Ocean Physics and Chemistry: From Concepts to Applications

Ocean Physics and Chemistry: From Concepts to Applications

Edited by **Theodore Roa**

SYRAWOOD
PUBLISHING HOUSE
New York

Published by Syrawood Publishing House,
750 Third Avenue, 9th Floor,
New York, NY 10017, USA
www.syrawoodpublishinghouse.com

Ocean Physics and Chemistry: From Concepts to Applications
Edited by Theodore Roa

International Standard Book Number: 978-1-68286-369-5 (Hardback)

Printed in China.

Contents

Preface

This book has been a concerted effort by a group of academicians, researchers and scientists, who have contributed their research works for the realization of the book. This book has materialized in the wake of emerging advancements and innovations in this field. Therefore, the need of the hour was to compile all the required researches and disseminate the knowledge to a broad spectrum of people comprising of students, researchers and specialists of the field.

There has been rapid advancement in the research focused on marine systems and oceanic processes. This book contains path-breaking studies on ocean physics and chemistry. It offers a thorough understanding of concepts such as ocean currents and eddies, air-sea fluxes, chemical, biochemical and physical interactions, etc. The researches and case studies collated in the text discuss latest developments in data collection mechanisms, remote sensing, numerical and computational measurements and observations, etc. This book is a vital tool for all researching and studying this field.

At the end of the preface, I would like to thank the authors for their brilliant chapters and the publisher for guiding us all-through the making of the book till its final stage. Also, I would like to thank my family for providing the support and encouragement throughout my academic career and research projects.

Editor

1

Enhancing the accuracy of automatic eddy detection and the capability of recognizing the multi-core structures from maps of sea level anomaly

J. Yi[1], Y. Du[1], Z. He[2,3], and C. Zhou[1]

[1]State Key Laboratory of Resources and Environmental Information System, Institute of Geographic Sciences and Natural Resources Research, Chinese Academy of Sciences, Beijing 100101, China
[2]State Key Laboratory of Tropical Oceanography (LTO), South China Sea Institute of Oceanology, Chinese Academy of Sciences, Guangzhou 510301, Guangdong, China
[3]College of Ocean and Earth Sciences, Xiamen University, Xiamen 361005, China

Correspondence to: Y. Du (duyy@lreis.ac.cn)

Abstract. Automated methods are important for automatically detecting mesoscale eddies in large volumes of altimeter data. While many algorithms have been proposed in the past, this paper presents a new method, called hybrid detection (HD), to enhance the eddy detection accuracy and the capability of recognizing eddy multi-core structures from maps of sea level anomaly (SLA). The HD method has integrated the criteria of the Okubo–Weiss (OW) method and the sea surface height-based (SSH-based) method, two commonly used eddy detection algorithms. Evaluation of the detection accuracy shows that the successful detection rate of HD is $\sim 96.6\,\%$ and the excessive detection rate is $\sim 14.2\,\%$, which outperforms the OW and those methods using SLA extrema to identify eddies. The capability of recognizing multi-core structures and its significance in tracking eddy splitting or merging events have been illustrated by comparing with the detection results of different algorithms and observations in previous literature.

1 Introduction

Mesoscale eddies play an important role in ocean circulation as well as in heat and mass transport (McWilliams, 2008; Nencioli et al., 2010). With the development of observation technology, automated detection methods are essential for understanding the complex movements and dynamic characteristics of mesoscale eddies in large volumes of altimeter data. In Eulerian schemes (distinguished from Lagrangian detection schemes using drifting trajectory data (Dong et al., 2011)), existing eddy detection methods can be categorized into three classes: (1) those based on a physical parameter, (2) those based on flow geometry, and (3) those based on sea surface height anomaly (SSHA).

Among the first category, the Okubo–Weiss (OW) approach (Okubo, 1970; Weiss, 1991) is the most widely used. The physical parameter W is computed from the horizontal velocity field as $W = \mathrm{Sn}^2 + \mathrm{Ss}^2 - \omega^2$, where Sn and Ss are the shear and strain deformation, respectively, and ω is the vertical component of vorticity. Applications of the OW method can be found in much literature (Chelton et al., 2007; Henson and Thomas, 2008; Isern-Fontanet et al., 2003b, 2006; Morrow et al., 2004; Xiu et al., 2010). Despite its popularity, three major weaknesses were reported about OW. First, it is difficult to decide on the optimal or appropriate threshold value of the parameter W that makes identification more accurate (Chelton et al., 2011). Recent studies (Williams et al., 2011; Petersen et al., 2013) have developed a new R^2 algorithm to eliminate the dependency on threshold choosing, which judges the quality of an eddy based on the similarity of certain functional fits with an idealized Gaussian vortex. Second, the derivation of this physical parameter may bring extra noise that increases falsely detected eddies (Ari Sadarjoen and Post, 2000; Chaigneau et al., 2008; Chelton et al., 2011;

Nencioli et al., 2010). Third, the physical criteria may fail to locate the eddies or underestimate the dimensions (Basdevant and Philipovitch, 1994; Doglioli et al., 2007; Henson and Thomas, 2008; Isern-Fontanet et al., 2003a).

To improve eddy detection, some novel approaches based on flow geometry characteristics were developed. The winding-angle (WA) approach first proposed by Ari Sadarjoen and Post (2000) is a representative that identifies eddies by clustering closed or spiral streamlines. Chaigneau et al. (2008) adapted this approach by using sea level anomaly (SLA) local extrema as potential eddy centers and the outermost streamlines as corresponding boundaries. Nencioli et al. (2010) argued that this adaption incorporated a physical quantity (SLA) and thus should be regarded as a hybrid approach. Consequently, they further developed a pure flow geometry-based, or vector geometry-based (VG) approach, which identifies eddies only by the geometry characteristics of velocity fields and is independent from parameters derived using velocity derivatives as well as from the SLA field (Nencioli et al., 2010). However, the specification and sensitivity test of two additional parameters that are used for searching eddy centers complicate the identification procedure.

The third category directly uses SSHA or SLA for eddy identification in which a threshold is nevertheless always required to delimit eddy dimensions. Fang and Morrow (2003) used a 10 cm threshold to identify the large eddies in the South Indian Ocean. Chaigneau and Pizzaro (2005) chose 6 cm to detect eddies in the region west of South America, and Wang et al. (2003) selected 7.5 cm for eddy identification in the South China Sea (SCS). In 2010, Chelton et al. (2011) proposed a sea surface height-based (SSH-based) method for global studies and eliminated the threshold dependency on SSHA. However, this method relies on other thresholds (e.g., area and horizontal scale) that were specified in its detection criteria to determine the eddy boundaries, and may yield eddies with more than one local extremum, which are referred to as multi-core structures in this paper.

Multi-core structures exist in the vicinity of mutual interactions between eddies, so that identifying and tracking them may allow one to understand how eddies interact with each other, how they split or merge, and how the energy transfers and exchanges. However, few studies have developed algorithms capable of detecting and tracking them. Chelton et al. (2011) tried splitting the multi-core structures, only to find extra induced problems in tracking them, so they eventually abandoned the splitting procedure. This paper presents a hybrid detection method (HD) that attempts to identify and characterize the multi-core structures and lays the foundation for further tracking the evolution processes correctly. A companion tracking method that addresses the difficulties in tracking multi-core structures will be elaborated in another paper. This paper is mainly focused on the discussion of the new detection method and its performance.

The detection accuracy defines the quality of eddy identification algorithms and is a necessary evaluation criterion for newly developed methods. Integrating different methods presents a promising avenue to improving the performance of detection algorithms, for it tries to make the best of the advantages and minimize the disadvantages of each single method. In previous literature, the adapted WA method (Chaigneau et al., 2008) incorporated SLA extrema to make the algorithm more simple and efficient. Morrow et al. (2004) combined additional SLA criteria to reduce excessive detections of the Q-parameter approach. In this paper, the HD method is developed with the integration of the OW method and the SSH-based method; it combines the W parameter criterion and SLA local extrema to detect eddy centers and uses SLA contours to delineate the dimensions of eddies. An objective validation protocol that has been used by Chaigneau et al. (2008) and Nencioli et al. (2010) is adopted to measure the detection accuracy of the HD method. Meanwhile, the detection results of different algorithms are compared to examine the performance of HD and the ability of characterizing multi-core structures. Nan et al. (2011)'s study of three anticyclonic eddies among which splitting and merging events had been observed is used to verify HD's detection results and to demonstrate the significance of recognizing multi-core structures.

The rest of the paper is organized as follows: Sect. 2 introduces the altimetry data and the detailed procedures of the HD method; Sect. 3 presents the results and discussion of HD's detection evaluation, method comparisons, and historical verifications; and the final section provides the conclusions.

2 Data and method

2.1 Altimetry data

As combining data from different satellite missions improves the estimation of mesoscale signals (Le Traon and Dibarboure, 1999; Chelton and Schlax, 2003; Pascual et al., 2006), the SLA data set (October 1992 to April 2012) of the AVISO (http://www.aviso.oceanobs.com) Reference Series delayed-time altimeter products merging TOPEX/Poseidon (T/P), Jason-1, ERS-1/2, and Envisat data (Ducet et al., 2000) was used to identify mesoscale eddies. This merged altimeter data set consists of weekly SLA maps computed with respect to a seven-year mean and resampled on a 1/3° × 1/3° Mercator grid.

2.2 Eddy detection methodology

Two major procedures are involved in conventional eddy identification: the first is to recognize the center of the eddy, the second is to extract the boundary. However, to identify the multi-core structures, HD includes an additional procedure to determine if the eddies constitute a multi-core structure of which the closed boundary contains more than one local extremum of the same polarity. The methodology of

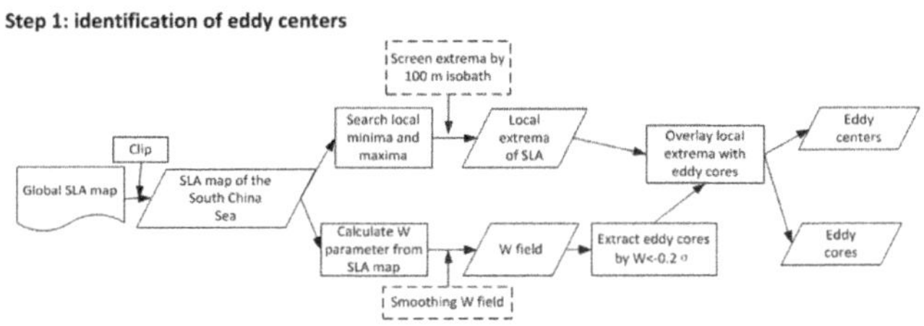

Fig. 1. Detailed procedures of step 1. Procedure in the box with dashed border is adaptable according to the data quality and the eddy characteristics of different study regions.

the new method thus has three steps: (1) identify eddy centers; (2) find eddy boundaries; and (3) distinguish multi-core structures.

2.2.1 Step 1: a hybrid way to identify eddy centers

The velocity field within an ocean eddy is dominated by rotation, corresponding to the negative W values in the OW method that can be used to recognize where eddies exist. Among the studies using the OW method, many adopted the criterion $W < -0.2\,\sigma_w$ (σ_w is the spatial standard deviation of W) for detecting ocean eddies from altimetry data, but this criterion usually brings undesirable noise into the data and causes false detections (Isern-Fontanet et al., 2006; Henson et al., 2008; Xiu et al., 2010).

Having an SLA maximum (or minimum) inside the eddy-dominant region is another essential character of ocean eddies, which has been adopted in the WA method, the SSH-based method as well as other eddy detection methods; hereafter they are all referred to as the SLA extremum-based methods. Following the definition that "a vortex exists when instantaneous streamlines mapped onto a plane normal to the vortex core exhibit a roughly circular or spiral pattern [...]" (Robinson, 1991), Chaigneau et al. (2008) improved the WA method by incorporating the SLA extrema criterion. Besides, the threshold-free SSH-based method developed by Chelton et al. (2011) for global eddy detection also required qualified eddies to contain at least one local minimum or maximum. However, it should be noted that noise on SSHA or SLA fields may form false maxima and minima and therefore could induce spurious detection of extrema.

Based on these definitions of an eddy and the corresponding eddy detection methods developed in previous studies, we argue that an eddy should have a core area where the velocity field is dominated by rotation and the streamlines show a circular or spiral pattern, and an SLA maximum or minimum inside. So, this study developed a hybrid detection algorithm that has integrated both the W parameter criterion and SLA extrema criterion to effectively reduce excessive detections and enhance the detection accuracy. Here, to avoid confusion, the local extrema of SLA in eddies are

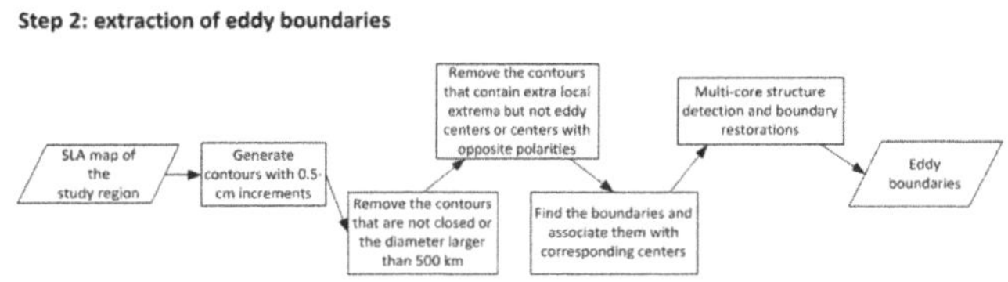

Fig. 2. Detailed procedures of step 2.

referred to as "eddy centers" while the connected regions where $W < -0.2\,\sigma_w$ are referred to as "eddy cores".

Figure 1 shows the details of step 1. Each global SLA map is first clipped to the study region (e.g., the South China Sea). Then, the program calculates the W parameter and searches for local extrema within that clipped SLA. Regions where $W < -0.2\,\sigma_w$ are extracted as eddy cores. Meanwhile, the local SLA extrema are searched out by running a 3×3 grid window (the smallest scale) on an SLA map. If the central point is a maximum or minimum of the window, then it represents a local extremum, and the extremum will be qualified as an eddy center only if it is located in a core area.

In this procedure, extra restrictions can be added based on the data quality of different study regions. In the South China Sea, for example, eddy centers located over water depths shallower than 100 m are removed due to alias from tides and internal waves contained in the SLA data over the shallow shelf area (Yuan et al., 2006). In addition, a smoothing algorithm, like a half-power filter (Chelton et al., 2011) or Hanning filter (Penven et al., 2005), can also be added to reduce the noise in the W field, but to avoid removing any physical information, this study applied no smoothing algorithms.

2.2.2 Step 2: definition and extraction of eddy boundaries

Existing algorithms lack a consistent definition of eddy boundaries, and detection results can vary significantly depending on the rules for extracting the boundary (Nencioli et al., 2010). The OW method defines enclosed regions as eddies where W values satisfy the criterion. However, this criterion is restricted to the core of the vortices and may underestimate eddy dimensions (Basdevant and Philipovitch, 1994). The WA method and the VG method represent eddy boundaries by streamlines, which make better approximation of eddy shapes, and given the good agreement with streamlines, the SSH-based method takes the outermost closed contours of SSHA as eddy boundaries. So, in the HD method, we also use SLA contours to represent eddy boundaries, but the "outermost" criterion is further refined to make the boundary extraction more sensible. The detailed procedures are shown in Fig. 2.

First, a series of contours of an SLA map is generated with increments of 0.5 cm, although 1 cm increments had been confirmed good enough to resolve the eddies with compact

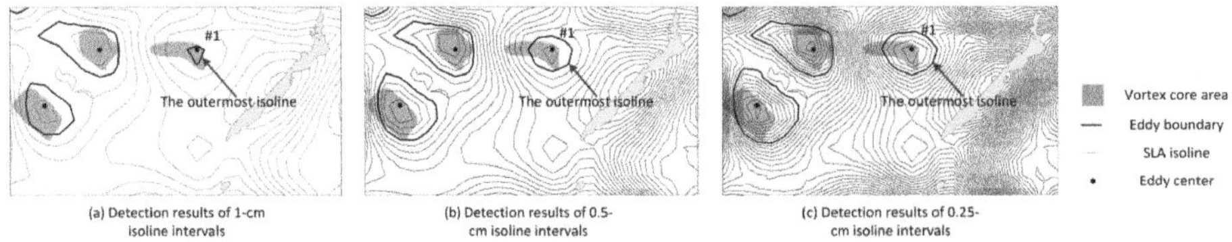

Fig. 3. Detection results of different contour intervals. The eddy boundary is determined by the HD algorithm that has combined the OW criterion and the SSH-based criterion.

interiors (Chelton et al., 2011). The smaller increment is preferred for it provides a better approximation of eddy dimensions. We carefully checked the detection results of different contour intervals and found that contours with 1 cm increments may underestimate the dimensions of eddies (Fig. 3a). We also found that 0.25 cm improves the boundary approximation, but the time consumption increases significantly when the increment becomes much smaller. So, the increment of 0.5 cm represents a compromise choice. Contours that are unclosed or have a diameter greater than 500 km are removed from the data set.

Second, the eddy centers and cores identified in step 1 are combined to decide which contours surrounding the eddy centers should be their corresponding boundaries. For each eddy center, the smallest contour that encompasses its core area is regarded as the qualified boundary. This definition improves the approximation of eddy shape while choosing the outermost closed contours, as boundaries tend to enlarge the dimensions. Figure 4a illustrates the difference between the two criteria.

It is noteworthy that some special situations may occur during the process of extracting eddy boundaries. For instance, it is possible that eddies may have no contours completely containing their core areas (Fig. 4b), or even no closed contours around the eddy center (Fig. 4c). In the former case, eddy boundaries are defined by the outermost closed contour that intersects the core areas, while in the latter case, the eddy dimensions are defined by the core, just as the OW method does. While some may argue about the existence of these weak eddies in the latter case, this study retains them during the identification process, as they may be in the immature or unstable stages (e.g., forming or disappearing) of their evolution processes and removing them will break a coherent process into discrete evolution pieces when they are tracked. Figure 4d illustrates another case that eddy boundaries contain extra local extrema but not qualified eddy centers. In such cases, the boundaries improperly enlarge the dimensions, and so are removed before the boundary extraction procedure. Moreover, contours containing two or more eddy centers of opposite polarities are also discarded in this study.

We summarize all the criteria or constraints used for detecting eddy centers and boundaries in the HD method as follows:

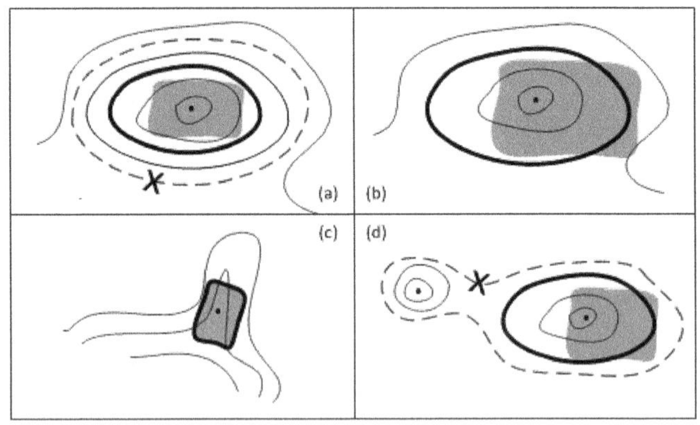

Fig. 4. Possible situations in extracting an eddy boundary. The contours are represented by the thin solid lines, while the eddy boundary is underlined by the thick solid line. Eddy core areas are symbolized by grey polygons, and the dashed lines denote the outermost closed contours.

1. $W < -0.2\,\sigma_w$ for detecting eddy core areas;
2. 3×3 grid moving window for searching SLA maxima and minima;
3. only those SLA maxima or minima that lie within core areas are qualified as eddy centers;
4. eddy centers located over shallow depths (taken to be 100 m in the SCS) are discarded;
5. the contours of SLA are generated with increments of 0.5 cm;
6. unclosed contours or diameter greater than 500 km are discarded;
7. the minimum closed SLA contours that encompass eddy core areas are defined as qualified boundaries;
8. if no closed contours completely contain the core areas, then take the outermost closed contours that intersect with the core areas as eddy boundaries;
9. if no closed contours intersect with the core areas, then use core areas to define eddy boundaries.

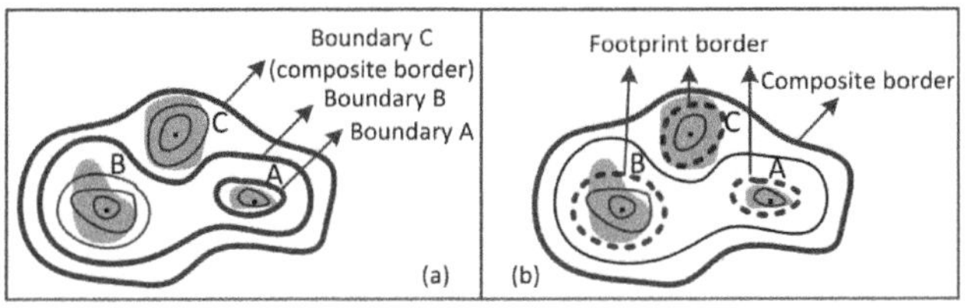

Fig. 5. Identification of multi-core structures and boundary restorations. The contours are represented by thin solid lines. The grey polygons represent eddy core areas.

2.2.3 Step 3: multi-core recognition and boundary restoration

The multi-core structures of eddies, which have two or more closed eddies of the same polarity within the boundaries, represent the important transitional stages of their lives in which component eddies may experience splitting, merging or other energy-transferring interactions. They were little mentioned and had no clear definition in previous studies. The SSH-based criteria can yield multi-core structures, but they fail to characterize their shapes and track their evolution processes. The HD method developed in this study attempts to recognize eddy multi-core structures fully so that we can address the difficulties they bring to tracking algorithms.

After step 2, every eddy center has been associated with a unique boundary. This step further defines that if the boundary of an eddy contains other eddy centers, it reveals a multi-core structure, and the boundary is taken as the "composite" border of the whole structure (Fig. 5a). If there are multiple composite borders, the algorithm just retains the outermost one. The borders of inner component eddies, which we called "footprint" borders, are defined by the outermost closed contours that only contain one eddy center. In brief, HD characterizes the spatial domain of multi-eddy structures and included eddies by the composite borders and the footprint borders, respectively.

After recognizing multi-core structures, the procedure needs to check every eddy's boundary and extract the composite borders and footprint borders. For example, in Fig. 5a, after step 2, sub-eddy C's boundary was found to be the outermost boundary containing sub-eddies A and B, and therefore it should be recognized as the composite border. Sub-eddy B's boundary conflicts with the definition of a footprint border for it contains eddy A, so that subsequent restorations are needed to assign the correct boundary. Figure 5b shows the correct borders of the multi-core structure and its component eddies after the boundary restoration based on the definitions of composite borders and footprint borders.

3 Results and discussion

3.1 Detection accuracy in the South China Sea

The accuracy is an important measure of the eddy identification quality. According to the validation protocol in Chaigneau et al. (2008), two quantities can be used to evaluate the accuracy of identification algorithms: the success of detection rate (SDR) and the excess of detection rate (EDR). Definitions are shown as follows:

$$\mathrm{SDR} = \frac{N_\mathrm{c}}{N_\mathrm{e}} \tag{1}$$

$$\mathrm{EDR} = \frac{N_\mathrm{om}}{N_\mathrm{e}} \tag{2}$$

Where N_c corresponds to the number of the eddies identified by both experts and the automated algorithm, N_e corresponds to the number of the eddies only identified by the experts, and N_om corresponds to the number of the eddies only identified by the algorithm.

This study randomly selected ten SLA maps of SCS (100° E–125° E; 5° N–26° N) from 1992 to 2012 to compute SDR and EDR of the HD method. Five experts were invited to manually detect eddies on these sample maps. On each map, only eddies recognized by at least three experts were counted in the detection result used for accuracy evaluation. As subjective bias usually exists among the experts, estimating the uncertainty of each derived manual detection result is necessary to understand how difficult it may be to correctly identify eddies from a specific SLA map, and to what extent the detection error of automatic algorithms can be tolerated. Here, we introduce the ratio of inconsistency (RI) to measure the uncertainty of expert results. The definition is as follows:

$$\mathrm{RI}_i = \frac{|N_i - N_\mathrm{ce}|}{N_\mathrm{ce}} \tag{3}$$

Where RI_i denotes the inconsistency rate of expert i. N_i is the number of eddies identified by expert i and N_ce is the number of eddies that are commonly recognized by at least three experts. N_i may be less than N_ce, and so to avoid negative values we only compute the absolute value of RI.

Figure 6a shows the comparison between expert detection (circles) and the HD result (dots) on 9 May 2007, SCS. There are 25 eddies ($N_\mathrm{e} = 25$) detected by the experts on this data set (12 cyclonic and 13 anticyclonic). Four eddies were overly detected ($N_\mathrm{om} = 4$) and one was missed ($N_\mathrm{c} = 24$) by the HD. So, according to the definition, SDR = 96.0 % and EDR = 16.0 % in this experiment.

Table 1 presents the evaluation results of the ten randomly selected SLA maps. About 22 eddies were identified in each SLA map by the experts on average. The mean inconsistency rate is ~20.7 %, indicating that such a number of eddies raises disagreement among experts, and when evaluating the detection accuracy, the rate value can be used as

Table 1. Accuracy evaluation results of the ten sample experiments on SLA maps.

Date		1 13/07/1994	2 03/07/1995	3 17/07/1996	4 21/01/1998	5 08/09/1999	6 16/05/2001	7 09/10/2002	8 12/11/2003	9 11/05/2005	10 09/05/2007	Average
expert	N_c	23	23	18	19	17	26	25	19	22	25	21.7 ± 3.2
	RI (%)	8.7	36.5	31.1	21.1	15.3	20.7	16.0	16.8	20.9	20.0	20.7 ± 8.0
OW	SDR (%)	100.0	100.0	94.4	89.5	100.0	96.2	100.0	89.5	100.0	96.0	96.6 ± 4.3
	EDR (%)	69.9	47.8	77.7	89.5	64.7	38.5	56.0	100	54.5	104.0	70.3 ± 22.2
SLA extrema	SDR (%)	100.0	100.0	100.0	100.0	100.0	100.0	100.0	100.0	100.0	100.0	100.0 ± 0.0
	EDR (%)	17.4	30.4	83.3	42.1	58.8	30.8	16.0	21.1	31.8	36.0	36.8 ± 20.6
HD	SDR (%)	100.0	100.0	94.4	89.5	100.0	96.2	100.0	89.5	100.0	96.0	96.6 ± 4.3
	EDR (%)	8.7	21.7	22.2	10.6	17.6	7.7	4.0	10.6	22.7	16.0	14.2 ± 6.7

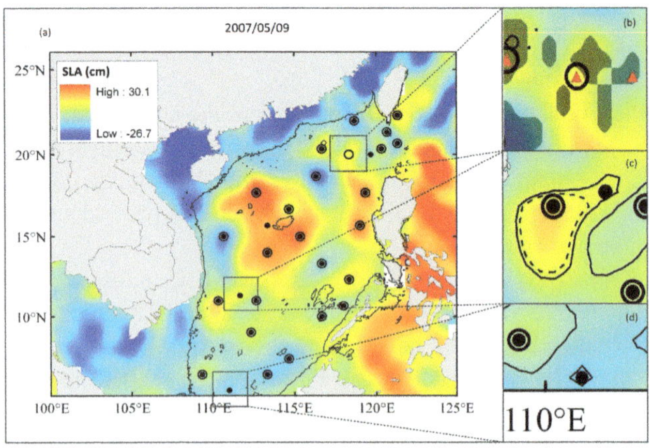

Fig. 6. (a) Comparisons between experts' detection result and HD's result on 9 May 2007. The 100 m depth isobath is delineated by the black solid line. The circles represent eddies identified by experts, while the dots represent the ones identified by HD. These detection results are superposed onto the SLA field of that date. **(b)** One missed eddy in HD's result. Red triangles represent the SLA local extrema, and the grey polygons represent eddy core areas. **(c)** One over-detected eddy that constitutes a multi-core structure. The composite borders are delineated by the solid lines and the footprint borders by the dashed lines. **(d)** The other over-detected eddy located near the border of the study region.

an upper bound of the acceptable tolerance for detection errors. The detection accuracy of the OW method and the SLA extremum-based method were also presented in Table 1 for comparison.

The results show that all the detection methods have relatively high SDRs. The methods using SLA local extrema to identify eddies maintain 100 % SDR on every map, which confirms that the SLA local extrema are necessary evidence for the existence of eddies. The SDRs of the OW method and the HD method are both equal to $\sim 96.6\,\%$ on average; $\sim 3.4\,\%$ eddies failed to be recognized by either OW or HD. Further examination reveals that these overlooked eddies were not located in the intense vorticity regions (where $W < -0.2\,\sigma_w$) but nearby (Fig. 6b), and were hence excluded from the qualified eddy centers identified by the HD method.

Second, EDR of different methods in the results varies wildly. The EDR of OW is high at $\sim 70.3\,\%$ on average and even exceeds 100 % in single experiments (e.g., on 19 May 2007), which is evidence of the over-detection weakness of the OW method. The average EDR of the SLA extremum-based method ($\sim 36.8\,\%$) is much lower than that of OW, which agrees with the validation results in Chaigneau et al. (2008) and confirms that the WA method is more accurate than OW. The HD method has the lowest EDR in single experiments and on average. This result substantiates that the integration of the OW and SSH-based criteria effectively improves the detection accuracy. In addition, the average EDR of HD ($\sim 14.2\,\%$) is under the tolerance bound (20.7 %) of acceptable detection errors, which further confirms the feasibility as an automatic algorithm in practical applications. In-depth investigation shows that the $\sim 14.2\,\%$ excessively detected eddies are mainly ascribed to two reasons: most of them (1) constitute a part of multi-core structures (Fig. 6c); or (2) are located near the boundary of the study region (Fig. 6d). Both cases tend to be overlooked in manual detection.

Another analysis using the receiver operating characteristic (ROC) graphs (Fawcett, 2006) was also performed to compare the detection accuracy of different methods, and the analysis result (Fig. 7) reconfirms the good performance of the HD method over the OW method and the SLA extremum-based method. Most of the ROC points of the HD method are located near the top-left corner, which represents perfect detection accuracy. The OW method's performance is the poorest as most of the ROC points are clustered in the top-right corner.

3.2 Detection results in the Eastern South Pacific Ocean

Although Sect. 3.1 has provided a quantitative evaluation of the detection accuracy of different methods in the SCS, this section intends to further examine the detection results on a specific SLA map in a different region and to observe the detection capability of different algorithms. We employed different detection methods to identify eddies in the Eastern South Pacific Ocean (3° S–20° S; 70° W–90° W) on the SLA map of 8 December 2004. Figure 8 shows the manual

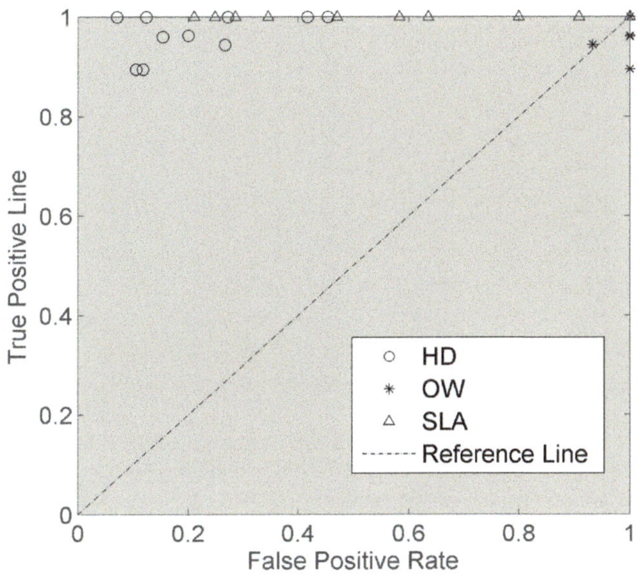

Fig. 7. The ROC analysis result of the detection methods in the SCS.

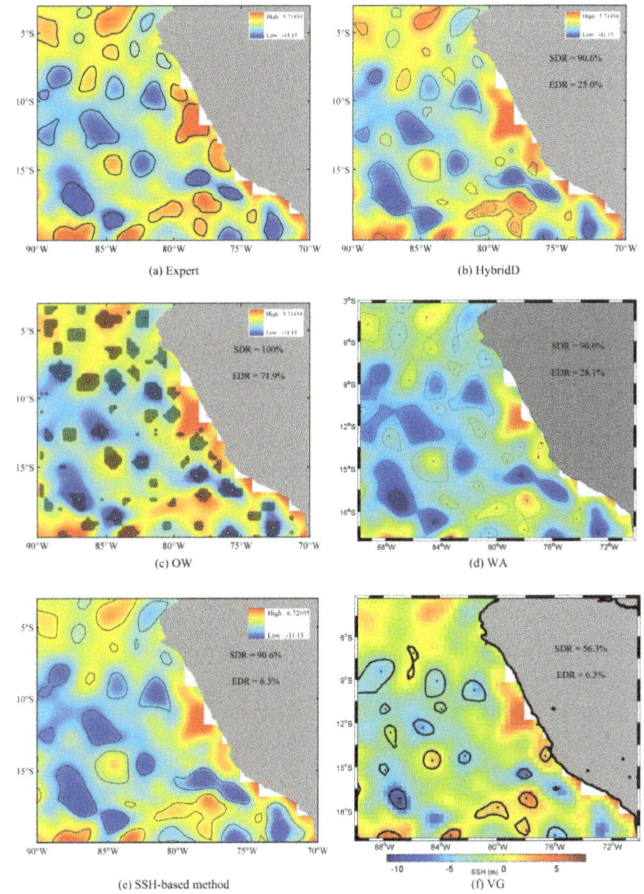

Fig. 8. Results of manual detection and automatic algorithms. The boundaries are delineated by black solid lines. Grey dark polygons in the OW detection denote where W parameter satisfies the criterion, and the white dots represent the SLA local extrema. The detection result of WA method is provided by A. Chaigneau. The SSH-based method result is generated according to the criteria in Chelton et al. (2011). The VG result is derived by the MATLAB code provided by F. Nencioli. The results in these subgraphs are superposed on SLA field of 8 December 2004.

detection result as well as the automatic methods' results on that map.

By comparing with the expert result, the detection performance of each method within our test can be sorted from good to poor: SSH-based method, HD, WA, OW, and VG. The OW method has a 100 % SDR, but the EDR is relatively high (71.9 %). The detection accuracy of VG (SDR = 50.0 %, EDR = 6.3 %) is not as good as it was in Nencioli et al. (2010) in which high-resolution data sets were used. The limited resolution of the SLA maps could be the main reason. The HD, WA and SSH-based methods all have a good SDR (90.6 %), but the SSH-based method outperforms the other two for its EDR (6.3 %) is minimal. However, considering the capability of detecting eddies in multi-core structures, only HD has developed algorithms to characterize the boundaries of component eddies and composite wholes. WA was unable to recognize such mixing structures, and the SSH-based method, although it had recognized two composite structures located around 17° S, 78° W, failed to distinguish the contained eddies.

In brief, this analysis has demonstrated that HD, WA, and the SSH-based method are more suitable for detecting ocean eddies from SLA maps than OW and VG. Moreover, HD has the advantage of characterizing eddies in the multi-core structures over other methods, which provides the opportunity to explore eddy interactions.

3.3 Detection verification with previous literature

Eddies that have been studied in previous literature provide good references for verifying the detection results of automatic algorithms. Among the extensive research on mesoscale eddies in the South China Sea (Wang et al., 2008; Wang et al., 2006; Li et al., 1998; Jia and Liu, 2004; Yi et al., 2013), Nan et al. (2011) investigated three long-lived anticyclonic eddies (AE) in the northern SCS in 2007, two of which (named ACE2 and ACE3) experienced splitting and merging events during their evolution processes. So, this paper cites the two AEs to verify the detection results of the HD method and discusses the roles that multi-core structures play in eddies' evolution changes.

Figure 9 shows four consecutive snapshots of ACE3 during the merging event. According to Nan et al. (2011), one small eddy lasted for about 2 weeks and then merged into ACE3 on 22 August. In Fig. 9, a two-core structure was found near Luzon Island on 8 August and the smaller component eddy close to the shore shrunk sharply but still remained in the composite structure after one week. On 22 August, the

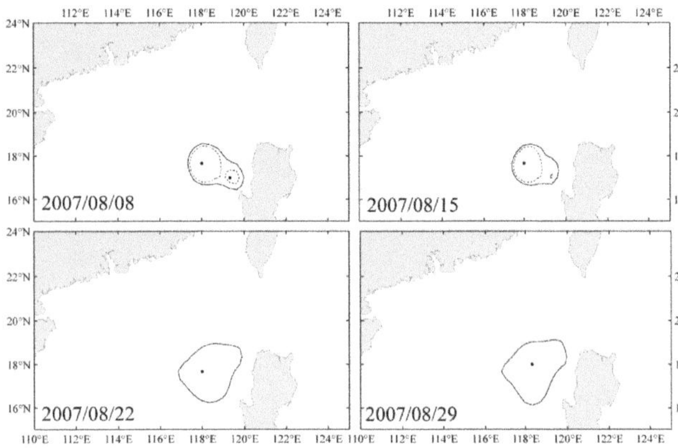

Fig. 9. Evolution snapshots of ACE3's merging event. The composite border is delineated by the solid line, while the footprint border is delineated by the dashed line.

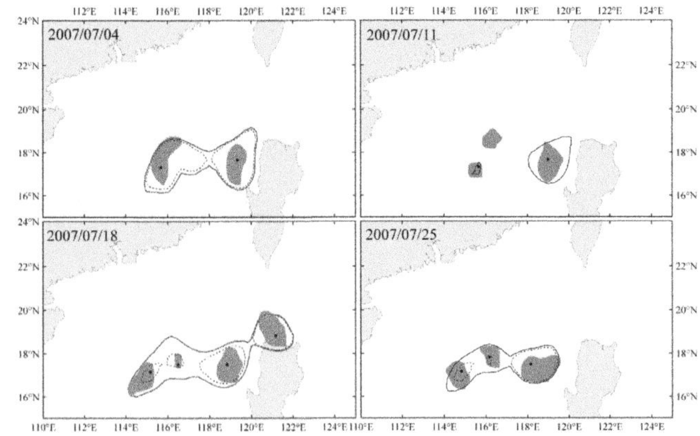

Fig. 10. Evolution snapshots of ACE2's splitting event. The dots represent eddy centers, while the dark grey polygons represent eddy core areas. The composite border is delineated by the solid line, while the footprint border is delineated by the dashed line.

shrinking eddy disappeared and the two-core structure eventually turned into a single eddy. Such an evolution event of ACE3 recognized by HD is basically consistent with the description in Nan et al. (2011). Additionally, the multi-core structure is delineated, providing detailed information (e.g., how the energy tranfers) about the merging effect.

The other long-lived anticyclonic eddy, i.e., ACE2, which split off a small eddy on 11 July according to Nan et al. (2011), is also examined, and Fig. 10 shows the identification results of HD during the split. On 4 July, ACE2 constituted a two-core structure in a loose connection with ACE3, but one week later, they separated and ACE2 shrunk significantly. Afterwards, ACE2 was strengthened judging by the core areas, and contained in a more complex four-core structure on 18 July. In addition, between ACE2 and ACE3 a small newborn eddy can be observed expanding rapidly in the multi-core structure, which suggests that the evolution of ACE2 and ACE3 had contributed much to the generation and the development of this internal newborn eddy.

Some differences can be observed, relative to Nan et al. (2011), especially that ACE2 was not found splitting off a small eddy but rather separated from ACE3 during 4–11 July. Careful examination reveals how this discrepancy results from the different detection criteria used. Nan et al. employed the OW method to identify eddies so that they observed the splitting event since ACE2's core was indeed divided into two parts during that period. However, one part after the separation contained no SLA local maximum and was consequently excluded from HD's detection results. Despite some differences between the results of our algorithm and the description of Nan et al. (2011), our results demonstrate the feasibility of the automated detection even when eddy structures contain multiple cores, presenting a means to algorithmically identify splitting and merging events, enabling one to locate an entire class of events involving multiple eddy cores and a change in the number of eddies.

4 Conclusions

This study proposed a new hybrid method, HD, which is designed to automatically identify mesoscale eddies from SLA data sets. The combination of detection criteria of the OW method and the SSH-based method effectively enhances the accuracy of eddy detection, and procedures are implemented in the HD method to recognize the multi-core structure of mesoscale eddies, which serves as the foundation of tracking eddies' dynamic events (e.g., splitting or merging) and eddies' mutual interactions during the evolution processes.

To evaluate the detection accuracy of the HD method objectively, this study adopted the validation protocol used by Chaigneau et al. (2008) and made a comparison with the manual results detected by 5 experts. Ten random experiments in the South China Sea showed that the average successful detection rate was $\sim 96.6\,\%$ and the excessive detection rate was $\sim 14.2\,\%$, which confirmed HD's improvement in detection accuracy over either OW or the SLA extremum-based method individually. Second, the comparisons between different detection algorithms in the ESP study area illustrated every method's performance as well as HD's capability of recognizing eddy multi-core structures. Finally, by comparing with two long-lived anticyclonic eddies investigated by Nan et al. (2011), this study verified HD's detection results and found that the merging event involving ACE3 was well recognized by HD, while the discrepancy of ACE2's splitting event was mainly due to the different detection criteria between OW and HD. This detection verification also demonstrated the additional capability of recognizing the multi-core structures, which facilitated the identification of the splitting and merging changes and unveiled details of eddy evolution that cannot be seen with other detection algorithms.

Limitations of the HD method are also obvious. The HD method can only detect ocean eddies that produce a sufficiently strong SLA. Although HD is able to address complex multi-eddy structures, the extent of mass or energy exchanges and interactions under the surface require further careful studies and validation with in situ data. So, having presented a method for the identification of multi-core structures, we expect and welcome subsequent refinement to our method.

This study has developed a hybrid method for identifying single eddies and composite structures containing multiple cores from an instantaneous map of SLA. In the next paper, we will present the companion tracking method that is able to extract the continuous evolution processes as well as complex dynamic changes and interactions of the identified eddies and composite structures.

Acknowledgements. We thank Matthew Hecht, Sean Williams, and other two anonymous reviewers for their constructive comments on this paper. We are grateful to Alexis Chaigneau for providing the detection results of their improved wind-angle method, and Francesco Nencioli for sharing the source code of the Vector-Geometry Eddy Detection Algorithm. We also thank the experts participating in the manual detection of eddies so that we can make method validations. This research was a contribution to research projects 41071250 and 41371378 of the National Science Foundation of China and 088RA500KA of the Innovation Projects of the State Key Laboratory of Resource and Environment Information System, Chinese Academy of Sciences. The altimeter products were produced by Ssalto/Duacs and distributed by Aviso, with support from Cnes (http://www.aviso.oceanobs.com/duacs/).

Edited by: M. Hecht

References

Ari Sadarjoen, I. and Post, F. H.: Detection, quantification, and tracking of vortices using streamline geometry, Comput. Graph., 24, 333–341, doi:10.1016/s0097-8493(00)00029-7, 2000.

Basdevant, C. and Philipovitch, T.: On the validity of the "Weiss criterion" in two-dimensional turbulence, Physica D, 73, 17–30, doi:10.1016/0167-2789(94)90222-4, 1994.

Chaigneau, A. and Pizarro, O.: Eddy characteristics in the eastern South Pacific, J. Geophys. Res., 110, C06005, doi:10.1029/2004jc002815, 2005.

Chaigneau, A., Gizolme, A., and Grados, C.: Mesoscale eddies off Peru in altimeter records: Identification algorithms and eddy spatio-temporal patterns, Prog. Oceanogr., 79, 106–119, doi:10.1016/j.pocean.2008.10.013, 2008.

Chelton, D. B. and Schlax, M. G.: The Accuracies of Smoothed Sea Surface Height Fields Constructed from Tandem Satellite Altimeter Datasets, J. Atmos. Ocean. Tech., 20, 1276–1302, doi:10.1175/1520-0426(2003)020<1276:taosss>2.0.co;2, 2003.

Chelton, D. B., Schlax, M. G., Samelson, R. M., and de Szoeke, R. A.: Global observations of large oceanic eddies, Geophys. Res. Lett., 34, L15606, doi:10.1029/2007gl030812, 2007.

Chelton, D. B., Schlax, M. G., and Samelson, R. M.: Global observations of nonlinear mesoscale eddies, Prog. Oceanogr., 91, 167–216, doi:10.1016/j.pocean.2011.01.002, 2011.

Doglioli, A. M., Blanke, B., Speich, S., and Lapeyre, G.: Tracking coherent structures in a regional ocean model with wavelet analysis: Application to Cape Basin eddies, J. Geophys. Res., 112, C05043, doi:10.1029/2006jc003952, 2007.

Dong, C., Liu, Y., Lumpkin, R., Lankhorst, M., Chen, D., McWilliams, J. C., and Guan, Y.: A Scheme to Identify Loops from Trajectories of Oceanic Surface Drifters: An Application in the Kuroshio Extension Region, J. Atmos. Ocean. Technol., 28, 1167–1176, doi:10.1175/jtech-d-10-05028.1, 2011.

Ducet, N., Le Traon, P. Y., and Reverdin, G.: Global high-resolution mapping of ocean circulation from TOPEX/Poseidon and ERS-1 and -2, J. Geophys. Res., 105, 19477–19498, doi:10.1029/2000jc900063, 2000.

Fang, F. and Morrow, R.: Evolution, movement and decay of warm-core Leeuwin Current eddies, Deep Sea Res.-Pt. II, 50, 2245–2261, doi:10.1016/s0967-0645(03)00055-9, 2003.

Fawcett, T.: An introduction to ROC analysis, Pattern Recogn. Lett., 27, 861–874, doi:10.1016/j.patrec.2005.10.010, 2006.

Henson, S. A. and Thomas, A. C.: A census of oceanic anticyclonic eddies in the Gulf of Alaska, Deep Sea Res.-Pt I, 55, 163–176, doi:10.1016/j.dsr.2007.11.005, 2008.

Isern-Fontanet, J., Garcia-Ladona, E., and Font, J.: Identification of marine eddies from altimetric maps, 5, American Meteorological Society, Boston, MA, ETATS-UNIS, 7 pp., 2003a.

Isern-Fontanet, J., Garcia-Ladona, E., and Font, J.: Identification of marine eddies from altimetric maps, J. Atmos. Ocean. Tech., 20, 772–778, doi:10.1175/1520-0426(2003)20<772:IOMEFA>2.0.CO;2, 2003b.

Isern-Fontanet, J., Garcia-Ladona, E., and Font, J.: Vortices of the Mediterranean Sea: An altimetric perspective, J. Phys. Oceanogr., 36, 87–103, doi:10.1175/JPO2826.1, 2006.

Jia, Y. and Liu, Q.: Eddy Shedding from the Kuroshio Bend at Luzon Strait, J. Oceanogr., 60, 1063–1069, doi:10.1007/s10872-005-0014-6, 2004.

Le Traon, P. Y. and Dibarboure, G.: Mesoscale Mapping Capabilities of Multiple-Satellite Altimeter Missions, J. Atmos. Ocean. Tech., 16, 1208–1223, doi:10.1175/1520-0426(1999)016<1208:MMCOMS>2.0.CO;2, 1999.

Li, L., Nowlin Jr., W. D., and Jilan, S.: Anticyclonic rings from the Kuroshio in the South China Sea, Deep Sea Res.-Pt. I, 45, 1469–1482, doi:10.1016/s0967-0637(98)00026-0, 1998.

McWilliams, J. C.: the nature and consequence of oceanic eddies, in: Eddy-Resolving Ocean Modeling, edited by: Hecht, M. and Hasumi, H., AGU Monograph, 5–15, 2008.

Morrow, R., Birol, F., Griffin, D., and Sudre, J.: Divergent pathways of cyclonic and anti-cyclonic ocean eddies, Geophys. Res. Lett., 31, L24311, doi:10.1029/2004gl020974, 2004.

Nan, F., He, Z., Zhou, H., and Wang, D.: Three long-lived anticyclonic eddies in the northern South China Sea, J. Geophys. Res., 116, C05002, doi:10.1029/2010jc006790, 2011.

Nencioli, F., Dong, C., Dickey, T., Washburn, L., and McWilliams, J. C.: A Vector Geometry–Based Eddy Detection Algorithm and Its Application to a High-Resolution Numerical Model Product and High-Frequency Radar Surface Velocities in the Southern California Bight, J. Atmos. Ocean. Tech., 27, 564–579, doi:10.1175/2009jtecho725.1, 2010.

Okubo, A.: Horizontal dispersion of floatable particles in the vicinity of velocity singularities such as convergences, Deep

Sea Research and Oceanographic Abstracts, 17, 445–454, doi:10.1016/0011-7471(70)90059-8, 1970.

Pascual, A., Faugère, Y., Larnicol, G., and Le Traon, P.-Y.: Improved description of the ocean mesoscale variability by combining four satellite altimeters, Geophys. Res. Lett., 33, L02611, doi:10.1029/2005gl024633, 2006.

Penven, P., Echevin, V., Pasapera, J., Colas, F., and Tam, J.: Average circulation, seasonal cycle, and mesoscale dynamics of the Peru Current System: A modeling approach, J. Geophys. Res., 110, C10021, doi:10.1029/2005jc002945, 2005.

Petersen, M. R., Williams, S. J., Maltrud, M. E., Hecht, M. W., and Hamann, B.: A three-dimensional eddy census of a high-resolution global ocean simulation, J. Geophys. Res.-Oceans, 118, 1759–1774, doi:10.1002/jgrc.20155, 2013.

Robinson, S. K.: Coherent Motions in the Turbulent Boundary Layer, Annu. Rev. Fluid Mech., 23, 601–639, doi:10.1146/annurev.fl.23.010191.003125, 1991.

Wang, D., Xu, H., Lin, J., and Hu, J.: Anticyclonic eddies in the northeastern South China Sea during winter 2003/2004, J. Oceanogr., 64, 925–935, doi:10.1007/s10872-008-0076-3, 2008.

Wang, G., Su, J., and Chu, P. C.: Mesoscale eddies in the South China Sea observed with altimeter data, Geophys. Res. Lett., 30, 2121, doi:10.1029/2003gl018532, 2003.

Wang, G., Chen, D., and Su, J.: Generation and life cycle of the dipole in the South China Sea summer circulation, J. Geophys. Res., 111, C06002, doi:10.1029/2005jc003314, 2006.

Weiss, J.: The dynamics of enstrophy transfer in two-dimensional hydrodynamics, Physica D, 48, 273–294, doi:10.1016/0167-2789(91)90088-q, 1991.

Williams, S., Petersen, M., Bremer, P.-T., Hecht, M., Pascucci, V., Ahrens, J., Hlawitschka, M., and Hamann, B.: Adaptive extraction and quantification of geophysical vortices, IEEE T. Vis. Comput. Gr., 17, 2088–2095, 2011.

Xiu, P., Chai, F., Shi, L., Xue, H., and Chao, Y.: A census of eddy activities in the South China Sea during 1993–2007, J. Geophys. Res., 115, C03012, doi:10.1029/2009jc005657, 2010.

Yi, J., Du, Y., Wang, X., He, Z., and Zhou, C.: A clustering analysis of eddies' spatial distribution in the South China Sea, Ocean Sci., 9, 171–182, doi:10.5194/os-9-171-2013, 2013.

Yuan, D., Han, W., and Hu, D.: Surface Kuroshio path in the Luzon Strait area derived from satellite remote sensing data, J. Geophys. Res., 111, C11007, doi:10.1029/2005jc003412, 2006.

2

Decadal variability of heat content in the South China Sea inferred from observation data and an ocean data assimilation product

Wei Song[1,2], Jian Lan[1], Qinyan Liu[2], Dandan Sui[2], Lili Zeng[2], and Dongxiao Wang[2]

[1]Physical Oceanography Laboratory, Ocean University of China, Qingdao, China
[2]State Key Laboratory of Tropical Oceanography, South China Sea Institute of Oceanology, Chinese Academy of Sciences, Guangzhou, China

Correspondence to: Dongxiao Wang (dxwang@scsio.ac.cn)

Abstract. Using an observation data set of temperature and the Simple Ocean Data Assimilation (SODA), the decadal variability of upper ocean heat content (0–400 m; hereafter, OHC) in the South China Sea (SCS) was investigated for the period from 1958 to 2007. Decadal variability was identified as the dominant mode of upper OHC besides the seasonal cycle. According to decreasing or increasing OHC, four periods were chosen to discuss detailed processes behind OHC variability in the SCS; the four periods are 1958–1968, 1969–1981, 1982–1992, and 1993–2003. Results show that advection was the major factor for decreasing (increasing) OHC during 1958–1968 (1968–1981). During 1982–1992 and 1993–2003, the net surface heat flux was the main contributor to the variability of OHC besides the advection. The OHC, advection and net surface heat flux had significant positive trends during 1992–2003. The spatial characteristics of OHC variability and heat budget in the Luzon Strait, west of Luzon Island, and in the Xisha warm eddy region are also discussed.

1 Introduction

The ocean has the greatest heat capacity in the climate system. Significant changes in the global ocean heat content were revealed for the past 50 yr (Levitus et al., 2000). To understand the ocean's roles in the climate system, the global ocean's ability to store and transport heat has been extensively studied by a number of researchers. Most of the recent studies focused on upper ocean warming (Levitus et al., 2000, 2005, 2009; Gouretski and Koltermann, 2007; Domingues et al., 2008) in association with the ocean's significant contribution to global warming and climate change.

The South China Sea (SCS) is the largest marginal sea in Southeast Asian waters (Fig. 1). It is connected to the East China Sea through the Taiwan Strait, to the Pacific Ocean through the Luzon Strait, and to the Sulu Sea via the Mindoro Strait. There are also channels connecting the SCS and the Indian Ocean: the Karimata Strait and the Malacca Strait. In this paper, we choose a cross section covering the two straits and call it the Sunda shelf (the location is given in Sect. 3). The interannual variability of latent and sensible heat fluxes in the SCS has been studied (Zong et al., 2010). He and Guan (1997) discussed the interannual and interdecadal variations in the upper ocean heat content (OHC) of the SCS, and pointed out that the vertically averaged temperature trend changed from negative to positive in the recent decades. OHC contributes interdecadal variability to ENSO events, the Asian winter monsoon and the tropical atmospheric circulation anomalies (He and Guan, 1997). The OHC in the northern SCS shows propagation, but that in the southern SCS is stationary (Yan et al., 2010). Studying interdecadal variability of the OHC itself is not enough, because ocean circulations play an important role as well. A few studies (Fang et al., 2003; Qu et al., 2004; Willis et al., 2004; Wang et al., 2006a, b; Liu et al., 2012) pointed out that the Luzon Strait allows the ENSO signal into the SCS, and therefore the role of advection should not be neglected. Horizontal advection is especially important to the OHC during the mature phase of ENSO events (Liu et al., 2012).

2 Data and methods

The data used in the present study include the expendable bathythermograph (XBT) data from the World Ocean Database 09 (WOD09) and the Simple Ocean Data Assimilation (SODA, Carton et al., 2000a, b; Carton and Giese, 2008), covering the period from 1958 to 2007. The ocean model used in making the SODA is based on Parallel Ocean Program (POP; Smith et al., 1992) physics with an average resolution of $0.25° \times 0.4°$ and 40 levels. The observations include virtually all available hydrographic profile data, ocean station data, moored temperature and salinity time series, surface temperature and salinity observations of various types, and night-time infrared satellite sea-surface temperature (SST) data. The SODA output is in monthly-averaged form, mapped onto uniform $0.5° \times 0.5°$ grids. Before the analysis, the XBT data had already been processed under quality control. We first selected the data by removing those profiles that were flagged as "bad" or as not passing the monthly, seasonal, and annual standard deviation checks. Detailed procedures are as follows. The data is flagged if it is more than n standard deviation (STD) from the climatological monthly mean at any one depth. For depths deeper than 50 m, $n = 3$ is assigned, while for depths shallower than 50 m, $n = 4$ is assigned to accommodate greater variability of upper ocean states (Locarnini et al., 2006). We then removed those profiles that did not extend to 50 m, thus eliminating profiles over the shallow continental shelf, and those profiles with obviously erroneous records (e.g., temperature < 10 °C) or exceptionally large vertical increment (> 20 m) between two samples in the top 100 m. The final data set contains a total of 23 356 temperature profiles in the SCS (Fig. 1), with irregular spatial and temporal distributions.

A heat budget calculation was carried out: the heat content change is equal to the summation of surface heat fluxes, advection flux, and diffusion as in Na et al. (2012). The equation is:

$$\mathrm{HCC} = \mathrm{QNET} + \mathrm{ADV} + R. \quad (1)$$

In Eq. (1), HCC denotes heat-content change in the upper 400 m. QNET (positive into the SCS) denotes the net surface heat flux. ADV (positive into the SCS) is the lateral ocean heat advection, and R is the residual heat flux (vertical advection, diffusion, etc.) The heat content used in the calculation is the integration of the top 400 m, which contain the thermocline process (Qu et al., 2004). The surface heat fluxes are from the NCEP/NCAR Reanalysis 1 data set, which provides four kinds of air–sea fluxes from the study period. To discuss the monthly anomaly, the climatological seasonal cycle is removed: every month in the time series minus the mean of corresponding month. For example, the anomaly in January 1958 is the HC minus the average for all Januaries from 1958 to 2007.

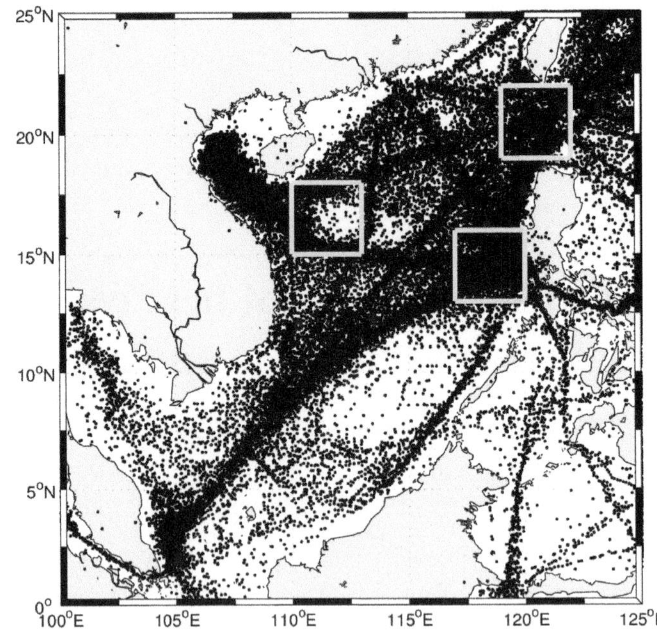

Fig. 1. Geographical distribution (asterisks) of the XBT data used in this study, which consist of more than 23 000 profiles. The three boxes are used for illustrating spatial differences: Box 1 for the Luzon Strait, Box 2 for west of Luzon Island, and Box 3 for the Xisha warm eddy.

3 Results

3.1 Decadal variability

Figure 2 shows the decadal variability of vertical temperature profiles and OHC in the SCS (90–121° E, 0–25° N). The results show that there are two dominant peaks and troughs from 1958 to 2007. The vertical temperature evolution averaged in the SCS inferred from the SODA data set shows obvious decadal variability (Fig. 2, middle), which is more evident than that inferred from the XBT data set. The strongest temperature anomalies in 100–200 m are likely to be related to the thermocline variability (Qu et al., 2004). The most notable cold and warm anomalies occurred in 1964 and 2000, respectively. In addition, the temperature appears to have risen in the recent decades.

For the comparison of XBT data and the SODA product, nine boxes are chosen where the XBT data are abundant. The results show that the two data sets are comparable in terms of heat content (not shown here), and an example (the entire SCS) is given in Fig. 2. The XBT data and the SODA product have similar variability, especially in terms of OHC time series (Fig. 2, bottom). So, only the SODA product will be used next for heat budget calculation.

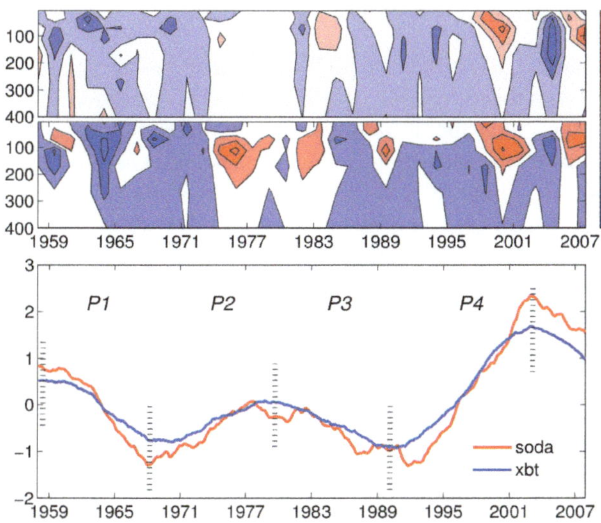

Fig. 2. Temperature anomalies of the whole South China Sea (contour interval: 0.2, units: K; seasonal cycle removed) of the XBT (top) and SODA (middle), and the heat-content time series of XBT and SODA (bottom), units: 10^{23} J. Also indicated are P1: 1958–1968; P2: 1969–1981; P3: 1982–1992; and P4: 1993–2003.

Table 1. The terms of heat budget for the climatological mean and in the four periods shown in Fig. 2. HCC denotes heat-content change in the upper 400 m. QNET (positive into the SCS) denotes the net surface heat flux. ADV (positive into the SCS) is the ocean heat advection, and R is the residual heat flux (vertical advection, diffusion, etc.). Units: TW, where 1 TW $= 10^{12}$ W.

	HCC	QNET	ADV	R
Clim	3.38	67.69	−51.10	−13.28
P1	−18.72	2.53	−14.92	−6.33
P2	14.36	−1.77	13.23	2.90
P3	−15.53	−7.24	−4.87	−3.42
P4	24.31	16.69	3.45	4.17

3.2 Heat balance

According to the decadal variability of OHC in the SCS (Fig. 2, bottom), four periods are chosen for studying the heat budget, which are 1958–1968, 1969–1981, 1982–1992, and 1993–2003 (Table 1). All terms in the heat equation except the diffusion are calculated using the SODA product as in Na et al. (2012). The advection term is the summation of fluxes through the four channels (Luzon Strait: 120.25° E, 18.25–22.25° N; Taiwan Strait: 117.25–120.25° E, 20.25° N; Sunda Shelf: 104.25–111.25° E, 1.75° N; Mindoro Strait: 117.25–120.25° E, 6.75–14.75° N). The Luzon Strait fluxes are positive every period and the other three fluxes are all negative every period, but between the different periods the value of each strait's flux varies; the contribution of ADV (four straits' summation) to HCC is different. The climatological heat budget is consistent with a previous study based on the results from an ocean general circulation model (Qu et al., 2004); climatological HCC, Q and ADV in Table 1 are each within 93 % of the model values.

Inferred from the terms in Table 1, the significant factors controlling the heat content change (HCC) in the SCS differ for the four chosen periods. During 1958–1968 (period P1), the OHC in the SCS decreased rate was −18.72 TW, and cannot be explained by the net surface heat flux, which was 2.53 TW into the ocean. So the advective cooling was the major contributor to the HCC in this period. Advection can also explain the increase in OHC during 1969–1981 (period P2). Advection was the major factor causing the heat content variation before the climate shift in 1976, explaining up to 78 % and 93 % of the change in periods P1 and P2, respectively. The situations are different after the climate shift during periods P3 and P4, when the net surface heat flux played an important role in regulating the decadal variability during 1982–1992 and 1993–2003, although the contribution of advection cannot be ignored, with 31 % and 14 %, respectively, for these periods. After the climate shift, the net heat flux and ocean heat advection can explain about 78 % and 83 % of the HCC during 1982–1992 and 1993–2003, respectively, and the rest of HCC may be caused by vertical flux and diffusion. There was an evident positive trend in OHC from 1992 to 2003, which was stronger than the HCC in the years from 1982 to 1992 and was mainly caused by the strong cold anomaly in 1992 and warm anomaly in 2003 (Fig. 2, bottom panel).

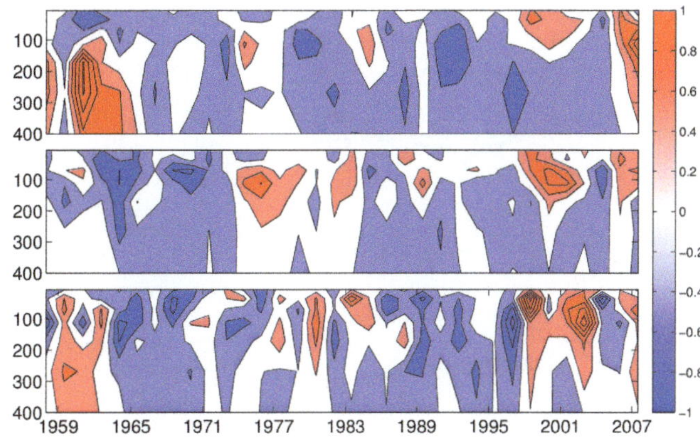

Fig. 3. Temperature anomalies (contour interval: 0.2, units: K; seasonal cycle removed) of Box 1 (top), Box 2 (middle), and Box 3 (bottom).

3.3 Spatial differences

When comparing XBT data and the SODA product, we found that the HCCs in different areas have a general resemblance (Figs. 3 and 4). Three key boxes (shown in Fig. 1) are chosen to study the spatial characteristics of OHC and its budget due to HCC, net heat flux and advection. The boxes are for the Luzon strait, west of Luzon Island, and

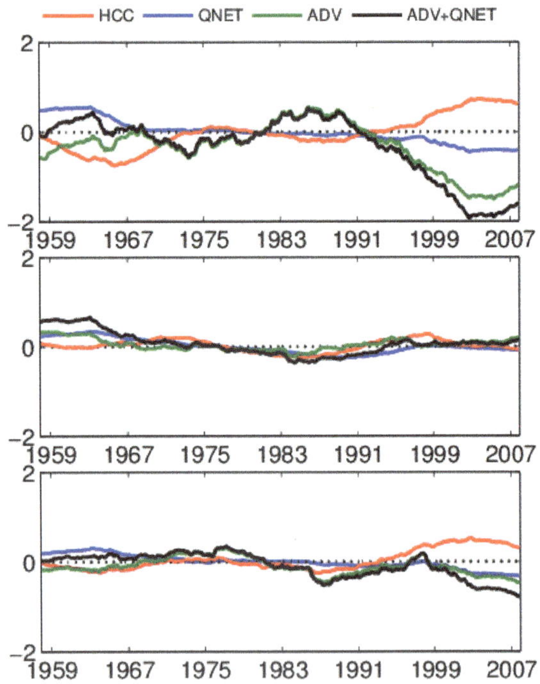

Fig. 4. Time series of 10 yr running mean HCC (red), QNET (blue) and ADV (green) in Box 1 (top), Box 2 (middle) and Box 3 (bottom); units: TW, where 1 TW = 10^{12} W.

the Xisha warm eddy region. It seems that the warming happened both in the subsurface and the surface near the Luzon Strait (Fig. 3, upper panel), but the data before the 1970s should be used carefully, so the detailed characteristics should be investigated further. The net heat flux and advection cannot explain the variability of OHC in the Luzon Strait, which is mainly affected by the Kuroshio Current; the dynamical processes here are complicated (such as a strong Kuroshio associated with meso-scale eddy and diffusion, etc.). Both boxes 2 and 3 are located in the interior SCS and influenced by the Pacific Ocean, but each has its own characteristics. The box we choose in the Xisha warm eddy is the pathway of the western boundary current of the South China Sea (SCS) and the SCS warm current; the heat content change here is mainly caused by the advection flux. The box we choose to the west of Luzon Island is connected to the Pacific Ocean through the Mindoro Strait and other marginal seas. Some research shows that the Pacific water and signal can get into the SCS through the Mindoro Strait (Zhuang et al., 2013). The temperature profile in Box 3 (Xisha warm eddy) has some similarity with that in Box 2 (west of Luzon Island), but the oscillation of temperature in Box 3 is more vigorous than that in Box 2 (Fig. 3, bottom and middle panels). The results of the heat budget show that advection is the major contributor to the HCC in boxes 2 and 3, with similar trends especially from the 1970s to 1990s (Fig. 4, middle and bottom panels). The net surface heat flux decreased during the study period, and advection seems to dominate the HCC in the SCS (Qiu and Chen, 2012).

4 Summary and discussion

Using the SODA, XBT, and NCEP/NCAR flux data sets, the decadal variability of OHC in the SCS was discussed. The results show that the decadal variability of OHC is evident, with two peaks and troughs during the study period. The OHC in the whole SCS decreased in 1958–1968 and 1982–1992, and increased in 1968–1981 and 1993–2003. Before the climate shift in 1976, the advection was the major factor in the heat content variation, with 78 % and 93 % contributions in periods P1 and P2, respectively. The situations were different after the climate shift, and the net surface heat flux played an important role in regulating the decadal variability during 1982–1992 and 1993–2003, although the contributions of advection cannot be ignored, at 31 % and 14 %, respectively.

In this paper, the spatial characteristics of the OHC in the SCS were also discussed. In the Luzon Strait, the net heat flux and advection could not account for the variability of OHC there, perhaps due to the complicated ocean dynamic processes, such as strong Kuroshio and associated meso-scale eddies and diffusion. The trends of HCC and net heat flux in the study period were negative, and it is difficult to explain the HCC with the net heat flux alone. In fact, the advection was the main factor to control the decadal variability of the OHC in the interior SCS, especially during the 1970s to 1990s.

The OHC time series actually indicates that the timing of the OHC minimum matched a volcanic eruption. Church et al. (2005) noted that after a volcanic eruption, the volcanic ash can be sent into the stratosphere and remain stable; thus, the solar radiation can be reduced, leading to decreased heat content in the ocean. For either the whole SCS or the boxes chosen, however, the local solar radiation does not match the OHC variability. Each volcanic eruption affects the global solar radiation, and then the global OHC; the effect on the SCS is in another form (i.e., advection dominated by atmospheric circulation) (Tozuka et al., 2009). The Mt. Pinatubo eruption that occurred near the SCS in 1992 may be partly responsible for the greatest cold anomaly in the OHC seen in Fig. 2 (bottom).

Acknowledgements. This research was supported by the National Basic Research Program of China (2011CB403502), by the National Nature Science Foundation of China under grants 41306014, 41206011, and 41221063, and by the Guangzhou Municipal Science and Technology Project under grant 12F084060027. Qinyan Liu is supported by the open program of The First Institute of Oceanography, SOA (grant: LDAA-2012-03). The authors wish to thank the reviewers and editors for their valuable constructive comments.

Edited by: J. M. Huthnance

References

Carton, J. A. and Giese B. S.: A Reanalysis of Ocean Climate Using Simple Ocean Data Assimilation (SODA), Mon. Weather Rev., 136, 2999–3017, 2008.

Carton, J. A., Chepurin, G., Cao, X., and Giese, B.: A simple ocean data assimilation analysis of the global upper ocean 1950–95, Part I: Methodology, J. Phys. Oceanogr., 30, 294–309, 2000a.

Carton, J. A., Chepurin, G., Cao, X., and Giese, B.: A simple ocean data assimilation analysis of the global upper ocean 1950-95. Part II: Results, J. Phys. Oceanogr., 30, 311–329, 2000b.

Church, J. A., White, N. J., and Arblaster, J. M.: Significant decadal-scale impact of volcanic eruptions on sea level and ocean heat content, Nature, 438, 74–77, doi:10.1038/nature04237, 2005.

Domingues, C. M., Church, J. A., White, N. J., Gleckler, P. J., Wijffels, S. E., Barker, P. M., and Dunn, J. R.: Improved estimates of upper-ocean warming and multi-decadal sea-level rise, Nature, 453, 1090–1093, doi:10.1038/nature07080, 2008.

Fang, G. H., Wei, Z. X., Choi, B. H., Wang, K., Fang, Y., and Li, W.: Inter-basin freshwater, heat and salt transport through the boundaries of the East and South China seas from a variable-grid global ocean circulation model, Sci. China, Ser. D, 46, 149– 161, 2003.

Gouretski, V. and Koltermann K. P.: How much is the ocean really warming?, Geophys. Res. Lett., 34, L01610, doi:10.1029/2006GL027834, 2007.

He, Y. and Guan, C.: Interannual and interdecadal variability in heat content of the upper ocean of South China Sea, Trop. Oceanol. 16, 23–29, 1997.

Levitus, S., Antonov, J., Boyer, T., and Stephens, C.: Warming of theworld ocean, Science, 287, 2225–2229, doi:10.1126/science.287.5461.2225, 2000.

Levitus, S., Antonov, J., and Boyer, T.: Warming of the world ocean, 1955–2003, Geophys. Res. Lett., 32, L02604, doi:10.1029/2004GL021592, 2005.

Levitus, S., Antonov, J. I., Boyer, T. P., Locarnini, R. A., Garcia, H. E., and Mishonov, A. V.: Global ocean heat content 1955–2008 in light of recently revealed instrumentation problems, Geophys. Res. Lett., 36, L07608, doi:10.1029/2008GL037155, 2009.

Liu, Q., Huang, R., and Wang, D.: Implication of the South China Sea throughflow for the interannual variability of the regional upper-ocean heat content, Adv. Atmos. Sci., 29, 54–63, 2012.

Locarnini, R. A., Mishonov, A. V., Antonov, J. I., Boyer, T. P., and Garcia, H. E.: World Ocean Atlas 2005, 3–6, in: NOAA Atlas NESDIS 61, Vol. 1, edited by: Levitus, S., Government Printing Office, Washington, DC, 2006.

Na, H., Kim, K.-Y., Chang, K.-I., Park, J. J., Kim, K., and Minobe, S.: Decadal variability of the upper ocean heat content in the East/Japan Sea and its possible relationship to northwestern Pacific variability, J. Geophys. Res., 117, C02017, doi:10.1029/2011JC007369, 2012.

Qiu, B. and Chen, S.: Multidecadal sea level and gyre circulation variability in the northwestern tropical Pacific Ocean., J. Phys. Oceanogr., 42, 193–206, 2012.

Qu, T., Kim, Y. Y., Yaremchuk, M., Tozuka, T., Ishida, A., and Yamagata, T.: Can Luzon Strait transport play a role in conveying the impact of ENSO to the South China Sea?, J. Climate, 17, 3644–3657, 2004.

Smith, R. D., Dukowicz, J. K., and Malone, R. C.: Parallel ocean general circulation modeling, Phys. D: Nonlinear Phenomena, 60, 38–61, 1992.

Tozuka, T., Qu, T., Masumoto, Y., and Yamagata, T.: Impacts of the South China Sea Throughflow on seasonal and interannual variations of the Indonesian Throughflow, Dyn. Atmos. Oceans, 47, 73–85, 2009.

Wang, D., Liu, Q., Huang, R., Du, Y., and Qu, T.: Interannual variability of the South China Sea throughflow inferred from wind data and an ocean data assimilation product, Geophys. Res. Lett., 33, L14605, doi:10.1029/2006GL026316, 2006a.

Wang, Y., Fang, G., Wei, Z., Qiao, F., and Chen, H.: Interannual variation of the South China Sea circulation and its relation to El Niño, as seen from a variable grid global ocean model, J. Geophys. Res., 111, C11S14, doi:10.1029/2005JC003269, 2006b.

Willis, J. K., Roemmich, D., and Cornuelle, B.: Interannual variability in upper ocean heat content, temperature, and thermosteric expansion on global scales, J. Geophys. Res., 109, C12036, doi:10.1029/2003JC002260, 2004.

Yan, Y., Qi, Y., and Zhou, W.: Interannual heat content variability in the South China Sea and its response to ENSO, Dynam. Atmos. Oceans, 50, 400–414, 2010.

Zhuang, W., Qiu, B., and Du, Y.: Lowfrequency western Pacific Ocean sea level and circulation changes due to the connectivity of the Philippine Archipelago, J. Geophys. Res. Oceans, 118, 6759–6773, doi:10.1002/2013JC009376, 2013.

Zong, H., Liu, Y., Xiu, P., Xu, Q., and Rong, Z.: Interannual variability of latent and sensible heat fluxes in the South China Sea, Chinese J. Ocean. Limnol., 28, 153–159, doi:10.1007/s00343-010-9241-9, 2010.

3

Roles of initial ocean surface and subsurface states on successfully predicting 2006–2007 El Niño with an intermediate coupled model

F. Zheng and J. Zhu

International Center for Climate and Environment Science (ICCES), Institute of Atmospheric Physics, Chinese Academy of Sciences, Beijing 100029, China

Correspondence to: F. Zheng (zhengfei@mail.iap.ac.cn)

Abstract. The 2006–2007 El Niño event, an unusually weak event, was predicted by most models only after the warming in the eastern Pacific had commenced. In this study, on the basis of an El Niño prediction system, roles of the initial ocean surface and subsurface states on predicting the 2006–2007 El Niño event are investigated to determine conditions favorable for predicting El Niño growth and are isolated in three sets of hindcast experiments. The hindcast is initialized through assimilation of only the sea surface temperature (SST) observations to optimize the initial surface condition, only the sea level (SL) data to update the initial subsurface state, or both the SST and SL data. Results highlight that the hindcasts with three different initial states can all successfully predict the 2006–2007 El Niño event 1 year in advance and that the hindcast initialized by both the SST and SL data performs best. A comparison between the various sets of hindcast results further demonstrates that successful prediction is more significantly affected by the initial subsurface state than by the initial surface condition. The accurate initial surface state can trigger the easier prediction of the 2006–2007 El Niño, whereas a more reasonable initial subsurface state can contribute to improving the prediction in the growth of the warm event.

1 Introduction

El Niño–Southern Oscillation (ENSO), which is one of the most striking interannual variabilities in the tropical Pacific Ocean, has been studied for several decades. Understanding the changes in its characteristics remains an important issue for worldwide environmental and socioeconomic interests (McPhaden et al., 2006; Ashok and Yamagata, 2009). During the past 2–3 decades, ENSO forecasts have made remarkable progress, reaching a stage in which skillful predictions can be made 6–12 months in advance. Many climate models from operational centers have routinely been used to make ENSO predictions in real time (Latif et al., 1998; Kirtman et al., 2002), although the skill of sea surface temperature (SST) forecasts in the equatorial Pacific is strongly model dependent and widely divergent across various prediction systems (Jin et al., 2008; Barnston et al., 2012). To improve the prediction skills, intensive effort has been made in both model improvements and better constraint of initial conditions. The latter is particularly important in the ocean because the memory for ENSO resides in the ocean (e.g., Neelin et al., 1998), and the importance of ocean data in making ENSO predictions has been demonstrated in a number of studies (e.g., Ji and Leetmaa, 1997; Ji et al., 2000; Alves et al., 2004; Keenlyside et al., 2005; Zheng et al., 2006, 2007; Zheng and Zhu, 2008; Yang et al., 2010; Zhu et al., 2012).

However, as demonstrated by Barnston et al. (2012), an apparent retrogression in skill exists for ENSO predictions in the 2002–2011 study period, compared with those in the 1981–2010 period, and the real-time ENSO prediction skill of models in the past decade is significantly lower than that of the less-advanced models of the 1980s and 1990s. For example, the 9-year sliding correlation for the hindcasts during the 1981–2010 period has an average score of 0.65 at a 6-month lead time, but the prediction skill of the correlation decreases to 0.42 for the 2002–2011 period (Barnston et al., 2012). Specifically, in the case of the 2006–2007 El Niño event, which was unusually weak (Hackert et al., 2007; McPhaden, 2008; Yang et al., 2010), most seasonal climate forecasts from operational centers only predicted the event after the warming had already become apparent and was basinwide (McPhaden, 2008). The composite forecast for December 2006 from the International Research Institute for Climate Prediction (IRI) indicates that the warm anomaly was underpredicted by more than 0.5 °C. (A review of forecast skills is available on the IRI website at http://iri.columbia.edu.)

The limited predictability may be attributed to factors such as errors in the oceanic initial conditions, state-dependent stochastic forcing, or model errors (Palmer, 2001). In this study, we investigated possible improvements in predicting this warm event by focusing on the role of the oceanic initial states. On the basis of the ensemble Kalman filter (EnKF) algorithm (e.g., Evensen, 2009), the roles of the initial ocean surface and subsurface states on the 2006–2007 El Niño predictions were examined in three sets of retrospective forecast experiments. In the first set, only the SST data was assimilated to provide accurate surface initial conditions, whereas only sea level (SL) data was assimilated to provide the updated information of the subsurface water temperature entrained into the mixed layer in the model. In the third set, both SST and SL data were assimilated. The forecast differences in the three sets of the retrospective experiments with initializations by the three separate data assimilation schemes were then examined to isolate the effects of the various initial states on predicting the warm event.

2 Model and initialization schemes

In this section, we briefly describe the model components and the designs of three initialization schemes on the basis of the EnKF algorithm.

2.1 Description of the model components

The ENSO-coupled model we used here is an intermediate coupled model (ICM) that was first developed by Keenlyside and Kleeman (2002) and Zhang et al. (2005); various components of the ICM are briefly described here (Fig. 1). Its ocean dynamical component consists of both linear and nonlinear components. The former was essentially a McCreary-type (1981) modal model but was extended to include horizontally varying background stratification. Different from the commonly used Zebiak–Cane model (1987), this ICM considers the realistic vertical structure of the upper ocean, spatially varying stratification, and partial nonlinear effects. For example, when 10 vertical modes are included in the calculation of dynamic ocean fields, along with a parameterization of the local Ekman-driven upwelling, this model yields realistic simulations of the mean equatorial circulation and its variability in the tropical Pacific (Keenlyside and Kleeman, 2002).

The ocean model domain extends from 33.5° S to 33.5° N and from 124° E to 70° W, covering the tropical Pacific basin

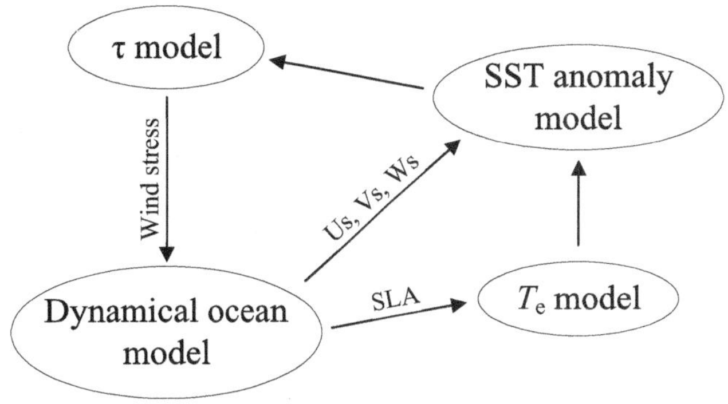

Figure 1. A schematic illustrating the ICM consisting of a dynamical ocean model, an SST anomaly model with an empirical parameterization for T_e in terms of SLA, and a statistical atmospheric wind stress (τ) model. Adopted from Zhang et al. (2005).

with a realistic representation of continents. The model has a 2° zonal grid spacing and a meridional grid stretching from 0.5° within 10° of the Equator to 3° at the northern and southern boundaries. Vertically, a 5500 m flat-bottom ocean is assumed; the linear component has 33 levels, chosen as in Levitus (1982), with 8 levels in the upper 125 m. The two layers used to simulate the nonlinear effects and high-order baroclinic modes span the upper 125 m and are divided by a surface mixed layer whose depth is prescribed from observations (Monterey and Levitus, 1997).

An SST anomaly model is embedded within this dynamical framework. The governing equation (i.e., Eq. 1) describes the evolution of surface mixed layer temperature anomalies, driven by ocean horizontal advection and entrainment associated with both specified mean and simulated anomalous currents.

$$\begin{aligned}\frac{\partial T'}{\partial t} =& -u'\frac{\partial \overline{T}}{\partial x} - (\overline{U}+u')\frac{\partial T'}{\partial x} - v'\frac{\partial \overline{T}}{\partial y} - (\overline{V}+v')\frac{\partial T'}{\partial y} \\ & -\left\{(\overline{W}+w')M(-\overline{W}-w') - \overline{W}M(-\overline{W})\right\}\frac{\overline{T}_e-\overline{T}}{H} \\ & -(\overline{W}+w')M(-\overline{W}-w')\frac{T'_e-T'}{H} \\ & +\frac{\kappa_h}{H}\nabla_h(H\nabla_h T') + \frac{2\kappa_v}{H(H+H_2)}(T'_e-T') - \alpha T'\end{aligned}, \quad (1)$$

where, T' and T'_e are anomalies of SST and the temperature of subsurface water entrained into the mixed layer; u', v' and w' are anomalies of ocean currents (horizontal and vertical) in the surface mixed layer; H is the depth of the mixed layer; H_2 is a constant (125 m); $M(x)$ is the Heaviside function; κ_h and κ_v are horizontal diffusion and vertical mixing parameters, respectively; and other variables are mean climatology fields. The surface heat flux is parameterized as being negatively proportional to local SST anomalies. These components are used to calculate SST changes from the equation given above.

As demonstrated by Zhang et al. (2005), having a realistic parameterization for the temperature of the subsurface water entrained into the mixed layer (T_e) is crucial to the performance of SST simulations in the equatorial Pacific. Based on the empirical orthogonal function (EOF) method, an empirical T_e model was constructed from a historical relationship between the interannual variations in T_e and SL during the period 1963–1996 and was demonstrated to be effective in improving the SST simulations. The ocean model is coupled with a statistical atmospheric model, which specifically relates the wind stress (τ) to SST anomaly fields (Zhang et al., 2005). The τ field was constructed from an SVD (singular value decomposition) of the covariance matrix that was calculated from the time series of observed monthly anomalies of SST, zonal, and meridional wind stress components during the period of 1963–1996. To achieve reasonable amplitudes, the first five SVD modes were retained in estimating τ fields from SST anomalies.

All coupled model components exchange simulated anomaly fields (Fig. 1). At each time step, the dynamical ocean component produces anomalous ocean pressure, mixed-layer averaged currents, and vertical velocity at the base of the mixed layer (entrainment). Then, from the SL anomaly, a T_e anomaly is calculated with the EOF-based T_e model, serving as an interface between the SST anomaly and dynamical model components. The SST anomaly model then takes the T_e and ocean circulation fields (prescribed mean and simulated), and the observed climatologies of mean SST and vertical temperature gradient to update the SST anomaly. The resultant SST anomaly is then used to calculate wind stress anomalies with the SVD-based τ model. These are then used to force the dynamical ocean model on the next time step. Information between the atmosphere (τ) and the ocean (SST) is exchanged once a day, and the T_e anomalies for the SST anomaly model are also updated once a day from the SL anomalies.

2.2 Initialization schemes

An EnKF data assimilation system for initializing the ICM was first developed by Zheng et al. (2006) and Zheng et al. (2007). In the system, the EnKF is implemented by using an ensemble square-root filter algorithm with no perturbation of observations (e.g., Evensen, 2004). This EnKF data assimilation scheme was further improved by using a balanced, multivariate model-error approach (Zheng and Zhu, 2008) and was upgraded to use mean preserving transformations in the square-root scheme (Sakov and Oke, 2008; Evensen, 2009). The observed interannual SST and SL anomalies can then be consistently assimilated into the coupled model to provide an accurate initial condition for real-time SST prediction (Zheng and Zhu, 2010). In this study, we performed three initialization schemes to isolate the effects of the various initial states on ENSO prediction (Table 1). The first scheme, which is known as Assim_SST, assimilates only the observed SST anomalies into the ICM during the initialization process. In the second scheme, referred to as Assim_SL, the model states of the ICM are periodically updated by assimilating only the SL anomaly data by using EnKF. Moreover, both the SST and SL observations are assimilated into the coupled model to initialize the ICM for the ENSO forecast by using the Assim_SST + SL scheme.

As previously mentioned, the ICM is formed by coupling the ocean model to a simple statistical atmospheric model for interannual wind stress anomalies (τ), and the atmospheric state is a slave only to the modeled SST based on the SVD τ model. Therefore, assimilation of SST observations can provide an accurate initial surface thermal state of ICM and supply a better estimation of the initial atmospheric state. Thus, the Assim_SST scheme in this study represents a better initialization scheme of the surface model states of ICM (Zheng et al., 2006). As described in Sect. 2.1, a distinguished feature added to the dynamical ocean model of ICM for a better performance of SST simulation in the equatorial Pacific is an empirical parameterization for T_e. Observations and modeling studies indicate that variations in SL and the thermocline are well correlated over the equatorial Pacific (e.g., Wyrtki, 1975; Wang and McPhaden, 2000). As such, subsurface temperature anomalies associated with thermocline displacements are closely related with SL variability, which provides a statistical basis for inferring subsurface temperature anomalies from SL variability. Moreover, T_e can be directly estimated from the modeled SL anomalies by the EOF T_e model developed from the historical relationship between the interannual variations in T_e and SL. Thus, assimilation of SL data can directly improve the accuracy of the subsurface thermal state (T_e) of ICM (Zheng et al., 2007), and the Assim_SL scheme represents a better initialization of the subsurface model states of ICM. The Assim_SST + SL scheme is the best initialization scheme of the three different schemes because it can perfect both the surface and subsurface initial states for ICM to provide an ENSO forecast (Zheng and Zhu, 2008). Moreover, to further quantify the relative effects of the accuracy of the surface and subsurface initial states on the 2006–2007 El Niño predictions, the forecast differences between the two single-observational initialization schemes, Assim_SST and Assim_SL, and the double-observational initialization scheme, Assim_SST + SL, were examined to illustrate the relative contributions of the various initial states to ENSO prediction.

2.3 Data sets

The observed monthly SST anomaly fields were from the extended reconstructed SST (ERSSTv3b; Smith et al., 2008), and the monthly-averaged altimeter products were produced by Ssalto/Duacs and were distributed by Aviso, with support from Cnes (http://www.aviso.oceanobs.com/duacs/). Both of the SST and SL data were assimilated into the coupled model once a month, and the SST data were also used to verify the model predictions.

Table 1. Summary of the initialization scheme design.

Initialization scheme	Assimilated data		Description
	SST	SL	
Assim_SST	√		Optimizing the initial surface conditions (i.e., SST and wind stress)
Assim_SL		√	Updating the initial subsurface states (i.e., SL and T_e)
Assim_SST + SL	√	√	Improving the accuracy of both the initial surface and subsurface states

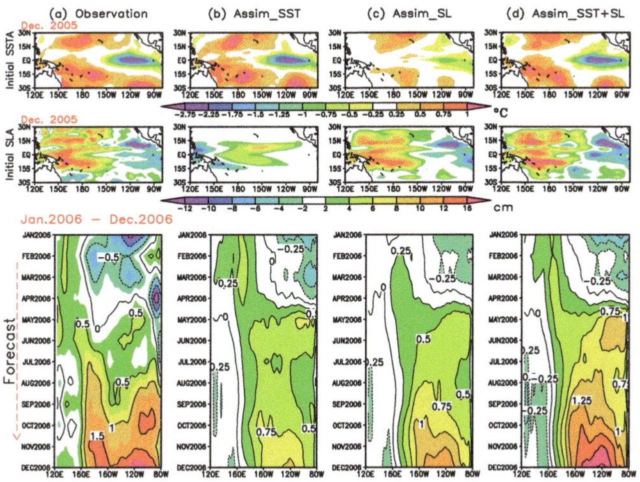

Figure 2. Initial conditions of anomalous SST (top row), SL (middle row), and forecasted SST anomalies (bottom row) from the **(a)** observations, **(b)** Assim_SST results, **(c)** Assim_SL results, and **(d)** Assim_SST + SL results, respectively. The initial fields are from December 2005, and the 12-month forecasts start in January 2006.

3 Isolating the impact of oceanic initial conditions on predicting the 2006–2007 El Niño

We initialized the coupled forecasts of the 2006–2007 El Niño event with the three oceanic analyses on 1 January 2006, and all forecasts had the same atmospheric initial condition. Thus, the outcome depended on the effectiveness of the oceanic initial condition for predicting the 2006–2007 El Niño. The initial conditions of anomalous SST and SL from the Assim_SST, Assim_SL, and Assim_SST + SL analysis results are compared with the observations in the top and middle rows in Fig. 2. As expected, for both the amplitude and spatial pattern of the observed SST, the Assim_SST scheme had a better analysis result than the Assim_SL scheme. However, the Assim_SL scheme had a much more similar assimilation result to the observed SL anomaly field than the Assim_SST scheme. Overall, the Assim_SST + SL scheme had a more reasonable analysis result in both the SST and SL and can provide a more dynamically consistent initial condition for both the surface and subsurface states (Zheng and Zhu, 2008).

The SST hindcasts are also compared with observations in Fig. 2. After initialization, all three hindcasts developed a warm anomaly event during the entire 12 months of the forecast. All of the hindcast experiments with the three different initial conditions successfully predicted the 2006–2007 El Niño event before 12 months, and the hindcast initialized by the Assim_SST + SL scheme performed best. The Assim_SST + SL hindcast exhibited a more realistic evolution during the developing and maturing stages of this El Niño and had the earliest and largest warming of the three experiments, with the first warming in June 2006 and a peak anomaly in December 2006. Moreover, this hindcast had a similar magnitude to the observed warming. The hindcast initialized from the Assim_SL analysis predicted a slightly weaker warming development of the event. Its peak anomaly was approximately 1.3 °C, weaker than that in the Assim_SST + SL forecast and in the observed SST. With the Assim_SST analysis, the warm anomaly developed much slower than the prediction initialized by the Assim_SL analysis or the Assim_SST + SL analysis. For example, in the Assim_SST hindcast, it takes 12 months for the forecasted SST anomalies increasing from 0.2 to 1.0 °C; while in the Assim_SL hindcast, it takes 11 months for the forecasted SST anomalies increasing from 0.2 to 1.3 °C; and in the Assim_SST + SL hindcast, it also takes 12 months for the forecasted SST anomalies increasing from 0.2 to 2.0 °C. The peak anomaly of the Assim_SST hindcast was the weakest among the three hindcast experiments.

When sea level data are assimilated, one major concern is how to project the surface information downward to subsurface layers (Fu and Zhu, 2011). Currently, satellite altimetric observations are assimilated into the coupled general circulation models (CGCMs) or oceanic general circulation models (OGCMs), either by projecting the surface signal to the subsurface due to the statistical relationships between SLA (SL anomalies) and subsurface temperatures (e.g., Fischer et al., 1997; Ji et al., 2000; Masina et al., 2001) or by the inherent multivariate relation derived from the ensembles in order to provide a straightforward way to assimilate SL data through some ensemble-based data assimilation methods (e.g., Oke et al., 2008; Counillon and Bertino, 2009). In this work, similar to the work of Fischer et al. (1997), the relationship between SLA and subsurface thermal state (T_e) is prescribed in a statistical EOF SL-T_e model, and the assimilated anomalous

SL information can be directly projected onto the anomalous subsurface temperature fields of the ocean. Then, the updated subsurface temperature will act to improve the accuracy of the SST prediction during the following forecast process (Zheng et al., 2007).

To illustrate the effect of ocean assimilation performance on the prediction skill for the 2006–2007 El Niño event, the differences between the three forecasts are shown in Figs. 3 and 4. In terms of the observed development of the warm event, the prediction was best when assimilating both the SST and SL observations into the coupled model, Assim_SST + SL. Therefore, the forecast differences between the two single-observational initialization schemes, Assim_SST and Assim_SL, with the double-observational initialization scheme, Assim_SST + SL, can help to isolate the relative effects of the accuracy of the surface and subsurface initial states on the 2006–2007 El Niño predictions.

Figure 3 shows the initial and forecast differences estimated between the Assim_SST + SL and Assim_SL forecasts for SST, zonal wind stress (τ_x), SL, and T_e, illustrating the effects of the initial surface states, SST and wind stress, on predicting the 2006–2007 El Niño event. For the initial period, a more reasonable zonal SST gradient was captured by assimilating the SST observational data into the model in which the SST is anomalously warm west of the dateline and cold to the east (Fig. 3a). This type of zonal SST gradient is convenient for generating stronger westerly winds over the equatorial Pacific (Fig. 3b) at the initial time, which results in the triggering of a considerable El Niño event. As indicated by the SL anomalies (Fig. 3c), the strong westerly wind stress anomalies in the lower atmosphere initiated an oceanic downwelling Kelvin wave, which has been alluded to as a facilitator of El Niño (Picaut et al., 2002; Hackert et al., 2007). Forced by the downwelling Kelvin wave, warm water accumulated in the western Pacific tended to sink and began to propagate eastward, resulting in a net warming at the subsurface layers in the eastern equatorial Pacific during the first half of 2006 (Fig. 3d). When the subsurface warm water reached the eastern tropical Pacific and warmed the surface water through vertical mixing in summer 2006, the warm SST activated the local coupled air–sea interactions to produce atmospheric–oceanic anomalies that developed and evolved with continuous warming during fall 2006. Finally, a 0.4 °C warmer El Niño condition at the end of 2006 was predicted when considering the accuracy of the initial surface ocean–atmosphere condition.

Similarly, the initial and forecasted differences estimated between the Assim_SST + SL and Assim_SST forecasts for SL, T_e, SST, and τ_x are shown in Fig. 4, illustrating the effects of the initial subsurface states, SL and T_e, on prediction of the 2006 El Niño event. Different from the discrepancy induced by the initial surface conditions, the Assim_SST + SL hindcast had an obviously stronger variation in SL than that of Assim_SST, which is associated with the propagation of two downwelling Kelvin waves during the forecast process.

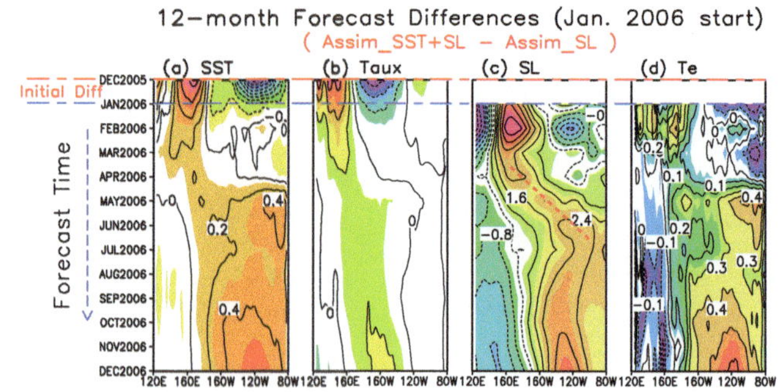

Figure 3. Initial and forecasted differences between the Assim_SST + SL hindcast and the Assim_SL hindcast for **(a)** SST, **(b)** zonal τ_x, **(c)** SL, and **(d)** T_e, illustrating the effects of the initial surface states, SST and τ on predicting the 2006–2007 El Niño event. The dashed red line in **(c)** represents the propagation of the downwelling Kelvin wave. The contour interval is 0.2 °C in **(a)**, 0.08 dyn cm^{-2} in **(b)**, 0.8 cm in **(c)**, and 0.1 °C in **(d)**. The initial differences in December 2005 for the surface states (SST and τ) are demonstrated in the left two panels.

As shown in Fig. 4c and b, the early warming at both the surface and subsurface layers over the eastern equatorial Pacific in the Assim_SST + SL hindcast corresponds to a weak Kelvin downwelling wave initiated east of the dateline before 2006 (Fig. 4a). An additional stronger Kelvin downwelling wave propagated eastbound in approximately January 2006 in the equatorial western Pacific (Fig. 4a, b) and reached the eastern tropical Pacific in mid-2006. These two eastward downwelling propagations resemble the observed events and agree with the timing of the large deepening anomalies of the thermocline depth in the equatorial eastern Pacific (not shown), the details of which have been reported by Hackert et al. (2007). Finally, a 0.8 °C warmer El Niño condition at the end of 2006 was predicted when considering the accuracy of the initial subsurface condition; additionally, much stronger coupled air–sea interaction and vertical mixing induced by the SL assimilation were evident during the second half of 2006 when compared to those induced by the SST assimilation.

As a result, the inclusion of more reasonable and accurate initial conditions during the 2006–2007 El Niño forecast process can lead to better predictions. Figure 5 compares the hindcasts initiated from the three initialization schemes for the 2006–2007 El Niño episode. The hindcasts initialized by the Assim_SST + SL scheme could successfully predict the onset, development, and decay of the 2006 El Niño at all times prior to the event, although some small errors remained in the onset and magnitude of the 2006–2007 El Niño event when forecasted 9 months ahead. The hindcasts initialized by the Assim_SL scheme did not perform as well as those initialized by the Assim_SST + SL scheme; however, they excelled over the those initialized by the Assim_SST scheme. The hindcasts initialized by the Assim_SST scheme diverged

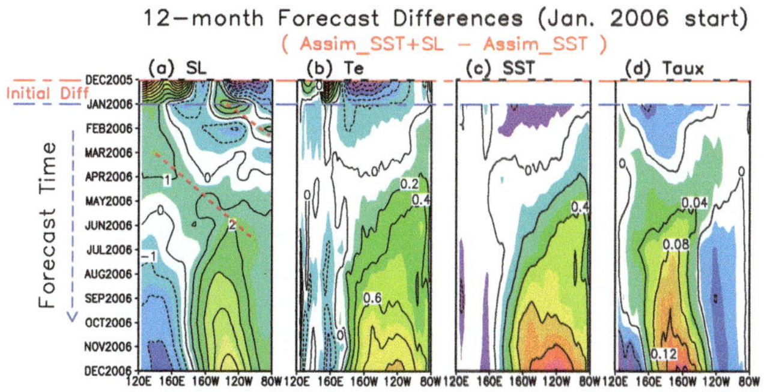

Figure 4. Initial and forecasted differences between the Assim_SST + SL hindcast and Assim_SST hindcast for **(a)** SL, **(b)** T_e, **(c)** SST, and **(d)** zonal τ_x, illustrating the effects of the initial subsurface states, SL and T_e on predicting the 2006–2007 El Niño event. The dashed red line in **(a)** represents the propagation of the downwelling Kelvin wave. The contour interval is 1.0 cm in **(a)**, 0.2 °C in **(b)** and **(c)**, and 0.04 dyn cm^{-2} in **(d)**. The initial differences in December 2005 for the subsurface states (SL and T_e) are demonstrated in the two panels on the left.

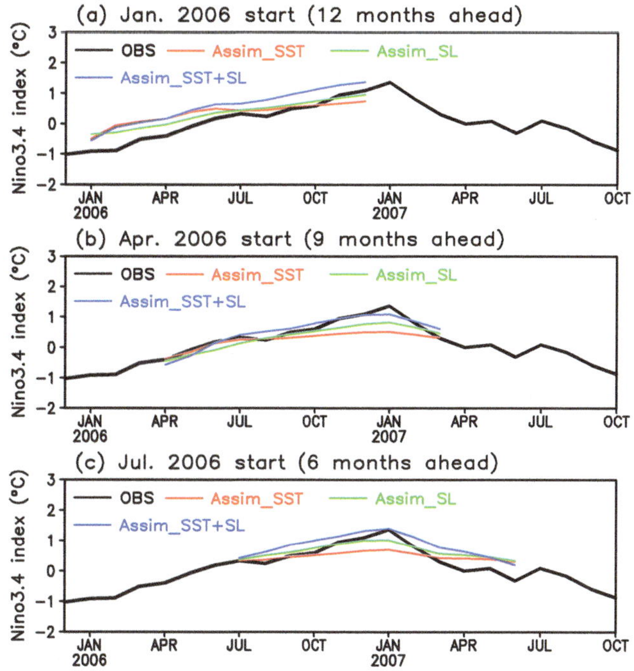

Figure 5. Comparisons of the hindcast results for the 2006–2007 El Niño event. The thick black curves are the observed Niño 3.4 SST anomalies, and the thin curves of red, green, and blue are the predictions initialized by the Assim_SST + SL, Assim_SST, and Assim_SL assimilation results, respectively. **(a)**, **(b)**, and **(c)** are hindcasts that started 12, 9, and 6 months, respectively, ahead of the peak of the 2006–2007 El Niño event.

more obviously from the actual observations when using 12-, 9-, and 6-month time points ahead of the event. However, all of these hindcasts were able to predict the development and decay trend of the event with little departure from the observations. Our methodology of isolating the roles of the initial oceanic states reproduced all of the major features of the 2006–2007 El Niño event. The successful prediction of this event demonstrates that benefits are gained from both the corresponding initial surface and subsurface conditions, although the initial subsurface conditions were more effective.

4 Concluding remarks

The IRI composite forecast table for the December 2006 validation showed that, in general, the amplitude of this 2006–2007 El Niño event was under-predicted by more than 0.5 °C prior to September 2006. By that time, warming in the east should have naturally triggered the dynamical and statistical coupled models to predict El Niño through Bjerknes feedback (McPhaden, 2008). That is, the models were deficient and began to predict El Niño only after the warming in the eastern Pacific had commenced. The results in this study demonstrate the unique ability of data assimilation on improving the initial ocean states to show warming in the coupled system 12 months prior to the event.

Moreover, the impact of the initial details in surface and subsurface states estimated from oceanic data assimilation on coupled forecasts of the 2006–2007 El Niño event was investigated and was isolated in this study. The conceptually simple technique of isolating the role of the initial ocean state through determining the differences in pairs of initialization experiments has been shown to increase the accuracy prediction of the 2006–2007 El Niño. This study further demonstrated that for the 2006–2007 El Niño event, the forecast skill is more significantly affected by the initial state of the subsurface field than by that of the surface field (Zheng et al., 2007; Zheng and Zhu, 2008), suggesting that advanced data assimilation methodology combined with accurate subsurface information should be used in conjunction with coupled forecasts to extend the lead time and accuracy of ENSO forecast systems (e.g., Zhu et al., 2012). Analysis of the results show that the initial conditions having optimal subsurface structures such as SL and T_e fields from the EnKF analysis mainly contributed to improving the intensity prediction in the growth of the 2006–2007 El Niño, although an accurate initial surface state such as SST or wind stress fields can help to easily trigger a warming tendency for predicting the 2006–2007 El Niño event.

Until now, initializing an ENSO-coupled model had mostly focused on generating more realistic ocean surface and subsurface initial states by advanced data assimilation methods (e.g., Alves et al., 2004; Behringer and Xue, 2004; Cazes-Boezio et al., 2010; Vidard et al., 2010). And the assimilation of subsurface temperature observations has been demonstrated to lead to improved forecasts (e.g., Ji and Leetmaa 1997; Zhu et al., 2012). However, over the tropical Pacific Ocean, the distribution of the subsurface data in space and time is quite irregular and sometimes very sparse, even Argo profiles have become available since the 2000s (Balmaseda and Anderson, 2009). Previous works indicated that the assimilation of SL anomalies yields comparable results to the case in which subsurface temperature anomalies were assimilated (e.g., Segschneider et al., 2001). Thus, an alternative data type is given by SL observations due to the availability of the altimetry data in near-real time and the advantage of high spatiotemporal resolution. In this work, we investigated the potential impact of SL assimilation on ENSO forecasts in the 2006–2007 El Niño case. The subsurface thermal state (T_e) can be directly updated by the assimilated SL anomalies through a pre-established statistical EOF SL-T_e model. At the same time, uncertainties in the parameterization of entrainment and vertical mixing continue to lower the SST prediction skill in most state-of-the-art CGCMs (e.g., Latif et al., 2001; Zhang et al., 2005). Our results indicate that there is substantial potential for improving ENSO prediction by minimizing the uncertainties of parameterizing subsurface thermal effects, from projecting the observed SL anomaly information onto subsurface layers based on the statistical relationship.

It should be noted that this work is limited to only one event, and it is difficult to make generic statements about all El Niño events. However, the conceptually simple and robust, technique used here will help coupled models to isolate the role of the initial state of the ocean and to evaluate the conditions favorable for improving El Niño predictions.

Acknowledgements. The authors wish to thank two anonymous reviewers for their helpful comments. This work was supported by the National Program for Support of Top-notch Young Professionals, the National Basic Research Program of China (grant no. 2012CB955202), the Chinese Academy of Sciences (WPOS: XDA10010405) project: Western Pacific Ocean System: Structure, Dynamics and Consequences.

Edited by: D. Stevens

References

Alves, O., Balmaseda, M., Anderson, D., and Stockdale, T.: Sensitivity of dynamical seasonal forecasts to ocean initial conditions, Q. J. Roy. Meteor. Soc., 130, 647–668, 2004.

Ashok, K. and Yamagata, T.: Climate change: The El Niño with a difference, Nature, 461, 481–484, 2009.

Balmaseda, M. A. and Anderson, D.: Impact on initialization strategies and observations on seasonal forecast skill, Geophys. Res. Lett., 36, L01701, doi:10.1029/2008GL035561, 2009.

Barnston, A. G., Tippett, M. K., L'Heureux, M. L., Li, S., and DeWitt, D. G.: Skill of real-time seasonal ENSO model predictions during 2002–2011: is our capability increasing?, B. Am. Meteorol. Soc., 93, 631–651, 2012.

Behringer, D. W. and Xue, Y.: Evaluation of the global ocean data assimilation system at NCEP: The Pacific ocean. Preprints, Eighth Symp. on Integrated Observing and Assimilation Systems for Atmosphere, Oceans, and Land Surface, Seattle, WA, Amer. Meteor. Soc., 2.3, 2004.

Cazes-Boezio, G., Menemenlis, D., and Mechoso, C. R.: Impact of ECCO ocean-state estimates on the initialization of seasonal climate forecasts, J.Climate, 21, 1929–1947, 2010.

Counillon, F. and Bertino, L.: Ensemble optimal interpolation: Multivariate properties in the Gulf of Mexico, Tellus A, 61, 296–308, 2009.

Evensen, G.: Sampling strategies and square root analysis schemes for the EnKF, Ocean Dynam., 54, 539–560, 2004.

Evensen, G.: Data Assimilation: The Ensemble Kalman Filter, 2nd Edn., Springer, Berlin, 2009.

Fischer, M., Flugel, M., Ji, M., and Latif, M.: The impact of data assimilation on ENSO simulations and predictions, Mon. Weather Rev., 125, 819–829, 1997.

Fu, W.-W. and Zhu, J.: Effects of sea level data assimilation by ensemble optimal interpolation and 3D variational data assimilation on the simulation of variability in a tropical Pacific model, J. Atmos. Ocean. Tech., 28, 1624–1640, 2011.

Hackert, E., Ballabrera-Poy, J., Busalacchi, A. J., Zhang, R.-H., and Murtugudde, R.: Role of the initial ocean state for the 2006 El Niño, Geophys. Res. Lett., 34, L09605, doi:10.1029/2007GL029452, 2007.

Ji, M. and Leetmaa, A.: Impact of data assimilation on ocean initialization and El Niño prediction, Mon. Weather Rev., 125, 742–753, 1997.

Ji, M., Reynolds, R. W., and Behringer, D. W.: Use of TOPEX/Poseidon sea level data for ocean analyses and ENSO prediction: Some early results, J. Climate, 13, 216–231, 2000.

Jin, E. K., Kinter III, J. L., Wang, B., Park, C.-K., Kang, I.-S., Kirtman, B. P., Kug, J.-S., Kumar, A., Luo, J.-J., Schemm, J., Shukla, J., and Yamagata, T.: Current status of ENSO prediction skill in coupled ocean–atmosphere models, Clim. Dynam., 31, 647–664, 2008.

Keenlyside, N. and Kleeman, R.: Annual cycle of equatorial zonal currents in the Pacific, J. Geophys. Res., 107, 3093, doi:10.1029/2000JC000711, 2002.

Keenlyside, N., Latif, M., Botzet, M., Jungclaus, J., and Schulzweida, U.: A coupled method for initialising ENSO forecasts using SST, Tellus A, 57, 340–356, 2005.

Kirtman, B. P., Shukla, J., Balmaseda, M., Graham, N., Penland, C., Xue, Y., and Zebiak, S. E.: Current status of ENSO forecast skill: A report to the Climate Variability and Predictability (CLIVAR) Working Group on Seasonal to Interannual Prediction, WCRP Informal Report No. 23/01, 31 pp., 2002.

Latif, M., Anderson, D., Barnett, T., Cane, M., Kleeman, R., Leetmaa, A., O'Brien, J., Rosati, A., and Schneider, E.: A review of the predictability and prediction of ENSO, J. Geophys. Res., 103, 14375–14393, 1998.

Latif, M., Sperber, K., Arblaster, J., Braconnot, P., Chen, D., Colman, A., Cubasch, U., Cooper, C., Delecluse, P., Dewitt, D., Fairhead, L., Flato, G., Hogan, T., Ji, M., Kimoto, M., Kitoh, A., Knutson, T., Le Treut, H., L., T., Manabe, S., Marti, O., Mechoso, C., Meehl, G., Power, S., Roeckner, E., Sirven, J., Terray, L., Vintzileos, A., Voß, R., Wang, B., Washington, W., Yoshikawa,

I., Yu, J., and Zebiak, S.: ENSIP: The El Niño simulation intercomparison project, Clim. Dynam., 18, 255–276, 2001.

Levitus, S.: Climatological Atlas of the World Ocean, NOAA Prof. Paper 13, 173 pp. and 17 microfiche, 1982.

Masina, S., Pinardi, N., and Navarra, A.: A global ocean temperature and altimeter data assimilation system for studies of climate variability, Clim. Dynam., 17, 687–700, 2001.

McCreary, J. P.: A linear stratified ocean model of the equatorial undercurrent, Philos. T. R. Soc. Lond., 298, 603–635, 1981.

McPhaden, M. J.: Evolution of the 2006–07 El Niño: The role of intraseasonal to interannual time scale dynamics, Adv. Geosci., 14, 219–230, 2008, http://www.adv-geosci.net/14/219/2008/.

McPhaden, M. J., Zebiak, S. E., and Glantz, M. H.: ENSO as an integrating concept in earth science, Science, 314, 1739–1745, 2006.

Monterey, G., and Levitus, S.: Seasonal variability of mixed layer depth for the world ocean, NOAA Atlas NESDIS 14, 96 pp., 1997.

Neelin, J. D., Battisti, D. S., Hirst, A. C., Jin, F.-F., Wakata, Y., Yamagata, T., and Zebiak, S. E.: ENSO theory, J. Geophys. Res., 103, 14262–14290, 1998.

Oke, P. R., Brassington, G. B., Griffin, D. A., and Schiller, A.: The Bluelink ocean data assimilation system (BODAS), Ocean Model., 21, 46–70, 2008.

Palmer, T. N.: A nonlinear dynamical perspective on model error: A proposal for non-local stochastic-dynamic parametrization in weather and climate prediction models, Q. J. Roy. Meteor. Soc., 127, 279–304, 2001.

Picaut, J., Hackert, E., Busalacchi, A. J., Murtugudde, R., and Lagerloef, G. S. E.: Mechanisms of the 1997–1998 El Niño–La Niña, as inferred from space-based observations, J. Geophys. Res., 107, 3037, doi:10.1029/2001JC000850, 2002.

Sakov, P. and Oke, P. R.: Implications of the form of the ensemble transformations in the ensemble square root filters, Mon. Weather Rev., 136, 1042–1053, 2008.

Segschneider, J., Anderson, D. L. T., Vialard, J., Balmaseda, M., Stockdale, T. N., Troccoli, A., and Haines, K.: Initialization of Seasonal Forecasts Assimilating Sea Level and Temperature Observations, J. Climate, 14, 4292–4307, 2001.

Smith, T. M., Reynolds, R. W., Peterson, T. C., and Lawrimore, J.: Improvements to NOAA's Historical Merged Land-Ocean Surface Temperature Analysis (1880-2006), J. Climate, 21, 2283–2296, 2008.

Vidard, A., Balmaseda, M., and Anderson, D.: Assimilation of altimeter data in the ECMWF ocean analysis system 3, Mon. Weather Rev., 137, 1393–1408, 2010.

Wang, W. and McPhaden, M. J.: The surface-layer heat balance in the equatorial Pacific Ocean. Part II: Interannual variability, J. Phys. Oceanogr., 30, 2989–3008, 2000.

Wyrtki, K.: El Niño—The dynamic response of the equatorial Pacific Ocean to atmospheric forcing, J. Phys. Oceanogr., 5, 527–584, 1975.

Yang, S.-C., Rienecker, M., ad Keppenne, C.: The Impact of Ocean Data Assimilation on Seasonal-to-Interannual Forecasts: A Case Study of the 2006 El Niño Event, J. Climate, 23, 4080–4095, 2010.

Zebiak, S. E. and Cane, M. A.: A model El NinoSouthern Oscillation, Mon. Weather Rev., 115, 2262–2278, 1987.

Zhang, R.-H., Zebiak, S. E., Kleeman, R., and Keenlyside, N.: Retrospective El Niño forecast using an improved intermediate coupled model, Mon. Weather Rev., 133, 2777–2802, 2005.

Zheng, F. and Zhu, J.: Balanced multivariate model errors of an intermediate coupled model for ensemble Kalman filter data assimilation, J. Geophys. Res., 113, C07002, doi:10.1029/2007JC004621, 2008.

Zheng, F. and Zhu, J.: Coupled assimilation for an intermediated coupled ENSO prediction model, Ocean Dynam., 60, 1061–1073, doi:10.1007/s10236-010-0307-1, 2010.

Zheng, F., Zhu, J., Zhang, R.-H., and Zhou, G.-Q.: Ensemble hindcasts of SST anomalies in the tropical Pacific using an intermediate coupled model, Geophys. Res. Lett., 33, L19604, doi:10.1029/2006GL026994, 2006.

Zheng, F., Zhu, J., and Zhang, R.-H.: The impact of altimetry data on ENSO ensemble initializations and predictions, Geophys. Res. Lett., 34, L13611, doi:10.1029/2007GL030451, 2007.

Zhu, J., Huang, B., Marx, L., Kinter III, J. L., Balmaseda, M. A., Zhang, R.-H., and Hu, Z.-Z.: Ensemble ENSO hindcasts initialized from multiple ocean analyses, Geophys. Res. Lett., 39, L09602, doi:10.1029/2012GL051503, 2012.

4

The instability of diffusive convection and its implication for the thermohaline staircases in the deep Arctic Ocean

S.-Q. Zhou[1], L. Qu[1,2], Y.-Z. Lu[1,2], and X.-L. Song[1,2]

[1]State Key Laboratory of Tropical Oceanography, South China Sea Institute of Oceanology, 164 West Xingang Road, Haizhu District, Guangzhou 510301, China
[2]University of Chinese Academy of Sciences, Beijing, 100049, China

Correspondence to: S.-Q. Zhou (sqzhou@scsio.ac.cn)

Abstract. In the present study, the classical description of diffusive convection is updated to interpret the instability of diffusive interfaces and the dynamical evolution of the bottom layer in the deep Arctic Ocean. In the new consideration of convective instability, both the background salinity stratification and rotation are involved. The critical Rayleigh number of diffusive convection is found to vary from 10^3 to 10^{11} in the deep Arctic Ocean as well as in other oceans and lakes. In such a wide range of conditions, the interface-induced thermal Rayleigh number is shown to be consistent with the critical Rayleigh number of diffusive convection. In most regions, background salinity stratification is found to be the main hindrance to the occurrence of convecting layers. With the new parameterization, it is predicted that the maximum thickness of the bottom layer is 1051 m in the deep Arctic Ocean, which is close to the observed value of 929 m. The evolution time of the bottom layer is predicted to be ~ 100 yr, which is on the same order as that based on ^{14}C isolation age estimation.

1 Introduction

Double diffusion is one of the most important non-mechanically driven mixing processes. It occurs in a fluid in which there are gradients of two (or more) properties with different molecular diffusivities and of opposing effects on the vertical density distribution. This phenomenon is of great interest to many disciplines in physical sciences and engineering (Turner, 1973), but the most active research area is exploration of thermocline staircases in oceans and lakes (Schmitt, 1994; Kelley et al., 2003). Basically, there exist two modes: salt finger (SF) and diffusive convection (DC). When cold and fresh water lies on top of warm and saline water, vertical mixing is triggered by different molecular diffusion rates of heat and salt, and DC forms (Kelley et al., 2003). DC is characterized by a series of thermohaline staircases: a stack of homogenous mixed layers of nearly constant temperature and salinity, separated by strongly stratified thin interfaces.

As sketched in Fig. 1, each convecting layer includes a well-mixed layer and is bounded by two adjacent interfaces. In each convecting layer of DC, the fluid properties and flow dynamics can be described by four dimensionless parameters. The first one is the thermal Rayleigh number $Ra_\mathrm{T} = \alpha g L^3 \Delta T / \nu \kappa_\mathrm{T}$, where g is the gravitational acceleration, and α, ν, κ_T, ΔT, and L being, respectively, the thermal expansion coefficient, the kinematic viscosity, the thermal diffusivity, the temperature difference and the typical length scale of the convecting layer. The second one is the salinity Rayleigh number, $Ra_\mathrm{S} = \beta g L^3 \Delta S / \nu \kappa_\mathrm{T}$, where ΔS is the salinity difference and β is the haline contraction coefficient. The other two parameters are the Prandtl number, $Pr = \nu / \kappa_\mathrm{T}$, and the Lewis number, $Le = \kappa_\mathrm{S} / \kappa_\mathrm{T}$, where κ_S is the salinity diffusivity. As the interface is the boundary of two adjacent convecting layers, the thermal Rayleigh number of the interface, Ra_TI, is proposed to be on the same order as the critical Rayleigh number of convection, which is on the order of 1000 (Turner, 1968, 1973). This argument has been found to work well in DC staircases in Lake Banyoles (Sánchez and Roget, 2007).

Recently, DC staircases have been observed in the deep Arctic Ocean (Timmermans et al., 2003; Timmermans and Garrett, 2006; Björk and Winsor, 2006; Carmack et al., 2012). These thermohaline staircases exhibit several unique

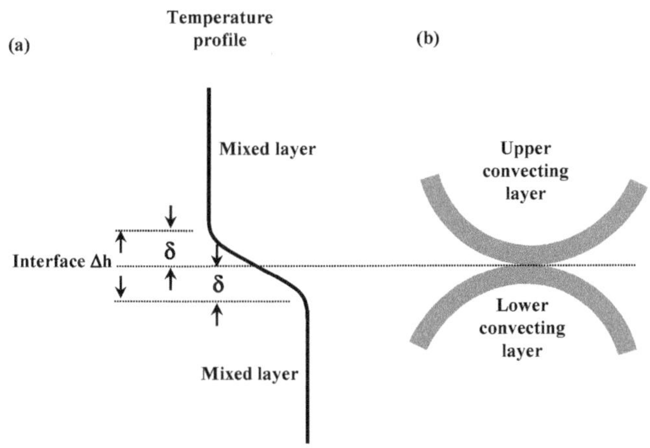

Fig. 1. **(a)** A sketch of temperature distribution of a diffusive interface in DC staircases; δ is the boundary layer thickness of a convecting layer. **(b)** A cartoon of flow pattern near the interface.

characteristics. One of them is the thick diffusive interface. It is about 5–8 m, which is much larger than those observed in laboratory experiments (Turner, 1968; Huppert and Linden, 1979; Fernando, 1987), lakes (Sánchez and Roget, 2007; Schmid et al., 2010) and other ocean regions (Voorhis and Dorson, 1975; Larson and Gregg, 1983; Padman and Dillon, 1987; Anschutz and Blanc, 1996). Associated with the temperature difference across the interface, $\Delta\theta \sim 1.3 \times 10^{-3}$ °C, and other fluid properties, as listed in Table 1 below, the thermal Rayleigh number of the interface, Ra_{TI}, is on the order of 10^9, which is much larger than the typical value reported in the literature (Turner, 1968, 1973; Sánchez and Roget, 2007). Therefore, additional understanding is needed to elucidate the difference between the result in the deep Arctic Ocean and those in previous studies.

Another prominent characteristic is that there exists a thick bottom mixed layer (Timmermans et al., 2003; Björk and Winsor, 2006; Timmermans and Garrett, 2006; Carmack et al., 2012). In the Canada Basin, this isothermal and isohaline bottom layer reaches to approximately 1000 m thick and extends about 1000 km across the basin (Timmermans et al., 2003). Björk and Winsor (2006) proposed a simple one-dimensional diffusive-convection model to understand the dynamics of the bottom layer. However, their proposed evolution process strongly depends on turbulent eddy diffusivity, which is unavailable at the present stage, and can only explain parts of the observation. Therefore, it is necessary to seek other mechanisms that may dominate in the deep-water evolution process.

In this paper, we explore the instability of the diffusive interface using hydrographic data measured by McLane Moored Profilers (MMP) at a fixed location in the Canada Basin. The classical description proposed by Turner (1968, 1973) is extended here by considering the influences of background salinity stratification and rotation. Then the

Table 1. Mean properties of four diffusive interfaces at Mooring A between 2 October 2009 and 9 August 2010. They are potential temperature difference, $\Delta\theta$, salinity difference, ΔS, thickness, Δh, density ratio, R_ρ, buoyancy frequency, N, thermal Rayleigh number, Ra_{TI}, salinity Rayleigh number, Ra_{SI}, and Taylor number, Ta_I.

Interface	1	2	3	4
$\Delta\theta$ (10^{-3} °C)	1.79	1.25	1.34	0.96
ΔS (10^{-4})	6.46	5.11	8.23	5.26
Δh (m)	7.79	6.29	8.61	5.32
R_ρ	2.13	2.46	3.67	3.27
N (10^{-4} s^{-1})	5.86	6.14	7.36	7.46
Ra_{TI} (10^9)	3.91	1.43	4.11	0.59
Ra_{SI} (10^9)	8.33	3.52	15.1	1.96
Ta_I (10^7)	2.13	0.91	3.18	0.46

instability of the interface is discussed in comparison with the onset thermal Rayleigh number of DC. With the new parameterization, we evaluate the thickness and evolution time of the bottom layer in the deep Arctic Ocean.

2 Data

Hydrographic data were mainly obtained from the Beaufort Gyre Exploration Project (BGEP) (Ostrom et al., 2004; Proshutinsky et al., 2009). We focus on the variances of temperature and salinity in the deep ocean at a fixed location in the Beaufort Sea of the Canada Basin. As marked in Fig. 2, Mooring A was deployed at 75° N, 150° W at the depth of 3825 m. From 2 October 2009 to 9 August 2010, the temperature and salinity between 2135 and 3080 dbar were measured by using an MMP at an interval of 8 or 11 h. This resulted in 791 profiles. Typical potential temperature, θ, and salinity, S, profiles at Mooring A are shown in Fig. 3. Note that these data measured by conductivity–temperature–depth (CTD), not by MMP, are used here because data of CTD covers a larger depth range. As reported in Zhou and Lu (2013), both CTD and MMP detected exactly the same staircase structures at the same depth range. The potential temperature, θ, profile, is shown in Fig. 3a. θ decreases with increasing depth till it reaches a minimum, θ_{min}, around a depth of 2400 m. When the depth increases further, θ increases and is accompanied by obvious staircases, where the mixed layer and the interface are well resolved. Between 2950 m and the sea floor, both θ and S are homogenous and uniform, and this range forms the bottom layer. Similar step structures were observed in the salinity profiles (Timmermans et al., 2003; Björk and Winsor, 2006; Timmermans and Garrett, 2006). These structures are less pronounced in the salinity profiles, as shown in Fig. 3b, due to the instrument resolution. As analyzed in our previous work (Zhou and Lu, 2013), the interface properties could be determined with an averaging technique.

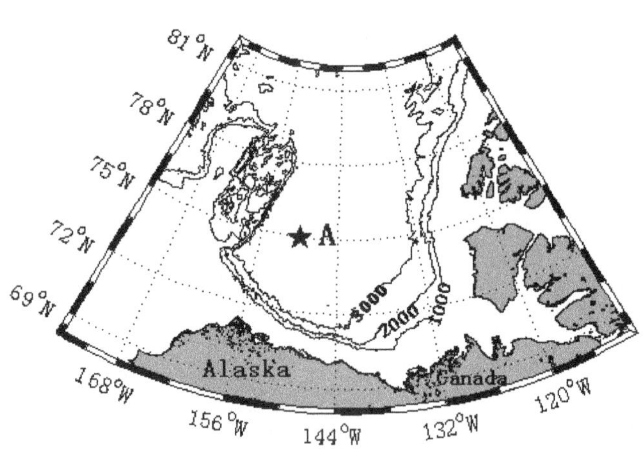

Fig. 2. Map of the Canada Basin in the Arctic Ocean. Isobaths are plotted using the ETOP data. The location of Mooring A is marked by a star.

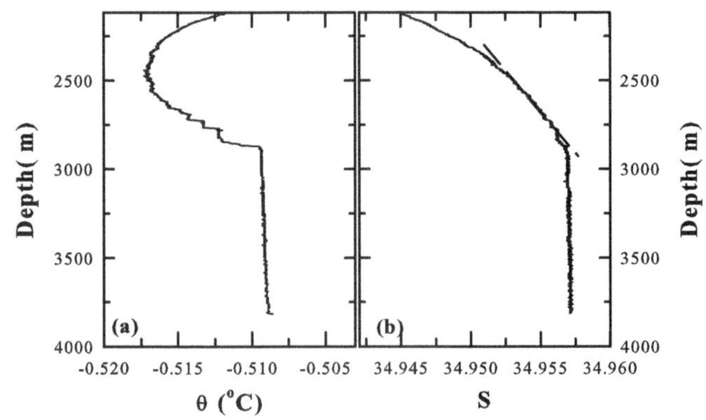

Fig. 3. Profiles of **(a)** potential temperature (θ) and **(b)** salinity (S) measured by conductivity–temperature–depth (CTD) at Mooring A on 28 Septermber 2010. The dashed line is the linear fit of the DC staircases in the salinity profile.

In Fig. 3a, four DC steps can be identified from the potential temperature, θ, profile, which are referred to as the 1st, 2nd, 3rd and 4th steps from the bottom to the top, respectively. Each step includes a mixed layer and its overlying interface. Note that the mixed layer of the first step is the bottom layer. Usually, two parameters are used to characterize the susceptibility of water columns to DC staircases. One is the density ratio R_ρ (Turner, 1965), which is expressed as

$$R_\rho = \frac{\beta \partial S/\partial z}{\alpha \partial \theta / \partial z}. \qquad (1)$$

R_ρ is the ratio of the stabilizing force due to the salinity gradient and the destabilizing force due to the temperature gradient. The other one is the buoyancy frequency, which has the form

$$N = [g(\alpha \partial \theta/\partial z - \beta \partial S/\partial z)]^{1/2}. \qquad (2)$$

N shows the stability of the stratification of water columns. It has been proposed that the shape of DC structures and the vertical heat transfer strongly depend on these parameters (see the review paper of Kelley et al., 2003). In the case of well-developed DC steps, where the mixed layer is homogeneous in temperature and salinity, $\partial S/\partial z$ and $\partial \theta/\partial z$ are reduced to the salinity and temperature gradients across the interface. According to the definitions of Ra_T and Ra_S, R_ρ can be rewritten as $R_\rho = Ra_\mathrm{S}/Ra_\mathrm{T}$. For detailed analysis of the determination of the diffusive interface properties, readers are referred to Zhou and Lu (2013). The mean properties of the four interfaces are listed in Table 1.

3 The classical description of DC instability

The instability of DC has been extensively studied based on the results in laboratory experiments (Veronis, 1965; Turner, 1973; Caldwell, 1974; Huppert and Moore, 1976; Pearlstein, 1981). Basically, two modes may occur at the onset of convection; one is steady convection and the other is oscillatory convection. In the case of the ocean ($Le \sim 0.01$), it has been found that the onset Rayleigh number of steady convection appears to be less than the value for oscillatory convection (Huppert and Moore, 1976). This implies that the steady convection mode is responsible for convective instability in DC staircases. In terms of linear stability analysis (Turner, 1973), the convection occurs when the thermal Rayleigh number, Ra_T, exceeds a critical value, Ra_c. Ra_c is written as

$$Ra_\mathrm{c} = \frac{Pr + Le}{Pr + 1} Ra_\mathrm{S} + (1 + Le)(1 + \frac{Le}{Pr})\frac{27\pi^4}{4}. \qquad (3)$$

In the case of homogeneous fluid, $Ra_\mathrm{S} = 0$ and $Le = 0$, and the equation is reduced to the onset Rayleigh number of Rayleigh–Bénard convection, $Ra_\mathrm{c} = \frac{27\pi^4}{4}$.

In the deep Arctic Ocean, the DC staircases are thick and the corresponding Coriolis frequency, f, is large; one may expect that the rotation influences the occurrence of convection. The rotation is characterized by the Taylor number, $Ta = 4(\frac{fL^2}{\nu})^2$. In linear stability analysis, it has been found that the rotation inhibits the onset of convective instability (Pearlstein, 1981). When the rotation strongly affects DC, the onset Rayleigh number, Ra_c, would be in the form of

$$Ra_\mathrm{c} = \frac{Pr + Le}{Pr + 1} Ra_\mathrm{S} + (1 + Le)(1 + \frac{Le}{Pr})(\frac{27\pi^4}{4})^{\frac{1}{3}} Ta^{\frac{2}{3}}. \qquad (4)$$

Equation (4) implies that the convection occurs when the heat-induced buoyancy force gradient overcomes the resistance produced by the salt stratification and rotation. In the ocean, $Le \sim 0.01$ and $Pr \sim 4 - 13$; Eqs. (3) and (4) are then simplified to

$$Ra_\mathrm{c} = \frac{Pr\, Ra_\mathrm{S}}{Pr + 1} + \frac{27\pi^4}{4}, \qquad (5)$$

for $Ta < 10^3$, and

$$Ra_c = \frac{Pr\,Ra_S}{Pr+1} + \left(\frac{27\pi^4}{4}\right)^{\frac{1}{3}} Ta^{\frac{2}{3}}, \quad (6)$$

for $Ta > 10^5$.

4 Results and discussions

There is not much work to address the instability of DC in the ocean because the involved convective flow state is typically far beyond the onset convection regime (Carpenter et al., 2012). However, findings about the instability mechanism of DC in previous laboratory experiments (Howard, 1964; Turner, 1968; Huppert and Linden, 1979) may actually help to understand the characteristics of diffusive interface and the bottom layer in the deep Arctic Ocean, both of which are discussed next.

4.1 Diffusive interface

For the diffusive interface in the deep Arctic Ocean, the corresponding thermal Rayleigh number, Ra_{TI}, is obtained from its definition, with the typical scale L and temperature difference ΔT being identified by the interface thickness Δh and temperature difference $\Delta\theta$. As Δh and $\Delta\theta$ vary with time (Zhou and Lu, 2013), the deduced Ra_{TI} is a function of time too. Its temporal distributions (shown in Fig. 4) indicate that the Ra_{TI} of each interface is distributed approximately log-normal, which suggests that Ra_{TI} is strongly intermittent. Similar distributions have been found in vertical heat flux, eddy diffusivity and other properties (Zhou and Lu, 2013). These results imply that the deep Arctic Ocean exhibits certain turbulent behaviors (Frisch, 1995). The mean Ra_{TI} of each interface, as listed in Table 1, is on the order of $\sim 10^9$.

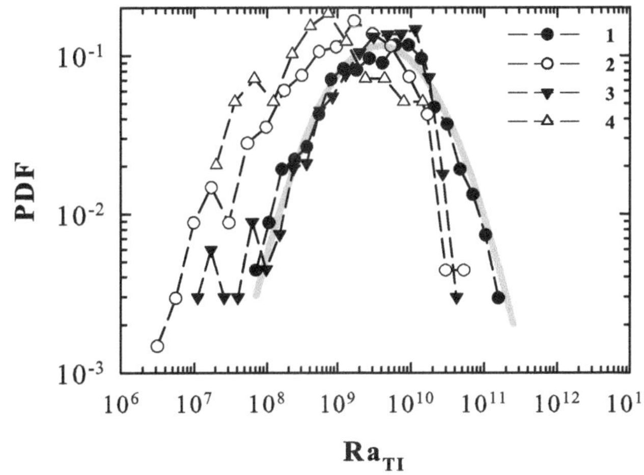

Fig. 4. Probability density functions (PDFs) of the thermal Rayleigh number, Ra_{TI}, of the four diffusive interfaces. Data of the first, second, third and fourth interfaces are plotted using the dashed lines with solid circles, open circles, solid down triangles and open triangles, respectively. The log-normal fitting of the PDF of the first interface is shown as a gray curve.

Although the diffusive interface is the internal boundary of two adjacent convecting layers, how to relate the interface to the boundary layer is arguable (Turner, 1973; Linden and Shirtcliffe, 1978; Padman and Dillon, 1989). Linden and Shirtcliffe (1978) proposed a boundary layer model. In this model, the boundary layer, δ, is defined as $\delta = (\Delta h_T - \Delta h_S)/2$, where Δh_T and Δh_S are the thicknesses of interface based on temperature and salinity. This model has been further employed in some studies, e.g., Worster (2004), and Carpenter et al. (2012).

On the other hand, many studies follow the classical boundary layer definition in diffusive convection (Turner, 1968, 1973; Padman and Dillon, 1989; Sánchez and Roget, 2007; Zaussinger and Spruit, 2013), which is similar to that in Rayleigh–Bénard convection. In Rayleigh–Bénard convection, two thermal boundary layers exist near the top and bottom ends. At the boundary, the temperature gradient is the largest. Away from the boundary, the temperature gradient becomes smaller and smaller till it reaches zero in the mixed layer. Such a temperature distribution is the result of convecting rolls. The thermal boundary layer thickness is defined as a small region between the boundary and the mixed layer (Lui and Xia, 1998). In DC, the temperature gradient is the largest around the middle of the interface, and it gets smaller and smaller as it approaches the edges of the interface. By analogy with Rayleigh–Bénard convection, each interface indeed consists of two boundary layers, as shown in Fig. 1, which are the top boundary layer of the lower convecting layer and the bottom one of the upper convecting layer. Here, we still use this definition of the boundary layer thickness. To the first-order approximation, the boundary layer thickness, δ, and the corresponding temperature difference, $\Delta\theta_\delta$, can be taken as $\delta \sim \Delta h/2$ and $\Delta\theta_\delta \sim \Delta\theta/2$. Consequently, the boundary layer thermal Rayleigh number, $Ra_{T\delta}$, is $Ra_{T\delta} \sim Ra_{TI}/16$. The same consideration is also applied to the salinity Rayleigh number, $Ra_{S\delta}$, and the Taylor number, Ta_δ, of the boundary layer. Typical values of $Ra_{T\delta}$ and Ta_δ in the deep Arctic Ocean are listed in Table 2.

In Rayleigh–Bénard convection, it is assumed that the Rayleigh number based on the boundary layer thickness, Ra_δ, is on the order of the critical value, namely, $Ra_{T\delta} \sim Ra_c$ (Howard, 1964; Siggia, 1994). This assumption has been confirmed to be marginally correct in experiments (e.g., Castaing et al., 1989; Lui and Xia, 1998). Analogously, the same argument has been employed to study the diffusive interface instability by Turner (1968, 1973). This argument has been found to work well in the DC staircases of the Arctic Ocean (Padman and Dillon, 1989) and Lake Banyoles (Sánchez and Roget, 2007). However, in the deep Arctic Ocean, Ra_{TI} of interfaces is on the order of $\sim 10^9$; then the deduced $Ra_{T\delta}$ is 10^8, which is much larger than the observed value ($\sim 10^3$)

Table 2. Data used in the analysis. The average values are listed when more than one data set are included in the data source. There are the temperature difference, $\Delta\theta$, salinity difference, ΔS, thickness, Δh, and Prandtl number, Pr, of the interface; saline buoyancy frequency, N_S, based on the background salinity profile; Taylor number, Ta_δ, thermal Rayleigh number, $Ra_{T\delta}$, and salinity Rayleigh number, $Ra'_{S\delta}$, of the boundary layer. Deep Arctic – data of the present work; Upper Arctic – average data measured by the Ice-Tethered Profiler (ITP) 2 in the Beaufort Gyre observation system (BGOS); Atlantis II 1 – Fig. 4 of Anschutz and Blanc (1996); Atlantis II 2 – Fig. 4 of Swift et al. (2012); Lake Kivu – Fig. 1 and Table 2 of Schmid et al. (2010); Lake Banyoles – Figs. 2 and 8 of Sánchez and Roget (2007); and the Bahamas – Fig. 2 of Larson and Gregg (1983).

Location	$\Delta\theta$ (°C)	ΔS	Δh (m)	Pr	N_S (s^{-1})	$Ra_{T\delta}$	$Ra'_{S\delta}$	Ta_δ
Deep Arctic	0.0012	0.00063	6.74	13.3	0.00031	1.3×10^8	6.2×10^7	9.5×10^5
Upper Arctic	0.044	0.016	0.62	13.3	0.0056	1.7×10^6	1.3×10^6	62.0
Atlantis II 1	4.082	30.08	1.70	3.8	0.14	6.3×10^{10}	9.4×10^{10}	3.7×10^3
Atlantis II 2	4.436	30.09	1.92	4.3	0.17	1.2×10^{11}	2.9×10^{11}	6.6×10^3
Lake Banyoles	0.360	0.073	0.018	6.2	–	1.7×10^3	55.1	8.9×10^{-5}
Lake Kivu	0.0045	–	0.19	6.2	0.0086	3.5×10^4	4.3×10^4	2.5×10^{-3}
Bahamas	0.047	0.026	0.069	6.4	0.018	2.2×10^4	3.8×10^3	6.8×10^{-3}

in Rayleigh–Bénard convection (Castaing et al., 1989) and in DC (Sánchez and Roget, 2007).

One may expect that the influences of salt stratification or rotation must be involved, that is, Eqs. (5) and (6) can be used to explain the results here. However, as listed in Table 1, to each interface, the salinity Rayleigh number Ra_{SI} is always larger than Ra_{TI}. Thus, when $Ra_{S\delta}$ ($\sim Ra_{SI}/16$) is employed in Eqs. (5) and (6), the obtained Ra_c would be larger than $Ra_{T\delta}$ ($\sim Ra_{TI}/16$), which may mean that the assumption of $Ra_{T\delta} \sim Ra_c$ does not work here. Carpenter et al. (2012) suggested that Eqs. (5) and (6) cannot apply to the instability of diffusive convection of large density ratio R_ρ ($R_\rho > 1.15$).

An alternative consideration can be found in a previous laboratory study by Turner (1968). In this study, Eq. (5) was used to interpret the onset of the convecting layer in DC. When the breakdown of the unstable boundary layer is described, it is further argued that the original salinity gradient remains unchanged because the salt diffuses much more slowly than heat. In other words, the salinity Rayleigh number, Ra'_S, is in the form of

$$Ra'_S = \frac{N_S^2 L^4}{\nu \kappa_T}. \tag{7}$$

In Eq. (7), $N_S = \sqrt{g\beta \frac{dS}{dz}}$ is the buoyancy frequency based on the initial salinity gradient before being heated. This argument has been supported by the measurement of the advancing boundary layer in another laboratory experiment (Huppert and Linden, 1979). In the laboratory, however, the calculated Ra'_S is found to be very small due to the thin boundary layer, and it is reasonable to neglect it. Consequently, Ra_c is reduced to the ordinary critical Rayleigh number based on heat alone (Turner, 1968).

It is expected that the same idea can be applied to the case of interest here. As it is impossible to obtain the initial salinity gradient of the deep Arctic Ocean, the background salinity gradient is alternatively used in Eq. (7). As shown in Fig. 3b, the background salinity gradient is obtained by the linear fitting to all DC staircases in salinity profiles, resulting in $\frac{ds}{dz} = 1.3 \times 10^{-5}\,m^{-1}$ and the corresponding $N_S = 3.1 \times 10^{-4}\,s^{-1}$. Meanwhile, more data in the oceans and lakes have been collected to check the applicability of the argument. These data include the field observations in the upper layer of the Canada Basin (BGOS, 2013), deep Red Sea (Anschutz and Blanc, 1996; Swift et al., 2012), Lake Kivu (Schmid et al., 2010), Lake Banyoles (Sánchez and Roget, 2007), and the Bahamas (Larson and Gregg, 1983). Typical interface properties of these data sources are listed in Table 2.

In terms of Eqs. (6) and (7), the thermal critical Rayleigh number, Ra_c, is obtained here from the δ-based salinity Rayleigh number, $Ra'_{S\delta}$, and the δ-based Taylor number, Ta_δ. Note that Eq. (5) is used when $Ta < 1000$. Within the collected data, the thermal critical Rayleigh number, Ra_c, is found to vary in a wide range from 10^3 in the Lake Banyoles to 10^{11} in the deep Red Sea, as shown in Fig. 5a. Figure 5a also shows that the calculated thermal critical Rayleigh number, Ra_c, is very close to the δ-based thermal Rayleigh number, $Ra_{T\delta}$, in almost all the collected data. As some data were captured from the published figures (Sánchez and Roget, 2007; Larson and Gregg, 1983), the limited accuracy would be the most probable reason responsible for the data scatters in Fig. 5a. When the set of data is large enough to have good statistics, e.g., those in the Upper Arctic (BGOS, 2013), Ra_c is in good agreement with $Ra_{T\delta}$. As far as we know, this is the first time that the classical description of DC has been applied to interpret instability of diffusive interfaces in the oceans.

In addition, we examine the influences of background salinity stratification and the rotation on the onset of convection. The comparison between the first and the second terms in Eq. (6) is plotted in Fig. 5b. In the deep Arctic Ocean, the influence of rotation is supposed to be the largest because of the thickest diffusive interface and the largest Coriolis

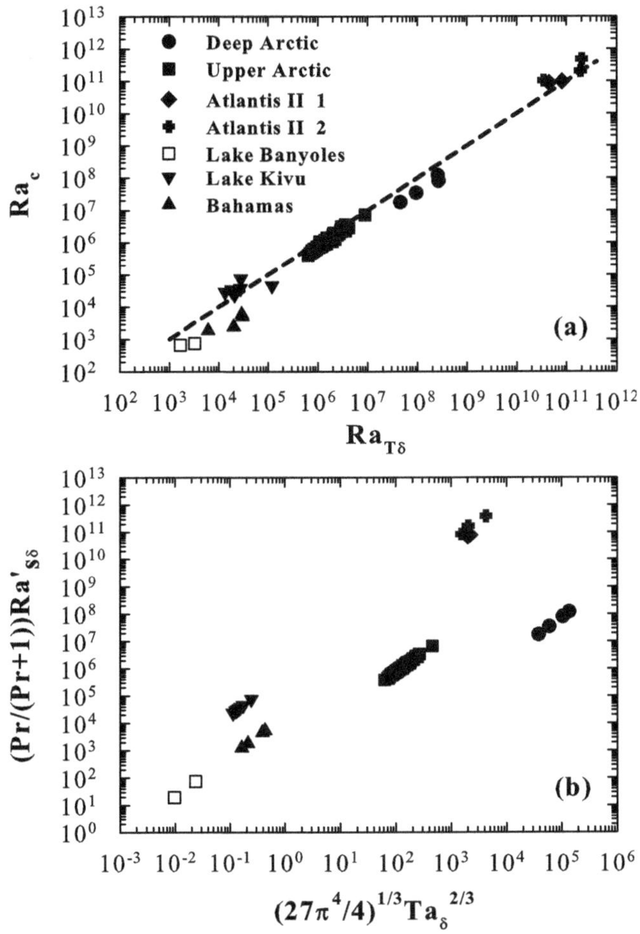

Fig. 5. **(a)** Comparison between the calculated onset thermal Rayleigh number, Ra_c, based on Eqs. (5) and (6) and the boundary layer thermal Rayleigh number, $Ra_{T\delta}$. **(b)** Comparison between the first term, $PrRa'_{S\delta}/(Pr+1)$, and the second term, $(27\pi^4/4)^{1/3}Ta_\delta^{2/3}$, in Eq. (6).

frequency in all the data. Even so, the contribution of rotation is found to be about 1000 times smaller than that of the salinity stratification. In other regions, the contribution of rotation is even smaller. Therefore, background salinity stratification would be the main hindrance to the occurrence of convective flow in the ocean. The influence of the salinity stratification, however, is not always important. In Lake Banyoles, as shown in Fig. 5b, it is found that $Ra'_{S\delta} \sim 55$; thus, the inherent mechanism of thermal convection plays a dominating role there, which results in $Ra_c \sim 1000$ (Sánchez and Roget, 2007).

4.2 Bottom homogenous layer

As introduced in Sect. 1, the thick homogenous bottom layer is a unique feature of the deep Arctic Ocean (Timmermans et al., 2003; Björk and Winsor, 2006; Timmermans and Garrett, 2006; Carmack et al., 2012; Zhou and Lu, 2013). As the model proposed by Björk and Winsor (2006) cannot explain the dynamics of the bottom layer well, we attempt to find other mechanisms instead. Recently, more evidence has been collected that geothermal heating plays an important role in the hydrographic configuration of deep water in the Arctic Ocean (Timmermans et al., 2003; Björk and Winsor, 2006; Timmermans and Garrett, 2006; Carmack et al., 2012; Zhou and Lu, 2013). The situation that the weakly stratified deep Arctic Ocean is being heated by geothermal heating is similar to those studied in the laboratory experiments and numerical simulations, where DC occurred in salt-stratified fluid heated from below (Turner, 1968; Huppert and Linden, 1979; Fernando, 1987). Thus, it is interesting to examine whether the results obtained from the laboratory experiments can be applied to the deep ocean and whether they can shed more light on the deep water dynamic process.

In the laboratory experiments where a salt-stratified fluid was heated from below (Turner, 1968, 1973; Huppert and Linden, 1979; Fernando, 1987), it has been found that the bottom layer is much thicker than the overlying staircases, which is similar to the observation in the deep Arctic Ocean. In Turner's (1968) theoretical work, he suggested that the homogeneous bottom layer reaches a critical thickness before a second convecting layer appears on top of it. This maximum thickness was proposed to be in the form of

$$h_{c1} = \left(\frac{\frac{1}{4}Ra_c \nu q_0{}^3}{\kappa_T^2 N_S^8}\right)^{\frac{1}{4}}. \quad (8)$$

Here q_0 is the buoyancy flux, which is derived as $q_0 = \alpha g F/\rho_0 c_p$, and F is the heat flux supplied from the bottom boundary. Eq. (8) has been verified in Turner's laboratory, except that the fitted Ra_c was about $\sim 2.4 \times 10^4$, which is larger than the ordinary value (~ 1000). As discussed in the last subsection, the instability of DC follows the arguments based on Eqs. (5), (6) and (7). Then we can evaluate the bottom layer thickness in the Arctic Ocean by using Eq. (8). In the deep Arctic Ocean, geothermal heating is taken as $F \sim 50\,\mathrm{m\,W\,m^{-2}}$ (Langseth et al., 1990). At Mooring A, the physical properties of sea water are respectively $\alpha = 1.27 \times 10^{-4}\,^\circ\mathrm{C}^{-1}$, $\beta = 7.5 \times 10^{-4}$, $\nu = 1.85 \times 10^{-6}\,\mathrm{m^2\,s^{-1}}$, $\kappa_T = 1.39 \times 10^{-7}\,\mathrm{m^2\,s^{-1}}$, $g = 9.8\,\mathrm{m\,s^{-2}}$, $\rho_o = 1041.3\,\mathrm{kg\,m^{-3}}$, and $c_p = 3899.1\,\mathrm{J\,kg^{-1}\,^\circ C^{-1}}$. With the related properties of the first interface, as listed in Table 1, h_{c1} is calculated to be 526 m, which is smaller than the observed value of 929 m (Zhou and Lu, 2013). However, as found in the laboratory experiment (Turner, 1968), the fitted Ra_c was about 20 times the ordinary value, and so was close to the interface thermal Rayleigh number Ra_{TI}. If Ra_c is replaced by Ra_{TI} in Eq. (8), h_{c1} is found to be 1051 m, which is much closer to the observed value.

In another laboratory study, Fernando (1987) found that the bottom mixed layer still grows upwards even after the second mixed layer forms. Based on the assumption that the kinetic and potential energies of turbulent eddies are balanced when the mixed layer grows to a critical height, an alternative expression for the layer thickness has been proposed,

$$h_{c2} = 41.5(\frac{q_0}{N_S^3})^{\frac{1}{2}}. \quad (9)$$

With the properties of the interface, the thickness h_{c2} is found to be 29 m, which is much smaller than the observation. This suggests that the assumption of energy balance cannot explain the formation of the bottom layer.

In a study of ^{14}C isolation age, the deep water of the Canada Basin is estimated to be about 500 yr old (Macdonald et al., 1993). According to the laboratory study of Turner (1968), the evolution time of the bottom layer can be predicted by

$$\tau = \frac{(N_S h)^2}{2q_0} \quad (10)$$

before the formation of the secondary convecting layers. Supposing that the bottom layer thickness, h, is about 1000 m, the time τ is calculated to be $\tau = 109$ yr. In observations, however, there are more than four thermohaline steps overlying the bottom layer, which means that the evolution time of the bottom layer must be longer than τ. In some sense, τ can be taken as the lower bound of evolution time of the bottom layer.

Based on the above results, it is suggested that the main evolution mechanism of the deep water in the deep Arctic Ocean is similar to that in the laboratory experiments (Turner, 1968, 1973; Huppert and Linden, 1979; Fernando, 1987). The homogenous bottom layer is the result of geothermal heating under salinity stratification. Its maximum thickness can be described by Eq. (8), where Ra_{TI} is alternatively used. The time, τ, is inferred to be on the same order as that estimated from the ^{14}C isolation age detection. Both results imply that the evolution of the bottom layer is mainly controlled by the convective instability mechanism.

5 Conclusions

In summary, the classical description of DC has been updated to interpret the instability of diffusive interfaces and the dynamical characteristics of the bottom layer in the deep Arctic Ocean. When both background salinity stratification and rotation are considered, the critical Rayleigh number, Ra_c, of DC is found to vary in a wide range from 10^3 to 10^{11} within the collected data, and the thermal Rayleigh number of the boundary layer, $Ra_{T\delta}$, is consistent with the critical Rayleigh number, Ra_c, of DC. It is expected that this new parameterization of Eqs. (5), (6) and (7) can be extensively applied to DC in other oceans and lakes. In most cases, the salinity stratification is found to be the main hindrance to the occurrence of convective flow, except for regions where the diffusive interface is thin, e.g., in Lake Banyoles. When the interface thermal Rayleigh number, Ra_{TI}, is alternatively employed in the old parameterization of Eq. (8), the predicted maximum thickness of the bottom layer is 1051 m, which is close to the observed value of 929 m. The evolution time of the bottom layer, τ, is predicted to be $\sim$ 100 yr, which is on the same order as the ^{14}C isolation age estimate. As several layers overlie the bottom layer, the evolution time of water in the deep Arctic Ocean must be longer than τ, and τ can be taken as the lower bound of the residence time.

In the formation of deep Arctic water, one perhaps cannot exclude the effects of other instabilities, e.g., thermobaric convection (Carmack et al., 2012) or topographic Rossby waves formed as a result of instability of a strong current (Timmermans et al., 2010). Nonetheless, according to the results in this study, the main evolution mechanism of the deep water would be similar to that in the laboratory experiments where salt-stratified fluid was heated from below (Turner, 1968, 1973; Huppert and Linden, 1979; Fernando, 1987). Therefore, when the classical description of DC is updated, it can be applied to interpret the flow dynamics of the deep Arctic Ocean.

Acknowledgements. We gratefully acknowledge Mary-Louise Timmermans for her support and generosity in sharing valuable communications of the data set in the Arctic Ocean. We thank Martin Schmid for sending the raw data of Lake Kivu. This work was supported by China NSF grants (41176027 and 11072253), the Strategic Priority Research Program of the Chinese Academy of Sciences (XDA11030302), and an LTO grant (LTOZZ1304). The data were collected and made available by the Beaufort Gyre Exploration Program based at the Woods Hole Oceanographic Institution (http://www.whoi.edu/beaufortgyre) in collaboration with researchers from Fisheries and Oceans Canada at the Institute of Ocean Sciences.

Edited by: J. M. Huthnance

References

Anschutz, P. and Blanc, G.: Heat and salt fluxes in the Atlantis II Deep (Red Sea), Earth Planet. Sc. Lett, 142, 147–159, doi:10.1016/0012-821X(96)00098-2, 1996.

BGOS: The hydrographic data measured by the Ice-Tethered Profiler 2 (ITP2) in Beaufort Gyre Observing System (BGOS) were used in the present work, available at: http://www.whoi.edu/itp (last access: 20 February 2014), 2013.

Björk, G. and Winsor, P.: The deep waters of the Eurasian Basin, Arctic Ocean: Geothermal heat flow, mixing and renewal, Deep Sea Res.-Pt. I, 53, 1253–1271, doi:10.1016/j.dsr.2006.05.006, 2006.

Caldwell, D. R.: Experimental studies on the onset of thermohaline convection, J. Fluid Mech., 64, 347–368, doi:10.1017/S0022112074002436, 1974.

Carmack, E. C., Williams, W. J., Zimmermann, S. L., and McLaughlin, F. A.: The Arctic Ocean warms from below, J. Geophys. Res., 39, L07604, doi:10.1029/2012GL050890, 2012.

Carpenter, J. R., Sommer, T., and Wuest, A.: Stability of a double-diffusive interface in the diffusive convection regime, J. Phys. Oceanogr., 42, 840–854, doi:10.1175/JPO-D-11-0118.1, 2012.

Castaing, B., Gunaratne, G., Heslot, F., Kadanoff, L., Libchaber, A., Thomae, S., Wu, X.-Z., Zaleski, S., and Zanetti, G.: Scaling of hard thermal turbulence in Rayleigh-Bénard convection, J. Fluid Mech., 204, 1–30, doi:10.1017/S0022112089001643, 1989.

Fernando, H. J. S.: The formation of layered structure when a stable salinity gradient is heated from below, J. Fluid Mech., 182, 525–541, doi:10.1017/S0022112087002441, 1987.

Frisch, U.: Turbulence: the Legacy of A. N. Kolmogorov, Cambridge University Press, UK, 1995.

Howard, L. N.: Convection at high Rayleigh number, Proc. 11th Cong. Applied Mech., edited by: Grtler, H., 1964.

Huppert, H. E. and Linden, P. F.: On heating a stable salinity gradient from below, J. Fluid Mech., 95, 431–464, doi:10.1017/S0022112079001543, 1979.

Huppert, H. E. and Moore, D. R.: Nonlinear double-diffusive convection, J. Fluid Mech., 78, 821–854, doi:10.1017/S0022112076002759, 1976.

Kelley, D. E., Fernando, H. J. S., Gargett, A. E., Tanny, J., and Özsoy, E.: The diffusive regime of double diffusive convection, Progr. Oceanogr., 56, 461–481, doi:10.1016/S0079-6611(03)00026-0, 2003.

Langseth, M. G., Lachenbruch, A., and Marshall, V. B.: Geothermal observations in the Arctic region, The Geology of North America: the Arctic Ocean region, The Geological Society of America, 1990.

Larson, N. G. and Gregg, M. C.: Turbulent dissipation and shear in thennohaline intrusions, Nature, 306, 26–32, doi:10.1038/306026a0, 1983.

Linden, P. F. and Shirtcliffe, T. G. L.: The diffusive interface in double-diffusive convection, J. Fluid Mech., 87, 417–432, doi:10.1017/S002211207800169X, 1978.

Lui, S. L. and Xia, K. Q.: Spatial structure of the thermal boundary layer in turbulent convection, Phys. Rev. E, 57, 5494–5503, doi:10.1103/PhysRevE.57.5494, 1998.

Macdonald, R., Carmack, E. C., and Wallace, D. W. R.: Tritium and radiocarbon dating of Canada basin deep waters, Science, 259, 103–104, doi:10.1126/science.259.5091.103, 1993.

Ostrom, W., Kemp, J., Krishfield, R., and Proshutinsky, A.: Beaufort Gyre freshwater experiment: Deployment operations and technology in 2003, Technical Report WHOI-2004-1, Woods Hole Oceanographic Institution, 2004.

Padman, L. and Dillon, T. M.: Vertical heat fluxes through the Beaufort Sea thermohaline staircase, J. Geophys. Res., 92, 10799–10806, doi:10.1029/JC092iC10p10799, 1987.

Padman, L. and Dillon, T. M.: Thermal microstructure and internal waves in the Canada Basin diffusive staircase, Deep Sea Res.-Pt. I, 36, 531–542, doi:10.1016/0198-0149(89)90004-6, 1989.

Pearlstein, A. J.: Effect of rotation on the stability of a doubly diffusive fluid layer, J. Fluid Mech., 103, 389–412, doi:10.1017/S0022112081001390, 1981.

Proshutinsky, A., Krishfield, R., Timmermans, M.-L., Toole, J., Carmack, E., McLaughlin, F., Williams, W. J., Zimmermann, S., Itoh, M., and Shimada, K.: Beaufort Gyre freshwater reservoir: State and variability from observations, J. Geophys. Res., 114, C00A10, doi:10.1029/2008JC005104, 2009.

Sánchez, X. and Roget, E.: Microstructure measurements and heat flux calculations of a triple-diffusive process in a lake within the diffusive layer convection regime, J. Geophys. Res., 112, C02012, doi:10.1029/2006JC003750, 2007.

Schmid, M., Busbridge, M., and Wüest, A.: Double diffusive convection in Lake Kivu, Limnol. Oceanogr., 55, 225–238, doi:10.4319/lo.2010.55.1.0225, 2010.

Schmitt, R. W.: Double diffusion in oceanography, Annu. Rev. Fluid Mech., 26, 255–285, doi:10.1146/annurev.fl.26.010194.001351, 1994.

Siggia, E. B.: High Rayleigh number convection, Annu. Rev. Fluid Mech., 26, 137–168, doi:10.1146/annurev.fl.26.010194.001033, 1994.

Swift, S. A., Bower, A. S., and Schmitt, R. W.: Vertical, horizontal, and temporal changes in temperature in the Atlantis II and Discovery hot brine pools, Red Sea, Deep Sea Res.-Pt. I, 64, 118–128, doi:10.1016/j.dsr.2012.02.006, 2012.

Timmermans, M. L. and Garrett, C.: Evolution of the deep water in the Canadian Basin of the Arctic Ocean, J. Phys. Oceanogr., 36, 866–874, doi:10.1175/JPO2906.1, 2006.

Timmermans, M. L., Garrett, C., and Carmack, E.: The thermohaline structure and evolution of the deep waters in the Canada Basin, Arctic Ocean, Deep Sea Res.-Pt. I, 50, 1305–1321, doi:10.1016/S0967-0637(03)00125-0, 2003.

Timmermans, M. L., Rainville, L., Thomas, L., and Proshutinsky, A.: Moored observations of bottom-intensified motions in the deep Canada Basin, Arctic Ocean, J. Mar. Res., 68, 625–641, doi:10.1357/002224010794657137, 2010.

Turner, J. S.: The coupled turbulent transport of salt and heat across a sharp density interface, Int. J. Heat Mass Tran., 8, 759–767, doi:10.1016/0017-9310(65)90022-0, 1965.

Turner, J. S.: The behavior of a stable salinity gradient heated from below, J. Fluid Mech., 33, 183–200, doi:10.1017/S0022112068002442, 1968.

Turner, J. S.: Buoyancy Effects in Fluids, Cambridge University Press, 1973.

Veronis, G.: On finite amplitude instability in thermohaline convection, J . Mar. Res., 23, 1–17, 1965.

Voorhis, A. D. and Dorson, D. L.: Thermal convection in the Atlantis II hot brine pool, Deep Sea Res., 22, 167–175, doi:10.1016/0011-7471(75)90056-X, 1975.

Worster, M. G.: Time-dependent fluxes across double-diffusive interfaces, J. Fluid Mech., 505, 287–307, doi:10.1017/S0022112004008523, 2004.

Zaussinger, F. and Spruit, H. C.: Semiconvection: numerical simulations, Astron. Astrophys., 554, A119, doi:10.1051/0004-6361/201220573, 2013.

Zhou, S.-Q. and Lu, Y.-Z.: Characterizations of Double Diffusive Convection Steps and Heat Budget in the Deep Arctic Ocean, J. Geophys. Res., 118, 1–15, doi:10.1002/2013JC009141, 2013.

5

Geostrophic currents and kinetic energies in the Black Sea estimated from merged drifter and satellite altimetry data

M. Menna and P.-M. Poulain

Istituto Nazionale di Oceanografia e Geofisica Sperimentale – OGS, Trieste, Italy

Correspondence to: M. Menna (mmenna@ogs.trieste.it)

Abstract. Drifter measurements and satellite altimetry data are merged to reconstruct the surface geostrophic circulation of the Black Sea in the period 1999–2009. This combined data set is used to estimate pseudo-Eulerian velocity statistics for different time periods. Seasonal and interannual variability of currents and kinetic energy fields are described with particular attention to the mesoscale and sub-basin coastal eddies. The mean currents are generally stronger in winter and enhanced speeds are observed in the period 2002–2006. The most intense activity of sub-basin Batumi Eddy occurs in summer with greater speeds and dimensions in 2006 and 2008. The sub-basin Sevastopol Eddy is generated in spring from a meander of the Rim Current. Mesoscale eddies located along the Anatolia, Caucasus and Crimea coasts are permanent, quasi-permanent or intermittent features and can interact and merge with each other, showing high values of kinetic energy.

1 Introduction

Surface currents contribute significantly to the distribution of hydrological properties, nutrients, pollution and other kinds of dissolved and suspended particles, among adjacent coastal regions and between coastal and open-ocean regions. Their assessment assume an important role in the densely inhabited coastal areas and semi-enclosed basins, where pollutants and contaminants are produced and discharged in large amounts. In this context, the Black Sea (BS), with a catchment area that cover extended areas of Europe and Asia (Stanev, 2005), represents a good example of marginal semi-enclosed basin, characterised by a limited water exchange with the other basins and an intense anthropic activity and river discharge. The upper-layer dynamics of the BS is conventionally described as the combination of a basin-scale cyclonic boundary current over the continental slope (e.g. Oguz et al., 1992), cyclonic gyres in the basin interior (Korotaev et al., 2001) and quasi-stationary or recurrent anticyclonic eddies along the coast (Korotaev et al., 2003). The interaction among these different and multi-scaled features contribute to the mixing of coastal water with the open sea and results in a high spatio-temporal variability of the current field.

The most exhaustive descriptions of the interannual, seasonal and mesoscale variability in the BS were obtained with satellite altimetry products (Stanev et al., 2000; Korotaev et al., 2001, 2003) and Lagrangian observations (Poulain et al., 2005). The BS surface circulation has a strong seasonal variability, with intensification of the cyclonic along-slope Rim Current (RC) in winter and attenuation in summer and fall (Stanev et al., 2000; Korotaev et al., 2001; Poulain et al., 2005). The organised winter circulation gradually disintegrates into a series of interconnecting eddies in summer and fall, when the RC exhibits more pronounced and complex mesoscale activity (Korotaev et al., 2003). The mesoscale variations are strongest in the RC and in large quasi-permanent nearshore sub-basin-scale anticyclonic eddies (Stanev et al., 2000), such as the Batumi Eddy (BE) and the Sevastopol Eddy (SE), and off the Crimean Peninsula (Korotaev et al., 2001). Mesoscale eddies induce the process of exchange between shelf and open sea waters (Shapiro et al., 2010). The BE has a seasonal modulation with intense cyclonic circulation in winter and weaker (sometimes anti-cyclonic) rotation in the summer–fall period (Stanev et al., 2000; Poulain et al., 2005). The SE is characterised by low seasonal variability and has large sea level oscillations at the inter- and intraannual scales. The interannual variability has the strongest signature in the southeastern BS (Stanev et al., 2000). Globally, the kinetic energy levels were found higher in winter–spring and lower in summer–fall periods (Poulain et al., 2005).

Both drifter and satellite data have been recognised as most useful to observe variation of sea-surface velocity fields (Uchida and Imawaki, 2003), although they present some drawbacks. Altimetry data are regularly sampled in space and time but they are generally low-pass filtered, interpolated and inaccurate in coastal areas and in the resolution of mesoscale features. In contrast, drifter data yield an accurate but irregular spatial and temporal sampling of the surface currents (Menna et al., 2012), therefore the resulting descriptions of surface circulation can be biased and strongly dependent on the specific data distribution (Poulain et al., 2012).

To overcome the aforementioned limitations, drifter and altimetry data can be combined to provide a more accurate representation of the mean circulation with respect to the computation performed from each data set alone (Centurioni et al., 2008; Poulain et al., 2012). The combination of these two independent data sets has already been used to produce quantitative descriptions of the circulation in several marine areas, such as the California Current System (Centurioni et al., 2008), the South China Sea (Centurioni et al., 2009), the Kuroshio Extension (Niiler et al., 2003), the North Pacific (Uchida and Imawaki, 2003) and the Mediterranean Sea (Menna et al., 2012; Poulain et al., 2012). The main advantages of this method are (1) the removal of biases that arise from the irregular sampling of drifters and (2) the enhancement of the accuracy of velocity estimates for coastal areas and mesoscale features.

In this work, the combination of the geostrophic velocities derived from drifter and satellite altimetry is implemented in the BS, so as to describe its surface circulation and its seasonal and interannual variability over the period 1999–2009. The combined method, applied to concurrent satellite and drifter observations, gives an estimation of the mean dynamic topography (MDT) of the BS in terms of absolute geostrophic velocity field.

A first attempt to estimate a MDT for the BS based on in situ and satellite data was made by Kubryakov and Stanichny (2011), applying the synthetic method described by Rio and Hernandez (2004) and Rio et al. (2007). This synthetic MDT (SMDT) was computed over the period 1993–1999 using measurements of drifter velocities, hydrological data and along-track sea level anomalies (SLA). The SMDT defines the basin scale RC, two separate cyclonic cells in the western and eastern parts of the BS and some mesoscale and sub-basin-scale features near the periphery of the basin. Anticyclonic mesoscale eddies are located between the continental slope and the coast and can feed the sub-basin-scale eddies (Stanev, 2005); the typical location and evolution of these structures are defined by Staneva et al. (2001) and Korotaev et al. (2003).

The goal of this paper is to produce a regularly sampled data set of the geostrophic currents in the BS, so as to describe quantitatively the pathways and the spatio-temporal variability of the surface circulation in different regions of the basin. The paper is organised as follows: information on drifter, remotely sensed altimetry data and wind products is given in Sect. 2, followed by a description of the methods used to compute the surface geostrophic field. Pseudo-Eulerian statistics, seasonal and interannual characteristics and eddy variability are presented and discussed in Sect. 3. The main conclusions are summarised in Sect. 4.

2 Data and methods

The Lagrangian data used in this study come from 89 satellite-tracked drifters deployed in the BS between 1999 and 2009. The majority are surface velocity program (SVP) drifters, consisting of a surface buoy tethered to a holey-sock drogue centred at a depth of 15 m. The others types are coastal ocean dynamic experiment (CODE) drifters, composed of a vertical 1 m-long negative buoyant tube with four radial vanes, and compact meteorological and oceanographic drifters (CMOD) that consist of a 60 cm-long aluminium cylindrical hull drogue with the sonobuoy case. They measure the near-surface currents in the first 15 m of the water column with various accuracies and errors induced by wind and waves (Poulain et al., 2009). The drifter position time series were edited to remove spikes and outliers and linearly interpolated at 2 h intervals using the kriging technique (Poulain et al., 2004). The interpolated positions were low-pass filtered (Hamming filter with cut-off period at 36 h) to remove high frequency current components and sub-sampled at 6 h intervals. Velocity components were then estimated from centred finite differences of the 6 h positions and finally archived in a database (http://doga.ogs.trieste.it/sire/medsvp/). The pseudo-Eulerian velocity statistics directly computed with the drifter data are defined as "biased" because the drifter data are usually non-uniformly sampled in space and time (Centurioni et al., 2008; Menna et al., 2012; Poulain et al., 2012).

Cross-calibrated, multi-platform (CCMP) ocean surface wind velocities (Atlas et al., 2009) were filtered and interpolated at the drifter locations and times, then used with the drifter velocities in a linear regression model in order to estimate the currents directly (slippage) and indirectly (Ekman currents) induced by the winds ($\boldsymbol{U}_{\text{wind-driven}}$). This method has been already applied to estimate the wind-driven currents in the tropical Pacific (Ralf and Niiler, 1999) and in the Mediterranean Sea (Menna et al., 2012; Poulain et al., 2009, 2012). Drifter velocities ($\boldsymbol{U}$) were used to estimate $\boldsymbol{U}_{\text{wind-driven}}$ using the following linear regression model (Poulain et al., 2012):

$$\boldsymbol{U} = \beta e^{i\theta} \boldsymbol{W} + \text{error} = \boldsymbol{U}_{\text{wind-driven}} + \text{error}, \quad (1)$$

Table 1. Results of the regression model (Eq. 1) applied to extract the wind-driven currents from the velocities of the different types of drifters. R^2 is the coefficient of determination and N is the number of observations considered.

Drifter Type	$\beta e^{i\theta}$	R^2 (%)	N
CMOD	$0.01 \exp(-25°i)$	20	1992
CODE	$0.008 \exp(-28°i)$	5	4813
SVP	$0.007 \exp(-31°i)$	4	25732

where β is a real constant and θ is the angle (positive anticlockwise) which represent, respectively, the estimations of intensity and direction of drifter wind-driven currents with respect to the wind velocity; $\boldsymbol{W}$ is the wind velocity interpolated at the drifter locations and times. The model was applied to the different drifter designs separately, over the whole BS, and the results are presented in Table 1. The SVP drifters measured the smallest wind-driven currents, amounting to 0.7 % of the wind speed and directed about 30° to the right of the wind; these currents represent 4 % of total velocity variance. The CODE drifters show wind-driven currents of 0.8 % of the wind speed with an angle of 28° to the right of the wind and an explained variance of about 5 %. The CMOD drifters move more downwind (about 25° to the right of the wind) with wind-driven currents amounting to 1 % of the wind speed and explain a significant portion of the velocity variance (about 20 %). The regression coefficients in the BS are similar to those estimated by Poulain et al. (2009, 2012) in the Mediterranean Sea. The mean intensity of the ageostrophic current estimated from Eq. (1) in $0.25° \times 0.25°$ bins is $1.5\,\mathrm{cm\,s^{-1}}$, with maximum values of $\sim 7\,\mathrm{cm\,s^{-1}}$ located in the interior of the basin and in the western region (not shown). Finally, the wind-driven currents estimated from Eq. (1) were subtracted from the drifter velocities to obtain an estimation of the geostrophic components $\boldsymbol{U}_{\mathrm{DG}}(\boldsymbol{U}_{\mathrm{DG}} = \boldsymbol{U} - \boldsymbol{U}_{\text{wind-driven}})$. These currents can obviously contain some contribution which is non geostrophic, especially near the coast and in small-scale eddies due to non-linear effects, but for simplicity and because mean currents will be mostly discussed, we will refer to these currents as geostrophic in this paper. To quantify the contribution of the wind-driven currents to the total drifter velocity, the percentage of total mean kinetic energy of the residuals per unit of mass (eddy kinetic energy; EKE) due to slippage and Ekman currents is computed on a $0.25° \times 0.25°$ resolution grid (not shown). The removal of Ekman component results in a reduction of the EKE computed from the drifter data set in 80 % of the cases; the mean reduction accounts for 30 % of the total EKE.

The altimetry data used in this work are gridded (one-eight of a degree Mercator projection grid) Ssalto/Duacs daily, multi-mission, delayed time products from AVISO (Ssalto/Duacs users handbook 2013). SLA data are defined with respect to a 7 yr mean (1993–1999). The high spatial resolution of these regional products allow one to discriminate the mesoscale patterns. Because of the intermittent temporal distribution and the scarcity of the drifter data (not shown), daily altimetry products were preferred to weekly products in order to have a larger number of drifter-satellite concurrent observations.

Drifter geostrophic velocities and satellite altimetry data were averaged in non-overlapping geographical bins of $0.25° \times 0.25° \times 1$ day .The bin-size was chosen as a compromise to resolve the basin and sub-basin (50–100 km) circulation of the BS and also to have a statistically significant number of drifter data inside each bin. Drifter and satellite data were combined in each spatial bin using the following regression model (according to the method described in Poulain et al., 2012):

$$\boldsymbol{U}_{\mathrm{DG}} = \boldsymbol{A}\boldsymbol{U}_{\mathrm{SLA}} + \boldsymbol{B} + \text{error}, \quad (2)$$

where the vector $\boldsymbol{U}_{\mathrm{SLA}}$ includes the bin-averaged anomalies of surface geostrophic velocities, concurrent with the bin-averaged drifter geostrophic velocities $\boldsymbol{U}_{\mathrm{DG}}$. The unknowns $\boldsymbol{A}$ and $\boldsymbol{B}$ are complex numbers or 2-D vectors. The slope $\boldsymbol{A}$ is the local adjustment of amplitude of $\boldsymbol{U}_{\mathrm{SLA}}$ (Niiler et al., 2003). Over the time period considered for the definition of SLA (1993–1999), the mean of $\boldsymbol{U}_{\mathrm{SLA}}$ is zero and $\boldsymbol{B}$ coincides with the time-mean drifter circulation; for any other period, the mean of $\boldsymbol{U}_{\mathrm{SLA}}$ is not zero and $\boldsymbol{B}$ is the offset between $\boldsymbol{U}_{\mathrm{SLA}}$ and $\boldsymbol{U}_{\mathrm{DG}}$ (Menna et al., 2012). The offset $\boldsymbol{B}$ represents the MDT expressed in terms of geostrophic velocities, partly referred to the SLA definition period (1993–1999) and partly referred to the drifter data period (1999–2009). Following Poulain et al. (2012), the slope $\boldsymbol{A}$ was subsequently low-pass filtered in $0.75° \times 0.75°$ overlapping bins to remove insignificant noise. The vector complex correlation (Kundu, 1976) between $\boldsymbol{U}_{\mathrm{DG}}$ and $\boldsymbol{U}_{\mathrm{SLA}}$ (Fig. 1a) is generally larger than 0.6 in the BS region, supporting the validity of the approach described by Eq. (2); correlation is low west of the Crimean Peninsula and off southern Bulgaria. The magnitude of the low pass filtered $|\boldsymbol{A}|$ and of $\boldsymbol{B}$ are shown in Fig. 1b and c, respectively. The magnitude of $|\boldsymbol{A}|$ varies mostly between 0.5 and 1.5; it exceeds 1.5 in the centre of the BS basin and off the northern coast of Bulgaria. Deviation of $\boldsymbol{A}$ from unity is mainly due to the oversmoothing of the satellite altimeter data, to the existence of residual wind-driven components, non-linear boundary currents and ageostrophic acceleration in the drifter velocities (Niiler et al., 2003; Poulain et al., 2012) and to the limited number of drifter data in the interior of the basin. The veering angle introduced by $\boldsymbol{A}$ (not shown) is not significantly different from zero. The offset $\boldsymbol{B}$ is as large as $30\,\mathrm{cm\,s^{-1}}$ in the fast RC, in particular along the Anatolian and Crimean coasts, and smaller than $5\,\mathrm{cm\,s^{-1}}$ in the interior of the basin. The relationship between drifter velocities and satellite SLA ($\boldsymbol{A}$ and $\boldsymbol{B}$), derived from 11 yr of concurrent data (1999–2009), can be used to estimate the mean unbiased absolute geostrophic currents $\langle \boldsymbol{U}_{\mathrm{G}} \rangle_{\mathrm{u}}$ for any

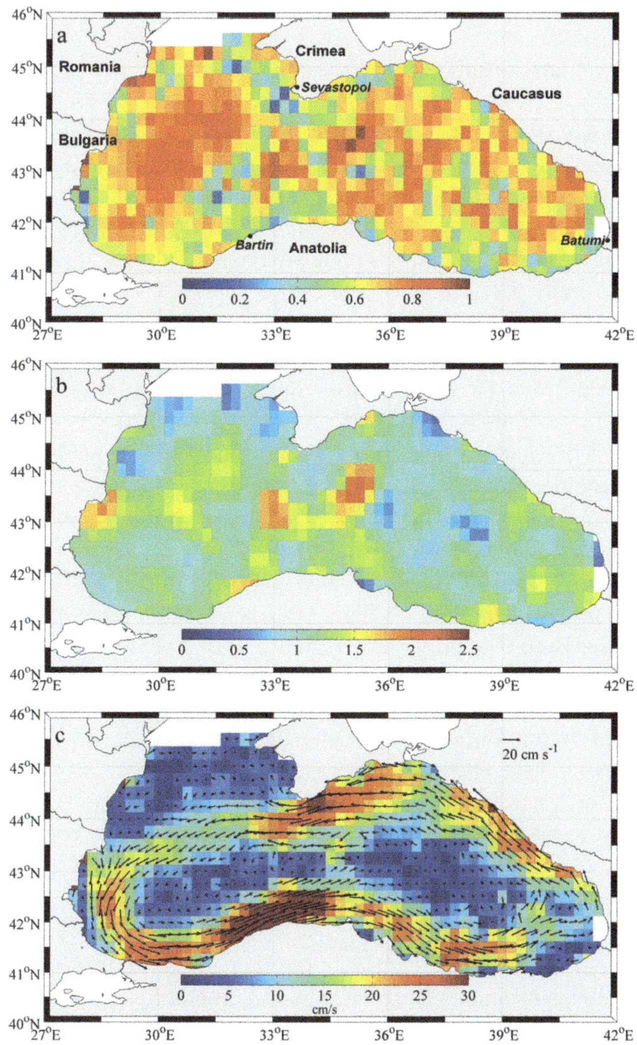

Fig. 1. Spatial distribution of the magnitude of **(a)** the complex correlation between drifter geostrophic velocities $\boldsymbol{U}_{\mathrm{DG}}$ and satellite anomalies of surface geostrophic velocities $\boldsymbol{U}_{\mathrm{SLA}}$; **(b)** the low pass filtered slope $|\boldsymbol{A}|$ and **(c)** the offset $\boldsymbol{B}$ depicted with vectors and colours.

time period in which $\boldsymbol{U}_{\mathrm{SLA}}$ is available, independently from the availability of drifter data:

$$\langle \boldsymbol{U}_{\mathrm{G}}\rangle_{\mathrm{u}} = \boldsymbol{A}\langle \boldsymbol{U}_{\mathrm{SLA}}\rangle_{\mathrm{u}} + \boldsymbol{B}, \qquad (3)$$

where $\boldsymbol{U}_{\mathrm{G}}$ are the surface geostrophic currents described by Eq. (2) and $\langle\rangle_{\mathrm{u}}$ indicates the "unbiased" temporal averages in each spatial bin. The pseudo-Eulerian statistics computed with these geostrophic currents are defined as "unbiased" because they are less biased with respect to those calculated directly with the drifter data (Menna et al., 2012; Poulain et al., 2012).

The interannual variability of the geostrophic circulation was assessed using the annual maps of kinetic energy of the mean flow per unit mass (MKE) and of EKE. The MKE is defined as

$$\mathrm{MKE} = \frac{1}{2}\left(\langle U_{\mathrm{G}}\rangle_{\mathrm{u}}^2 + \langle V_{\mathrm{G}}\rangle_{\mathrm{u}}^2\right). \qquad (4)$$

The residuals of geostrophic velocity field were evaluated by removing the mean unbiased velocity $\langle \boldsymbol{U}_{\mathrm{G}}\rangle_{\mathrm{u}}$ from each daily value of $\boldsymbol{U}_{\mathrm{G}}$ ($\boldsymbol{U}'_{\mathrm{G}} = \boldsymbol{U}_{\mathrm{G}} - \langle \boldsymbol{U}_{\mathrm{G}}\rangle_{\mathrm{u}}$) and the EKE is defined as the following annual mean:

$$\mathrm{EKE} = \frac{1}{2}\left(\left\langle U_{\mathrm{G}}'^2\right\rangle_{\mathrm{u}} + \left\langle V_{\mathrm{G}}'^2\right\rangle_{\mathrm{u}}\right). \qquad (5)$$

The seasonal and interannual variability of mesoscale and sub-basin-scale eddies in the BS was investigated in terms of kinetic energy of geostrophic velocity residuals (KE) per unit of mass, evaluated for each daily value of $\boldsymbol{U}'_{\mathrm{G}}$ and defined as

$$\mathrm{KE} = \frac{1}{2}\left(U_{\mathrm{G}}'^2 + V_{\mathrm{G}}'^2\right); \qquad (6)$$

Hovmoller diagrams of the KE in the period 1999–2009 were computed along the Anatolia and Crimean–Caucasian coasts.

Finally, the combined data set was used to reconstruct the relative vorticity (ζ) field associated with the surface circulation in the region of BE and SE. The relative vorticity was evaluated as the vertical component of the velocity field curl (Pedlosky, 1987):

$$\zeta = \frac{\partial \boldsymbol{V}_{\mathrm{G}}}{\partial x} - \frac{\partial \boldsymbol{U}_{\mathrm{G}}}{\partial y}. \qquad (7)$$

3 Results and discussion

The unbiased pseudo-Eulerian velocity statistics (Fig. 1c) depict the well known characteristics of the mean circulation in the upper layer of the BS suggested by several authors in the past (e.g. Korotaev et al., 2001; Poulain et al., 2005), including the RC, BE and SE. However they add new information about the seasonal and interannual characteristics of the surface circulation in the BS. This analysis is focused on the boundary regions characterised by high density of drifter data (Anatolia and Crimea coasts, Batumi and Sevastopol areas).

3.1 Seasonal variability

Seasonal variations are emphasised by removing the mean velocity field (Fig. 1c) from the mean seasonal current fields; the resulting anomalies are shown in Fig. 2. Anomalies larger than 8–10 cm s^{-1} are observed in the BE (41–43° N, 39.5–42° E) and SE (43.5–45° N, 30.5–32.5° E) regions and east of the Crimean Peninsula. The cyclonic RC is stronger in winter (mean anomalies of about 4–5 cm s^{-1}; maximum values larger than 10 cm s^{-1} along the Anatolia and Crimea coasts and in the BE; Fig. 2a) and weaker in spring/fall (mean anomalies of about 2 cm s^{-1}; Fig. 2b, d) and summer (the anticyclonic anomalies of RC indicate that the summer currents

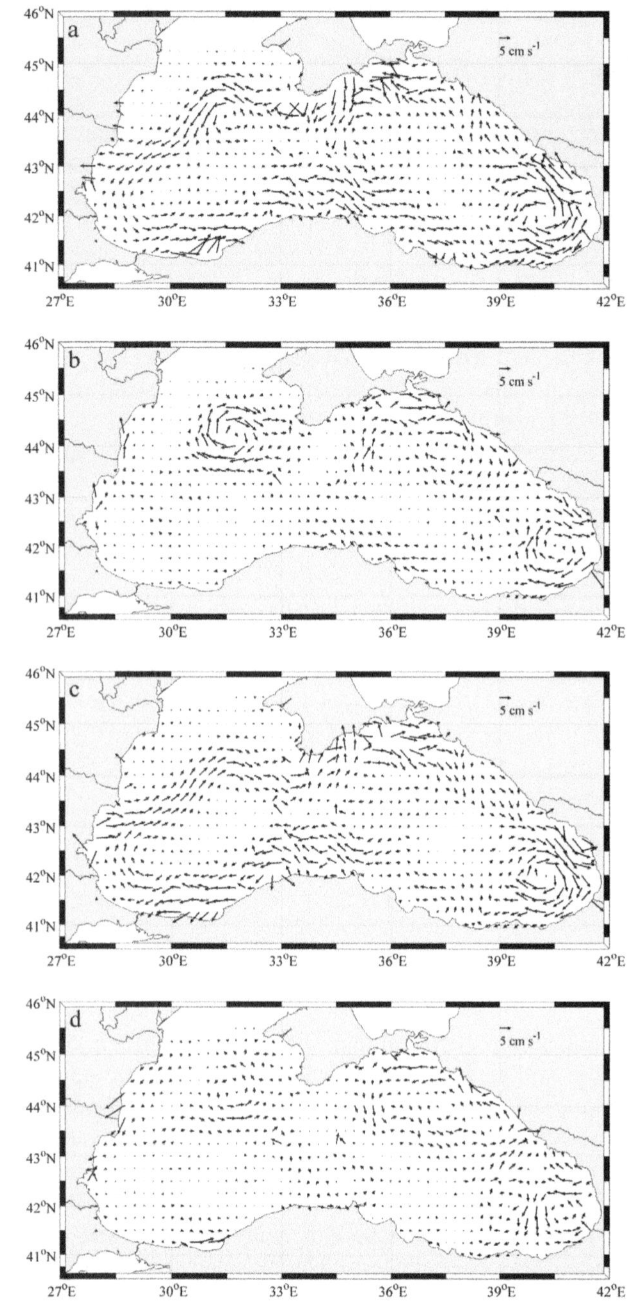

Fig. 2. Unbiased estimates of the anomalies of the mean surface geostrophic circulation in the Black Sea during **(a)** winter (January–March), **(b)** spring (April–June), **(c)** summer (July–September), **(d)** fall (October–December), in spatial bins of $0.25° \times 0.25°$ and for the period 1999–2009.

are 4–5 cm s^{-1} smaller than the mean currents; Fig. 2c). The BE is characterised by an intense cyclonic circulation in winter and an anticyclonic circulation in summer; the SE is most intense in spring and shows a less periodic nature than the BE. These results agree well with Stanev et al. (2000) and Poulain et al. (2005).

3.2 Interannual variability

3.2.1 Mean flow

The MKE computed on a yearly basis is used to describe the annual variation of the mean circulation. The annual maps (Fig. 3) depict a more intense RC circuit in the period 2002–2006 with speeds of 20–40 cm s^{-1} and MKE larger than 250 $cm^2\ s^{-2}$; maximum values are observed along the Anatolia (600–700 $cm^2\ s^{-2}$) and Crimea (500 $cm^2\ s^{-2}$) coasts. Generally, the northwest branch of the Rim Current is weaker with speeds of 5–15 cm s^{-1} and values of MKE smaller than 150 $cm^2\ s^{-2}$; larger intensities in this area are observed in 2002 and in the period 2005–2006 (speeds of 20–25 cm s^{-1} and MKE larger than 200 $cm^2\ s^{-2}$). The RC surrounds two cyclonic structures (described in the previous studies as the Western and Eastern gyres), that show the same characteristics over the whole study period (speeds smaller than 10 cm s^{-1} and MKE smaller than 50 $cm^2\ s^{-2}$).

The annual maps of EKE (Fig. 4) reach higher values in the Batumi region (larger than 250 $cm^2\ s^{-2}$) in 2002, in the Sevastopol region (larger than 150 $cm^2\ s^{-2}$) in 2000 and 2005, and in both these regions in 2003, 2004, 2006 and 2008; in 2003 high energy values are reached also along the western coast of the BS and along the Crimean coast. The regions characterised by the most intense interannual variations are the same as those estimated in Stanev et al. (2000).

3.2.2 Mesoscale eddies

Variability of the geostrophic component of the mesoscale (5–10 km) features in the BS was estimated using the time series of the KE. Mesoscale velocities are mostly included in the velocity variance and in the eddy kinetic energy. In particular, the KE can be used as a good indicator to follow the seasonal and interannual evolution of the mesoscale structures, as its large and variable values coincide with the main locations of the mesoscale eddies along the coasts of the BS (Fig. 5). The Hovmoller diagram in Fig. 5a, computed by meridionally averaging each KE value between 41 and 42.5° N, illustrates the interannual changes in intensity and location of the mesoscale eddies off the Anatolia coast (28.5–39° E). The Bosphorus Eddy is a recurrent structure located near 28–29° E with values of KE generally smaller than 200 $cm^2\ s^{-2}$; its energy increases without a specific periodicity (as in March 2001, July 2006, November–December 2007); sometimes it moves eastward and interacts with the close Sakarya Eddy (August–October 2003, October 2005, January 2006, December 2008–January 2009) reaching KE values larger than 500 $cm^2\ s^{-2}$. The interaction between the Bosphorus and the Sakarya eddies confirms the description of Korotaev et al. (2003). The Sakarya Eddy, located around 31° E, prevails only in some years (2000, 2002, 2003, 2005, 2006, 2009) with KE values larger than 250 $cm^2\ s^{-2}$. Figure 5a shows the presence of another eddy

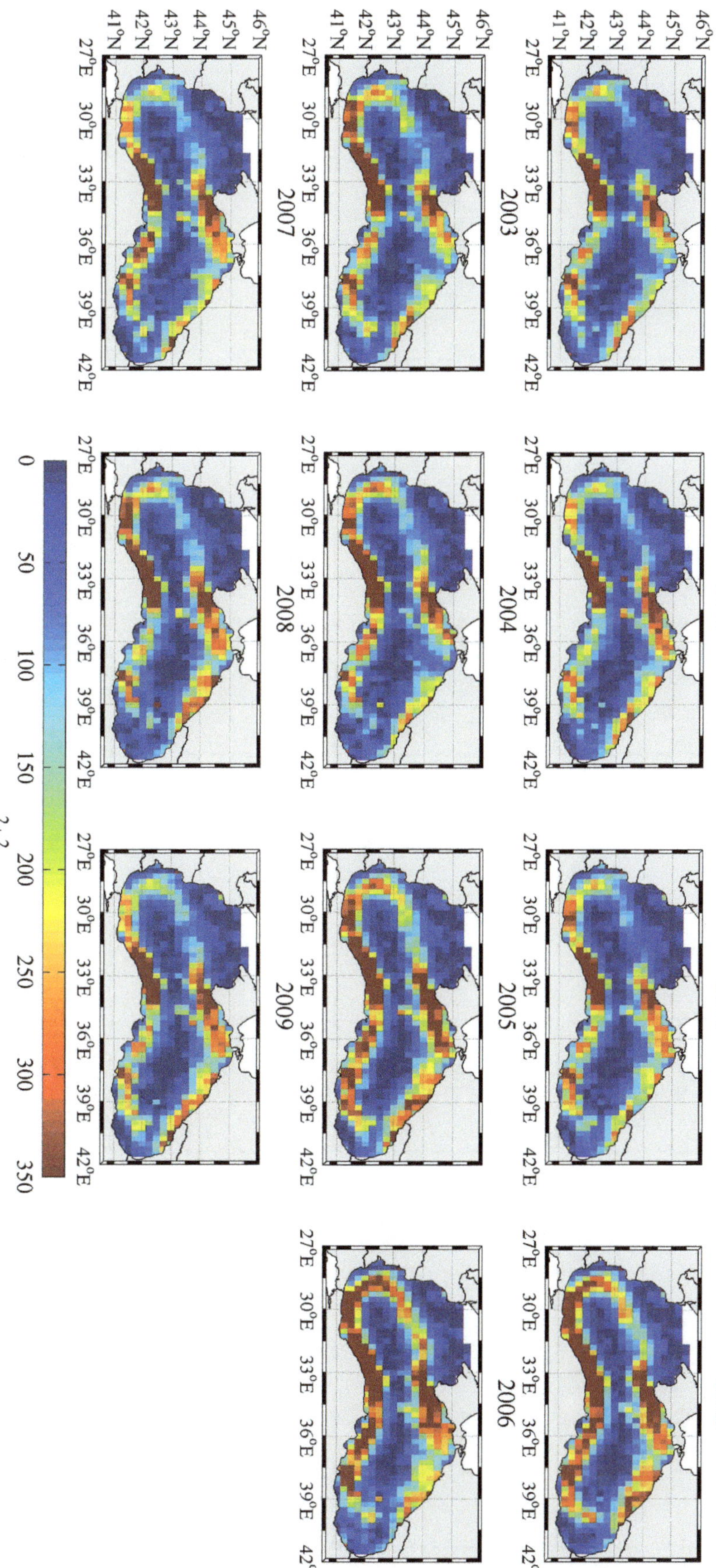

Fig. 3. Unbiased annual statistics of mean kinetic energy of the surface geostrophic currents in the period 1999–2009.

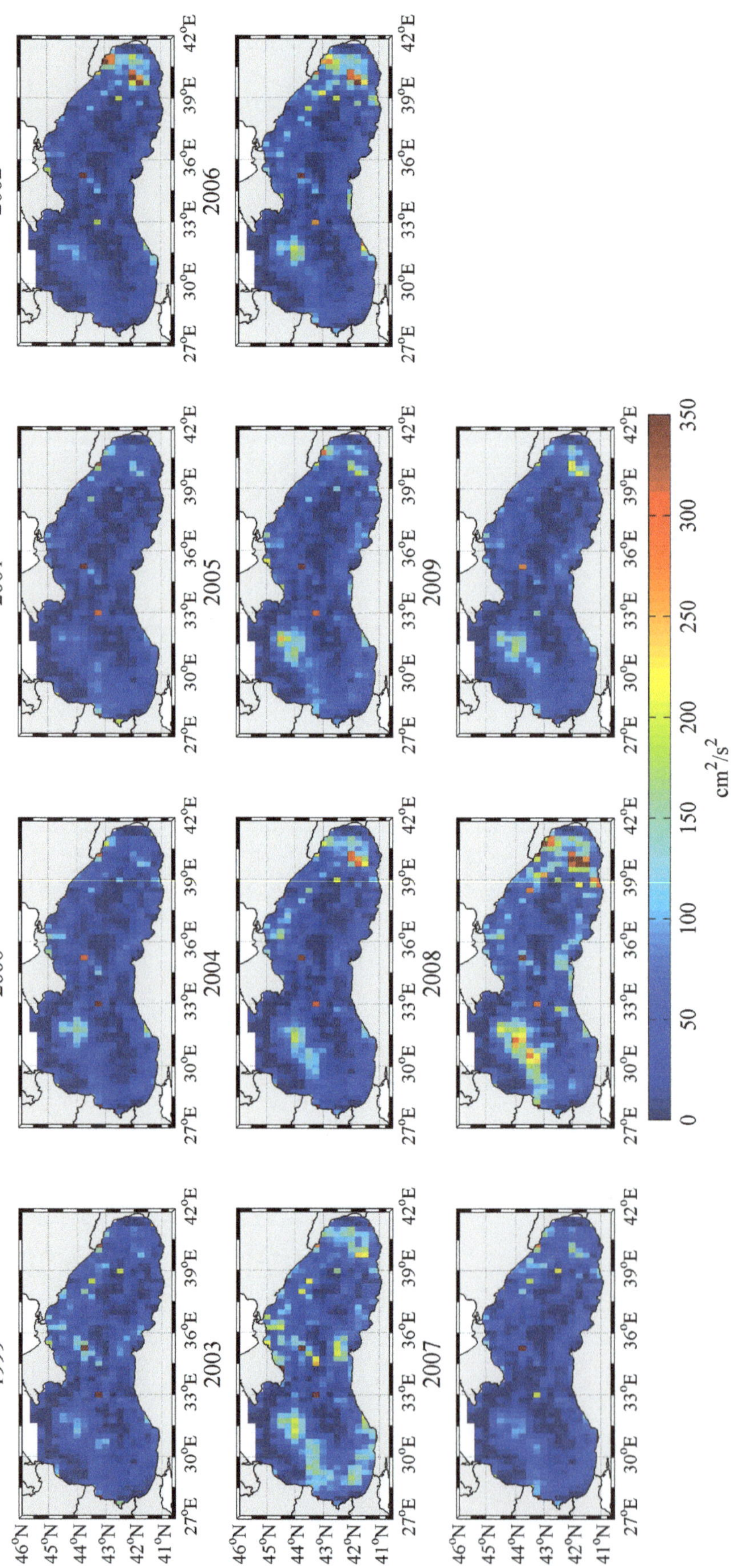

Fig. 4. Unbiased annual statistics of eddy kinetic energy of the surface geostrophic currents in the period 1999–2009.

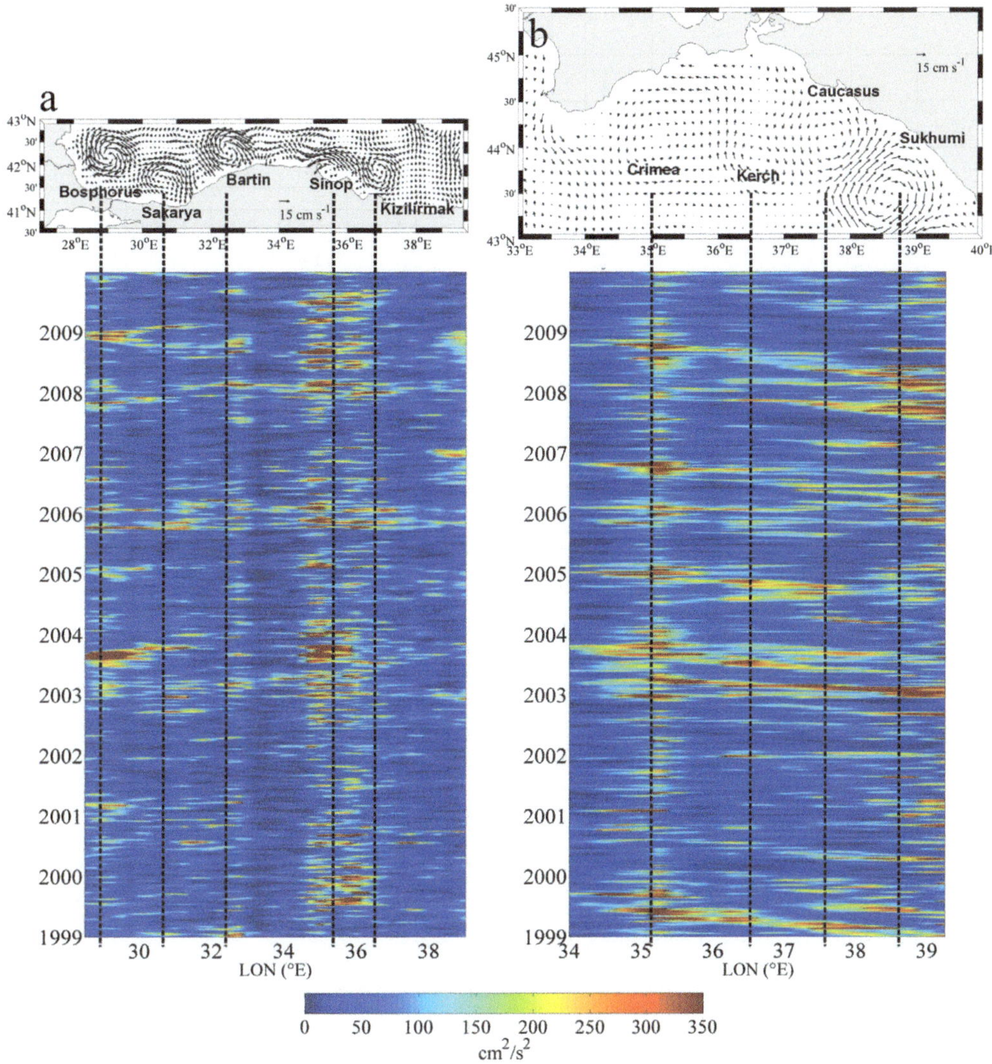

Fig. 5. Hovmoller diagrams of KE over the **(a)** Anatolia and **(b)** Crimea–Caucasus coasts. In the upper panels daily snapshots of the anomalies of surface geostrophic velocities (3 May 2001 and 15 January 1999, respectively) indicate the location of the main eddies along the BS coasts; these anomalies are depicted with the original spatial resolution of AVISO products (of one-eighth of a degree) in order to better discriminate the mesoscale structures. Dashed lines mark the locations of mesoscale eddies.

between 32 and 33° E with a quasi-persistent nature and KE signal between 150 and 250 $cm^2\ s^{-2}$; this structure can move westward and interact with the Sakarya Eddy (2005–2006). A mesoscale feature in this area was already recognised by Korotaev et al. (2003); according to its geographical location along the coast of the Bartin Province we define this structure as the Bartin Eddy. The Sinop and Kizilirmak Eddies, located between 34.5 and 37° E, show a quasi-permanent nature, in contrast with the results of Korotaev et al. (2003), and high values of KE (200–600 $cm^2\ s^{-2}$) throughout the period considered; the highest KE signal is observed in the period August–October 2003 in concomitance with the maximum intensities of the Bosphorus Eddy.

The diagram in Fig. 5b, computed by meridionally averaging each KE value between 43.5 and 45° N, includes the Caucasus coast and the eastern side of the Crimean Peninsula (34–39.25° E). The Crimea Eddy is a quasi-permanent mesoscale feature, located between 34.5 and 35.5° E, characterised by a pronounced interannual variability in terms of KE and a zonal extension of the eddy. The KE signal shows generally higher values (larger than 500 $cm^2\ s^{-2}$) when the Crimea Eddy becomes larger (1999, 2003, 2006, 2008). Another mesoscale eddy, identified by Korotaev et al. (2003) as the Kerch Eddy, is located between 36 and 37° E (KE values of 250–400 $cm^2\ s^{-2}$); it occurs sporadically and persists for a few months. The Caucasus Eddy, located around 37–38° E, is a sporadic structure with a KE larger than 200 $cm^2\ s^{-2}$ that occurs in the same periods as the Kerch Eddy, with which it sometimes interacts (2003, 2004, 2006, 2007). The interactions among the Kerch Eddy, the Caucasus Eddy and the

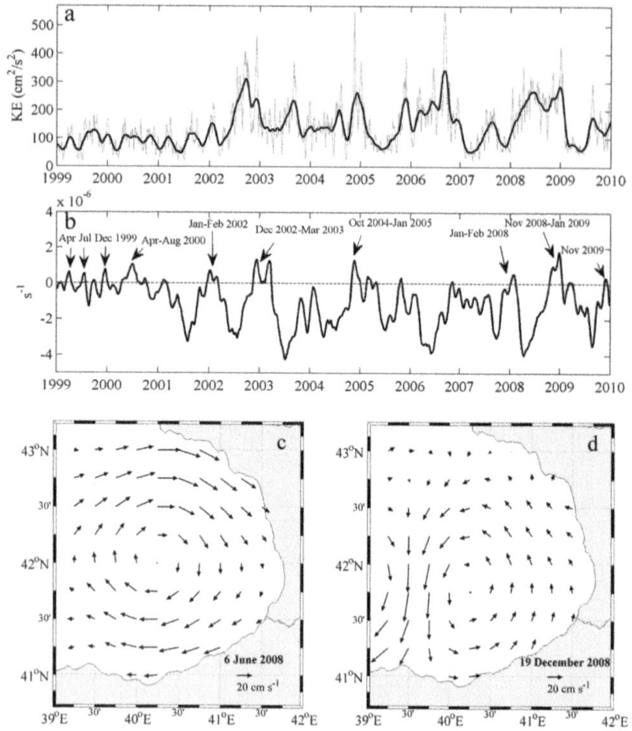

Fig. 6. **(a)** Time series of KE spatially averaged in the region of BE (41–43° N, 39–42° E); grey line represents the daily time series while the black line is the 90 day moving average of KE. **(b)** Time series of the spatially averaged relative vorticity of the surface geostrophic currents in the region of BE; black arrows emphasise the main periods of cyclonic circulation. Snapshots of the **(c)** anticyclonic and **(d)** cyclonic circulation ($\boldsymbol{U}_{\mathrm{G}}$) over the Batumi area.

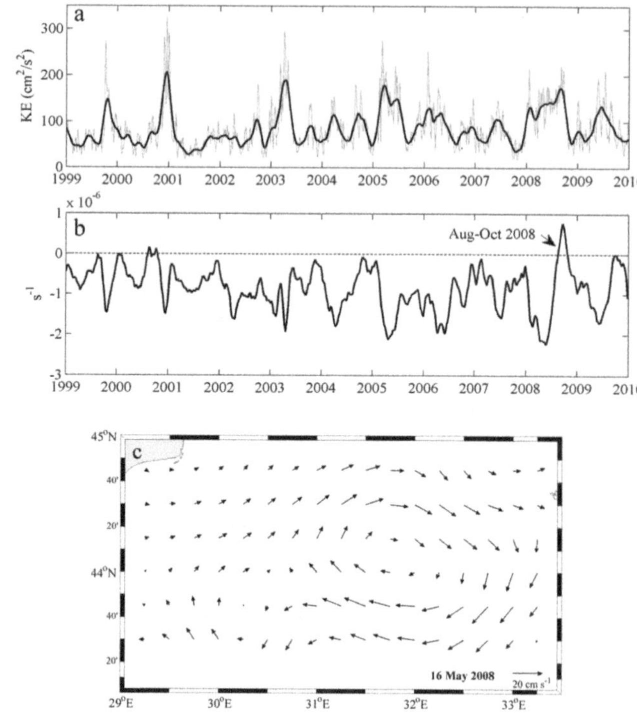

Fig. 7. **(a)** Time series of KE spatially averaged in the region of SE (43.75–44.75° N, 30–33° E); grey line represents the daily time series while the black line is the 90 day moving average of KE. **(b)** Time series of the spatially averaged relative vorticity of the surface geostrophic currents in the region of SE; black arrow indicate the period characterised by cyclonic circulation. **(c)** Snapshot of geostrophic currents ($\boldsymbol{U}_{\mathrm{G}}$) in the region of SE.

RC confirm the previous results of Korotaev et al. (2003) while their temporal intermittence contradicts Korotaev et al. (2003). The Sukhumi Eddy (38.5–39.25° E) reveals a quasi-permanent nature with greater intensities during the winter and fall months (KE values of 200–400 $\mathrm{cm^2\,s^{-2}}$); it merges with the close Caucasus Eddy in winter 2003, fall 2007 and winter–spring 2008. The continual occurrence of the Sukhumi Eddy contradicts the results of Korotaev et al. (2003).

3.2.3 Sub-basin-scale eddies

The variability of the major sub-basin structures in the BS (the BE and the SE) is investigated using the spatially averaged time series of KE (Figs. 6a, 7a) computed in the eddy regions. The BE is more intense in 2002–2006 and 2008 with typical values of KE between 200 $\mathrm{cm^2\,s^{-2}}$ and 300 $\mathrm{cm^2\,s^{-2}}$ (Fig. 6a); this structure is usually characterised by a diameter of $\sim$ 100 km and is located in the southeast corner of the BS. The activity of BE is weaker in the period 1999–2001 and 2007 (Fig. 6a). The pronounced interannual variability in the Batumi region confirms the results of the model simulations of Staneva et al. (2001). Reversals of the surface circulation in the Batumi region are documented using the time series of the spatially averaged ζ (Fig. 6b). The typical anticyclonic pathway, characterised by negative vorticity values, is sometimes interrupted by a cyclonic regime (positive vorticity). The main cyclonic circulation events are observed in spring–summer 2000, winters 2002 and 2003, fall 2004–early winter 2005, fall 2008–early winter 2009 (black arrows in Fig. 6b). High values of KE correspond to anticyclonic rotation. Examples of anticyclonic and cyclonic regimes are shown in Fig. 6c and d, respectively.

The SE is characterised by low energy levels with respect to the BE (Fig. 7a); the maximum mean value of KE ($\sim$ 300 $\mathrm{cm^2\,s^{-2}}$) is observed in fall 2000. During the period 2003–2007 the SE shows a pronounced seasonal variability, with higher KE in spring and fall and lower KE in winter and summer. The activity of SE is weaker in 2001–2002 and more intense in 2008. The time series of the spatially averaged ζ (Fig. 7b) show that the SE is always anticyclonic, except for a cyclonic event in summer–fall 2008 probably due to the intrusion of an RC cyclonic meander in the Sevastopol region. The generation of SE is related to the meandering of the RC in the Sevastopol area and is generally observed in spring (speeds of about 15–20 $\mathrm{cm\,s^{-1}}$; Fig. 7c).

4 Conclusions

The surface geostrophic circulation in the BS has been described combining 11 yr (1999–2009) of drifter and satellite altimetry data in order to construct a regularly sampled data set of observations. Velocity statistics computed from this data set are defined as "unbiased" because they are less biased with respect to those calculated directly with the drifter data. The unbiased pseudo-Eulerian statistics are used to define seasonal and interannual variations of the mean field and of the sub-basin and mesoscale eddies. Mean geostrophic currents obtained in the present work (Fig. 1c) are based on a larger number of in situ drifter observation compared with those estimated from the SMDT of Kubryakov and Stanichny (2011).

Seasonal maps confirm the results of Stanev et al. (2000) and Poulain et al. (2005) and add new information about the SE, which is generated in spring from a meander of the RC (Fig. 2b). The most pronounced seasonal variabilities are located in the RC and in the BE areas, with anomalies of 8–10 cm s^{-1} with respect to the mean geostrophic field (Fig. 2).

The most intense interannual variations are located in the Batumi and Sevastopol regions (EKE levels larger than 200 $cm^2 s^{-2}$), in good agreement with the results of Stanev et al. (2000) and Staneva et al. (2001). Reversals of the surface circulation in the Batumi region, from anticyclonic to cyclonic regimes, are usually observed in fall–winter; the cyclonic pattern generally persists for periods of 2–4 months.

The analysis of mesoscale eddies along the Anatolia and Crimea–Caucasus coasts expands the results of Korotaev et al. (2003), as describes the interaction between the Sinop and the Kizilirmak eddies (Anatolia) and among the Kerch, the Caucasus and the Sukhumi eddies (Caucasus) and resolves a new quasi-persistent eddy generated along the Anatolia coast and defined as the Bartin Eddy.

Acknowledgements. The authors would like to thanks all the people who deployed drifters in the Black Sea. The altimeter products were produced by SSALTO/DUACS and distributed by AVISO, with support from CNES (http://www.aviso.oceanobs.com/duacs/).

We thank E. Stanev, A. A. Kubryakov and the two anonymous reviewers for their constructive comments on the discussion paper.

Edited by: J. M. Huthnance

References

Atlas, R., Ardizzone, J. V., Hoffman, R., Jusem, J. C., and Leidner, S. M.: Cross-calibrated, multi-platform ocean surface wind velocity product (MEaSUREs Project), Guide Document, Physical Oceanography Distributed Active Archive Center (PO.DAAC), JPL, Pasadena, California, Version 1.0., 26 pp., 2009.

Centurioni, L. R., Ohlmann, J. C., and Niiler, P. P.: Permanent meanders in the California Current System, J. Phys. Oceanogr., 38, 1690–1710, doi:10.1175/2008JPO3746.1, 2008.

Centurioni, L. R., Niiler, P. N., and Lee, D.-K.: Near surface circulation in the South China Sea during the winter monsoon, Geophys. Res. Lett., 36, L06605, doi:10.1029/2008GL037076, 2009.

Korotaev, G. K., Saenko, O. A., and Koblinsky, C. J.: Satellite altimetry observations of the Black Sea level, J. Geophys. Res., 106, 917–933, doi:10.1029/2000JC900120, 2001.

Korotaev, G., Oguz, T., Nikiforov, A., and Koblinsky, C.: Seasonal, interannual, and mesoscale variability of the Black Sea upper layer circulation derived from altimeter data, J. Geophys. Res., 108, 3122, doi:10.1029/2002JC001508, 2003.

Kubryakov, A. A. and Stanichny, S. V.: Mean Dynamic Topography of the Black Sea, computed from altimetry, drifter measurements and hydrology data, Ocean Sci., 7, 745–753, doi:10.5194/os-7-745-2011, 2011.

Kundu, P. K.: Ekman veering observed near the ocean bottom, J. Phys. Oceanogr., 6, 238–242, 1976.

Menna, M., Poulain, P.-M., Zodiatis, G., and Gertman, I.: On the surface circulation of the Levantine sub-basin derived from Lagrangian drifter and satellite altimetry data, Deep-Sea Res., 65, 46–58, doi:10.1016/j.dsr.2012.02.008, 2012.

Niiler, P. P., Maximenko, N. A., Panteleev, G. G., Yamagata, T., and Olson, B. D.: Near-surface dynamical structure of the Kuroshio Extension, J. Geophys. Res.-Oceans, 108, 3193, doi:10.1029/2002JC001461, 2003.

Oguz, T., La Violete, P. E., and Unluata, U.: The upper layer circulation of the Black Sea: Its variability as inferred from hydrographic and satellite observations, J. Geophys. Res., 97, 12569–12584, 1992.

Pedlosky, J.: Geophysical Fluid Dynamics, 2nd Edn., 710 pp., Springer, New York, 1987.

Poulain, P.-M., Barbanti, R., Cecco, R., Fayes, C., Mauri, E., Ursella, L., and Zanasca P.: Mediterranean surface drifter database: 2 June 1986 to 11 November 1999, Rel. 75/2004/OGA/31, OGS, Trieste, Italy (CDRom), 2004.

Poulain, P.-M., Barbanti, R., Motyzhev, S., and Zatsepin, A.: Statistical description of the Black Sea near-surface circulation using drifter in 1999–2003, Deep-Sea Res.-Pt. I, 52, 2250–2274, doi:10.1016/j.dsr.2005.08.007, 2005.

Poulain, P.-M., Gerin, R., Mauri, E., and Pennel, R.: Wind Effects on Drogued and Undrogued Drifters in the Eastern Mediterranean, J. Atmos. Ocean Tech., 26, 1144–1156, 2009.

Poulain, P.-M., Menna, M., and Mauri, E.: Surface geostrophic circulation of the Mediterranean Sea derived from drifter and satellite altimeter data, J. Phys. Oceanogr., 42, 973–990, doi:10.1175/JPO-D-11-0159.1, 2012.

Ralph, E. A. and Niiler, P. P.: Wind driven currents in the tropical Pacific, J. Phys. Oceanogr., 29, 2121–2129, 1999.

Rio, M.-H. and Hernandez, F.: A mean dynamic topography computed over the world ocean from altimetry, in-situ measurements, and a geoid model, J. Geophys. Res., 109, C12032, doi:10.1029/2003JC002226, 2004.

Rio, M.-H., Poulain, P.-M., Pascual, A., Mauri, E., Larnicol, G., and Santoleri, R.: A Mean Dynamic Topography of the Mediterranean Sea computed from altimetric data, in-situ measurements and a general circulation model, J. Mar. Syst. 65, 484–508, doi:10.1016/j.jmarsys.2005.02.006, 2007.

Shapiro, G. I., Stanichny, S. V., and Stanychna, R. R.: Anatomy of shelf-deep sea exchanges by a mesoscale eddy in the North West

Black Sea as derived from remotely sensed data, Remote Sens. Environ., 114, 867–875, 2010.
Stanev, E. V.: Understanding Black Sea Dynamics: overview of recent numerical modelling, Oceanography, 18, 52-71, 2005.
Stanev, E. V., Le Traon, P. Y., and Peneva, E. L.: Sea level variations and their dependency on meteorological and hydrological forcing: Analysis of altimeter and surface data for the Black Sea, J. Geophys. Res., 105, 17203–17216, 2000.
Staneva, J. V., Dietrich, D. E., Stanev, E. V., and Bowman, M. J.: Rim current and coastal eddy mechanisms in an eddy-resolving Black Sea general circulation model, J. Mar. Syst., 31, 137–157, 2001.
Uchida, H. and Imawakki, S.: Eulerian mean surface velocity field derived by combining drifter and satellite altimetry data, Geophys. Res. Lett., 30, 1229, doi:10.1029/2002GL016445, 2003.

Argo data assimilation into HYCOM with an EnOI method in the Atlantic Ocean

D. Mignac[1,2], C. A. S. Tanajura[2,3,4], A. N. Santana[2], L. N. Lima[2], and J. Xie[5]

[1]Graduate Program in Geophysics, Physics Institute and Geosciences Institute, Federal University of Bahia, Salvador, Brazil
[2]Oceanographic Modeling and Observation Network, Center for Research in Geophysics and Geology, Federal University of Bahia, Salvador, Brazil
[3]Physics Institute, Federal University of Bahia, Salvador, Brazil
[4]Ocean Sciences Department, University of California, Santa Cruz, CA, USA
[5]Institute of Atmospheric Physics, Chinese Academy of Sciences, Beijing, China

Correspondence to: D. Mignac (davi.mignac@gmail.com)

Abstract. An ocean data assimilation system to assimilate Argo temperature (T) and salinity (S) profiles into the HYbrid Coordinate Ocean Model (HYCOM) was constructed, implemented and evaluated for the first time in the Atlantic Ocean (78° S to 50° N and 98° W to 20° E). The system is based on the ensemble optimal interpolation (EnOI) algorithm proposed by Xie and Zhu (2010), especially made to deal with the hybrid nature of the HYCOM vertical coordinate system with multiple steps. The Argo T–S profiles were projected to the model vertical space to create pseudo-observed layer thicknesses (Δp_{obs}), which correspond to the model target densities. The first step was to assimilate Δp_{obs} considering the sub-state vector composed by the model layer thickness (Δp) and the baroclinic velocity components. After that, T and S were assimilated separately. Finally, T was diagnosed below the mixed layer to preserve the density of the model isopycnal layers. Five experiments were performed from 1 January 2010 to 31 December 2012: a control run without assimilation, and four assimilation runs considering the different vertical localizations of T, S and Δp. The assimilation experiments were able to significantly improve the thermohaline structure produced by the control run. They reduced the root mean square deviation (RMSD) of T and S calculated with respect to Argo independent data in 34 and 44 %, respectively, in comparison to the control run. In some regions, such as the western North Atlantic, substantial corrections in the 20 °C isotherm depth and the upper ocean heat content towards climatological states were achieved. The runs with a vertical localization of Δp showed positive impacts in the correction of the thermohaline structure and reduced the RMSD of T (S) from 0.993 °C (0.149 psu) to 0.905 °C (0.138 psu) for the whole domain with respect to the other assimilation runs.

1 Introduction

Many ocean circulation models include highly sophisticated numerical schemes and a large set of physical parameterizations. However, these are approximations of the governing equations and therefore they are sources of errors or discrepancies with respect to observations. Also, errors may be produced due to inaccuracies of the initial conditions, atmospheric forcing and lateral boundary conditions (Kalnay et al., 1996; Chassignet et al., 2009). For these reasons, data assimilation methods are important scientific tools in oceanography and other fields. They combine model outputs with observational data in a mathematically optimal or sub-optimal way and produce an objective analysis with smaller errors than the model output alone (Daley, 1991; Kalnay, 2003). The analysis is used as the model initial condition for weather and climate forecasts (Kalnay, 2003), and more recently for ocean weather forecasts under the framework of operational oceanography (Dombrowsky et al., 2009; Chassignet et al., 2009; Schiller and Brassington, 2011). Data assimilation methods are also applied to produce long-term series of analyses for climate diagnostics studies and they contribute to a better understanding of the physical mechanisms that are responsible for the ocean variability. For example, the depth of the mixed layer and the heat content can be better represented by analyses than by model simulations without assimilation (Carton and Giese, 2008).

A major obstacle in ocean data assimilation is the relatively small volume of observed ocean data available for assimilation and validation. Most of the spatial and high-frequency temporal variability from the ocean surface is acquired by satellite measurements such as the sea surface height (SSH) and sea surface temperature (SST). However, these observations are available for only a few decades and they are insufficient to determine the sub-surface variability (Ezer and Mellor, 1994; Chassignet et al., 2006). Therefore, the implementation of the Argo network with more than 3300 profilers freely reporting temperature and salinity data down to 2000 m has transformed the in situ ocean observing system in the new millennium (Schiller and Brassington, 2011). For the first time it is possible to have continuous measurements of the temperature, salinity and velocity of the upper ocean which makes the Argo network indispensable for any global or regional data assimilation system (Chassignet et al., 2007; Oke et al., 2008; Xie and Zhu, 2010). For example, Oke and Schiller (2007) showed that the assimilation of SST and SSH should be complemented with Argo profiles since the assimilation of this component plays a crucial role in improving the model thermohaline state, especially for salinity.

Among all data assimilation methods, the ensemble-based methods use a set of different model states to estimate the model errors (Evensen, 2003; Oke et al., 2005). One widely used is the ensemble optimal interpolation (EnOI) (Oke et al., 2005) in which the ensemble members are derived from an existing model run. This reduces the computational cost of the assimilation and makes this method suitable for operational purposes. It is already verified that the EnOI is able to effectively constrain the model towards observations. For instance, it was successfully applied to assimilate Argo data in the Pacific Ocean (Xie and Zhu, 2010), sea level anomaly (SLA) data in the South China Sea (Xie et al., 2011) and in the Gulf of Mexico (Counillon and Bertino, 2009), and SLA, SST and Argo data in the Australian region (Oke et al., 2008) and in the Indo-Pacific Ocean (Yan et al., 2010).

Considering the importance of the model in the construction of the analysis, several state-of-the-art, publically available ocean circulation models should be evaluated in order to develop an operational ocean forecasting system with data assimilation. One choice is the HYbrid Coordinate Ocean Model (HYCOM). It is formulated in terms of target densities and employs a hybrid vertical coordinate system to combine the best features of each vertical coordinate in specific oceanic regions (Bleck, 2002). The model has fixed z-levels to better represent the mixed layer, isopycnal layers to discretize the deep stratified ocean and σ-levels to better reproduce the bathymetry in shallow areas. Because of the hybrid nature of the model, the best way to assimilate vertical profile data into HYCOM is still an open question. A choice of the model prognostic variables should be made a priori since only two of the three state variables – temperature, salinity and potential density – are independent. Furthermore, the layer thickness, which is a key model variable, varies spatially and temporally according to the evolution of temperature, salinity and density.

Taking into account these characteristics, Thacker and Esenkov (2002; hereafter TE) proposed a method to assimilate expendable bathythermographic (XBT) data into HYCOM with a three-dimensional variational scheme (3-D Var). In this work, the XBT profiles are converted into observed layer thicknesses that respect the target densities of each model layer, and the temperature and salinity of the XBTs are projected to each observed layer thickness. Xie and Zhu (2010; hereafter XZ) used an EnOI scheme to assimilate Argo data, and showed that the TE approach produced significant improvements in relation to simpler schemes, in which the innovation is calculated in the observational space.

Very little has been published to evaluate the impact of in situ profile data assimilation into HYCOM with a focus on the Atlantic Ocean (e.g., Thacker et al., 2004; Belyaev et al., 2012). In the present work, a data assimilation system for HYCOM was constructed, implemented and realized for the Atlantic Ocean for the first time. The data assimilation algorithm follows the EnOI scheme suggested by XZ very closely. The present system was developed within the Brazilian Oceanographic Modeling and Observation Network (REMO) to be a component of an operational ocean forecasting system for the Atlantic Ocean (www.rederemo.org) (Tanajura and Belyaev, 2009; Lima et al., 2013; Tanajura et al., 2013). In this paper, the focus is on the impact of Argo data assimilation over the Atlantic Ocean and on a sensitivity study of the analysis run, considering different vertical localizations of the model error co-variance matrix involving temperature, salinity and especially the layer thickness. The REMO forecasting system uses a nested model approach based on HYCOM. The present work deals with the construction of large-scale analyses over almost the whole Atlantic Ocean. This domain was conceived and configured to provide reasonable boundary conditions to higher-resolution grids of greater interest to REMO over the South Atlantic. In the near future, the present data assimilation methodology will be used in the Atlantic domain, and in the higher-resolution grids over the Metarea V (from 35.5 °S to 7 °N, west of 20 °W to Brazil) and sub-regions off the Brazilian coast of particular interest to the Brazilian Navy and the active petroleum industry located there.

This paper is organized as follows. In Sect. 2, HYCOM and the configuration used in this study are briefly described. In Sect. 3, the EnOI scheme to assimilate Argo data is presented. Section 4 shows the design of the assimilation experiments and Sect. 5 presents their results. Section 6 contains discussion and conclusions.

2 HYCOM and its configuration

HYCOM is a primitive equation general circulation model, which has evolved from the Miami Isopycnic Coordinate Ocean Model (MICOM) (Bleck and Smith, 1990). The main advantage of the isopycnal coordinate is its ability to maintain the properties of water masses which do not communicate directly with the mixed layer. In HYCOM, with the advection of layer thicknesses by the continuity equation, the isopycnal coordinates smoothly transform into the z-coordinate in the weakly stratified upper ocean mixed layer and into the terrain-following σ-coordinate in the shallow water regions (Bleck, 2002; Chassignet et al., 2007). The freedom to adjust the vertical spacing of the coordinate surfaces in HYCOM simplifies the numerical implementation of several physical processes, such as turbulence in the mixed layer, detrainment and convective adjustment. Also, the capability of assigning additional coordinate surfaces to the oceanic mixed layer in HYCOM allows the option of implementing sophisticated vertical mixing turbulence closure schemes (Halliwell, 2004). Hence, HYCOM is considered to be a suitable model for operational ocean forecasting systems and climate studies (Chassignet et al., 2007, 2009). In this work, the version 2.2.14 of HYCOM was used.

The model grid in the present configuration has 760×480 horizontal grid points, with a spatial resolution of 0.25°, which remains constant in longitude, but varies in latitude attaining higher resolution towards the poles. The computational model domain covered almost all the Atlantic Ocean from 78 °S to 50 °N and from 100 °W to 20 °E, excluding the Pacific Ocean, the Mediterranean Sea and the North Atlantic subpolar region. The choice for the northern limit at 50 °N was based on two facts. First, the surface currents are primarily zonal at this latitude (Gabioux et al., 2013). Second, the purpose of this grid is to provide reasonable boundary conditions to higher-resolution grids that will soon be configured along the Brazilian coast, which is the area of main interest to REMO and far away from the northern boundary. On the lateral boundaries, relaxation to monthly climatological temperature and salinity from Levitus (1982) was applied considering the outermost 10 grid cells and the timescale of 30 days. This approach attempts to preserve climatological shear through geostrophic adjustment and has been successfully used in previous works (e.g., Paiva and Chassignet, 2001; Gabioux et al., 2013). Constant barotropic volume fluxes were imposed: zero flux in the north, eastward flux of 110 Sv in the Drake Passage, westward flux of 10 Sv in 12 grid points adjacent to South Africa along 20 °E, and eastward flux of 120 Sv further south to Antarctica.

The model was configured to use surface pressure as reference for potential density, aiming for improved representation of near-surface fields, at the cost of not representing accurately the Antarctic Bottom Water (Chassignet et al., 2003). The vertical domain was discretized in 21 vertical layers. The chosen target potential densities were 19.50, 20.25, 21.00, 21.75, 22.50, 23.25, 24.00, 24.70, 25.28, 25.77, 26.18, 26.52, 26.80, 27.03, 27.22, 27.38, 27.52, 27.64, 27.74, 27.82 and 27.88. To obtain the volumetric density in kg m^{-3}, 1000 should be added to each target density. The first layers have a few light target density values that ensure a minimum of three fixed-depth layers near the surface of the ocean.

The vertical mixing scheme is the K-profile parameterization (KPP) (Large et al., 1994). The model bathymetry was interpolated from the Earth Topography 1 (ETOPO1) with 1 min resolution. The model was initialized from the state of rest with climatological thermohaline structure, and a 30-year spin-up was performed using monthly climatological forcing fields from the Comprehensive Ocean and Atmosphere data set (COADS) (Woodruff et al., 1987). Then, from January 1995 to December 2009, the model was forced on the ocean surface with 6-hourly atmospheric reanalysis version 1 by the National Centers for Environmental Prediction/National Centers for Atmospheric Research (NCEP/NCAR) (Kalnay et al., 1996), including precipitation, wind speed at 10 m, short- and long-wave radiation fluxes at the surface, air temperature and humidity at 2 m. The result of this simulation was used as the initial condition for the Argo data assimilation experiments starting on 1 January 2010.

Figure 1 shows the mean state of temperature and salinity simulated by HYCOM from 1 January 1997 to 31 December 2008 and its comparison with the World Ocean Atlas 2013 Climatology (WOA13) (Boyer and Mishonov, 2013) along 25 °W for the upper 1000 m. WOA13 climatology shown in Fig. 1 spans the period from 1995 to 2012. In general, the pattern of the simulated temperature and salinity is similar to the WOA13 climatology, particularly in the South Atlantic, which is the main target area for REMO. In the North Atlantic large differences are seen between 25 and 50 °N below 400 m. The Mediterranean water (MW) is more saline, warmer and found further north in comparison with WOA13. The Mediterranean Sea is not simulated in this grid and is imposed as a relaxation towards monthly climatological temperature and salinity at the boundary without mass flux. In addition, the simulated temperature is higher than WOA13 in the upper 300 m of the equatorial region, while the values of high-salinity cores in the subtropical gyres are smaller than the values found in the climatology. It is worth noting that this resolution of 1/4° is not enough for HYCOM to properly solve the Gulf Stream and its associated dynamics (Hulburt and Hogan, 2000), which may also contribute to some of the model discrepancies in comparison to WOA13 in the mid-latitudes of the North Atlantic. It is expected that assimilation of Argo data will improve the model state and reduce the existing differences in the thermohaline structure with respect to the WOA13 climatology.

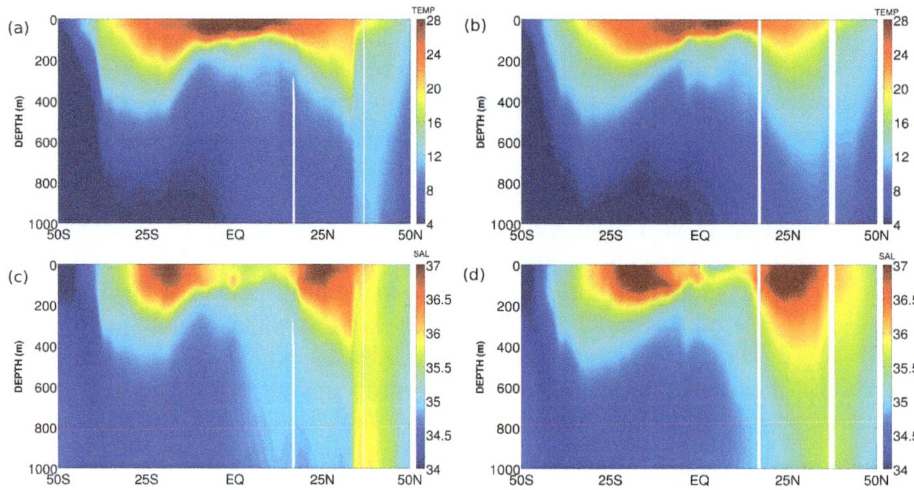

Figure 1. Mean temperature (°C) and salinity (psu) in the upper 1000 m along 25° W for **(a)** averaged simulated temperature from January 1997 to December 2008, **(b)** climatological temperature from WOA13, **(c)** averaged simulated salinity from January 1997 to December 2008, and **(d)** climatological salinity from WOA13.

3 The data assimilation scheme

The analysis ($\boldsymbol{X}^{\mathrm{a}}$) according to the EnOI scheme is given by the formula (Evensen, 2003)

$$\boldsymbol{X}^{\mathrm{a}} = \boldsymbol{X}^{\mathrm{b}} + \mathbf{K}(\boldsymbol{Y} - \mathbf{H}\boldsymbol{X}^{\mathrm{b}}), \tag{1}$$

where $\boldsymbol{X}^{\mathrm{b}} \in \mathbb{R}^{N}$ is the model background state or the prior, $\mathbf{K}$ is the gain matrix, $\boldsymbol{Y}$ is the vector of observations, $\boldsymbol{Y} \in \mathbb{R}^{N_{\mathrm{OBS}}}$, and $\mathbf{H}\boldsymbol{X}^{\mathrm{b}}$ is the projection of the prior onto the observational space by the observational operator $\mathbf{H}$. The term $(\boldsymbol{Y} - \mathbf{H}\boldsymbol{X}^{\mathrm{b}})$ is called the innovation vector and the term $\mathbf{K}(\boldsymbol{Y} - \mathbf{H}\boldsymbol{X}^{\mathrm{b}})$ is the analysis increment. The gain matrix $\mathbf{K}$ is calculated from the equation

$$\mathbf{K} = \alpha(\boldsymbol{\sigma} \circ \mathbf{B})\mathbf{H}^{\mathrm{T}}[\alpha\mathbf{H}(\boldsymbol{\sigma} \circ \mathbf{B})\mathbf{H}^{\mathrm{T}} + \mathbf{R}]^{-1}, \tag{2}$$

where $\mathbf{B}$ is the co-variance matrix of the model errors and $\mathbf{R}$ is the diagonal variance matrix of the observational error. The term $\alpha \in (0, 1]$ is a scalar that can tune the magnitude of the analysis increment and $\boldsymbol{\sigma}$ denotes the localization operator applied over $\mathbf{B}$ by a Schur product represented by the symbol $\circ$. Equation (2) is used by many data assimilation schemes such as the optimal interpolation, the ensemble Kalman filter (EnKF) or the EnOI. In the EnOI scheme, $\mathbf{B}$ is estimated from the equation

$$\mathbf{B} = \frac{\boldsymbol{A}'\boldsymbol{A}'^{\mathrm{T}}}{(M-1)}, \tag{3}$$

where $\boldsymbol{A}' = [A'^{1}A'^{2}, \ldots, A'^{M}]$, $A'^{k} = (\boldsymbol{X}^{k} - \frac{1}{M}\sum_{m=1}^{M}\boldsymbol{X}^{m})$, $\boldsymbol{X}^{k} \in \mathbb{R}^{N}$ is the model state vector of the kth ensemble member with k varying from 1 to M, and $M = 132$ is the number of ensemble members used in all assimilation steps in this study. This ensemble of model anomalies can be taken from a long-term model run (Evensen, 2003) or a spin-up run (Oke et al., 2008) in order to capture the model variability at certain scales. Thus, even being stationary in time, this ensemble of model anomalies allows for the description of the spatial correlations and the anisotropic nature of ocean circulation, keeping the analysis dynamically consistent and substantially reducing the computational cost. Details on how the $\mathbf{B}$ matrix was calculated here regarding the high-frequency variability of the model are described below.

3.1 Calculation of the innovation vector

Basically, there are two ways to calculate the innovation vector. The first one projects the model state vector into the observational space. In this case, the temperature, salinity and layer thicknesses of HYCOM are projected into the vertical levels of the Argo profiles, which provide almost vertically continuous measurements of temperature and salinity, ranging from near surface to 2000 m. This procedure makes the $\mathbf{H}$ operator become complex and non-linear. Since Eq. (1) is linear, the use of a non-linear operator may cause problems in the linear analysis update and may contribute to a sub-optimal assimilation performance (Xie and Zhu, 2010). A second way to calculate the innovation follows the strategy adopted by TE and by XZ for HYCOM. In this approach, temperature (T) and salinity (S) data profiles are projected into the model vertical space to create pseudo-observed layer thicknesses (Δp_{obs}). This is done following the hybrid nature of the model's layers: each layer is required to have a minimum thickness and, after that requirement is satisfied, it should be as close as possible to its specified target value of potential density. Thus, each Argo profile is processed as follows. Based on a pair of profiles of potential temperature and salinity, the profile of potential density can be calculated

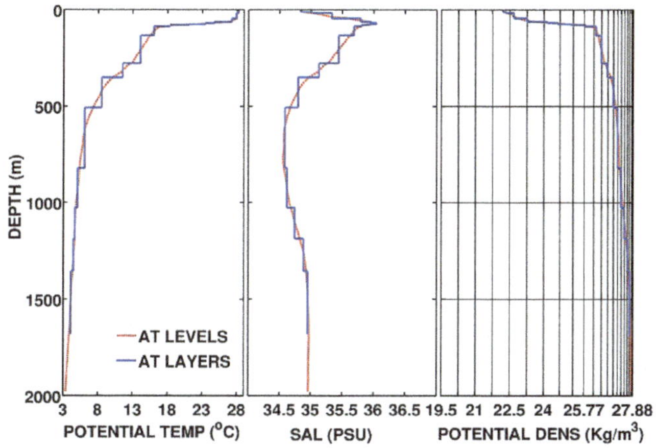

Figure 2. Profile of potential temperature (°C), salinity (psu) and potential density ($\mathrm{kg\,m^{-3}}$) from an Argo float located at 4.04° N and 23° W on 1 January 2010 plotted against its approximation for HYCOM as layer averages. The model target density is indicated by the solid black vertical lines.

by an equation of state for seawater (Brydon et al., 1999). The estimated surface density from the Argo profile is compared to the top layer target density to decide whether any sufficiently low-density water was observed. If not, the minimum thickness is assigned to the layer and the procedure is repeated for the layer below. Once water with the target density is encountered, the remainder of the potential density profile can be partitioned and the layer averages will correspond to the model target densities down to the maximum depth of the Argo profile.

Figure 2 shows the vertical profiles of potential temperature, salinity and potential density at z-levels and the new synthetic observations defined at model layers for an Argo float located at 4.04° N and 23° W on 1 January 2010. In Fig. 2, each Δp_{obs} respects the target densities of the model as soon as the first target density is found in the potential density profile of the Argo data. Also, the averages for all the observational variables are computed for each layer, as shown by the discretized profiles of potential temperature and salinity in the model layers. The step functions of T, S and Δp_{obs} are the data that will be actually assimilated by the EnOI scheme.

3.2 The modified EnOI to assimilate profile data into HYCOM

After the observation is defined in the HYCOM layers, the observed layer thicknesses are assimilated in a first step by Eqs. (1) and (2), and the analysis update is carried out for the model control state vector $(\Delta p_i, \mathrm{U_i}, \mathrm{V_i})$; $i = 1, \ldots, nz$, where Δp_i, $\mathrm{U_i}$ and $\mathrm{V_i}$ are the layer thicknesses and the baroclinic velocity components, respectively, defined at the nz model layers for a single time step. To avoid the analyzed layer thicknesses occasionally becoming negative, a computationally efficient scheme based on Sakov et al. (2012) is used. If the thickness of a layer becomes negative, it is reset to zero and the thickness deficit is added to the neighboring layers. The layers are traversed twice, once from top to bottom, and a second time from bottom to top. Finally, the sum of the layer thicknesses should be equal to the initial bottom pressure (or local depth).

In the next step, temperature and salinity are assimilated separately and in a univariate way also according to Eqs. (1) and (2), but now with the previously adjusted layer thicknesses. Finally, T or S is diagnosed below the mixed layer by the seawater equation of state. According to TE, "within the context of HYCOM, when correcting temperature, it is necessary to decide whether to move interfaces, keeping potential densities of the layers unchanged, or to correct the densities, leaving the interfaces unchanged". Therefore, when the assimilation of layer thicknesses is performed, the potential density should be kept constant in the ocean simulated with isopycnal coordinates. Considering that most of the T corrections in the experiments of XZ were due to changes in the layer thicknesses by the assimilation of Δp_{obs}, T was chosen to be diagnosed below the mixed layer in the present work, instead of S.

3.3 Generation of a running ensemble

Many authors have shown how sensitive the EnOI and EnKF schemes are to the ensemble size (Mitchell et al., 2002; Evensen, 2003; Oke et al., 2007). In the EnOI scheme, the propagation of the observational information is highly dependent on the size and the quality of the ensemble, because the final analysis can be regarded as a combination of the ensemble anomalies whose relative weight is determined by the co-variances. In this work, 132 ensemble members were used. They were selected from the model free run for each assimilation day, regarding the intra-seasonal variability and the high-frequency model dynamics. This number of ensemble members was chosen after a few sensitivity experiments considering a reasonable representation of the model's anomalies without high computational cost, and is in agreement with the numbers used in recent works (e.g., Counillon and Bertino, 2009; Xie and Zhu, 2010; Xie et al., 2011).

The long-term model run that was used to select the ensemble members corresponded to the 12-year period from 1 January 1997 to 31 December 2008. For each assimilation step, a different model co-variance matrix was calculated. Considering the assimilation day, 11 ensemble members for each year of the 12-year period were selected around the date of the corresponding assimilation day. For instance, to perform assimilation on 15 March 2010, 11 members centered on 15 March of each year from 1997 to 2008 were taken 8 days apart, which gives a time window of 80 days for each year. However, the computational code developed here is flexible to use another number of ensemble members and to select different intervals between each ensemble member.

3.4 Localization

The localization technique is a feasible solution to reduce the effect of the sampling error in the ensemble-based methods, especially when the ensemble size is small (Hamill et al., 2001; Oke et al., 2007). The significance range of a measurement is a critical question in assimilation. In the present case, it should be unreasonable that a measurement in the Gulf Stream contributes to resolving the mesoscale features of the circulation of the Brazil Current. Therefore, the localization aims to delete long-distance correlations that may appear in the gain matrix and to limit the influence of a single observation by the Kalman update equation within a fixed region around the observation location. However, a drawback of the localization is that it can breakdown the geostrophic balance. Oke et al. (2007) showed that localization conserves the geostrophic balance when the radius of the localization is equal to or larger than the radius of decorrelation, which is the scale in which the correlations become negligible.

For many EnKF and EnOI schemes, localization is only applied in the horizontal direction (Oke et al., 2008; Sakov et al., 2012). However, some works have already investigated the vertical localization and its impact on the analysis, such as in XZ. They presented a vertical localization scheme for T and S when assimilating Argo profiles into HYCOM. Here, localization in the vertical direction is also investigated, but differently from XZ, as the focus is on Δp.

The operator $\boldsymbol{\sigma}\circ$ in Eq. (2) defines the implementation of the localization by a Schur product, i.e., a product between elements with the same index in the arrays. The notation $\boldsymbol{\sigma}\circ\mathbf{B}$ denotes the Schur product of a correlation matrix $\boldsymbol{\sigma}$ with the $\mathbf{B}$ matrix, and this approach is used in many works (Oke et al., 2007, 2008; Xie and Zhu, 2010). Here, the localization operator is separated into a horizontal component ($\boldsymbol{\sigma}_{\mathrm{h}}$) and a vertical component ($\boldsymbol{\sigma}_{\mathrm{v}}$), and it is defined as $\boldsymbol{\sigma} = \boldsymbol{\sigma}_{\mathrm{h}}\boldsymbol{\sigma}_{\mathrm{v}}$.

3.4.1 Horizontal localization

In order to define the horizontal correlation matrix $\boldsymbol{\sigma}_{\mathrm{h}}$, a fifth-order function is used as in Gaspari and Cohn (1999):

$$\boldsymbol{\sigma}_h\left(I_{ij},L\right) = \tag{4}$$
$$\begin{cases} -\frac{1}{4}\left(\frac{I_{ij}}{L}\right)^5+\frac{1}{2}\left(\frac{I_{ij}}{L}\right)^4+\frac{5}{8}\left(\frac{I_{ij}}{L}\right)^3-\frac{5}{3}\left(\frac{I_{ij}}{L}\right)^2+1, & 0\le I_{ij}\le L \\ \frac{1}{12}\left(\frac{I_{ij}}{L}\right)^5-\frac{1}{2}\left(\frac{I_{ij}}{L}\right)^4+\frac{5}{8}\left(\frac{I_{ij}}{L}\right)^3+\frac{5}{3}\left(\frac{I_{ij}}{L}\right)^2-5\left(\frac{I_{ij}}{L}\right)+4-\frac{2}{3}\left(\frac{I_{ij}}{L}\right)^{-1}, & L< I_{ij}\le 2L \\ 0, & I_{ij}>2L \end{cases}$$

In this function, I_{ij} is defined as the Euclidean distance between any two arbitrary points in the horizontal space and L is the horizontal scale of influence defined as 150 km for all the assimilated variables. It is similar to a Gaussian function in physical space but more compact. The correlation function $\boldsymbol{\sigma}_{\mathrm{h}}$ forces the model error co-variance matrix B to decrease to zero when I_{ij} reaches 300 km. Thus, the radius of localization is defined as 300 km.

The horizontal scale of influence was chosen to be 150 km because this number roughly matches the scales of boundary currents (Ezer and Mellor, 1994; Carton and Giese, 2008). Since the present work was the first REMO effort to implement an EnOI scheme to assimilate Argo data into HYCOM, the horizontal scale of influence was fixed for the entire model domain and for all depths. In this case, the choice of 150 km was conservative in order to avoid capturing co-variances that could damage the analysis in high-variability regions, such as the Gulf Stream and the Brazil–Malvinas Confluence.

3.4.2 Vertical localization

Concerning the vertical localization, XZ found considerable off-diagonal correlations for temperature and salinity between different HYCOM layers. However, there was no significant impact in the analysis when those co-variances were filtered and no significant differences in the experiments with and without vertical localization were reported. However, in the present work the vertical localization of co-variances between the layer thicknesses was investigated. They showed considerable off-diagonal correlations of above 0.4 or below −0.4 from the seventh layer to the bottom (Fig. 3). In the formulation of the vertical localization operator, the vertical distance between the layers was measured by the water column stratification, rather than by the Euclidian distance. Thus, in order to define the correlation matrix $\boldsymbol{\sigma}_v$, the following function was used:

$$\boldsymbol{\sigma}_{\mathrm{v}(i,j)} = \exp\left[-\left(\Delta\rho_{(i,j)}/L_\rho\right)^2\right], \tag{5}$$

where $\Delta\rho_{(i,j)}$ is the density difference between the HYCOM layers i and j, and L_ρ is a vertical-scale factor defined as $0.5\,\mathrm{kg\,m^{-3}}$ according to XZ. As shown by Fig. 3, when applying the vertical localization, the elements a few entries away from the diagonal in the co-variance matrix are almost canceled, but the correlations between the adjacent layers remain.

3.5 Observational errors

Since the observations that are actually assimilated are defined in the model vertical space, the observational errors of T and S in the model layers are calculated as a function of the depth D in meters, respectively, as in XZ:

$$\mathrm{SD_T}(D) = 0.05 + 0.45\exp(-0.002D), \tag{6}$$
$$\mathrm{SD_S}(D) = 0.02 + 0.10\exp(-0.008D). \tag{7}$$

The standard deviations of the observational errors of T vary in the vertical from 0.5 °C on the surface to 0.05 °C in the deep ocean. For S, they vary from 0.12 to 0.02 psu. According to Eqs. (6) and (7), the observational errors are assumed Gaussian with zero mean and uncorrelated.

In case of layer thickness, according to TE, the standard deviation of Δp_{obs} is calculated depending on the oceanic region. For example, in the mixed layer, the layer thickness

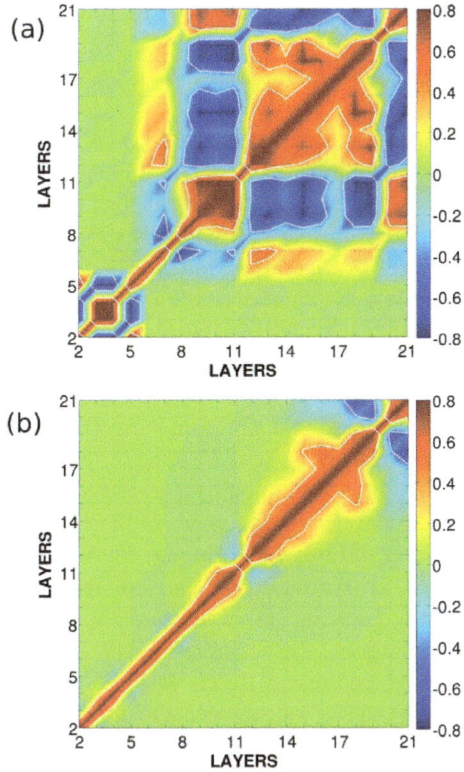

Figure 3. Vertical correlations of layer thickness between different layers at the location of an Argo float (69.95° W and 30.14° N) in the model domain: **(a)** calculated directly from the model ensembles ($M = 132$) and **(b)** after applying the vertical localization of layer thickness according to Eq. (5) and L_ρ defined as $0.5\,\mathrm{kg\,m^{-3}}$. The white lines denote correlations above 0.4 and below −0.4.

is assigned to the minimum layer thickness allowed by the model configuration, and the standard deviation is defined as $0.05\Delta p_k$, where Δp_k represents the layer thickness at the kth layer calculated from the observed profiles. In the isopycnal layers, the standard deviation is defined by the formula

$$\mathrm{SD}(\Delta p_k) = \max \left\{0.5\delta p_k, \max\left[0.05\Delta p_k, \Delta p_k\left(0.05 + (0.5-0.05)\frac{\mathrm{sd}_{\sigma(k)}}{\mathrm{SD}_{\sigma(k)}}\right)\right]\right\}, \quad (8)$$

where δp_k is the minimum layer thickness specified by the model configuration for the kth layer, $\mathrm{sd}_{\sigma(k)}$ is the minimum standard deviation of the potential density defined as $0.001\,\mathrm{kg\,m^{-3}}$, and $\mathrm{SD}_{\sigma(k)}$ is the standard deviation of the potential density from observations. The latter should be small when the potential density from the observed layer thickness has values close to its target density.

4 Assimilation experiments

4.1 Argo data and quality control

The Argo data employed in the assimilation were collected from a global data center (available at: ftp://ftp.ifremer.fr/ifremer/argo/geo/atlantic_ocean/). All those observations were required to step into a data quality control (QC) procedure, which is an essential part to any oceanic data assimilation system since spurious data can compromise the analysis quality and introduce artificial trends in the assimilation results (Yan and Zhu, 2010). The QC used in this work was both developed by REMO together with the Brazilian Navy and tests the date, the location and the temperature and salinity of each Argo profile previously collected. The validation of T and S profiles was made according to all criteria and references established by the Global Temperature and Salinity Pilot Program from the Intergovernmental Oceanographic Commission (IOC, 1990) and from the database of the National Oceanographic Data Center (NODC).

From January 2010 to December 2012, 47 999 valid Argo profiles were assimilated into HYCOM in the Atlantic Ocean. These profilers covered almost the entire model domain, and were especially dense in the North Atlantic, as shown in Fig. 4.

4.2 Configuration and evaluation of the assimilation experiments

Five integrations were performed from 1 January 2010 to 31 December 2012 to evaluate the impact of the Argo data assimilation and the vertical localization in the correction of the ocean state and circulation. The first one was a control run without assimilation (CTL). The other runs were assimilation runs, namely, (i) assimilation without vertical localization (ASSIM), (ii) assimilation with vertical localization of layer thickness (VL_DP), (iii) assimilation with vertical localization of temperature and salinity (VL_TS) and (iv) assimilation with vertical localization of all the assimilated variables (VL_DPTS). In all experiments with data assimilation, a 3-day observational window was considered in order to select all the valid profiles collected 3 days before the assimilation day. The interval between each assimilation step was also 3 days and the scalar α in Eq. (2) was defined as 0.3.

The Argo daily data were also used to validate the results of the experiments on a daily basis, despite the fact that assimilation was performed only every 3 days. However, the prior state was always considered in the evaluation of the assimilation runs. The prior is the model state immediately before the assimilation step. Considering each assimilation cycle, the ocean states at 24, 48 and 72 h after assimilation were assessed in the assimilation runs. Therefore, the validation of the prior is done with independent data, since all the Argo profiles employed in the validation were used only in the next assimilation cycle. This procedure is analogous

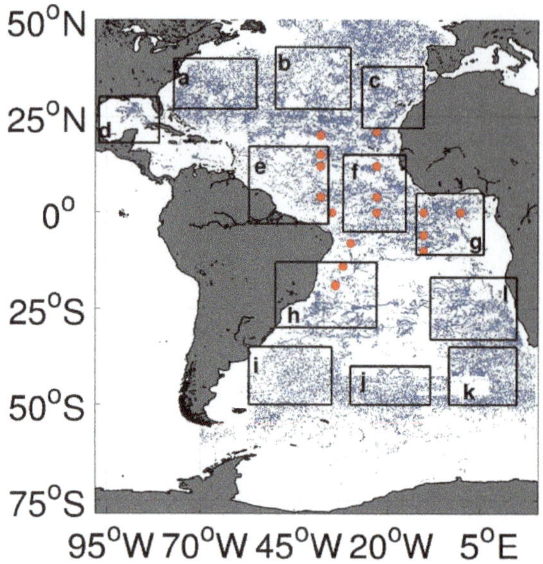

Figure 4. Locations of the 47 999 valid Argo profiles (blue dots) assimilated and also used to validate the prior state of each experiment from 2010 to 2012 in the model domain (100° W–20° E, 78° S–50° N), which excludes the Pacific Ocean, the Mediterranean Sea, and the North Atlantic subpolar region. The 16 fixed moorings from PIRATA used in the experiment's validation are represented by red dots. The numbers of valid Argo profiles in the 12 sub-regions (a–l) were 3085, 2352, 2522, 540, 1463, 2750, 2475, 1849, 1560, 1597, 2536 and 3067.

to the evaluation of forecasts. Moreover, 16 fixed moorings from the Prediction and Research Moored Array in the Tropical Atlantic (PIRATA) were used as another independent data set for validation. Their locations are represented by red dots in Fig. 4. Also, the domain was split into 12 sub-regions from a to l, as shown in Fig. 4, in order to evaluate the regional impact of the assimilation. These sub-regions and their coordinates were selected taking into account the spatial distribution of the surface mean kinetic energy and the mean standard deviations of SSH, SST and sea surface salinity (SSS) from 1997 to 2008. The numbers of Argo profiles assimilated in the experiments in the 12 sub-regions a–l were 3085, 2352, 2522, 540, 1463, 2750, 2475, 1849, 1560, 1597, 2536 and 3067.

Outputs from the HYCOM Navy Coupled Ocean Data Assimilation (HYCOM + NCODA) (Chassignet et al., 2007, 2009) system available in z-levels and fields from the Ocean Surface Current Analyses Real Time (OSCAR) (Johnson et al., 2007) were employed to compare the velocity fields produced by the assimilation runs. Also, WOA13 climatology for the period from 2005 to 2012 was used to evaluate the mean state of T, S and heat content of the upper 300 m (HC300) in all experiments. In this period, WOA13 includes the global coverage of Argo floats to compose the climatological gridded fields of T and S in 1/4° of spatial resolution.

5 Results

5.1 Comparison of mean states

The first comparison is conducted with the WOA13 climatology along 25° W for the upper 1000 m (Fig. 5). It was already verified that the free model run mean from 1997 to 2008 showed substantial differences with respect to the WOA13 climatology (Fig. 1). According to Chen et al. (2000), ocean circulation models can naturally present large and systematic biases in T and S due to their initialization and configuration. This is also shown by the control run, particularly in the North Atlantic. As mentioned before, to simulate the MW in the Atlantic, relaxation of T and S in the boundary condition is imposed without mass flux. Moreover, the resolution of 1/4° could also be a limitation to accurately solve the Gulf Stream and its associated dynamics in the mid-latitudes of the North Atlantic (Hulburt and Hogan, 2000). The main purpose of this grid is to provide boundary conditions for higher-resolution grids focusing on the Metarea V. However, even with these limitations, the assimilation schemes are able to substantially reduce these differences and correct the ocean state towards WOA13. For example, the positive temperature bias up to 3 °C and the negative salinity bias up to 0.4 psu in the upper 300 m of the control run are not produced in any assimilation run. Moreover, all the discrepancies found in the MW are remarkably diminished in the data assimilation runs, which are able to decrease the differences in 0.6 psu and more than 2 °C towards WOA13 between 20 and 40° N in the sub-surface, especially below 400 m. This correction is very effective in the VL_DP and VL_DPTS runs, which adopt the vertical localization of Δp. The VL_DPTS run is able to more efficiently improve the mean T and S states in the North Atlantic along the entire water column by further reducing the differences of 1 °C and 0.1 psu found in the other assimilation runs with respect to WOA13. Also, at 300 m near 25° S, only the VL_DPTS run decreases the differences towards WOA13 from more than 3 to almost 1 °C, while the ASSIM and VL_TS runs are not able to reduce these differences with respect to the control run. But, south of 30° S, the improvements of T and S in the VL_DPTS are considerably smaller than those obtained by the other assimilation runs.

Near 50° S, the control run does not well represent the formation of the Antarctic Intermediate Water Mass (AAIW). The experiments with data assimilation can correct the negative biases shown by the control run by up to 2 °C and 0.3 psu, except for in experiment VL_DPTS. Since this is a region of water mass formation, it seems that the vertical correlations between model layers are very important in order to represent physical processes. In the experiments with vertical localization of Δp, much of this information between layers is lost (Fig. 3), and the experiment VL_DPTS was not efficient at improving the model state for this region in particular.

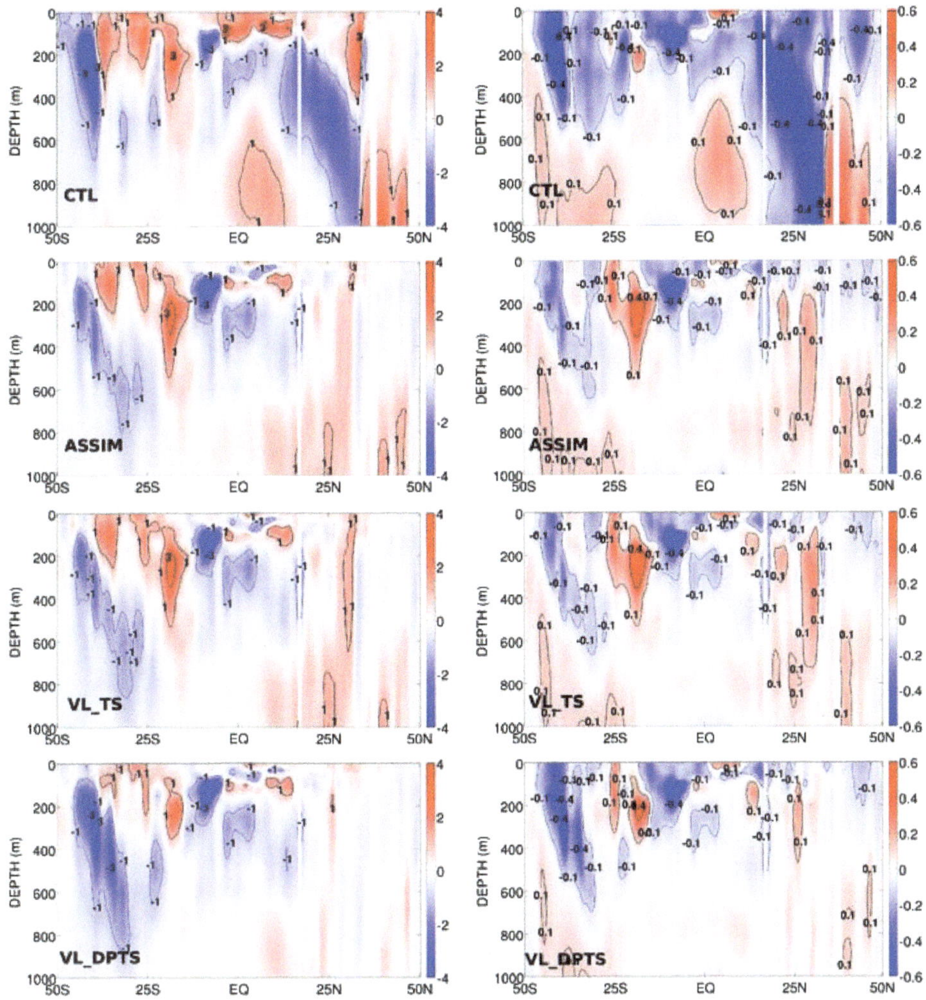

Figure 5. The model mean minus WOA13 along 25° W in the upper 1000 m for temperature (°C) on the left and salinity (psu) on the right from 1 January 2010 to 31 December 2012 according to the CTL run and the prior state of the ASSIM, VL_TS and VL_DPTS runs.

In order to investigate how stratification was modified by Argo data assimilation in the upper 300 m, Fig. 6 shows the meridional section along 25° W of the mean potential density and the position of the layers interfaces for the experiments CTL, ASSIM and VL_DPTS. In general, as already shown in Fig. 5, the assimilation runs tend to decrease the temperature and increase the salinity in the upper 300 m, driving the model towards WOA13. This is associated with a general increase of the mean potential density in the sub-surface in all the assimilation experiments, especially between 30° S and 30° N. This increase of the potential density can reach more than 0.5 $kg\,m^{-3}$ near the equator and in the subtropical gyres. According to Bleck (2002), the implementation of the isopycnal vertical coordinate in HYCOM follows the theoretical formulation that each isopycnal layer will move to be as close as possible to its target density. Therefore, in the assimilation experiments the deeper layers will move towards the surface to encompass the density increase produced by the T and S increments and satisfy their target densities. For example, near the equator and in the subtropical gyre of the South Atlantic, there is a displacement of 25 m or more of the deeper layer interfaces towards the surface. In the North Atlantic, this displacement is even larger and it can reach 150 m, as can be seen by the behavior of the lowermost white dashed line in Fig. 6 that represents the interface between the 11th and 12th layer. Large et al. (1997) showed that the warm temperature bias near the equator is a common feature of coarse-resolution models and it is related to weak zonal currents and weak zonal slopes of the isotherms. However, all the data assimilation experiments are able to force the present low-resolution model to stack the layer interfaces in the upper 300 m and correct the model bias of temperature and salinity not only at the equator but also at all latitudes, especially between 30° S and 30° N.

The depth of the 20 °C isotherm is also evaluated for the experiments and the WOA13 climatology in two longitudinal sections, one along the equator and the other along 30° N (Fig. 7). In the WOA13, the depth of 20 °C isotherm along equator is about 120 m in the west and it gets shallower in the east with a value of almost 50 m. The control run is clearly warmer than WOA13 in the upper ocean with temperatures higher than 28 °C in the first 100 m, so that the position of 20 °C isotherm is deeper, especially in the east of the section with values close to 100 m. All the assimilation experiments reduce the warm bias in the upper ocean, and the depth of 20 °C isotherm is lifted to a shallower position in much better agreement with the WOA13 climatology. In the North Atlantic, the control run has an even stronger bias of more than 4 °C in the upper 300 m. Hence, the depth of 20 °C isotherm of the control run is deeper than WOA13, attaining more than 300 m in the west of the section. Again, the experiments with data assimilation are able to significantly reduce the warm bias by moving the position of the 20 °C isotherm more than 150 m towards the surface. This agrees very well with the large displacement of the layer interfaces in the North Atlantic shown in Fig. 6, and also with the WOA13 climatology. Along 30° N, between 75 and 65° W and between 40 and 25° W, the experiment VL_DPTS is able to bring the 20 °C isotherm up to 50 m closer to the surface in comparison with the other runs without the vertical localization of layer thickness. On the other hand, the experiment VL_DPTS is the only assimilation experiment that does not represent very well the upwelling in the east margin of the Atlantic Ocean at 30° N in Fig. 7. Probably, this is because there is a strong vertical correlation among several model layers in the upwelling region that is not considered by the vertical localization of layer thickness in the VL_DPTS run. Similar behavior was observed in the region of AAIW formation shown in Fig. 5.

To further assess the impact of the assimilation in the upper ocean thermal structure, the spatial distribution of the HC300 between 50° S and 50° N is presented in Fig. 8 for the WOA13 together with the differences (model mean minus WOA13) for each run. In general, the control run has a larger HC300 in comparison with the WOA13 climatology. Vast areas of the North and South Atlantic have more than 50 MJ m^{-2} of HC300 excess over the climatology. The largest differences are attained in the Gulf Stream region (about 400 MJ m^{-2}), since it is misplaced northward with respect to observations, and in the southwest Atlantic around 40° S, associated with a misrepresentation of the Brazil–Malvinas Confluence. When Argo data are assimilated, the model temperature is constrained towards WOA13 and there is a strong impact on the HC300. All the assimilation experiments decrease the HC300 difference with respect to WOA13 by more than 150 MJ m^{-2} for large areas of the subtropical North Atlantic and 50 MJ m^{-2} in the equatorial and South Atlantic. However, large differences still remain in the simulated Gulf Stream region. No Argo data were available in this shelf region – to the north of the observed Gulf Stream – to constrain the model (Fig. 4); therefore, the assimilation runs could not correct this discrepancy. Also, no SST and SLA were assimilated and the assimilation of Argo only could not completely modify this important circulation feature.

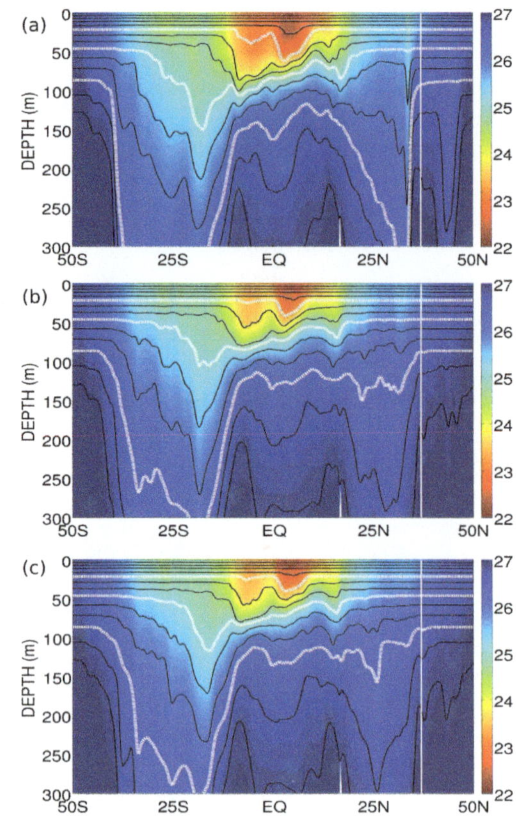

Figure 6. Potential density (kg m^{-3}) and the position of the layer interfaces along 25° W in the upper 300 m for the **(a)** CTL run and the prior state of the **(b)** ASSIM and **(c)** VL_DPTS runs from 1 January 2010 to 31 December 2012. The white dashed lines represent the interface between the 5th and the 6th, between the 8th and the 9th, and between the 11th and the 12th isopycnal layers, from top to bottom.

5.2 Comparison with Argo profilers and PIRATA moorings

The root mean square deviation (RMSD) is calculated on a daily basis against Argo and PIRATA observations to objectively evaluate the temperature and salinity produced by the experiments. This comparison is done with independent data, since the 24 h, the 48 h and the 72 h simulations after each analysis cycle is assessed. The 72 h simulations correspond to the prior or background states used in the assimilation steps.

The depth averaged RMSD of T and S above 2000 m against the Argo profilers in the model domain from 1 January 2010 to 31 December 2012 is shown in Fig. 9. The RMSDs of the ASSIM and VL_DPTS runs are substantially reduced from 1.75 °C and 0.30 psu in the beginning of the runs to 1 °C and 0.14 psu by the end of the runs. In Fig. 9,

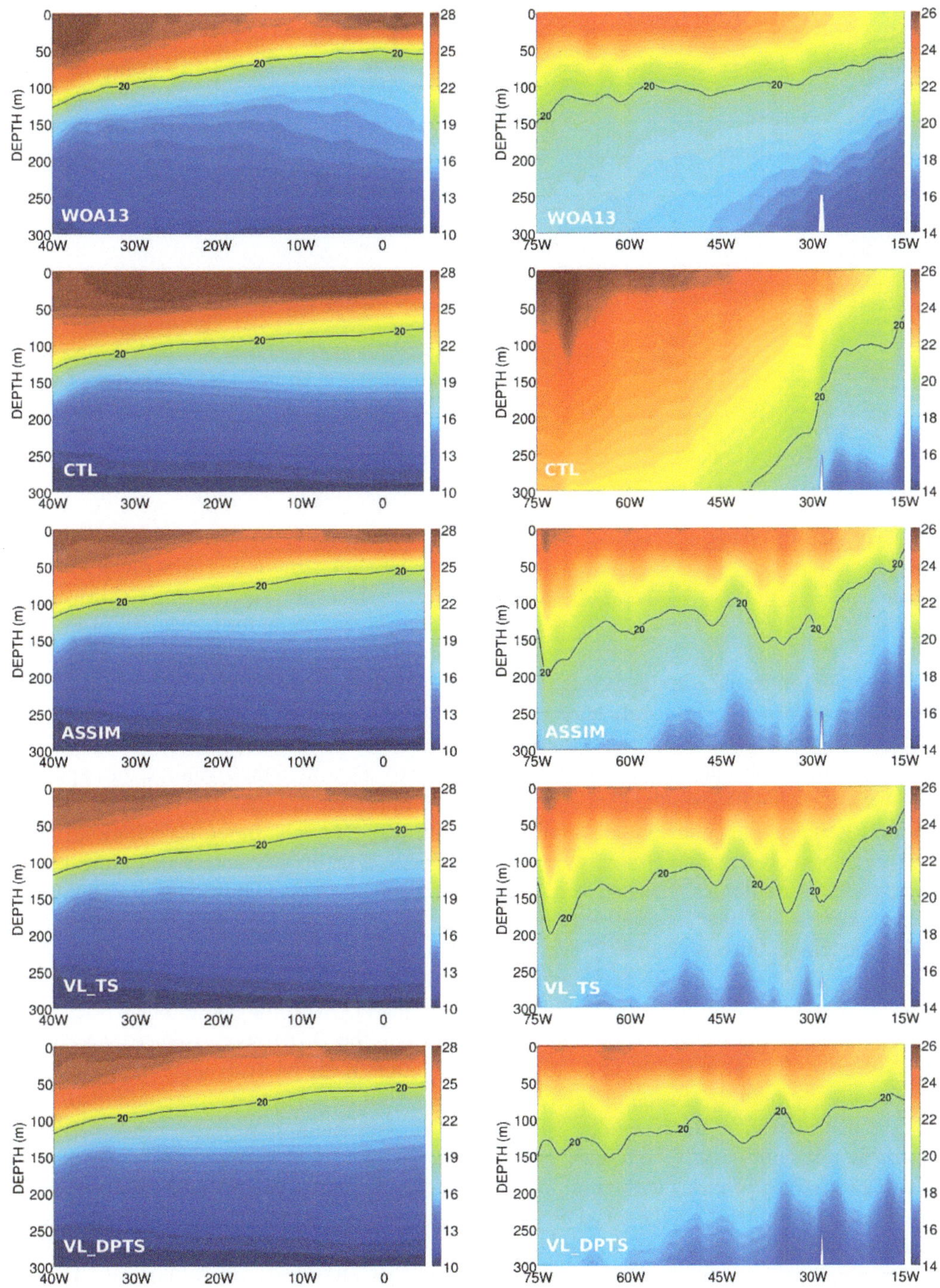

Figure 7. Temperature (°C) along the Equator (left) and along 30° N (right) in the upper 300 m for the WOA13 climatology, the CTL run and the prior state of the ASSIM, VL_TS and VL_DPTS runs from 1 January 2010 to 31 December 2012.

the only assimilation runs presented are the ASSIM and VL_DPTS, since the other two assimilation runs produced similar curves. The first year of integration has the greatest RMSD reduction for the assimilation experiments. From the second year on, especially in the last year, the RMSD reduction is smaller and the RMSD values tend to oscillate around a stable mean. This indicates that the analysis error is close to saturation and that changes in the assimilation parameters such as α may be necessary to better constrain the model towards observations. Here, α is equal to 0.3. However, care should be taken when increasing α. The relatively small value chosen here considered two aspects. First, a small

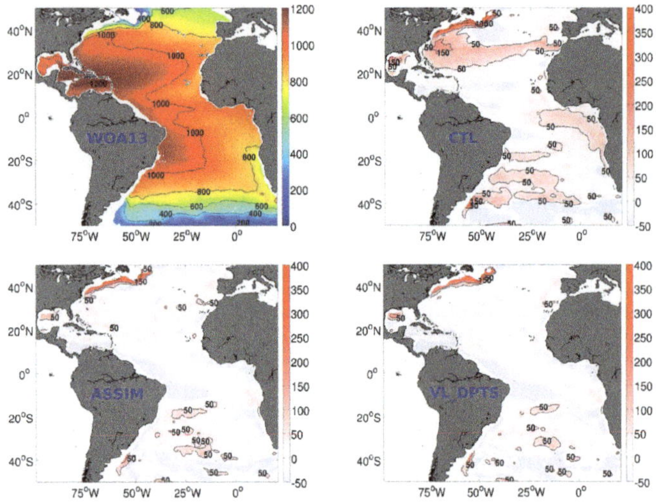

Figure 8. Heat content (MJ m^{-2}) to 300 m of the WOA13 climatology and the model mean minus WOA13 for the CTL run and the prior state of the ASSIM and VL_DPTS runs over the period from 1 January 2010 to 31 December 2012. Regions shallower than 300 m were not considered.

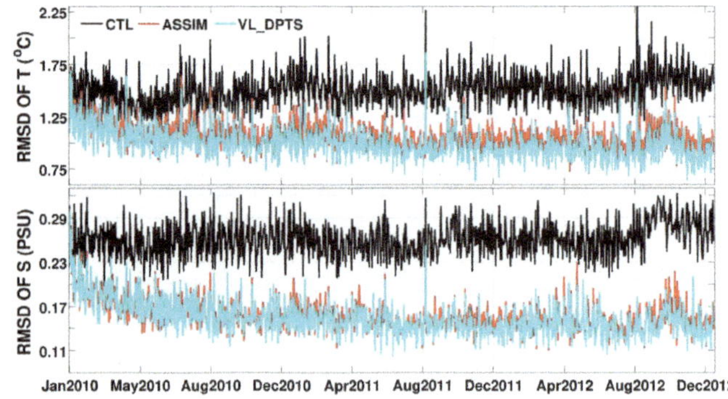

Figure 9. Depth-averaged RMSD of temperature (°C) and salinity (psu) to 2000 m against 47 999 Argo profilers in the model domain (100° W–20° E, 78° S–50° N) from 1 January 2010 to 31 December 2012 according to the control run (black) and the prior state of the experiment ASSIM (red) and VL_DPTS (light blue).

value of α helps avoiding abrupt changes in the model state and numerical instability considering the relatively large discrepancies between the model background and observations in certain regions. Also, the model resolution is relatively low and the use of high values of α could overfit the data and try to introduce mesoscale patterns that cannot be resolved by the model.

The VL_DPTS run produces smaller RMSDs than the ASSIM run. The RMSD of the VL_DPTS is about 86 % (69 %) of the ASSIM RMSD for temperature (salinity). However, this improvement is mainly due to the vertical localization of the layer thickness. It was found that when applying the vertical localization of the layer thickness, the number of grid points with negative Δp after assimilation and before the post-processing was greatly reduced. In the 365 assimilation steps from 1 January 2010 to 31 December 2012, more than 235 000 grid points with negative Δp were generated and needed post-processing in the ASSIM run. This number was decreased to less than 80 000 in the VL_DP and VL_DPTS runs, which means a reduction of 66 %. The improvement caused by the vertical localization of Δp, rather than the vertical localization of T and S, will be discussed in detail below.

The vertical profiles of the mean RMSD of all runs against Argo and PIRATA data over the whole model domain are presented in Fig. 10. The largest temperature and salinity RMSD of the control run are attained in the top 700 m with respect to Argo data and in the top 200 m with respect to PIRATA data. This is associated with difficulties that models have with representing the thermocline and pycnocline regions of sharp vertical gradients (Oke and Schiller, 2007; Xie and Zhu, 2010). When Argo data are assimilated, the vertical thermohaline structure is very much improved. With respect to Argo data, the experiment ASSIM decreases the RMSD of the control run in the top 600 m from 2 to 1.3 °C and from almost 0.4 to 0.19 psu. Regarding PIRATA data, the experiment ASSIM reduces the RMSD values of the control run by up to 1 °C and 0.13 psu in the first 120 m. Gains with the vertical localization of Δp are seen from 200 to 800 m and from 1200 to 2000 m. In these ranges, the experiment VL_DPTS is able to further decrease the RMSD by 0.15 °C and 0.025 psu with respect to the other assimilation runs. The vertical localization of only T and S had almost no positive impacts on the RMSD, and contributed to slightly degrading S near the surface when the runs were compared to PIRATA data. This is mainly because the most important HYCOM variable to determine the thermohaline structure is the model layer thickness (Thacker and Esenkov, 2002; Thacker et al., 2004; Xie and Zhu, 2010). For instance, according to HYCOM formulation, if the analysis increments of T and S decrease the density of a specific layer in comparison to its target density, the model attempts to move the lower interface downward, so that the flux of denser water across this interface increases the layer density (Bleck, 2002). The opposite happens if there is an increase of the layer density compared to its target density. Therefore, the analysis increments of T and S will try to indirectly change the model layer thicknesses in order to adjust the layer densities to their target densities. However, in the EnOI scheme proposed here, the assimilation of Δp_{obs} is realized before the assimilation of T and S, and is very effective in changing the model layer thicknesses. This may explain why the model thermohaline structure is much more sensitive to the vertical localization of Δp rather than to the vertical localization of T and S. The latter was studied by XZ and they also did not find any significant impact in their assimilation experiments.

According to Fig. 10, the vertical localization of Δp seems to be a good approach to better improve the model state,

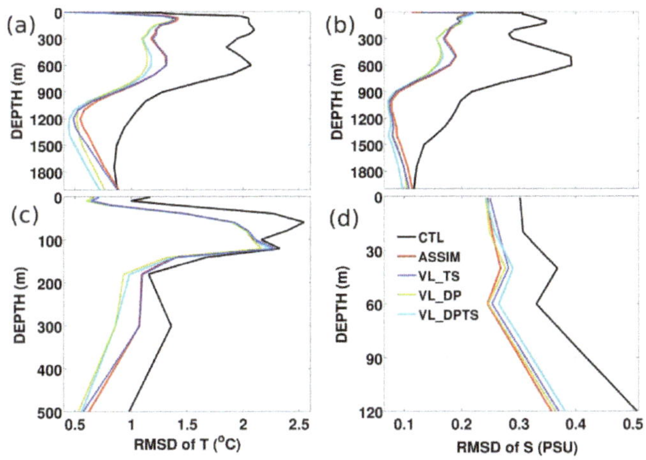

Figure 10. Vertical distribution of the RMSD of temperature (°C) and salinity (psu) against 47 999 Argo profilers and 16 PIRATA fixed moorings in the model domain (100° W–20° E, 78° S–50° N) from 1 January 2010 to 31 December 2012 according to the control run (black) and the prior state of the experiment ASSIM (red), VL_TS (dark dashed blue), VL_DP (dark dashed yellow) and VL_DPTS (light blue). **(a)** and **(b)** correspond to the RMSD of T and S with respect to Argo observations while **(c)** and **(d)** correspond to the RMSD of T and S with respect to PIRATA data, respectively.

especially when correlations of the ensemble members come from a free model run that contains large discrepancies in certain regions in comparison with climatology and observations. Since the co-variance matrix of the model errors in the EnOI scheme is calculated by many snapshots of the model state, the performance of the assimilation is quite dependent on the accuracy of the error co-variances (Oke et al., 2005; Xie and Zhu, 2010). Due to the biases of the ensemble members used in this work, many vertical correlations might not be well represented, especially between distant layers. Thus, since the vertical localization of layer thickness mostly keeps the correlations between adjacent layers, this strategy avoids the generation of many grid points with negative Δp during the assimilation process and makes this approach a safer way to produce a more physically consistent and reliable analysis.

To investigate the spatial distribution of the RMSD of T and S with respect to Argo data above 2000 m, Table 1 contains the deviations for all the 12 sub-regions previously defined in Fig. 4. In the control run, the largest RMSDs of T and S are attained in the sub-regions a, b and d with values greater than 1.6 °C and 0.27 psu. All those sub-regions are found in the North Atlantic, where the model has the largest differences in comparison with the WOA13 climatology and with Argo observations. This is particularly clear for sub-region a corresponding to the Gulf Stream region with RMSD of 2.3 °C, and for sub-region d in the Gulf of Mexico with RMSD of 0.368 psu. It should be reinforced that this grid configuration was prepared to provide boundary conditions for higher-resolution grids focusing on the Metarea V. Hence, the control run is better adjusted with smaller RMSDs in the South Atlantic than in the North Atlantic as shown in Table 1. The maximum RMSD in the South Atlantic is attained in the sub-region i corresponding to the Brazil–Malvinas Confluence with values of 1.29 °C and 0.207 psu. For all the sub-regions, the different EnOI runs are able to reduce the RMSD values of T and S in comparison with the control run. The greatest impact of data assimilation is in the North Atlantic, where approximately 60 % of the assimilated Argo profiles are found and where the control run has its largest RMSDs. For example, the experiment ASSIM decreases the RMSD of the control run about 1 °C in the sub-regions a and b, and more than 0.17 psu for the sub-regions b, c and d. In the Gulf of Mexico, this reduction is up to 0.2 psu. On other hand, the RMSD reductions by the ASSIM run in the South Atlantic are only about 0.2 °C and 0.04 psu. Even with the large impact of the Argo data assimilation, some regions still remain with large RMSDs. This is the case with the Gulf Stream, the Gulf of Mexico and the Brazil–Malvinas Confluence in the sub-regions a, d and i, respectively, where the experiments still have RMSDs greater than 1 °C and 0.145 psu. These are regions of strong gradients, variability and mesoscale eddy activity. The Gulf Stream is highly influenced by mesoscale circulation and the growth of baroclinic instability after its separation from the North American coast (Robinson et al., 1989; Spall and Robinson, 1990; Lee and Mellor, 2003). Also, the Brazil–Malvinas Confluence is characterized by a weak and warm southward Brazil Current meeting a strong, cold and less saline northward Malvinas Current, which results in large contrasts in stratification and strong eddy activity (Gordon, 1989; Garzoli and Garrafo, 1989; Goni et al., 1996). Thus, the large RMSD still found in the assimilation runs in these regions can be due to lack of data, inaccuracies in the ensemble members, and limitation of the model resolution to solve the mesoscale circulation patterns. Similarly, XZ found that the assimilation of Argo profiles did not capture the mesoscale activities in the Pacific Ocean due to the coarse model resolution, and the Kuroshio System remained with large RMSDs in their assimilation run.

In Table 1, the assimilation experiments with the best performances are the ones with vertical localization of layer thickness. This is particularly clear in the sub-regions a, b and c, where the experiments VL_DPTS and VL_DP are able to reduce the RMSD values up to 0.12 °C and 0.018 psu with respect to the ASSIM and VL_TS runs. These three sub-regions are located in the North Atlantic, where the model free run has its largest bias in comparison with WOA13 (Fig. 1). They are sub-regions where the ensemble vertical correlations might not be well represented. Therefore, relaxing the vertical constraints by vertical localization should improve the model state. Small gains can also be seen in the South Atlantic when applying the vertical localization of layer thickness, for example, in the sub-region h corresponding to the Brazil Current. The only regions where

Table 1. RMSD of temperature (°C) and salinity (psu) to 2000 m with respect to Argo data for the prior state of each experiment from 1 January 2010 to 31 December 2012. The calculations were performed for the whole model domain (100° W–20° E, 78° S–50° N) and for each one of the 12 sub-regions previously defined in Fig. 4.

	RMSD									
	CTL		ASSIM		VL_DPTS		VL_DP		VL_TS	
	T (°C)	S (psu)	T (°C)	S (psu)	T (°C)	S (psu)	T (°C)	S (psu)	T (°C)	S (psu)
A	2.302	0.279	1.300	0.166	1.180	0.148	1.153	0.147	1.268	0.157
B	1.722	0.283	0.728	0.113	0.647	0.107	0.674	0.110	0.707	0.109
C	1.214	0.287	0.813	0.116	0.665	0.112	0.652	0.111	0.815	0.142
D	1.663	0.368	1.089	0.160	0.995	0.175	1.016	0.170	1.034	0.175
E	0.972	0.227	0.678	0.119	0.655	0.115	0.633	0.114	0.662	0.119
F	1.177	0.163	0.761	0.097	0.729	0.094	0.729	0.094	0.765	0.098
G	1.139	0.161	0.800	0.107	0.759	0.105	0.759	0.105	0.764	0.106
H	0.756	0.166	0.633	0.125	0.550	0.115	0.554	0.109	0.651	0.132
I	1.290	0.207	1.048	0.179	1.035	0.179	1.038	0.178	1.036	0.180
J	0.920	0.158	0.670	0.124	0.700	0.127	0.709	0.129	0.673	0.124
K	0.845	0.151	0.594	0.106	0.596	0.113	0.599	0.112	0.638	0.110
L	0.786	0.140	0.575	0.090	0.549	0.082	0.548	0.079	0.595	0.092
All	1.507	0.264	0.993	0.149	0.905	0.139	0.915	0.138	0.972	0.147

the experiments VL_DP and VL_DPTS are just a little bit worse than the experiment ASSIM is the Gulf of Mexico, sub-region d, for salinity, and the sub-regions j and k in the mid-latitudes close to the AAIW formation. In general, the pairs of experiments ASSIM and VL_TS, and VL_DP and VL_DPTS, produced similar results in all sub-regions, and the strategies with vertical localization of layer thickness attained smaller RMSDs of T and S. It corroborates the conclusion that the vertical localization of Δp was more important than the vertical localization of T and S for the present experiments.

5.3 Adjustment of the altimetry and velocity fields

As shown above, the EnOI scheme used in this work adjusts the baroclinic velocity fields via their co-variance with layer thickness in the analysis increments. This is a physically consistent adjustment, since the horizontal velocity fields are dominated by the slope of the isopycnal layers, which in turn generates the pressure gradient force. XZ showed that around the assimilated Argo profiles in the mid-latitudes, cyclonic and anti-cyclonic circulations develop, consistent with the large local modifications of the model layer thicknesses and geostrophy. This is due to the nature of the co-variances, which come from an ensemble of model states and allow for the description of the anisotropic patterns of the circulation (Oke et al., 2005, 2008; Xie and Zhu, 2010). Figure 11 shows the analysis increment of layer thickness and the circulation originating in the 12th layer around an Argo profile located in the mid-latitudes of the North Atlantic on 1 January 2010 regarding the experiments ASSIM

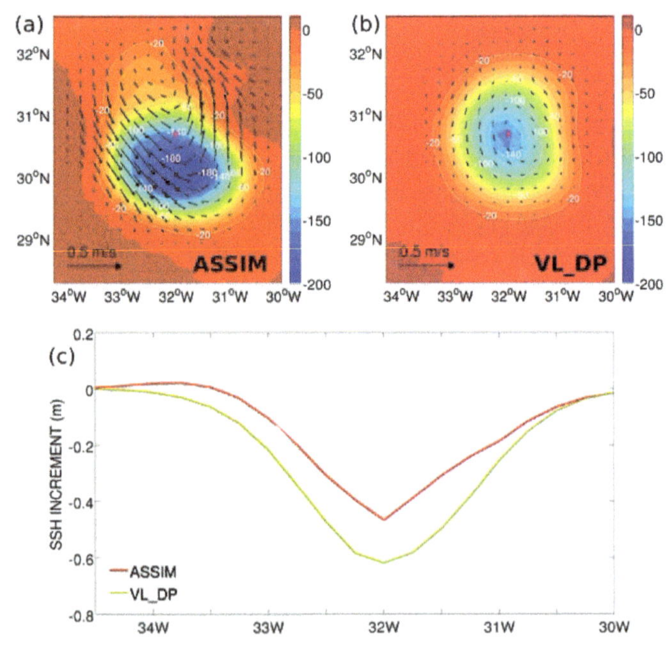

Figure 11. Analysis increments after assimilating an Argo profile located at 30.72° N and 31.98° W on 1 January 2010 of: **(a)** layer thickness (unit: m) and current components (unit: m s^{-1}) in the 12th layer for the experiment ASSIM, **(b)** layer thickness (unit: m) and current components (unit: m s^{-1}) in the 12th layer for the experiment VL_DP and **(c)** SSH (unit: m) along 30.72° N for the experiment ASSIM (red) and VL_DP (dark dashed yellow). The pink dot represents the location of the Argo float.

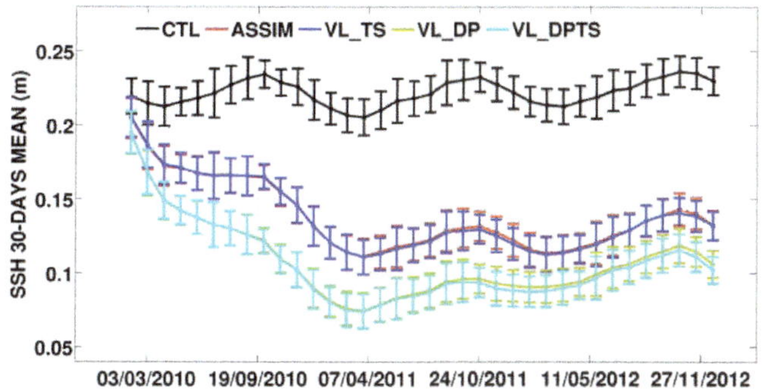

Figure 12. 30-day running SSH mean (m) of the control run (black) and the prior state of the experiment ASSIM (red), VL_TS (dark blue), VL_DP (dark yellow) and VL_DPTS (light blue) from 1 January 2010 to 31 December 2012. The markers represent the 30-day running mean values of SSH, while the bars correspond to the standard deviations of each run.

and VL_DP. The experiment ASSIM robustly induces thinning of layer thickness up to 200 m, and then a stronger and more anisotropic cyclonic pattern is imposed with velocity increments reaching $0.5\,\mathrm{m\,s^{-1}}$. Both experiments develop a cyclonic circulation around this mid-latitude Argo profile. This is coherent with the negative analysis increment of layer thickness and the negative SSH increment also shown in Fig. 11. Thus, the geostrophic balance was preserved in both experiments. Since the experiment VL_DP has limited influence between distant layers, the analysis increments are smoothed and smaller velocity increments of about $0.1\,\mathrm{m\,s^{-1}}$ are produced. The cyclonic gyre is better represented in the VL_DP and VL_DPTS (not shown) runs, which show a more negative and more continuous increment of SSH than the ASSIM run. The analysis constructed by this increment may be more suitable to initialize a model than the analysis by the ASSIM strategy considering the present background state and ensemble members. However, further adjustment of the velocity fields is made by the model itself during further integration to balance the analysis increments after assimilation of Δp, T and S.

The trend of SSH reduction by Argo data assimilation in the present experiments is confirmed in Fig. 12. It shows the 30-day SSH running mean of all experiments considering the model domain between 50° S and 50° N. From the beginning of the integration to December 2012, SSH is reduced from 0.22 to 0.12 m in the experiments ASSIM and VL_TS, which is compatible with the largest impact found in the first year of assimilation in Fig. 9. The VL_DP and VL_DPTS runs produce a larger reduction from 0.22 to 0.08 m. From this period on, the SSH of all assimilation experiments stabilize around these values and different SSH means are achieved in the experiments with and without vertical localization of layer thickness, but with variability very close to the control run. The diagnostic model SSH varies due to the barotropic pressure mode and especially due to the deviations in temperature and salinity caused by changes in the structure of the layer thicknesses (Chin et al., 2002). Hence, this lower SSH mean in the assimilation experiments reflects the new thermohaline state and stratification achieved when Argo data are assimilated. The thermal expansion of seawater constitutes a very important component of SSH and then the reduction of the temperature and the heat content in the assimilation experiments contribute strongly to the SSH decrease. For example, the heat content is positively correlated with SSH (Chambers et al., 1998; Willis et al., 2004; Dong et al., 2007). In this work, mean correlation values of 0.802, 0.815 and 0.812 between the HC and SSH are found for the experiments CTL, ASSIM and VL_DPTS, respectively. When comparing the monthly means of all the experiments with the WOA13 monthly means over the entire domain (not shown), the experiments VL_DP and VL_DPTS are able to decrease the HC by more than $7\,\mathrm{MJ\,m^{-2}}$ towards the climatology. This could explain why there is a larger decrease of the mean SSH in the experiments with vertical localization of layer thickness.

In order to evaluate how the local changes in layer thickness, SSH and velocity fields affect the large-scale circulation, the zonal velocity along 25° W in the upper 300 m for all the experiments and HYCOM + NCODA analysis is shown in Fig. 13. Some observed features are clearly represented in the HYCOM + NCODA analysis, such as the South Atlantic Current (SAC) south of 40° S, the branches of the westward South Equatorial Current (SEC), the eastward equatorial undercurrent (EUC) and the North Equatorial Countercurrent (NECC). Modifications with respect to the control run were imposed by assimilation. However, some can be considered improvements towards the HYCOM + NCODA analysis, and some cannot. For example, there is a strong zonal current in the control run up to $0.6\,\mathrm{m\,s^{-1}}$ between 30 and 40° N associated with the Gulf Stream recirculation path, which is much more intense than in the HYCOM + NCODA analysis. This zonal current is reduced to 0.1–$0.2\,\mathrm{m\,s^{-1}}$ in the experiments VL_TS and VL_DPTS. However, it is displaced further south in comparison with HYCOM + NCODA analysis, especially in the VL_TS run. In the experiment ASSIM, this recirculation path is even weaker and merges in the broader eastward flow. Also, in all assimilation experiments the EUC velocity decreases from 0.6 to $0.4\,\mathrm{m\,s^{-1}}$ and gets closer to the magnitude of the HYCOM + NCODA analysis, but the shape of its core remains the same as in the control run and is not as elongated as in the HYCOM + NCODA analysis. On the other hand, all data assimilation runs constrain the SAC near to the surface, while it reaches more than 300 m in the HYCOM + NCODA analysis. Finally, assimilation is not able to reduce the magnitude of the SEC and to increase the intensity of the NECC simulated by the control run.

XZ could observe modifications in the equatorial Pacific current system due to Argo data assimilation, but they were not always positive when compared to the Tropical Atmosphere Ocean (TAO) array data. For instance, a too deep

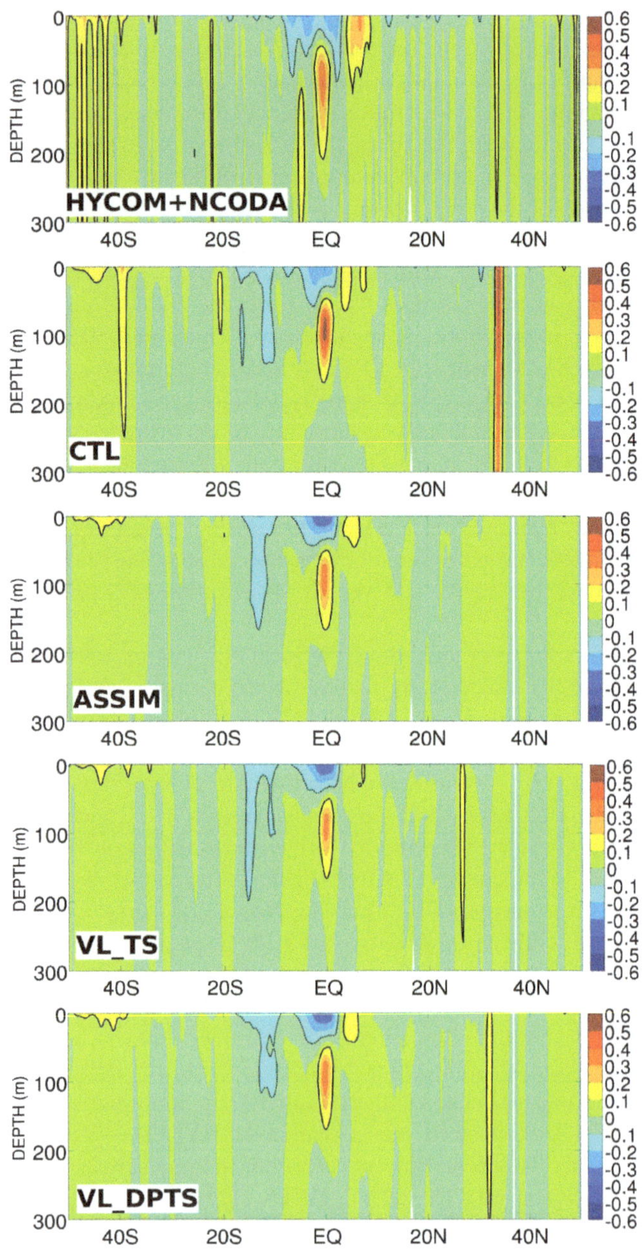

Figure 13. Zonal velocity fields ($m\,s^{-1}$) in the upper 300 m along 25° W of the HYCOM + NCODA analysis, the CTL run and the prior state of the ASSIM, VL_TS and VL_DPTS runs from 1 January 2010 to 31 December 2012. The black solid and black dashed lines denote the contours of $0.1\ m\,s^{-1}$ and $-0.1\ m\,s^{-1}$, respectively.

undercurrent maximum in the east and a too thick and strong westward current in the west were produced. In the present work, a few modifications were observed in the equatorial Atlantic region as well, but with smaller intensity than in the mid-latitudes. It should be noted that the geostrophic balance that holds in most of the ocean interior loses importance close to the equator. Therefore, the equatorial current system should be less sensitive to analysis increments by Argo data assimilation, so that changes in this region may be mostly attributed to remote impacts from the assimilation in extra-equatorial regions.

In the present work, the discrepancies of the model free run and its variability with respect to climatology and observations are an important limitation for the assimilation to produce the correct analysis increments, particularly for the large-scale circulation. For this reason, many works point out that the model biases should be considered during the assimilation process (Reynolds et al., 1996; Dee and Silva, 1998; Bell et al., 2004; Dee, 2005; Xie and Zhu, 2010). Also, improving the co-variances of the ensemble members may lead to more accurate analyses (Oke et al., 2008; Xie and Zhu, 2010). In addition to the comparison with the HYCOM + NCODA analysis, the surface velocity fields of the assimilation runs were compared to the OSCAR fields. Again, the results did not provide a clear signal about the impact of Argo data assimilation on the surface currents. The RMSD of U and V with respect to OSCAR were reduced by less than 5 % in comparison to the control run.

6 Conclusions and discussion

In this work, an EnOI scheme to assimilate Argo data into HYCOM was successfully constructed and implemented. The first results were evaluated against observations and analyses over the Atlantic Ocean. The EnOI scheme was based mostly on the work of XZ. A key variable in the assimilation algorithm was the observed model layer thickness, Δp_{obs}, constructed from the Argo temperature (T) and salinity (S) profiles, according to TE. First, Δp_{obs} was assimilated and the analysis state vector considered not only the model layer thickness, but also the baroclinic velocities. This procedure is physically consistent, since the horizontal velocity fields are dominated by the difference between the layer depths, which will be the responsible for the pressure gradient force. After that, with the previously adjusted model layer thicknesses, T and S were assimilated in separate steps. Finally, T was diagnosed below the mixed layer through the seawater equation of state following the best results found in XZ. Thus, this EnOI scheme respects the hybrid nature of the HYCOM vertical coordinate system and allows for the restructuring of the isopycnal layer thicknesses and velocities with the assimilation of Δp_{obs}. Also, a sensitivity study was performed considering different vertical localizations of the model error co-variance matrix, involving the variables T, S and especially Δp. Five integrations were realized from 1 January 2010 to 31 December 2012. The study of the impact of the vertical localization of Δp is an original contribution of this work.

The thermohaline structure of the experiments with assimilation was significantly improved above 2000 m, the maximum depth of the Argo data. The RMSDs with respect to Argo observations in the assimilation runs were reduced in at least 34 % (44 %) for T (S), regarding the control run over the whole domain. Spatially, the RMSD of T and S decreased for all the 12 selected sub-regions of the domain with remarkable corrections in the depth of the 20 °C isotherm and heat content in the upper 300 m towards WOA13, particularly in the North Atlantic. Also, the reorganization of the isopycnal layers by the assimilation of Δp_{obs} has provided a large reduction of the diagnosed SSH in the model, reflecting the new thermohaline state achieved by Argo data assimilation. Indeed, the correction of layer thickness played a role in correcting the model thermohaline structure and stratification, as stated in previous data assimilation works with HYCOM (Thacker and Esenkov, 2002; Chin et al., 2002; Thacker et al., 2004; Xie and Zhu, 2010).

The vertical localization of T and S did not produce any significant impact in the assimilation experiments, which is consistent with the results found in XZ. However, the experiments VL_DP and VL_DPTS with vertical localization of Δp were able to decrease the RMSD of T (S) from 0.993 °C (0.149 psu) to 0.905 °C (0.138 psu) in the whole model domain with respect to the other assimilation runs. This improvement was especially seen in the North Atlantic, where the ensemble members and background had their largest biases with respect to observations and climatology. The experiments with vertical localization of Δp decreased by 66 % the number of grid points with negative Δp generated during the assimilation process by constraining the vertical co-variances between more distant layers, and then reducing the degrees of freedom offered by the ensemble members. This result reinforces how important the quality of the ensemble members is in order to improve the performance of the assimilation, particularly in the EnOI scheme, in which the ensemble is stationary in time and does not evolve with the model integration as in the EnKF (Evensen, 2003; Oke et al., 2005, 2008). In future works with EnOI, the quality of the ensemble members should be taken into account. For example, long reanalysis with Argo data assimilation should be performed first to provide new and better ensemble members and, therefore, improve the accuracy of the model error co-variance to be employed in a new assimilation run. If more accurate ensemble members are employed, it is expected that the strong vertical localization of Δp will not lead to improvements in the analysis.

Despite the evidence that the assimilation experiments caused a strong SSH reduction in the model domain and that the analysis increments of Δp and velocities were locally consistent, there was not a clear improvement in the large-scale circulation. The assimilation schemes were simply bias blind here. The model biases should be considered in future studies to better analyze the impact of Argo data assimilation in the large-scale circulation. According to Oke and Schiller (2007), the role of Argo data assimilation is mainly to constrain the thermohaline structure of the model, especially for salinity, and this was obtained in the present work. To constrain the SST and T in the mixed layer, the assimilation of SST should be performed, and to correct the upper mesoscale circulation, the assimilation of SLA is needed.

The present work was a key step forward in two major directions. First, it was the basis for the future implementation of assimilation in the other REMO higher-resolution grids. For the current domain with 760 × 480 horizontal grid points, the Argo data assimilation code took around 12 min of CPU on 32 2 GHz processors to assimilate approximately 150 Argo T–S profiles. Therefore, it can be easily used in operational mode. Second, it served as the backbone to construct the assimilation code for along-track satellite altimetry data and SST analyses that will form the REMO Ocean Data Assimilation System into HYCOM (RODAS_H) for operational and research purposes (Tanajura et al., 2014).

Acknowledgements. This work was financially supported by PETROBRAS and Agência Nacional de Petróleo, Gás Natural e Biocombustíveis (ANP), Brazil, via the Oceanographic Modeling and Observation Network (REMO). The authors of this manuscript would like to acknowledge the support of the National Research Council (CNPq), Ministry of Science, Technology and Innovation of Brazil (proc. 130032/2012-3), and the support of the CAPES Foundation, Ministry of Education of Brazil (proc. BEX 3957/13-6).

Edited by: N. Wells

References

Bell, M. J., Martin, M. J., and Nichols, N. K.: Assimilation of data into an ocean model with systematic errors near the equator, Q. J. Roy. Meteor. Soc., 130, 873–893, 2004.

Belyaev, K. P., Tanajura, C. A. S., and Tuchkova, N. P.: Comparison of Methods for Argo Drifters Data Assimilation into a Hydrodynamical Model of the Ocean, Oceanology, 52, 593–603, 2012.

Bleck, R.: An oceanic general circulation model framed in hybrid isopycnic-Cartesian coordinates, Ocean Modell., 4, 55–88, 2002.

Bleck, R. and Smith, L. T.: A wind-driven isopycnic coordinate model of the north and equatorial atlantic ocean. 1. Model development and supporting experiments, J. Geophys. Res., 95, 3273–3285, 1990.

Boyer, T. and Mishonov, A.: World Ocean Atlas 2013 Product Documentation, Technical Ed. NOAA Atlas NEDSIS, 14 pp., 2013.

Brydon, D., Sun, S., and Bleck, R.: A new approximation of the equation of state for sea water, suitable for numerical ocean models, J. Geophys. Res., 104, 1537–1540, 1999.

Carton, J. A. and Giese, B. S.: A Reanalysis of Ocean Climate Using Simple Ocean Data Assimilation (SODA), Mon. Weather Rev., 136, 2999–3017, 2008.

Chambers, D., Tapley, B., and Stewart, R.: Measuring heat storage changes in the equatorial Pacific: A comparison between TOPEX altimetry and tropical atmosphere-ocean buoys, J. Geophys. Res., 103, 591–597, 1998.

Chassignet, E. P., Smith, L. T., and Halliwell, G. R.: North Atlantic Simulations with the Hybrid Coordinate Ocean Model (HYCOM): Impact of the vertical coordinate choice, reference pressure, and termobaricity, J. Phys. Oceanogr., 33, 2504–2526, 2003.

Chassignet, E. P., Hulburt, H. E., Smedstad, O. M., Halliwell, G. R., Hogan, P. J., Wallcraft, A. J., and Bleck, R.: Generalized Vertical Coordinates For Eddy-Resolving Global and Coastal Ocean Forecasts, Oceanography, 19, 20–31, 2006.

Chassignet, E. P., Hurlburt, H. E., Smedstad, O. M., Halliwell, G. R., Hogan, P. J., Wallcraft, A. J., Baraille, R., and Bleck, R.: The HYCOM (HYbrid Coordinate Ocean Model) data assimilative system, J. Marine Syst., 65, 60–83, 2007.

Chassignet, E. P., Hurlburt, H. E., Metzger, E. J., Smedstad, O. M., Cummings, J., Halliwell, G. R., Bleck, R., Baraille, R., Wallcraft, A. J., Lozano C., Tolman, H. L., Srinivasan, A., Hankin, S., Cornillon, P., Weisberg, R., Barth, A., He, R., Werner, F., and Wilkin, J.: US GODAE: Global Ocean Prediction with the HYbrid Coordinate Ocean Model (HYCOM), Oceanography, 22, 64–75, 2009.

Chen, D., Cane, M. A., Zebiak, S. E., Cafiizares, R., and Kaplan, A.: Bias correction of an ocean-atmosphere coupled model, Geophys. Res. Lett., 27, 2585–2588, 2000.

Chin, T. M., Haza, A. C., and Mariano, A. J.: A Reduced-Order Information Filter for Multilayer Shallow-Water Models: Profiling and Assimilation of Sea Surface Height, J. Atmos. Ocean. Technol., 19, 517–533, 2002.

Counillon, F. and Bertino, L.: Ensemble Optimal Interpolation: multivariate properties in the Gulf of Mexico, Tellus A, 61, 296–308, 2009.

Daley, R.: Atmospheric Data Analysis, Cambridge University Press, Cambridge, UK, New York, USA, 457 pp., 1991.

Dee, D. P.: Bias and data assimilation, Q. J. Roy. Meteor. Soc., 131, 3323–3343, 2005.

Dee, D. P. and da Silva, A. M.: Data assimilation in the presence of forecast bias, Q. J. Roy. Meteor. Soc., 124, 269–295, 1998.

Dombrowsky, E., Bertino, L., Brassington, G. B., Chassignet, E. P., Davidson, F., Hurlburt, H. E., Kamachi, M., Lee, T., Martin, M. J., Mei, S., and Tonani, M.: GODAE systems in operation, Oceanography, 22, 80–95, 2009.

Dong, S., Hautala, S., and Kelly, K.: Interannual variations in the upper-ocean heat content and heat transport convergence in the Western North Atlantic, J. Phys. Oceanogr., 37, 682–697, 2007.

Evensen, G.: The ensemble Kalman filter: theoretical formulation and practical implementation, Ocean Dyn., 53, 343–367, 2003.

Ezer, T. and Mellor, G. L.: Continuous assimilation of Geosat altimeter data into a three-dimensional primitive equation Gulf Stream model, J. Phys. Oceanogr., 24, 832–847, 1994.

Gabioux, M. G., Costa, V. S., Souza, J. M. A. C., Oliveira, B. F., and Paiva, A. M.: Modeling the South Atlantic Ocean from medium to high-resolution, Rev. Bras. Geo., 31, 229–242, 2013.

Garzoli, S. L. and Garrafo, Z.: Transports, frontal motions and eddies at the Brazil-Malvinas Currents Confluence, Deep-Sea Res., 36, 681–703, 1989.

Gaspari, G. and Cohn, S. E.: Construction of correlation functions in two and three dimensions, Q. J. Roy. Meteor. Soc., 125, 723–757, 1999.

Goni, G., Kamholz, S., Garzoli, S., and Olson, D.: Dynamics of the Brazil-Malvinas Confluence based on inverted echo sounders and altimetry, J. Geophys. Res., 101, 16273–16289, 1996.

Gordon, A. L.: Brazil-Malvinas Confluence-1984, Deep-Sea Res., 36, 359–384, 1989.

Halliwell, G. R.: Evaluation of vertical coordinate and vertical mixing algorithms in the HYbrid-Coordinate Ocean Model (HYCOM), Ocean Modell., 7, 285–322, 2004.

Hamill, T. M., Whitaker, J. S., and Snyder, C.: Distance-dependent filtering of background error co-variances in an ensemble Kalman Filter, Mon. Weather Rev., 129, 2776–2790, 2001.

Hulburt, H. E. and Hogan, P. J.: Impact of 1/8° to 1/64° resolution on Gulf Stream model-data comparisons in basin-scale subtropical Atlantic Ocean models, Dynam. Atmos. Oceans., 32, 283–329, 2000.

Intergovernmental Oceanographic Commission (IOC): GTSPP REAL-TIME Quality Control Manual, IOC Manuals and Guides 22, 128 pp., 1990.

Johnson, E. S., Bonjean, F., Lagerloef, G. S. E., Gunn, J. T., and Mitchum, G. T.: Validation and Error Analysis of OSCAR Sea Surface Currents, J. Atmos. Ocean. Tech., 24, 688–701, 2007.

Kalnay, E.: Atmospheric Modeling, Data Assimilation and Predictability, Cambridge University Press, Cambridge, UK, New York, USA, 341 pp., 2003.

Kalnay, E., Kistler, R., Collins, W., Deaven, D., Gandin, L., Iredell, M., Saha, S., Wollen, J., Zhu, Y., Leetmaa, A., Reynolds, R., Chelliah, M., Ebisuzaki, W., Higgins, W., Janowiak, J., Mo, K. C., Ropelewski, C., Wang, J., Jenne, R., and Joseph, D.: The NCEP/NCAR 40-year reanalysis project, B. Am. Meteorol. Soc., 77, 437–472, 1996.

Large, W. G., McWilliams, J. C., and Doney, S. C.: Oceanic vertical mixing: a review and a model with a nonlocal boundary layer parameterization, Rev. Geophys., 32, 363–403, 1994.

Large, W. G., Danabasoglu, G., and Doney, S. C.: Sensitivity to Surface Forcing and Boundary Layer Mixing in a Global Ocean Model: Annual-Mean Climatology, J. Phys. Oceanogr., 27, 2418–2447, 1997.

Lee, H. and Mellor, G. L.: Numerical Simulation of the Gulf Stream and the Deep Circulation, J. Oceanogr., 59, 343–357, 2003.

Levitus, S. E.: Climatological atlas of the world ocean, NOAA Professional Paper, 13, 1982.

Lima, J. A., Martins, R. P., Tanajura, C. A. S., Paiva, A. M., Cirano, M., Campos, E. J. D., Soares, I. D., França, G. B., Obino, R. S., and Alvarenga, J. B. R.: Design and implementation of the Oceanographic Modeling and Observation Network (REMO) for operational oceanography and ocean forecasting, Rev. Bras. Geofis., 31, 209–228, 2013.

Mitchell, M. J., Houtekamer, P., and Pellerin, G.: Ensemble size, balance and model-error representation in an ensemble Kalman filter, Mon. Weather Rev., 130, 2791–2808, 2002.

Oke, P. R. and Schiller, A.: Impact of Argo, SST, and altimeter data on an eddy-resolving ocean reanalysis, Geophys. Res. Lett., 34, L19601, doi:10.1029/2007GL031549, 2007.

Oke, P. R., Schiller, A., Griffin, D. A., and Brassington, G. B.: Ensemble data assimilation for an eddy-resolving ocean model of the Australian region, Q. J. Roy. Meteor. Soc., 131, 3301–3311, 2005.

Oke, P. R., Sakov, P., and Corney, S. P.: Impacts of localization in the EnKF and EnOI: experiments with a small model, Ocean Dyn., 57, 32–45, 2007.

Oke, P. R., Brassington, G. B., Griffin, D. A., and Schiller, A.: The Bluelink ocean data assimilation system, Ocean Modell., 21, 46–70, 2008.

Paiva, A. M. and Chassignet, E. P.: The Impact of Surface Flux Parameterizations on the Modeling of the North Atlantic Ocean, J. Phys. Oceanogr., 31, 1860–1879, 2001.

Reynolds, C., Gelaro, R., and Murphree, T.: Observed and simulated Northern Hemisphere intraseasonal circulation anomalies and the influence of model bias, Mon. Weather Rev., 124, 1100–1118, 1996.

Robinson, A. R., Glenn, S. M., Spall, M. A., Walstad, L. J., Gardner, G. M., and Leslie, W. G.: Forecasting Gulf Stream meanders and rings, Am. Geophys. Union, 70, 1464–1473, 1989.

Sakov, P., Counillon, F., Bertino, L., Liseter, K. A., Oke, P. R., and Korablev, A.: TOPAZ4: an ocean-sea ice data assimilation system for the North Atlantic and Arctic, Ocean Sci., 8, 633–656, 2012,
http://www.ocean-sci.net/8/633/2012/.

Schiller, A. and Brassington, G. (Eds.): Operational Oceanography in the 21st Century, Springer Science & Business Media, 745 pp., 2011.

Spall, M. A. and Robinson, A. R.: Regional Primitive Equation Studies of the Gulf Stream Meander and Ring Formation Region, J. Phys. Oceanogr., 20, 985–1016, 1990.

Tanajura, C. A. S. and Belyaev, K.: A sequential data assimilation method based on the properties of a diffusion-type process, Appl. Math. Model., 33, 2165–2174, 2009.

Tanajura, C. A. S., Costa, F. B., da Silva, R. R., Ruggiero, G. A., and Daher, V. B.: Assimilation of sea surface height anomalies into HYCOM with an optimal interpolation scheme over the Atlantic Ocean Metarea V, Rev. Bras. Geofis., 31, 257–270, 2013.

Tanajura, C. A. S., Santana, A. N., Mignac, D., Lima, L. N., Belyaev, K., and Xie, J.: The REMO Ocean Data Assimilation System into HYCOM (RODAS_H): General Description and Preliminary Results, Atmos. Ocean. Sci. Lett., 7, 464–470, 2014.

Thacker, W. C. and Esenkov, O. E.: Assimilating XBT Data into HYCOM, J. Atmos. Ocean. Technol., 19, 709–724, 2002.

Thacker, W. C., Lee, S. K., and Halliwell, G. R.: Assimilating 20 years of Atlantic XBT data into HYCOM: a first look, Ocean Modell., 7, 183–210, 2004.

Willis, J., Roemmich, D., and Cornuelle, B.: Interannual variability in upper ocean heat content, temperature and thermosteric expansion on global scales, J. Geophys. Res., 109, C12037, doi:10.1029/2003JC002260, 2004.

Woodruff, S. D., Slutz, R. J., Jenne, R. L., and Steurer, P. M.: A comprehensive ocean-atmosphere data set, B. Am. Meteorol. Soc., 68, 1239–1258, 1987.

Xie, J. and Zhu, J.: Ensemble optimal interpolation schemes for Assimilating Argo profiles into a hybrid coordinate ocean model, Ocean Modell., 33, 283–298, 2010.

Xie, J., Counillon, F., Zhu, J., and Bertino, L.: An eddy resolving tidal-driven model of the South China Sea assimilating along-track SLA data using the EnOI, Ocean Sci., 7, 609–627, 2011,
http://www.ocean-sci.net/7/609/2011/.

Yan, C. and Zhu, J.: The Impact of "Bad" Argo Profiles on Ocean Data Assimilation, Atmos. Ocean. Sci. Lett., 3, 59–63, 2010.

Yan, C., Zhu, J., and Xie, J.: An Ocean Reanalysis System for the Joining Area of Asia and Indian-Pacific Ocean, Atmos. Ocean. Sci. Lett., 3, 81–86, 2010.

7

Physical forcing and physical/biochemical variability of the Mediterranean Sea: a review of unresolved issues and directions for future research

P. Malanotte-Rizzoli[1], V. Artale[2], G. L. Borzelli-Eusebi[3], S. Brenner[4], A. Crise[3], M. Gacic[3], N. Kress[6], S. Marullo[2], M. Ribera d'Alcalà[8,*], S. Sofianos[11], T. Tanhua[12], A. Theocharis[13], M. Alvarez[14], Y. Ashkenazy[15], A. Bergamasco[10], V. Cardin[3], S. Carniel[10], G. Civitarese[3], F. D'Ortenzio[16], J. Font[5], E. Garcia-Ladona[5], J. M. Garcia-Lafuente[17], A. Gogou[13], M. Gregoire[18], D. Hainbucher[19], H. Kontoyannis[13], V. Kovacevic[3], E. Kraskapoulou[13], G. Kroskos[13], A. Incarbona[20], M. G. Mazzocchi[8], M. Orlic[21], E. Ozsoy[7], A. Pascual[22], P.-M. Poulain[3], W. Roether[9], A. Rubino[23], K. Schroeder[10], J. Siokou-Frangou[13], E. Souvermezoglou[13], M. Sprovieri[24], J. Tintoré[22], and G. Triantafyllou[13]

[1]Massachusetts Institute of Technology, Cambridge, MA, USA
[2]ENEA, Agenzia nazionale per le nuove tecnologie, l'energia e lo sviluppo economico sostenibile, Centri ricerca Frascati e Roma, Rome, Italy
[3]Istituto Nazionale di Oceanografia e di Geofisica Sperimentale – OGS, Trieste, Italy
[4]Department of Geography and Environment, Bar llan University, Ramat Gan 52900, Israel
[5]Institut de Ciencies del Mar – CSIC, Barcelona, Spain
[6]National Institute of Oceanography, Israel Oceanographic & Limnological Research, Haifa, Israel
[7]Middle East Technical University, Institute of Marine Sciences, Erdemli, Turkey
[8]Stazione Zoologica Anton Dohrn, Naples, Italy
[9]University of Bremen, Bremen, Germany
[10]CNR ISMAR, Istituto di Scienze Marine, Venice, Italy
[11]University of Athens, Division of Environmental Physics, Ocean Physics and Modeling Group, Athens, Greece
[12]GEOMAR Helmholz Center for Ocean Research, Kiel, Germany
[13]Institute of Oceanography, Hellenic Centre for Marine Research, Sounio Ave., P.O. Box 712, Athens, Greece
[14]Instituto Espanol de Oceanografia, Calle del Corazon de Maria 8, Madrid 28020, Spain
[15]Department of Solar Energy & Environmental Physics, The Jacob Blaustein Institutes for Desert Research, Ben-Gurion University of the Negev, Israel
[16]Laboratoire d'Oceanographie de Villefranche (LOV), CNRS and Université Pierre et Marie Curie, Paris 06, UMR7093, Villefraneche-sur-mer, France
[17]Physical Oceanography Group University of Malaga, Malaga, Spain
[18]Universite de Liege – Laboratoire d'Oceanologie – Centre MARE, Liege, Belgium
[19]University of Hamburg – Institute of Oceanography, Hamburg, Germany
[20]Universita di Palermo, Dipartimenti di Geologia & Geodesia, Palermo, Italy
[21]University of Zagreb, Faculty of Science Andrija Mohoroviviv Geophysical Institute, Zagreb, Croatia
[22]Instituto Mediterraneo de Estudio Avanzados, IMEDEA (CSIC-UIB), Balearic Islands, Spain
[23]Universita Ca' Forscari di Venezia, Venice, Italy
[24]CNR IAMC, Unita Operativa di Capo Granitola, Trapani, Italy

Correspondence to: P. Malanotte-Rizzoli (rizzoli@mit.edu)

Abstract. This paper is the outcome of a workshop held in Rome in November 2011 on the occasion of the 25th anniversary of the POEM (Physical Oceanography of the Eastern Mediterranean) program. In the workshop discussions, a number of unresolved issues were identified for the physical and biogeochemical properties of the Mediterranean Sea as a whole, i.e., comprising the Western and Eastern sub-basins. Over the successive two years, the related ideas were discussed among the group of scientists who participated in the workshop and who have contributed to the writing of this paper.

Three major topics were identified, each of them being the object of a section divided into a number of different subsections, each addressing a specific physical, chemical or biological issue:

1. Assessment of basin-wide physical/biochemical properties, of their variability and interactions.
2. Relative importance of external forcing functions (wind stress, heat/moisture fluxes, forcing through straits) vs. internal variability.
3. Shelf/deep sea interactions and exchanges of physical/biogeochemical properties and how they affect the sub-basin circulation and property distribution.

Furthermore, a number of unresolved scientific/methodological issues were also identified and are reported in each sub-section after a short discussion of the present knowledge. They represent the collegial consensus of the scientists contributing to the paper. Naturally, the unresolved issues presented here constitute the choice of the authors and therefore they may not be exhaustive and/or complete. The overall goal is to stimulate a broader interdisciplinary discussion among the scientists of the Mediterranean oceanographic community, leading to enhanced collaborative efforts and exciting future discoveries.

1 Introduction

The motivation for this paper stems from a workshop held in Rome in November 2011, on the occasion of the 25th anniversary of the POEM (Physical Oceanography of the Eastern Mediterranean) program. The objectives of the workshop were however rather more ambitious than simply having a memorial. First, the workshop was meant to provide a synopsis of the state-of-the-art of research and present knowledge of the Mediterranean Sea's physical, chemical, and biological properties. Second, it wanted to offer the opportunity to scientists working in different regions of the sea, both the Western and Eastern basins, to meet and share ideas, and hence foster pan-Mediterranean collaborations.

The importance of the Mediterranean Sea for the world ocean has long been recognized. First, the Mediterranean Sea has a profound impact on the Atlantic Ocean circulation and, consequently, on the global thermohaline conveyor belt. Maps of the Mediterranean salty water tongue exiting from the Gibraltar Strait at intermediate depths and spreading throughout the Atlantic interior have been well known since the 1950s. Through direct pathways to the Atlantic's polar regions or through indirect mixing processes, the salty Mediterranean water preconditions the deep convection cells of the polar Atlantic. There the North Atlantic Deep Water is formed that successively spreads throughout the world ocean, constituting the core of the global thermohaline circulation.

Even more importantly, the Mediterranean Sea is a laboratory basin for the investigation of processes of global importance, being much more amenable to observational surveys because of its location at mid-latitude and its dimensions. Both the western and eastern basins in fact possess closed thermohaline circulations analogous to the global conveyor belt. A unique upper layer open thermohaline cell connects the eastern basin to the western basin and, successively, to the North Atlantic through the Gibraltar Strait. In it, the Atlantic water entering into Gibraltar in the surface layer, after travelling to the easternmost Levantine Basin, is transformed into one of the saltiest water masses through air–sea heat and moisture fluxes. This is the salty water that, crossing the entire basin in the opposite direction below the surface Atlantic water, finally exits from the Gibraltar Strait at mid-depths.

Both the western and eastern basins are endowed with deep/intermediate convection cells analogous to the polar Atlantic deep convection cells or to the intermediate mode water ones. Deep and intermediate water masses are therefore formed in different sites of the entire basin. Because of their easily accessible locations, these convection cells are much more amenable to direct observational surveys and mooring arrays.

A ubiquitous, energetic mesoscale and sub-mesoscale eddy field is superimposed onto and interacts with the sub-basin scale, wind-driven gyres that characterize the upper thermocline circulation. Three different scales of motion are therefore superimposed, producing a richness of interaction processes which typify similar interactions in unexplored ocean regions.

Both wide and narrow shelves are at present separated by steep continental slopes from the deep interiors. Cross-shelf fluxes of physical as well as biogeochemical parameters are crucial in determining the properties of the shallow vs. deep local ecosystems and their trophic chain. Most importantly, the Mediterranean Sea is a basin of contrasting ecosystems, from the strongly oligotrophic deep interiors to the fully eutrophic northern Adriatic characterized by recurrent, anomalous algal blooms and related anoxia events.

This review focuses on the identification of major unresolved scientific issues and provides directions for future research to address these issues both theoretically and observationally. However, the unresolved issues discussed here constitute the selection of the group of authors who have contributed to the review. It cannot therefore be either exhaustive or complete. Furthermore, a different group of scientists might have provided a different set of unresolved issues for physical/biogeochemical properties. The major purpose in presenting the present choices is to induce the Mediterranean oceanographic community to a broader, more inclusive and fully interdisciplinary discussion to move forward in the theoretical and observational investigation of the Mediterranean Sea.

2 Assessment of basin-wide physical/biochemical properties and their variability and interactions

2.1 Scales of variability of the Mediterranean circulation

2.1.1 State of knowledge

It is of paramount importance to know which are the dominant scales of variability of the circulation characterizing a specific oceanic basin. External wind forcing determines the large barotropic circulation pattern while stratification determines the internal baroclinic deformation radius, the scale of the ubiquitous, energetic mesoscale eddy field. Furthermore, oceanic motions cover the entire space and time spectrum, from the global thermohaline conveyor belt to microturbulence. Therefore, to capture the variability of a specific oceanic region, knowledge of the dominant scales of motion and their variability and interactions is crucial, both theoretically and for the design of sampling strategies.

The Mediterranean circulation is characterized by three predominant interacting spatial scales, which represent the wide variety of forcings that characterize the basin: basin, sub-basin and mesoscale. The basin-scale circulation is broadly described in terms of a surface flow from the Atlantic Ocean entering through the Strait of Gibraltar and proceeding to the eastern basin, and a return flow of intermediate water, originating in the Levantine Basin, proceeding towards Gibraltar and finally exiting into the Atlantic. This basin-scale open cell is mainly driven by thermohaline forcing: an east–west density gradient, associated with enhanced heat and moisture fluxes in the Levantine Sea, drives the eastward flow of surface Atlantic water. In the Levantine Basin the ocean releases buoyancy into the atmosphere through heat loss and an evaporation/precipitation deficit. The buoyancy loss reduces the stability of the water column, with loss of potential energy in the Levantine. This energy deficit is compensated for by a buoyancy gain associated with the inflow of the fresh surface Atlantic water. For this open cell, the forcings of the Mediterranean basin-scale circulation are the inflows through the Gibraltar and Sicily straits. Sannino et al. (2009) discuss the importance of including variations of the Atlantic water inflow through the Gibraltar Strait to describe decadal variations in the western Mediterranean circulation pattern. Similarly, Malanotte-Rizzoli and Bergamasco (1991), investigating the relative contribution of wind forcing, thermohaline surface fluxes and boundary conditions to the EMed circulation, showed that flow changes in the Sicily Strait can induce variability in the circulation pattern on the basin scale. Pierini and Rubino (2001) modeled the remotely forced dynamics in the areas surrounding the Strait of Sicily and, imposing steady fluxes along the open boundaries, obtained in the absence of meteorological forcing, quasi-stationary circulations representing the local manifestation of the large-scale Mediterranean conveyor belt. It is interesting to note that Demirov and Pinardi (2002), discussing changes in the Ionian circulation in relation to the variability in the thermohaline properties of the water mass formed in the Levantine, provided evidence that variations in the latter one could be associated with changes in the Ionian circulation. Gačić et al. (2010, 2011) showed that these changes may be responsible for modifications of the eastern Mediterranean thermohaline cell, thus impacting the overall structure of the eastern Mediterranean circulation. This finding constitutes an example in which changes in the sub-basin scale alter the basin-scale circulation pattern.

Air–sea heat and moisture fluxes also drive the eastern and western convection cells through the formation of dense water in the southern Adriatic and the Gulf of Lion, respectively. In these cases, the related downslope flow modifies the vertical structure of the water column and determines changes in the horizontal scales of the circulation (Winters et al., 1995; CIESM, 2009; Winters and Young, 2009). Wintertime Dense Shelf Water Cascades (DSWC) can transport large amounts of water and sediment through submarine canyons, reshaping the canyons and affecting the deep water environment in the process. An excellent example was presented by Canals et al. (2006) from observations in the Cap de Creus Canyon in the Gulf of Lions in the Mediterranean Sea (one of his figures shows measurements made during a gravity/turbidity current event, which shows current speeds as high as 80 cm and high sediment concentrations). Similar events can be expected to occur in the Bari Canyon when North Adriatic Deep Water (NADW) formed in the northern parts reaches the Bari Canyon System (BCS) in a few months. The BCS intercepts sediments derived from the Po and southern Apennine rivers and funnels this material into the deep South Adriatic Basin.

A useful approach to describing scales of oceanic circulation relates to the vertical partitioning of oceanic energy into barotropic and baroclinic components (Wunsch and Stammer, 1995; Stammer, 1997; Wunsch, 1997). This allows us to identify the dominant components of the motion and estimate its characteristic scales in terms of the Rossby deformation radius (Lermusiaux and Robinson, 2001). However, the quantification of the relative importance of the barotropic and baroclinic mode(s) in determining the horizontal circulation is difficult, due to the fact that, for a faithful description of a given region of the ocean, nearly continuous sampling over wide areas is necessary. Another possible approach is based on satellite remote sensing, which allows data to be gained over wide areas of the ocean nearly continuously. An example is given in Fig. 1.

This approach implies direct evaluation of the scales of variability of a specific passive scalar, assumed to be a tracer of the surface circulation (Barron and Kara, 2006; Borzelli and Ligi, 1999a, b; Borzelli, 2008), or evaluation of the scales of variability of the sea level anomaly (SLA, Ioannone et al., 2011) obtained by orbiting satellite altimeters. In practice, this technique relies upon the estimation of the autocorrelation function of the parameter represented by satellite remotely sensed data and the evaluation of characteristic scales of the autocorrelation function, which are assumed to represent the scales of the sea surface structure. Figure 2 shows examples of SLA autocorrelation functions computed in the Ionian Sea in the different seasons. Figure 3 gives the time variability of the scales of the same autocorrelation function.

This approach, however, leads to great uncertainties since, first, through direct evaluation of scales, it is difficult to determine which is the dominant vertical component determining the horizontal circulation and, second, interacting components due to nonlinearities and active over short scales are filtered out.

2.1.2 Unresolved issues

1. Stratification, stratification changes and variability in the horizontal circulation scales

   The horizontal circulation of the sea can be represented as the superposition of barotropic and baroclinic modes (normal modes). This representation is convenient, since it allows us to express the characteristic scales of the horizontal motion in terms of Rossby radii of deformation. In the Mediterranean, the overturning circulation concurs in determining considerable variability over time in the stratification regime and producing changes both in the structure of individual modes and in the way in which these modes compete with each other to determine the horizontal motion. This requires an analysis of how, in the Mediterranean, stratification changes modify horizontal scales of motion.

2. Energy exchanges and excitation of normal dynamical modes (i.e., barotropic and baroclinic modes)

   In the available literature (see, e.g., Malanotte-Rizzoli and Bergamasco, 1989) it has been pointed out that, in the Mediterranean, the wind excites primarily the barotropic mode, which is a long, fast and transient component of the circulation. Over the longer

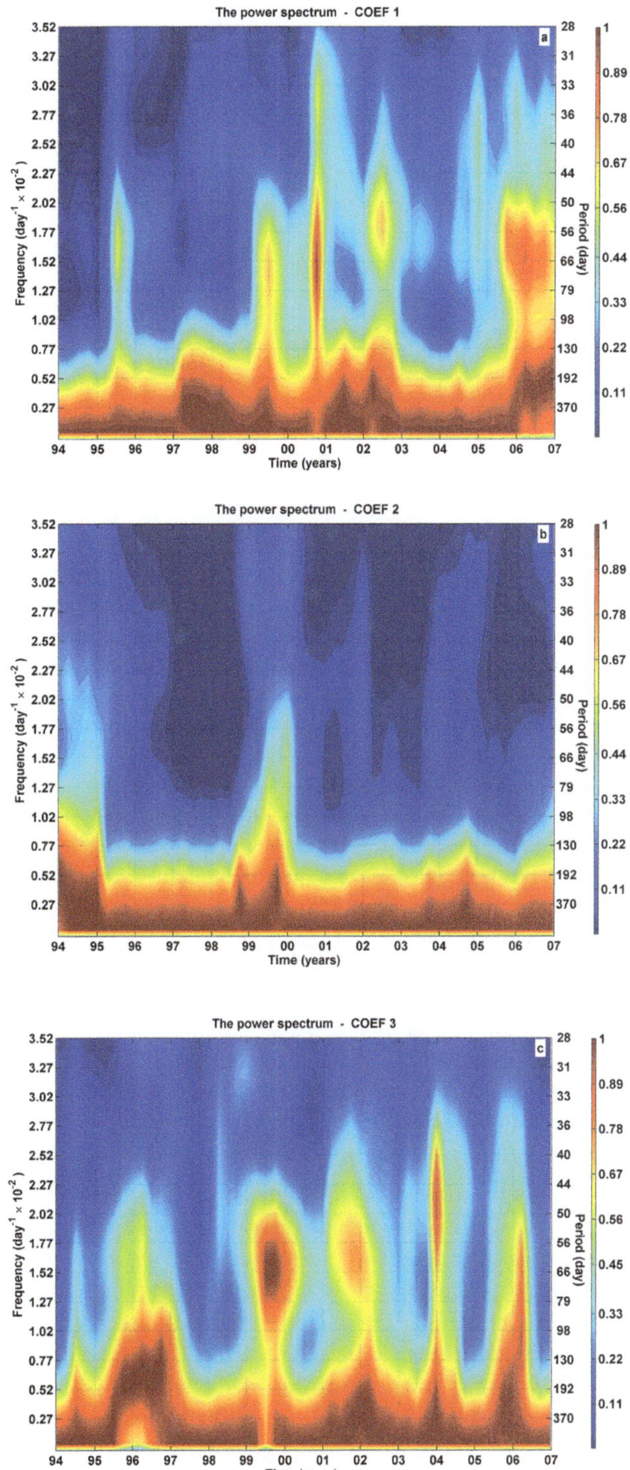

Fig. 1. Normalized time-dependent frequency spectra of empirical orthogonal function coefficients of the SLA. **(a–c)** Spectra of coefficients 1–3, respectively. Left and right vertical scales indicate frequencies and periods, respectively (adapted from Ioannone et al., 2011).

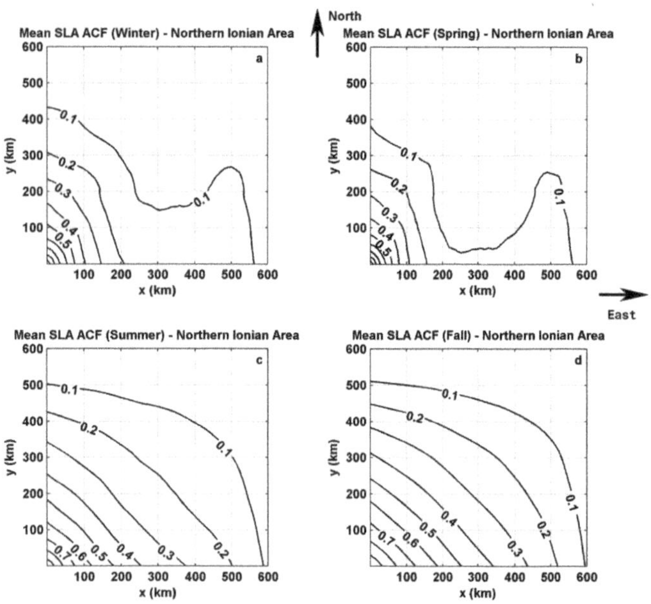

Fig. 2. Contours of the SSH autocorrelation function over the Ionian Sea in the period 1993–2007 averaged over the different seasons. **(a)** Winter; **(b)** spring; **(c)** summer; **(d)** fall (adapted from Ioannone et al., 2011).

timescales (annual and inter-annual) the circulation is essentially determined by baroclinic components. This implies that, in order to maintain the long-term circulation, energy transferred from the wind to the barotropic component has to be transferred from the barotropic component to baroclinic ones. In order to clarify this mechanism an analysis of the way different modes interact with each other is necessary.

3. Mediterranean overturning circulation and deep mixing rates

   Throughout the world ocean, stratification is maintained by mixing in the deep layers that, counteracting upwelling processes, provide the counterflow of dense water necessary for restoring the stability of the water column. In the Mediterranean, although scales of the horizontal circulation have been intensively investigated during POEM and PRIMO, the relationship between the intensity of the Mediterranean overturning circulation and deep mixing rates is not yet understood. Indeed, mixing is generally believed to occur through tidal dissipation over shelf breaks and/or breaking of internal gravity waves. In the Mediterranean, energy associated with these processes is too low compared to the energy required for maintaining the overturning circulation.

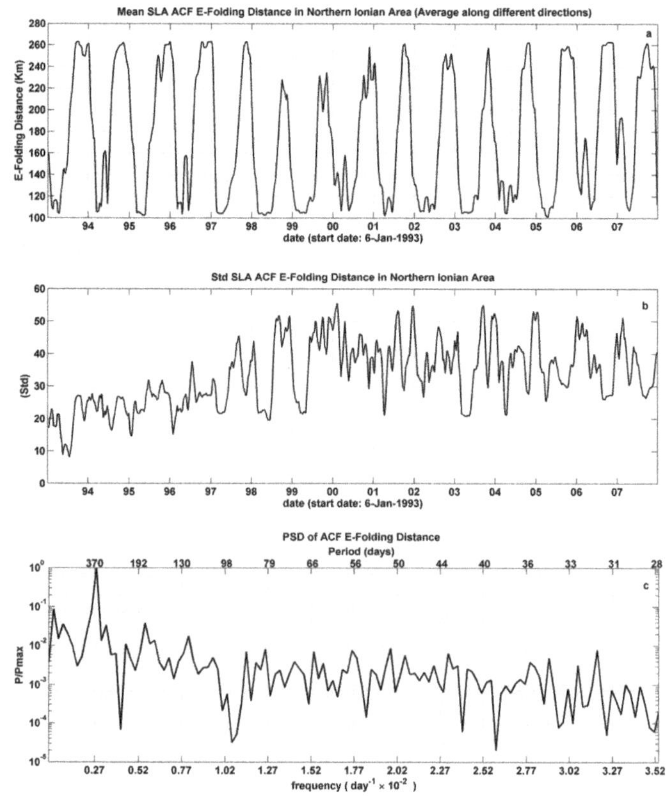

Fig. 3. Temporal variability and frequency spectra of the correlation length scale of the SSH over the Ionian in the period 1993–2007. The correlation length is computed as the average of 17 e-folding distances obtained by fitting the SSH autocorrelation with a Gaussian along 17 different radial directions. **(a)** Average over 17 radial directions between 0 and 90°; **(b)** standard deviation of the mean of Fig. 5a; **(c)** power spectrum of Fig. 5a (adapted from Ioannone et al., 2011).

2.2 Differences and similarities between circulation, forcing and water mass conversion in the WMed and EMed and interactions between the basins

2.2.1 Present knowledge

The Western and Eastern Mediterranean (WMed, EMed) are connected by the Strait of Sicily (sill depth 500 m). Forcing is dominated by exchange with the Atlantic Ocean through the Strait of Gibraltar in the west (sill depth about 300 m) and, to a lesser degree, also with the Black Sea through the Strait of Dardanelles in the northeast. There are also a number of straits and channels that connect both the WMed and EMed with the marginal seas. They play a crucial role in determining the water mass exchanges and related properties (Astraldi et al., 1999). The other factor is atmospheric forcing, primarily in the form of net evaporation. Local climate variability exists due to both the different impacts of large-scale teleconnection patterns (NAO, EA/WR, etc.) in the WMed and EMed (Krichak and Alpert, 2005; Josey et al., 2010) and the

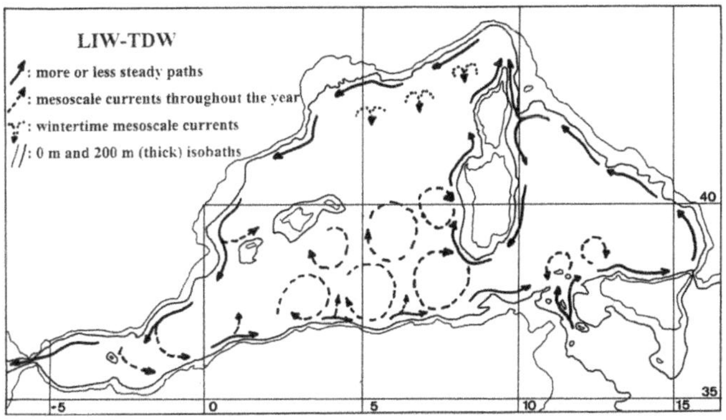

Fig. 4. Western Mediterranean circulation in the upper layer. In the northern part of the WMed, the westward rim current is permanently well defined, even if affected by seasonal variability. This AW flow is a recirculating branch that detaches AW from southern latitudes in the southwestern area of Sicily (Millot and Taupier-Letage, 2005). In the southern part of the WMed the Algerian Current is unstable and generates (anticyclonic) meanders and eddies of ~ 50–100 km (up to 200) in diameter. These eddies can detach from the Algerian current and propagate into the interior of the Algerian and Tyrrhenian basins. The pattern of Fig. 4 has recently been extended to the entire Mediterranean by Millot and Taupier-Letage (2005), who have inferred it from SST imagery. In the EMed, however, and especially in the Levantine Basin (or Levantine Sub-Basin, LSB, as in Menna et al., 2012), the AW is topped by the more saline Levantine surface waters during most of the year and the use of SST imagery to infer circulation patterns is rather controversial.

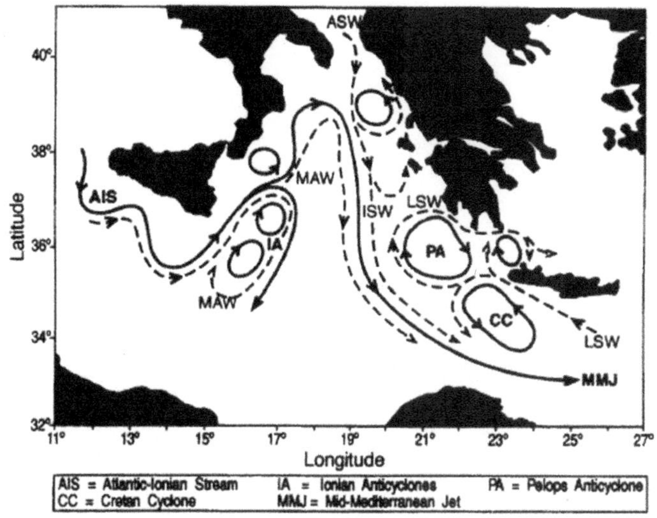

Fig. 5a. Schematic of the upper thermohaline circulation and water mass pathways in the central Mediterranean. Dashed lines mark water mass pathways. Solid lines denote permanent dynamic features, i.e., currents and sub-basin gyres, for which names are given in the table.

regional characteristics (Xoplaki et al., 2003, 2004). The induced eastward salinity increase in the upper waters results in two kinds of thermohaline cells.

The open conveyor belt consists of an upper layer eastward flow of low-salinity Atlantic Water (AW) and a subsurface westward spreading of the warm and saline Levantine Intermediate Water (LIW) in 200–400 m (i.e., at depths shallower than the Sicily and Gibraltar Strait sills; Schroeder et al., 2012) formed in the Levantine Basin, enriched sporadically by Cretan Intermediate Water (CIW) formed in the Cretan Sea. This Eastern Mediterranean Intermediate Water (EIW) enters the WMed, where part of it participates in the DWF processes. A further ingredient in the EIW is the Western (formerly Winter) Intermediate Water (WIW), cooler and fresher than the LIW, formed in the WMed (Send et al., 1999). The main volume of the EIW along with the WIW exit at depth at Gibraltar. The parts of the Intermediate Waters that cannot outflow directly through the Channel of Sicily and the Strait of Gibraltar continue recirculating in both Mediterranean basins (Millot, 2013).

Secondly, there exist internal or quasi-closed thermohaline cells in both Mediterranean basins, driven by deep water formation processes in the northern regions of the seas. These deep waters are partly involved in the Sicily and Gibraltar overflows. The WMDW is formed in the open region of the NW Mediterranean, while the EMDW originates in marginal seas, regularly mostly in the Adriatic but also in the Aegean and, sporadically, in the NW Levantine (Kontoyiannis et al., 1999).

A marked difference exists in the bathymetry between the two basins. A highly structured bathymetry characterizes the EMed compared to the WMed, which has a virtually flat bottom east and south of the Balearic Islands. This contributes to the differences in the circulation, especially the depth, of the two basins. In the WMed the eastward flow of AW is concentrated in the southern part of the basin, along the continental slope, i.e., the Algerian Current (Fig. 4, adapted from Millot, 1985, 1999).

Figure 5 shows the circulation pattern for the EMed, specifically Fig. 5a for the Ionian Sea (adapted from Malanotte-Rizzoli et al., 1997) and Fig. 5b for the LSB (adapted from Menna et al., 2012). The anticyclonic pattern of AW depicted in Fig. 5a synthesizes the hydrographic data set collected during POEM-Phase I, especially during 1987. It corresponds to the anticyclonic circulation mode of the recently discovered Bimodal Oscillation System (BIOS) (Gačić et al., 2010, 2011). During the anticyclonic mode there is increased spreading of the AW northeastward, thus freshening the northern Ionian Sea (see Sect. 3.1).

For the LSB, there has been an ongoing controversy regarding the pathway of surface waters propagating eastward from the Ionian Sea. Two pictures have been proposed and contrasted. The first one was proposed by Millot and Taupier-Letage (2005), Alhammoud et al. (2005), and Hamad et al. (2005, 2006), and is similar to the WMed pattern of Fig. 4, based again on SST imagery. The assumed instabilities of

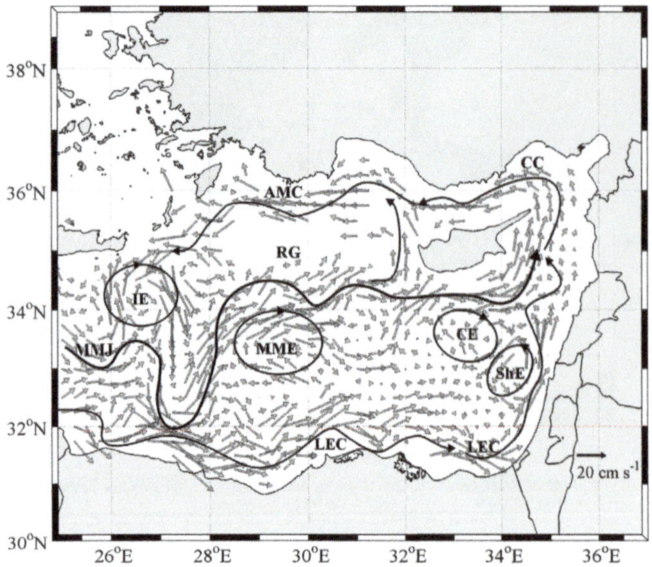

RG: Rhodes Gyre	CE: Cyprus Eddy	IE: Ierapetra Eddy	LEC: Libyo–Egyptian Current	AMC: Asian Minor current
MME: Mersa Matruh Eddy	ShE: Shikmona Eddy	CC: Cilician Current	MMJ: Mid Mediterranean Jet	

Fig. 5b. Mean absolute geostrophic velocities in the period 1992–2010 (dark grey). Black lines emphasize the main currents and sub-basin eddies for which names are given in the table.

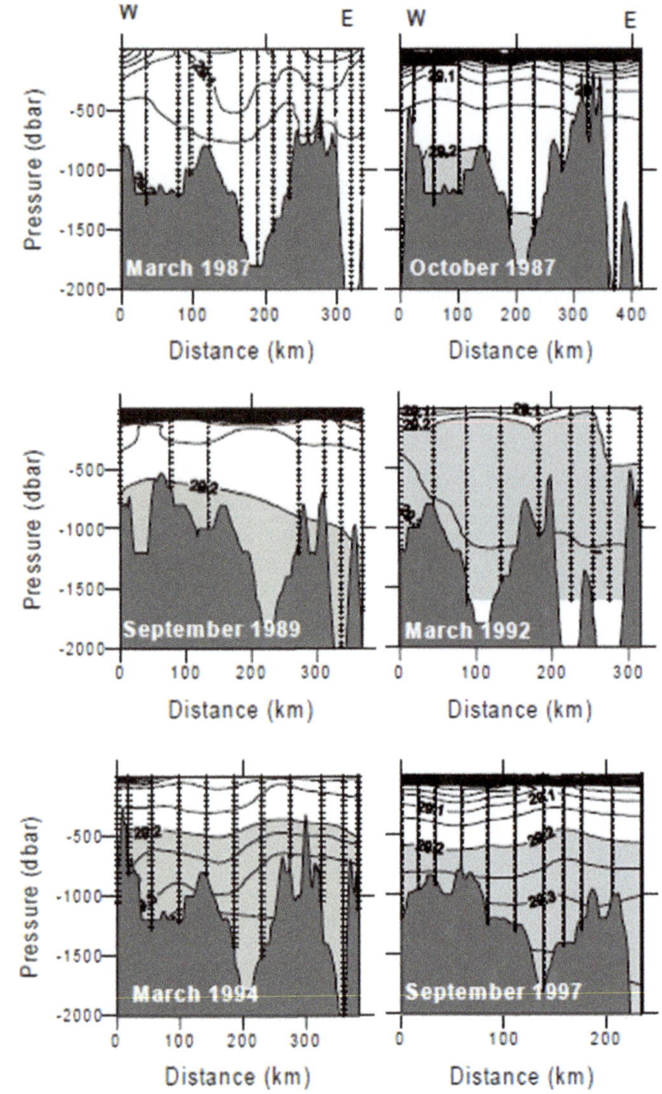

Fig. 6. West–east σ_θ cross-sections in the Cretan Sea showing the evolution of the 29.2 kg m^{-3} surface during the decade 1987–1997.

the Libyo–Egyptian current are responsible for the generation of anticyclonic eddies propagating the AW into the LSB interior. This picture is in sharp contrast to the picture originally proposed by the POEM group (1992), in which the surface/sub-surface waters carry the AW into the Levantine as an open sea central eastward current. This continues as the Mid Mediterranean Jet (MMJ) splitting into several branches (Robinson et al., 1991, 2001; Robinson and Golnaraghi, 1994; Malanotte-Rizzoli et al., 1997, 1999; Malanotte-Rizzoli and the LIWEX group, 2003). Furthermore, observational work comprising several oceanographic surveys carried out in the CYCLOPS experiment (Zodiatis et al., 2005) and the recent works by Schroeder et al. (2012) and especially Menna et al. (2012) demonstrate unambiguously the existence of the MMJ. Figure 5b, reproduced from Menna et al. (2012, their Fig. 5b) is based on the analysis of 18 years of drifter data and satellite-derived sea level anomalies. It clearly shows the major currents and eddies of the LSB, thus confirming and refining the old POEM pattern. The reader is referred to Menna et al. (2012) for a detailed description of the AW flow, the major eddies characterizing the region, and the mechanism of periodic generation of the Shikmona eddy (see also Brenner, 1993).

The classical view that the thermohaline circulation of the Mediterranean was quasi-stable has more recently been overruled. During the late 1980s to the early 1990s the Aegean deep water formation took over from the Adriatic (Fig. 6, adapted from Theocharis et al., 1999a, b).

Huge amounts of very dense waters characterized by enhanced salinity and temperature were released for a few years, forming the Eastern Mediterranean Transient (EMT; Roether et al., 1996, 2007; Theocharis et al., 1999, 2002a), significantly influencing the thermohaline structure and stratification of the entire EMed Basin (Fig. 7, adapted from Roether et al., 1996).

A specialty is that the Aegean dense-water outflow feeds into the Hellenic Trench region before being transferred into the EMed at large. However, deep water formation in the Aegean has been reported for the 1970s (Lascaratos et al., 1999; Theocharis et al., 2002b; Beuvier et al., 2010), but it is not known if the Aegean dense water observed in the deep layers of the EMed was enough to produce a substantial change in the deep thermohaline cell as in the 1990s.

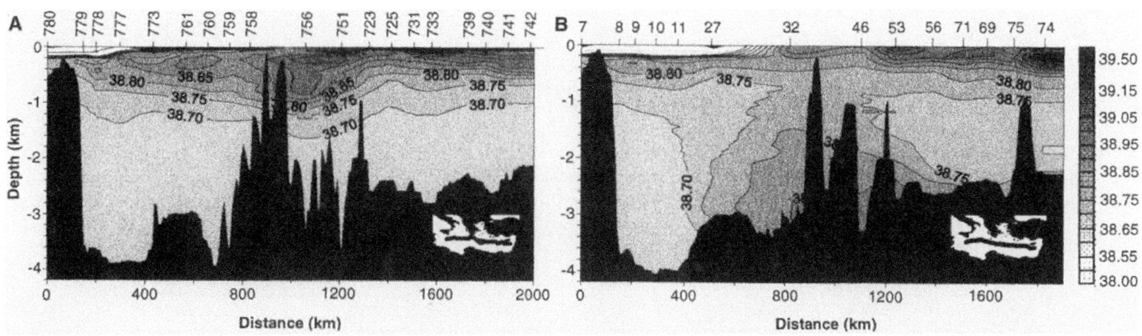

Fig. 7. Salinity isolines on sections across the Ionian Basin (see inset maps) for Meteor cruises M5/6 (1987) **(A)** and M31/1 (1995) **(B)** – stations are at the top.

Recently it was also suggested that the EMT could be a recurrent phenomenon (Borzelli et al., 2009; Pisacane et al., 2006). EMT-induced changes have been communicated through the Sicily Strait to the WMed (Fig. 8, adapted from Schroeder et al., 2006), with the role of the Tyrrhenian becoming enhanced for some years (Gasparini et al., 1995; Roether and Lupton, 2011).

More specifically, in the Tyrrhenian Sea, the WMDW and the overflow from the Eastern Mediterranean mix, lifting the former component to allow it to take part in the Sicily Strait outflow (Millot and Taupier-Letage, 2005), and, at the same time, forming a deep salinity source for the WMDW. Indeed, a significant warming and salinification of the whole water column has been observed also in the Western Mediterranean, comparable to the EMT, both in terms of intensity and observed effects (Schroeder et al., 2008). This event of high production of anomalously warm and salty new deep water during the winters of 2004/2005 and 2005/2006 (Fig. 9, adapted from Schroeder et al., 2008) is now known as the Western Mediterranean Transition (WMT) (Schroeder et al., 2008; Herrmann et al., 2010).

Currently, thus, the subsurface distributions of temperature and salinity, as well as of most other properties in the entire Mediterranean, are far from a steady state. Moreover, the changes observed in the circulation during the last decades, such as the reversals in the Ionian circulation (Gačić et al., 2011; Borzelli et al., 2009) and the transport of the EMT effects westwards (Gasparini et al., 2005) dictate a more thorough study on water mass spreading pathways and their variability and on water mass conversion, as well as temporal and spatial variability in the marginal seas of the EMed (Adriatic/Aegean). Only if these items have been resolved will it be possible to come up with quantitative answers on the geochemistry and ecology of the Mediterranean Sea.

In the last two decades there has existed an increased effort for the development of observational systems in the Mediterranean in the framework of programs or initiatives such as Med-GOOS, MFSTEP (Mediterranean forecasting system), ECOOP (European Coastal/Shelf seas operational observing and forecasting), MYOCEAN/GMES (Global Monitoring and Environmental Security) and EUROARGO. These observational systems are designed to transmit real-time data to operational centers with the ultimate goal of producing accurate model predictions. Recent developments in numerical modeling that assimilate available in situ observations are a promising step towards more reliable ocean predictions and long-term reanalysis simulations in the Mediterranean within the scope of operational oceanography (Adani et al., 2011; Dobricic et al., 2007, 2010; Pinardi and Coppini, 2010; Nilsson et al., 2011).

There is furthermore a need for clarification of the role of the characteristic atmospheric circulation patterns (west vs. east), also in the context of large climatic variations and trends that are associated with variability of the circulation and with deep water formation events (Josey et al., 2010; Kontoyiannis et al., 2012).

2.2.2 Unresolved issues

1. Long-term variability

   A principal concern is to improve the understanding of the long-term variability and evolution of the Mediterranean circulation. A special item that must urgently be amended is scarce data in the southern part of the EMed in consequence of the political situation in the past.

2. Influence of lateral forcing

   One need is to monitor the path and characteristics of the AW after it enters the Mediterranean at Gibraltar and after passing the Sicily Channel, to find the causes of its path variability and its contribution in water mass transformations. Furthermore, one needs to monitor the characteristics of the overflows at the Gibraltar and Sicily straits and to relate their variability to that observed in both Mediterranean basins. Finally, the interaction between subsurface Eastern and Western Mediterranean waters must be clarified further.

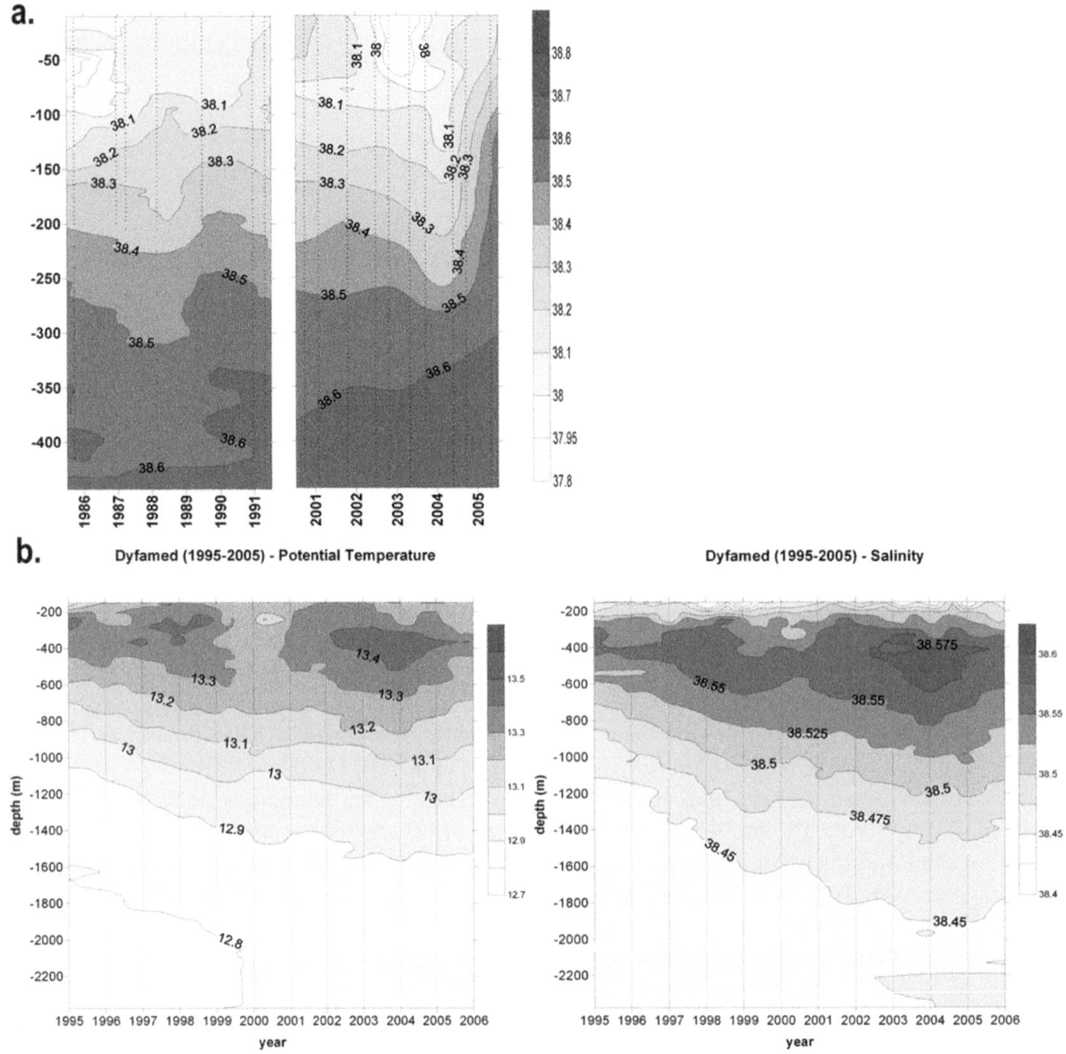

Fig. 8. **(a)** Temporal evolution of salinity in the Corsica Channel; **(b)** temporal evolution of potential temperature and salinity at the DYFAMED site showing the propagation of the EMT towards the WMed.

3. Sub-basin and mesoscale variability

   Additionally, we need to improve our understanding of the eddies, their nature (permanent/transient) and lifetimes and their role in water mass transport and distribution that have different influences in each sub-basin.

4. EMT and its influence in the WMed

   A further subject is the changes relevant to the EMT since the 1990s and the corresponding changes in the WMed, and the related variability over decadal or longer timescales in the entire Mediterranean, also in relation to the ongoing climatic changes (IPCC report). This analysis is straightforward for water properties based on historical hydrographic data after the early 50s, but the question of the past circulation patterns is more difficult, so that one has to employ model simulations.

5. Joint observational strategies and modeling studies

   Apart from the existing European observational systems mentioned above, field observations from ships are of great value because they provide unique information with respect to the circulation and the thermohaline processes. One strategy in this direction would be to reapply past multi-ship multi-national hydrographc surveys, but these surveys will have to (1) be on a decadal timescale jointly for both Mediterranean basins, (2) extend to the near-Africa areas and (3) employ presently available technology on direct current measurements (ADCPs, LADCPs) so as to produce near-synoptic realizations of the 3-D surface-to-bottom mass and flow fields. These observations will provide indisputable pictures of the Mediterranean circulation that can not be achieved by

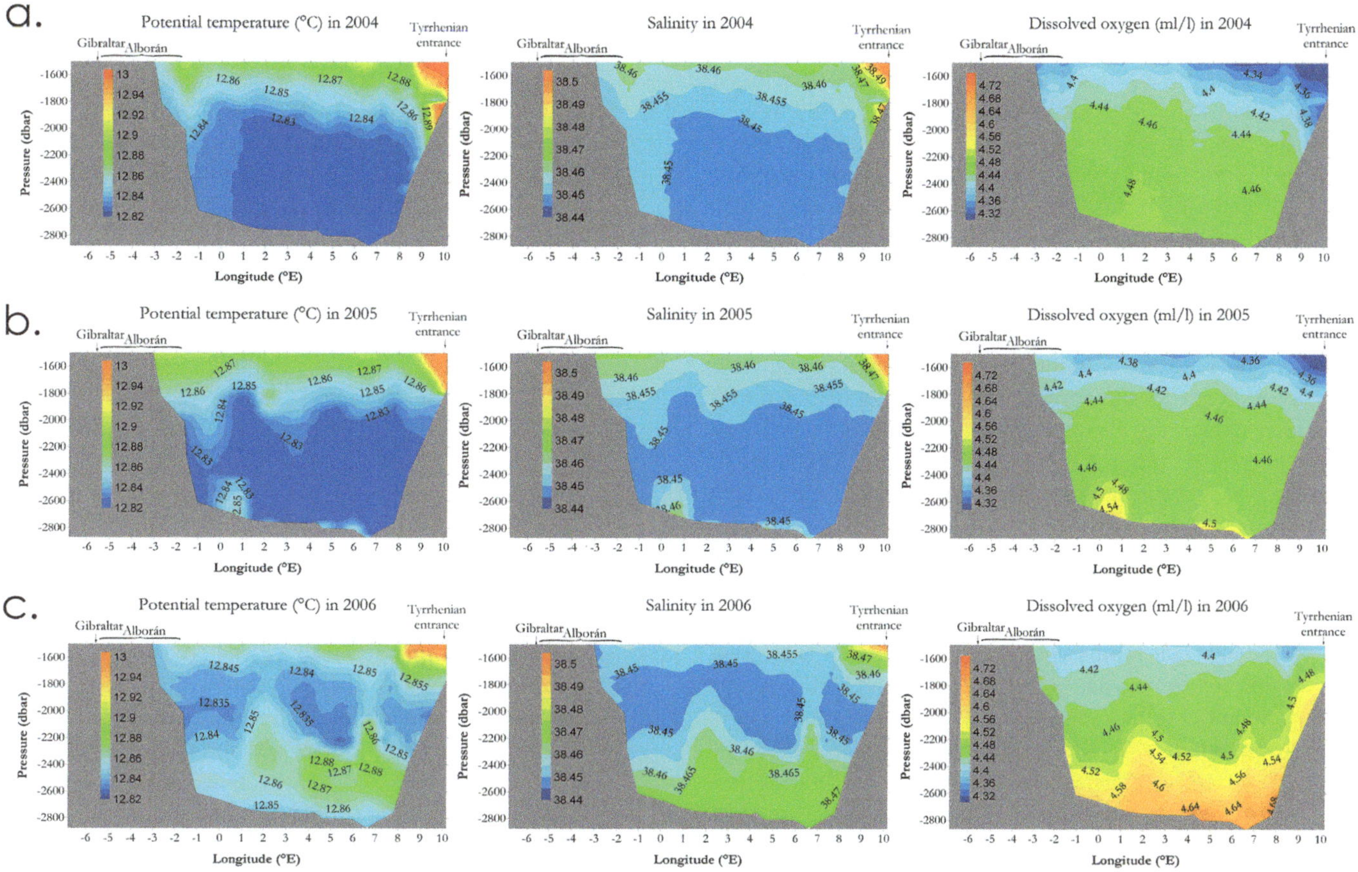

Fig. 9. Vertical θ, S and O_2 distributions below 1500 m depth along transects in the WMed showing the evolution of the WMT.

the operational observing systems. The other direction would be to continue monitoring certain parameters at key locations, such as straits and water formation areas. It is of importance to continue the effort towards an ocean–atmosphere interaction direction, which combines models with data, with schemes assimilating all different parameters of direct field measurements, from temperature and salinity to flow velocities throughout the water column.

2.3 Forcings and variability in the stock of nutrients of the Mediterranean Basin

2.3.1 Present knowledge

In this section we use the term biogeochemical in a restrictive sense, focusing only on the distributions of nutrients and the key processes affecting them. Based on nutrient concentrations, the western and the eastern Mediterranean are classified as oligotrophic or extremely oligotrophic. However, the trophic regimes of the basin cannot be characterized only by the low nutrient concentrations of Mediterranean waters. Atmospheric deposition, riverine inputs, exceptional cross-shelf transports and exchanges with the Black Sea suggest a less extreme lack of nutrient availability in the photic zone, with a quick utilization and channeling through the food web (e.g., Thingstad et al., 2005).

Estimates of atmospheric fluxes are generally based on a few stations located around the perimetry of the basin (e.g., Guerzoni et al., 1999; Markaki et al., 2008) or, in the case of Saharan dust deposition, inferred from remote-sensing observations (e.g., Volpe et al., 2009). Despite the obvious uncertainties, most of the estimates support the view that atmospheric contributions provide an amount of nutrients to the photic zone comparable to other sources. Krom et al. (2004) presented a detailed nutrient budget for the eastern Mediterranean and showed that atmospheric input of dissolved inorganic nitrogen (DIN) and dissolved inorganic phosphorus (DIP) may account for 61 % and 28 % of the total budget of nitrogen (N) and phosphorus (P), respectively. This contribution is the most difficult to track in surface concentrations due to the low values of instantaneous fluxes and rapid uptake. A parallel uncertainty exists for cross-shelf fluxes due to extreme and hard-to-monitor events . The importance of external sources as compared to internal stocks and the rapid turnover of the basin is also at the origin of the non-Redfieldian N : P ratio in the stocks. N : P ratio in the atmospheric deposition, in dissolved or soluble inorganic forms (Ridame et al., 2003), is considered the natural candidate for explaining the anomalous high N : P ratio typical of the Mediterranean. Indeed, it displays an eastwardly increasing trend, from about 20 : 1–24 : 1 in the western basin to 28 : 1 in the eastern one (Markaki et al., 2008).

In the Mediterranean, river loads, even relatively important ones, have limited shelves (with the noticeable exception of the Po and West Adriatic rivers) and release their nutrient content directly into the open ocean. They also display a N : P ratio higher than Redfield, especially after the reduction in phosphorus utilization on land. Despite this evidence the relative weight of the different processes is still elusive.

The internal stock is homogenized by general circulation processes, such as the recurrent meandering of the northern current over the Provencal–Catalan shelf, that facilitate the export of nutrients from the coastal region offshore. Irregular bathymetric features near the coastal region, such as canyons, further enhance the export of water properties and suspended material offshore, contributing to an efficient transfer of coastal signals to the open ocean (e.g., Gomez Gesteira et al., 2003; de Madron et al., 2011).

All the external inputs contribute to mitigate, or to compensate for, the imbalance in the phosphate and nitrate budget estimated at Gibraltar (surface waters with low nutrient concentration, about 3–4 11 M N, enter the Alboran Sea in the upper layer of -200 m, while roughly twice as much exits in the lower layer). On the other hand, the large differences in water fluxes at the Gibraltar Strait (0.81 Sv in, 0.76 out, CANIGO group) and the exchanges at the Sicily Channel (around 1.2 Sv) suggest that in the western Mediterranean a relevant recirculation (with an associated supply of nutrients) must take place between the intermediate and upper layers, even if wind-driven upwellings are relatively small and not permanent. This is testified to by the relatively steady concentration of the nutrients in the deep layers of the Mediterranean.

2.3.2 Unresolved issues

1. Understanding the functioning of the Gibraltar valve

   The Mediterranean is connected to the Atlantic through a shallow sill at Gibraltar, which is expected to decrease the turnover of tracers passing through the Gibraltar Strait. Despite its small section the strait is still responsible for the largest flux of water and elements into the basin. In the last decade observations have become more systematic and data are, therefore, increasing (e.g., Huertas et al., 2012). However, what is crucial is the net flux, which, as mentioned above, is significantly negative. Budget estimates have existed since the pioneering age of Mediterranean oceanography (e.g., McGill, 1961), but they generally assume a steady state. A dynamical reconstruction of the exchange with the Atlantic is still lacking. An analysis of pluriannual or long-term variations in fluxes, which may depend on the reorganization of Mediterranean and Central Atlantic circulation, impacting on Mediterranean stocks, is not available. A robust estimate of the role of entrainment and mixing among different water masses because of the peculiar functioning of the Strait has only recently been made (e.g., Huertas et al., 2012; Bruno et al., 2013), but should be better refined. Models have also provided important insight, e.g., Sannino et al. (2009), but, to date, no single term of the nutrient budget of the basin is really solid, which is striking considering the size and the closeness of the basin. A better understanding of the functioning of the Gibraltar valve at different timescales (from decadal to centennial) is a necessary step for analyzing the Mediterranean internal variability and changes in nutrient stocks.

2. Turnover rates and exchanges between Mediterranean sub-basins

   The turnover rates of Mediterranean nutrient stocks in different basins depend not only on advective processes and exchanges through the straits, but also on water mass transformation rates and biogeochemical processes. Constraining the physical drivers with classical approaches based on passive tracers and modeling would allow one to determine the coupling between surface biology and internal remineralization, which would also help in building scenarios of future trends in the basin. Furthermore, biogeochemical processes in filaments, fronts and eddies may be very important in coupling the surface and deep layers.

3. Timescales of internal variability

   To assess the scales of internal variability a careful analysis of possible longer term oscillations/trends is required. Existing databases, even in presently reviewed forms, display high internal noise and dispersion of data. Despite this an updated assessment of pluri-decadal trends in nutrient concentrations for the eastern and western Mediterranean must be carried out starting from the existing database.

4. Data on stable isotopes

   Stable isotopes have been recently used to differentiate the different atmospheric inputs in the eastern basin. Less is known for the western basin and, overall, the data available are still too few. ^{15}N has also been used to infer trophic levels of copepods or top predators (e.g., Koppelman et al., 2009), still without covering the whole basin. Future surveys should consider those parameters, especially to approach the problem of the external vs. internal contribution of elements to the basin production.

5. Silicon dynamics

   De Madron et al. (2011) noted that decrease in terrestrial inputs may decrease the Si content of the basin. Ribera d'Alcala et al. (2003) pointed out that balancing the silicon budget in the Mediterranean is difficult with the existing data. The silicon distribution in the Mediterranean shows a different gradient with respect to nitrogen and phosphorus, indicating that a better representation of the silicon sources is necessary to quantify its variability in the open sea. In addition, there are sediment trap data that show a large proportion of biogenic silica, in contrast to the low primary production (at least in the eastern basin).

6. Mutual feedbacks between biogeochemical functioning and community structure

   These feedbacks display differences between the subbasins of the Mediterranean that need to be addressed better (e.g., Moutin et al., 2012). Furthermore, the functioning of the biological pump should be investigated, combining estimates of pCO_2, satellite PP, Chi-a data and POC fluxes measured by sediment traps at various depths along with the role of community structure (D'Ortenzio and Ribera d'Alcala, 2009).

7. Biogeochemical processes in intermediate and deep layers

   These processes should be analyzed in relation to the higher respiration rates and the very low values of dissolved organic carbon (DOC, Santinelli et al., 2010).

2.4 Modeling and assessing ecosystems in the Mediterranean Sea

2.4.1 Present knowledge

The objective of this section is to identify some research priorities in physical–biological interactions limited to basin-wide seasonal-to-decadal scales, in order to be consistent with the overarching objectives of this article. The interested reader can refer for the climatic scale to a comprehensive review of functioning and responses of Mediterranean marine biogeochemical cycles and ecosystems to key natural and anthropogenic drivers, recently published by the MerMex Group (2012). To elicit the nonlinear impact of dynamical variability on the ecosystem on the above-mentioned scales, basic questions, opportunities offered by innovative observational networks, and scientific challanges in numerical models are discussed to underpin the emerging research themes in a crosscutting perspective. A holistic approach has been followed in this section, prioritizing those themes that have been neglected up to now both for conceptual and practical reasons. The key questions have been set, retaining the "Mediterranean flavor", i.e., the specificity that makes this basin unique and vulnerable. The peculiarity of the Mediterranean Sea however has indeed been recognized but still not understood in its full complexity.

The whole Mediterranean Sea was characterized as a single bio-province by Longhurst (1998) by using an intermittent, variable-quality data set produced by CZCS. This conclusion has been questioned more recently by other papers based on different approaches: in situ data (Bianchi and Morri, 2000), remotely sensed surface chlorophyll (D'Ortenzio and Ribera d'Alcala, 2009), and decadal simulation of hydrodynamical–biogeochemical models (Lazzari et al., 2012). In this last paper an analysis of the dynamics of the physical–biogeochemical seasonal cycle has been produced, showing contrasting behaviors according with the different forcing. In Fig. 10 the opposite responses of the ecosystems to the seasonal cycles are represented using a sub-basin-integrated Longhurst diagrams (see the caption for details): in the NWMed (left plate) deep convection plays a major role in delaying the spring bloom by a month, while, to the other extreme, the model predicts a deepening of mixed layer depth (MLD), relatively high daily irradiance even in winter and a concomitant gentle increase in primary production and autotrophic biomass. Thus, two dominant regimes have been identified: "bloom" and "no bloom". In fact, despite being smaller, the WMed displays a larger number of trophic regimes than the eastern basin, which is known to be oligotrophic with the exception of the shallower part of the Adriatic Sea (D'Ortenzio and Ribera d'Alcala, 2009).

This is largely due to the interplay between physical and biological processes (inverse estuarine thermohaline circulation, biological pump, limited river loads, ubiquitous atmospheric inputs, etc.), the different forcings and their modulation by the morphology of boundaries, which may in turn produce a different internal dynamics (Crispi et al., 1998). Interesting responses to the interplay between primary and secondary producers have been identified in the case of episodic nutrient inputs. Paradigmatic is the impact of atmospheric input, which has long been recognized as a major source of nutrients in the Mediterranean surface waters. The response of the planktonic community to this episodic input in the case of nutrient-depleted surface water has proved elusive: during the CYCLOPS experiments, as a response to an artificial additional phospate addition, the total chlorophyll declined in the surface waters and there was an increase in bacterial production and, lately, copepod egg abundance. Thus the nutrient supply to the higher trophic levels may have taken place by phosphorus uptake in heterotrophic bacteria and by "tunnelling" to copepod through the heterotrophic nanoflagellate predation (Thingstad et al., 2005). This pathway may partially explain the relatively high carry capacity of the eastern basin compared with its primary productivity.

The link between differences in multiannual through multi-decadal variability in the internal dynamics of the two basins and the biotic dynamics is still an unexplored trait of the Mediterranean Sea. This issue deserves even more

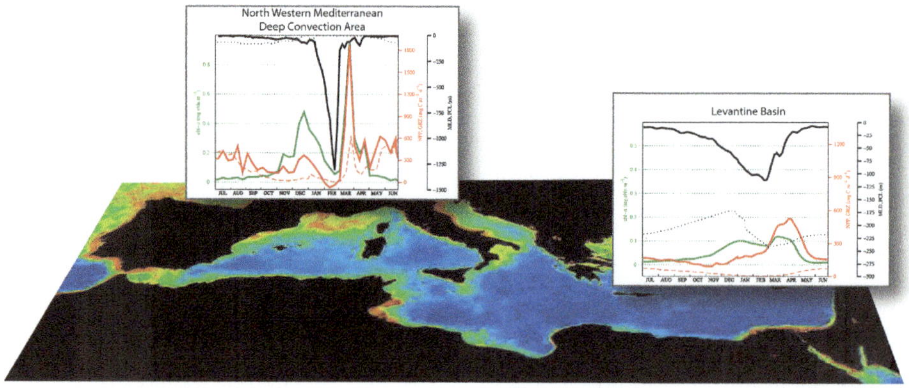

Fig. 10. Seasonal cycles of two contrasting ecosystem dynamics. The seasonal cycles of Chi-a concentration (green), MLD (black), net primary production (red solid), phosphocline (green dotted) and grazers (red dotted) are been modeled for the deep NWMed convection area (left) and for the Levantine Basin (right). The first reproduces very well the post-convection bloom (as seen in the Chi-a satellite images), while the latter follows a semi-tropical dynamics. Note that the vertical scales have been adjusted to accommodate the different intensities in the responses (after Lazzari et al., 2012, modified).

attention considering the recent finding of multiannual oscillations in the eastern Mediterranean (Gačić et al., 2011; Roether et al., 1996). On the other end of the drivers of changes, uncontrolled anthropic activities such as overfishing are impacting top predators of high commercial values. The expected impact on the food web is not confined to the upper trophic levels, but also influences the whole food chain (Parsons and Lalli, 2002). The biomass continuously supplied by primary producers was once channelled towards the top predators; now it is supposed increasingly to feed jellyfish growth (more generally the macro-zooplankton). "If jellyfish prevail, they are a further blow to fish populations, as they prey upon the prey of fish larvae (crustacean plankton) and on the larvae themselves, acting as both competitors and predators of fish" (Boero et al., 2008). The understanding of this potential regime shift should also be connected with the above dynamic variability and the introduction of non-indigenous species.

This fact, in turn, highlights that processes occurring in the basin should be approached through their appropriate space and timescales, spanning from the synoptical to the decadal (and longer) variability. The need to investigate the impacts of the internal and forced variability on longer timescales in the Mediterranean community structures and composition requires better, qualified, problem-oriented models and model systems (Arhonditsis and Brett, 2004).

The considerable efforts made up to now in trying to model the vertical dynamics and circulation processes and their impact on biogeochemical cycles on basin scale started approximately 20 years ago. A number of 3-D models were developed on the basin scale using a simplified food web (NPD) (Crise et al., 1998; Crispi et al., 1998) to study/simulate the east–west gradient in nutrient limitation and DCM depth, as well as the impact of the general circulation on the ecosystem dynamics. Many subregional studies have been carried out with intermediate complexity biogeochemical-coupled models in both the WMed and EMed, in particular in the NWMed (see Herrmann et al., 2013 and the references reported therein), the Adriatic Sea (Polimene et al., 2006; Cossarini et al., 2012) and the Alboran and Levantine basins. Despite its local validity, the regional approach to modeling is generally not useful for sub-basin intercomparison because of the forcing mismatches and the lack of spatio–temporal synopticity. In addition, even if the open boundary conditions derive from a coarser model, the solution is boundary condition dependent (in particular if the flushing timescale $L_{\text{basin}}/U_{\text{OBC}}$ is less than the internal adjustment time to local forcing). In this case sub-regional simulations largely mimic the coarse model solutions on scales longer than the seasonal one.

More complex 1-D (water column) configurations based on the ERSEM and BFM (Fig. 11) model were used to study the ecosystem functioning in more detail, although mostly on the regional scale, addressing the dynamics of the microbial loop, the bacterial dynamic and primary production (Allen et al., 1998; Anderson and Turley, 2003; Polimene et al., 2007; Lazzari et al., 2012). A significant effort has focused on the data assimilation of biophysical parameters into an ecosystem model of the eastern Mediterranean (Triantafyllou et al., 2005, 2007).

Only recently have food web models and ecological network analysis applications been developed for Mediterranean marine ecosystems. Because of the intrinsic formulation of this class of models, the studies have focused on data-rich areas of the basin (Coll and Libralato, 2011). Very few studies (if any) are available for the southern side of the Mediterranean Sea. Few comparisons of trophic networks from different Mediterranean sub-basins have been published (Coll et al., 2008; Barausse and Palmieri, 2013), even if they can be proved essential in understanding the differences in

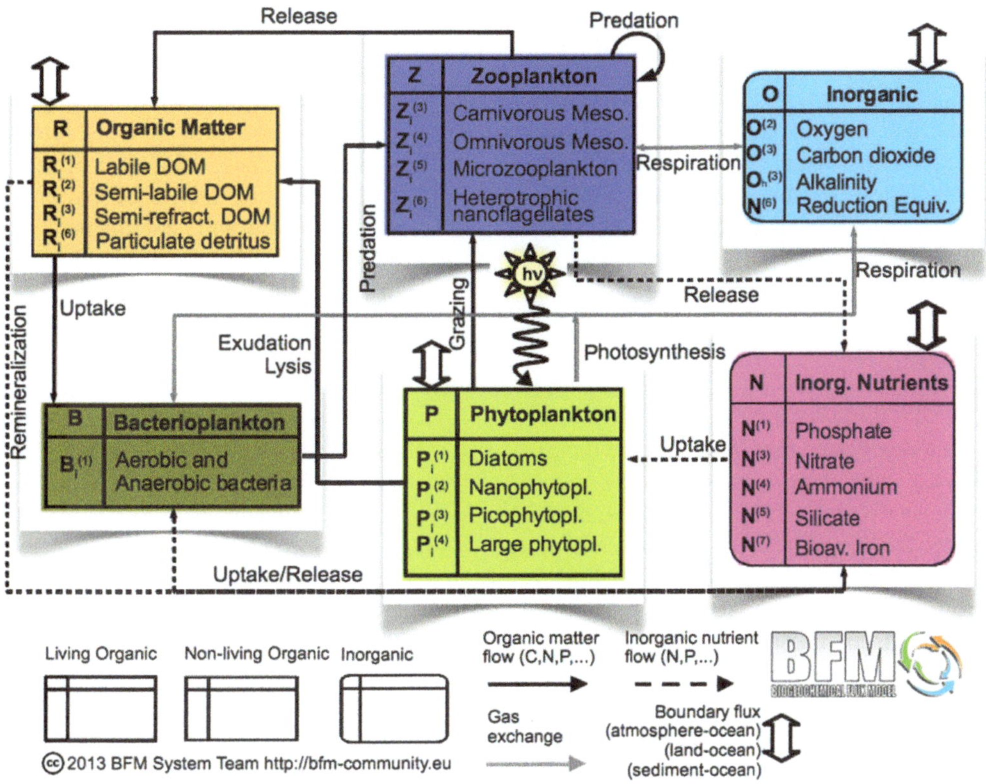

Fig. 11. State-of-the-art biogeochemical model schematics. The biogeochemical flux model is used extensively in Mediterranean biogeochemical studies. It describes the trophic interactions of the marine ecosystems by describing the cycles of the organic and inorganic carbon as well as the macro-nutrients. Major features are the plastic intracellular N : P : Si : C : Chl *a* ratios and the explicit description of the microbial loop.

the functioning of Mediterranean ecosystems. Such ecosystem models, if properly validated, can also provide a management tool for exploitation of renewable resources (Coll and Libralalto, 2012) and constitute a component of end-to-end implementation of the marine ecosystems (Libralato and Solidoro, 2009). Experiments in producing decadal-scale re-analyses provided by state-of-the-art biogeochemical models coupled with data-assimilation schemes are carried out in the framework of some ongoing EC FP7 projects (MyOCEAN2, OpEc) in order to be able to describe the dominant modes of variability of the Mediterranean ecosystem(s).

In recent years, substantial improvements in the science-to-policy interface in the marine sector have been achieved, driven by practical requirements such as the ecosystem approach to the management of natural resources. This cross-cutting approach also includes a socio-economical adaptive component to be connected in a seamless way to the ecosystem models. This approach is the backbone of ongoing projects aimed at connecting marine science findings with socio-economical models and management tools (e.g., the EC FP7 PERSEUS project).

2.4.2 Unresolved issues

The necessarily brief summary of the state of the art in ecosystem studies in relation to the physical forcing can generate many open questions. In the following, a selection of key questions focuses on basic understanding and modeling of the Mediterranean marine ecosystem to scales from synoptic to decadal.

1. Non-indigenous species (NIS)

   The Mediterranean ecosystem structure(s) is(are) clearly influenced by the overall physical/chemical processes and exchanges active in the basin that favored the immigration of many species of Atlantic origin that dominate the Mediterranean communities and the invasion of alien species from either Gibraltar, Suez, and anthropogenic transport (Galil et al., 2007; Galil, 2000). The short time needed by several new non-indigenous species to colonize vast areas of the Mediterranean Sea hints at the presence of favorable conditions that may contribute to NIS success related to resources, natural enemies and the physical environment (Shea and Chesson, 2002). The foreseen temperature and salinity positive trends summed to the anthropogenic pressures create conditions to observe and understand the evolution of NIS in a changing ocean. The relative importance of resources, enemies and environment in an evolutionary perspective has to be investigated, and their impacts on the interactions between them (resource and escape opportunity, interaction between natural enemies and competition in the native range of the invader and the invaded community, community maturity and invasion resilience). The Mediterranean can be an ideal natural laboratory for improved understanding of coexistence mechanisms in an evolving marine realm.

2. Mesopelagic community structure

   It has long been recognized that the dissolved oxygen vertical structure in the Mediterranean shows only a slightly pronounced oxygen minimum layer (if any) compared with other bio-provinces (i.e., the equatorial Pacific) (McGill, 1961). This highlights a relatively small export likely due to the peculiar community structure and composition. In addition, possible ventilation spots can occur in anticyclonic structures. This, in turn, further reduces the concentration of macronutrients in the ocean interiors, which suggests that the community efficiently recycles most of the nutrients in the surface layer. Conversely, dissolved organic carbon exhibits a well-shaped minimum with values comparable with those typical of the oceans (Santinelli et al., 2006). The mesopelagic trophic regime and respiration dynamics therefore deserve specific studies in light of these peculiar conditions.

3. Habitat mapping

   The Regional Activity Centre for Specially Protected Areas showed that 39 % of habitats and associated species considered in the list are scarcely covered by scientific knowledge and that the quantitative information about the geographical distribution of selected habitats and associated species is largely unbalanced toward the shallow western Mediterranean Sea (UNEP-MAP-RAC/SPA, 2006). There is a need to improve the classification protocols and the inventories of marine habitats (Fraschetti et al., 2011) in a novel perspective of continuously evolving ocean conditions. The classification would also include dynamical processes and rather new concepts such as ecological coherence and connectivity to allow better estimates of biodiversity distribution, and to reverse habitat loss rates through the systematic implementation of proper measures and instruments.

4. Mesoscale impact on primary producers

   The mesoscale has a typical length shorter than the major oceans (about 15–20 km according to Pinardi and Masetti, 2000), but the coherent structures found in the satellite images are roughly three to four times the Rossby internal deformation radius, since two-dimensional geostrophic turbulence is evolving toward larger scales. Large, long-lasting anticyclonic eddies are regularly generated in the Algerian Current, most probably because of the baroclinic instability of this current flowing along the African coast. A number of large, recurrent, continuously evolving structures connected by jets and separated by fronts (Robinson and Golnaraghi, 1994) have been observed in the surface waters of both sub-basins. In recent years the mesoscale activity in the Mediterranean has been a matter of scientific discussion, and is not the subject of this section. What is worth noting here is that the response in the surface chlorophyll images is not clearly connected to the vorticity polarity, so the simple equivalence cyclone (anticyclone) = more(less) nutrient supply in the surface layer = more(less) primary production seems not to hold in general. More intriguingly, during the stratified season no Chl *a* features are evident in the EMed. Since mesoscale, as tracked by the SSt and SSHA satellite-borne sensors, is ubiquitous, this unexpected result may be related to the well-known E–W nutricline slant. Alternatively (or in combination), a different penetration of the mixed layer, possibly modified by baroclinic processes, can be confined in the proximity of the surface, being unable to puncture the (deep) nutricline. This in turn can be responsible for the different responses in the standing crops and primary production of the two sub-basins. The origin of the differences in this response has still to be investigated thoroughly.

5. Impact of decadal variability on the western Mediterranean food web

   While recent analyses hint at a relevant role of interannual variability in the EMed (10–20 % of the seasonal cycle), existing data show that the seasonal cycle is the dominant component of variation. However, recent 50-year Mediterranean climatology provides evidence of consistent temperature changes in the WMed and the North Atlantic, explained by similarities in the atmospheric heat flux anomalies strongly correlated with NAO (Rixen et al., 2005). Conversely, the copepod community composition and abundance (even with a clear seasonal cycle) exhibits in the Ionian Sea a reduced variability on the interannual scale showing a remarkable resilience to the interannual variability. More generally, an assessment of the long-term atmospheric variability impact on the food web has still to be made.

6. Multiple stressor impacts on Mediterranean marine ecosystem functioning

   Global increases in atmospheric CO_2 and temperature are associated with changes in ocean chemistry and circulation, altering light, pH and nutrient regimes. Resulting changes in phytoplankton community structure are expected to have a cascading effect on primary and export production, food web dynamics and the structure of the marine food web as well as the biogeochemical cycling of carbon and bio-limiting elements in the sea. A review of the current literature indicates that cell size and elemental stoichiometry often respond predictably to abiotic conditions and follow biophysical rules that link environmental conditions to growth rates, and growth rates to food web interactions, and consequently to the biogeochemical cycling of elements. This suggests that cell size and elemental stoichiometry must be monitored to allow modeling and tracking changes in phytoplankton community structure in response to climate change. In turn, these changes are expected to have further impacts on phytoplankton community structure through as yet poorly understood secondary processes associated with trophic dynamics (Finkel et al., 2010).

7. Modeling and assessing the budgets of the relevant ocean variables for Mediterranean sub-basins

   The Mediterranean physiography as well as its circulation dynamics are composed of a limited number sub-basins connected by straits. In their interior, sub-basins have internal (in some ways independent) dynamics, whereas the impact of the other sub-basins is ruled by the exchanges at the connecting straits. In this context, the Mediterranean nutrient dynamics as depicted with full-blown 3-D coupled BGC hydrodynamical models can be represented as averaged 1-D or 0-D variable vectors, each representing a sub-basin interconnected by laterally averaged fluxes. This simplified analysis should allow the testing of scenarios (i.e., modifications of boxes' internal functioning, changing fluxes at the connecting boundaries, steady state vs. interannual/decadal variability, connections/feedbacks/linkages with climate alteration), directly focusing on the impact of the internal vs. external forcing on the nutrients, and spatio–temporal repartition between sub-basins.

8. Observational needs and perspectives

   The issues unresolved above often require closely interconnected state-of-the-art observational capacity and cutting edge technologies. Here some basic requirements are spelled out, in terms of observations, data management and modeling. Autonomous observational techniques profiting from the advancement of observational techniques should be adopted extensively because they provide new relevant information on distributions and internal dynamics of the mesopelagic and bathypelagic ecosystems at high space and time resolution. The integration with advanced technologies (metagenomics, automated image analysis and recognition, innovative acoustic measurements) on autonomous vehicles such as AUV is still a challenge, but will lead in the near future to a totally different approach to observing ocean interiors. There is a need for a network of sustained truly multidisciplinary deep sea observatories able to complement the "diffuse" and often problem-/project-dependent observing initiatives and experiments. The data integration and harmonization of observing stategies lead to the further development of a sustained ocean observing system prioritizing the Mediterranean at European level and taking advantage of what has been developed and consolidated within Euro-GOOS and, in particular, MONGOOS. There is a need for a sustained pan-Mediterranean cyber-infrastructure based on existing components (MyOCEAN, SeaDataNet, national initiatives) to link, qualify, preserve and disseminate the observations to facilitate the data and metadata interoperability in the framework of the GEO and GEOSS programs. An open and free data policy should be adopted by all the major actors in Mediterranean marine research acknowledging the EC INSPIRE directive, in order to enable easy access to data and their reuse.

3 Relative importance of external forcing functions (wind stress, heat/moisture fluxes, forcing through straits) vs. internal variability

3.1 The role of salinity decadal oscillations in triggering the thermohaline circulation and the Mediterranean conveyor belt

3.1.1 Present knowledge

The Mediterranean Sea open circulation cell is driven by the salinity differences between the inflowing low-salinity Atlantic Water (AW) and the highly saline Eastern Mediterranean waters, mainly the Levantine Intermediate Water (LIW). The salinity differences established between the two water masses are maintained due to the prevalence of evaporation over precipitation. Deep closed circulation cells are driven by the air–sea heat losses at specific locations, resulting in vertical convection and dense-water formation. This average circulation pattern is subject to interannual and decadal variability due to both external (meteorological) and internal forcings. Temporal variability of the closed circulation cells is primarily determined by the intensity of the dense-water formation, which in turn depends on the air–sea heat fluxes and the preconditioning. Deep circulation was thought for a long time to be invariable, and based on that the residence time of the Eastern Mediterranean was estimated at about 100 years. Recent long-term simulations gave evidence that the alteration of the two sources of dense waters (Adriatic and Aegean Seas) in the EMed could be a recurrent phenomenon, connected to salinity out-of-phase decadal oscillations between the two basins (Borzelli et al., 2009), given the necessary intense local atmospheric forcing. The salt content in each formation site, along with the intensity of the atmospheric forcing, determines the depth of the convection and thus the density of the newly produced water. In the early 70s a high-salinity event was documented in the Levantine and Aegean, but with lower formation rates and thus without a significant signature of Aegean dense water in the adjacent sub-basins (Beuvier et al., 2010). Theocharis et al. (2002) noted that the 70s episode was evident in all three basins, Levantine, Ionian and Aegean, with a maximum signal in the Levantine. The total volume of the produced waters was lower than that produced during the EMT, thus its effect vanished quickly.

In the mid-1990s, experimental evidence on the Eastern Mediterranean Transient (EMT) was presented and it was shown that the abyssal circulation is not in a steady state but can be subject to episodic sudden changes (Roether et al., 1996). The EMT involves the passage of the Eastern Mediterranean dense-water formation site from the Adriatic to the Aegean. This then resulted in important changes in the abyssal circulation of the entire basin. Changes were evidenced over the upper part of the water column as well. Generally, the most prominent variations in the upper layer circulation have taken place in the northwestern Ionian, where they manifested as the circulation reversal in 1997 reinforced by the EMT. In addition, evidence of two other events in 1987 and 2006 (Gačić et al., 2011) was presented. These reversals of the upper-layer circulation pattern in the Ionian denominated Bimodal Oscillation System (BiOS) (Gačić et al., 2010) determine the salt redistribution between the Adriatic and the Levantine/Cretan seas (Fig. 12, adapted from Gačić et al., 2010). This is due to a preferential pathway of the relatively fresh AW; in the cyclonic circulation pattern in the Ionian the AW pathway is towards the Levantine/Cretan seas, while during the Ionian anticyclonic circulation mode the northern Ionian and the southern Adriatic are subject to freshening due to the increased spreading of the AW northeastward.

Therefore, the salt content and buoyancy in the upper layers of the Adriatic and the Levantine/Aegean are out of phase, suggesting that one or the other sub-basin is more prone to the convective mixing. This suggests that the EMT, i.e., the Cretan Sea undertaking the role of the main dense-water source for the Eastern Mediterranean, is potentially a recurrent phenomenon if the winter air–sea heat losses are strong enough. Important decadal variability in the thermohaline properties and the deep circulation were evidenced in the Western Mediterranean as well. Again, salinity forcing seems to play an important role, since on decadal scale there is important building-up of the highly saline water spreading over the entire bottom layer and increasing continuously in

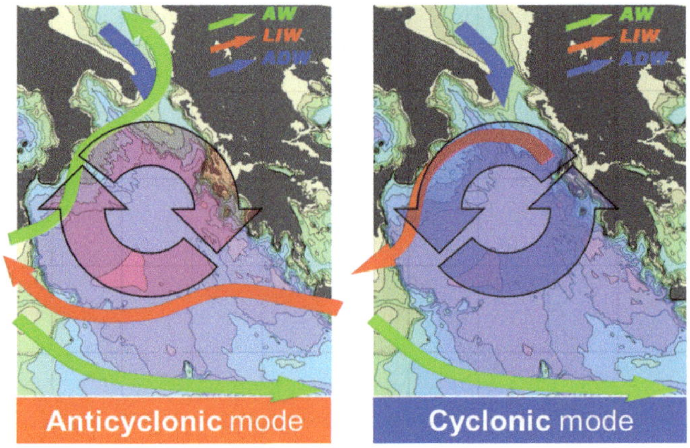

Fig. 12. Schematic presentation of the BiOS decadal inversions of the Ionian circulation and pathways of the principal water masses.

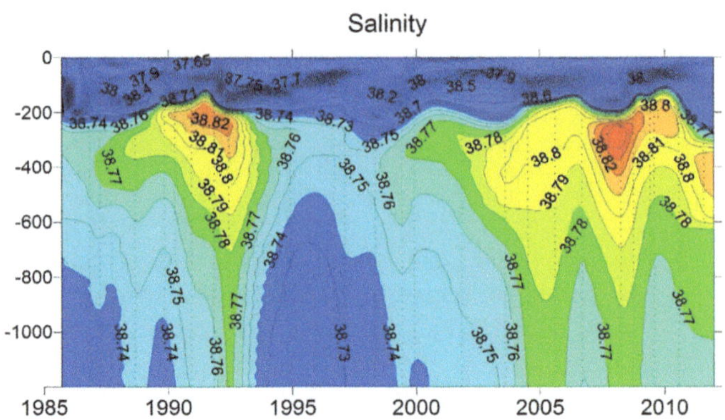

Fig. 13. Time–depth diagram of the salinity variations in the Sicily Channel. Two episodes of the salinity maximum at the depth of the LIW are evident in the early 1990s and in the second half of the 2000s associated with BiOS and impacting the WMDW and Western Mediterranean Transition pre-conditioning (modified from Gačić et al., 2013).

thickness from 2004 up to the present day. The phenomenon is called the Western Mediterranean Transition and can presumably be connected to the inflow of the LIW of varying salinity from the Eastern Mediterranean (Schroeder et al., 2009). In fact, it has been shown recently (Gačić et al., 2013) that this salinity increase can be explained by up to 60 % by the advective input of the highly saline LIW and thus related to BiOS and circulation reversals in the Ionian subbasin (Fig. 13). The replenishment of the deep basin is by newly formed Western Mediterranean Deep Water that, depending on its density, can either uplift old resident waters or lay above them, leaving in any case a cold signature in the temperature series (Garcia-Lafuente et al., 2009). This mechanism then determines interannual variability of thermohaline properties of the outflowing Mediterranean waters, possibly changing their impact on the global conveyor belt.

The Mediterranean outflow is a constant source of warm and salty intermediate water, and it has even been suggested that it influences the Atlantic Meridional Overturning Circulation (Candela, 2001). Thus, understanding the interannual and long-term variability of the Mediterranean Sea appears to have a broader significance than previously thought (Millot, 2007).

3.1.2 Unresolved issues

1. Completion of a more reliable data set for studies of long-term variability in the Mediterranean

   During the last few years considerable effort has been put into identifying past changes in the water mass characteristics of the Mediterranean Sea from historical hydrographic records related to salinity oscillations. The investigation of this long-term variability by using historical databases such as MEDAR-MEDATLAS has been proven to be inadequate due to spatial inhomogeneity of data and high noise levels (Schroeder et al., 2012). Important efforts have been made to construct new and extend existing data sets (SeaDataNet, EMODNET, MyOcean) that pass screening quality control procedures, but cannot resolve the inefficiencies of past measurements and spatial inhomogeneity. Long-term reanalysis simulations that assimilate in situ measurements and exploit the significant amount of information from observational systems in the Mediterranean are an important step towards reliable reconstructions for "short-term" climatic analysis. Moreover, studies of DWF and circulation on decadal and interdecadal scales must take into account the variability of the inflowing AW and in particular its observed increasing salt content (Lauzier and Sindlinger, 2009).

2. Relative importance of internal processes and external large-scale climatic forcing

   The salinity oscillations show a prevailing decadal timescale modulated by internal mechanisms, but could also be influenced by atmospheric forcing. This is especially related to the fact that the deep closed circulation cells are subject to the air–sea heat fluxes. The large-scale atmospheric circulations such as the NAO (North Atlantic Oscillation), which largely determines Mediterranean winter climatic conditions (Xoplaki et al., 2003) and EA (East Atlantic) and plays an important role over most of the region (Krichak et al., 2002; Josey, 2010), also exhibits variability on decadal timescales. Therefore, it is important to investigate the combined effects and relative importance of internal processes and external forcing.

3. Decadal and interdecadal variability studies in the Mediterranean as a whole

   The intensity of the EMT compared to similar past events is a major issue. Both the increase in salinity and the huge amount of dense waters (8 Sv years, Roether et al., 2007) produced in the Aegean were exceptional. Further study is needed to clarify whether the EMT can be attributed to coincidence of the salinity preconditioning of the area with intense atmospheric forcing. Diagnostic long-term simulations give evidence of internal modes of variability even on interdecadal timescales (Pisacane et al., 2006). It is important to find out whether the Mediterranean as a whole exhibits internal variability modes, similar to or inferred by the decadal oscillations evidenced in the EMed.

4. Studies of long-term trends

   It is of great interest to study the variability and trends in the EMED in response to changes and improve knowledge of the functioning of internal feedback mechanisms and related processes.

5. Biogeochemical responses in the Mediterranean to circulation changes

   Finally, biogeochemical responses in both the Eastern and Western Mediterranean to circulation changes on the decadal scale should be addressed in more detail (Civitarese et al., 2010).

3.2 Residence times and ventilation of water masses in the Mediterranean Sea: implications for dynamical and biogeochemical processes

3.2.1 Present knowledge

Water mass characteristics, structure and distribution in the Mediterranean, as well as the associated circulation patterns, are complex and show significant variability and sensitivity. This can be attributed to the complex topography of the basin, the diversity of the mixing processes involved, the variability of atmospheric forcing on various timescales and the variety of water mass formation processes present in the basin. Regional and local observational efforts and modeling studies of the Mediterranean and its sub-basins reveal a complete picture of the water mass structure and evolution. Even in the same sub-basin and nearby depressions, water mass characteristics and tracer concentrations of the deep waters present a remarkable diversity, as can be seen in Fig. 14, depicting the spatial and temporal variability in the Aegean Sea (Gertman et al., 2006; Zervakis et al., 2000; Vervatis et al., 2011). At the same time, significant and sometimes abrupt changes are encountered in the deep and intermediate layers (Roether et al., 1996; Schroeder et al., 2010). An example

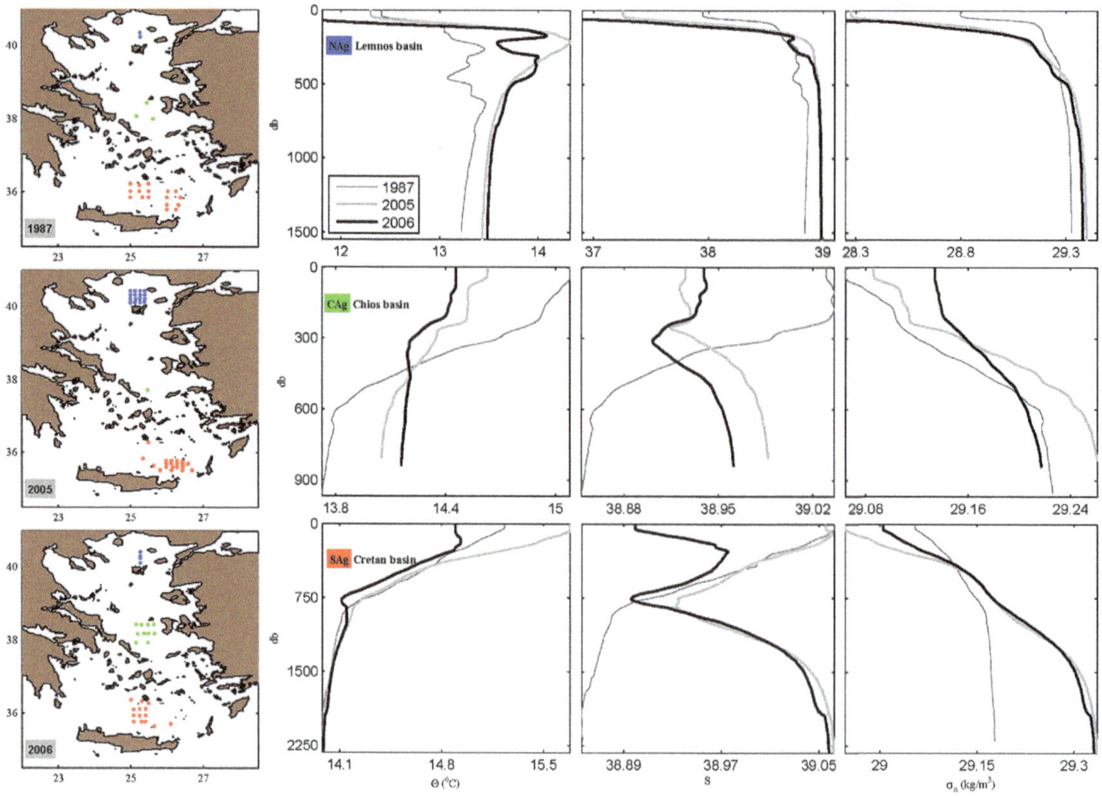

Fig. 14. Map panels: Aegean Sea winter hydrographic stations for the years 1987 (pre-EMT), 2005 and 2006 (post-EMT); North Aegean (NAg): blue dots at the Lemnos basin; Central Aegean (CAg): green dots at the Chios basin; South Aegean (SAg): red dots at the central–eastern Cretan basin. Profile panels: averaged profiles of Q (°C), S and σ_θ (kg m^{-3}), for all three major sub-basins; adapted from Vervatis et al. (2011).

of the abrupt change in the heat and salt content of the deep layers in the Western Mediterranean is illustrated in Fig. 15.

If we want to understand the dynamical and biogeochemical functioning of the Mediterranean Sea, we need to construct a clear picture of the water mass structure and characteristics, monitor their variability and relate it to internal and external forcing mechanisms. Evaluation of the renewal timescales and the associated processes of the Mediterranean water masses is crucial for understanding the variability of the thermohaline circulation and its response to the atmospheric variability. Progress in our understanding and quantification of these processes will also enhance the investigation of biogeochemical cycles on basin-wide or regional/local scales. Using a variety of observational and modeling approaches, the evaluation of residence times and the investigation of aging processes of the various water masses can be achieved.

3.2.2 Unresolved issues

1. Investigation of spatial and temporal variability of deep water masses in the Mediterranean Sea, using historical data and modeling reanalysis

A series of research projects was devoted to constructing oceanographic databases on the global and regional scales (e.g., SESAME, SeaDataNet, EMODNET, MyOcean). Today a wealth of data exists that can be used to analyze the spatio–temporal variability of physical and biogeochemical parameters in the Mediterranean Sea. Modeling techniques and computer power have been greatly improved during the last decades, enabling an ever increasing accuracy of results that can be applied on the Mediterranean and subbasin scales for the investigation of long-term variability. The contribution of operational oceanography in the Mediterranean Sea was important, developing and implementing new and advanced modeling tools and techniques in the Mediterranean Sea in the framework of research projects and initiatives at the Mediterranean and European levels (e.g., MFSPP, MFSTEP, ECOOP, MERSEA, MyOcean, MOON/MONGOOS). Blending the two approaches through data assimilation procedures can produce useful results for filling spatio–temporal gaps. It is important that additional quality control of existing data collections for, in particular, biogeochemical parameters is performed

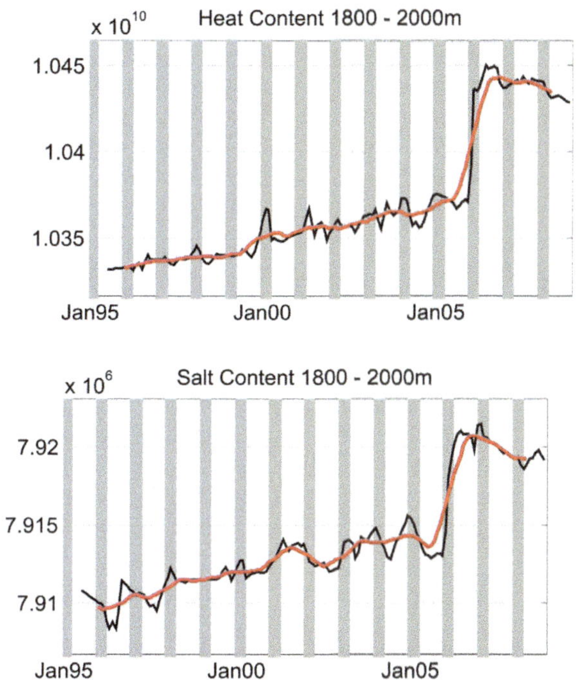

Fig. 15. **(a)** Temporal evolution of heat content ($J\,m^{-2}$) of the 1800–2000 m layer at the Dynamique des Flux Atmosphériques en Méditerranée (DYFAMED) site in each 200 m layer from 0 to 2000 m (black line, observed; red line, moving average over 13 months). Shaded regions indicate the "convection period" (December–March); **(b)** same as **(a)** except for salt content ($10^3\,kg\,m^{-2}$); (adapted from Schroeder et al., 2010).

to ensure that the data sets are internally consistent and that measurement biases are removed so that temporal trends can be estimated correctly.

2. Investigation of ventilation processes in water masses and monitoring of spatio–temporal variability

   Based on past observations, numerical model techniques and a strategy for observational process-orientated campaigns, renewal patterns and mixing processes that affect the characteristics and ventilation of the water masses can be revealed and explained. Places like deep depressions, straits and water-mass formation areas are key sites for understanding renewal processes. Monitoring of the spatio–temporal variability can also take advantage of national and international observational networks, among which the HydroChanges CIESM program deserves special mention (http://www.ciesm.org/marine/programs/hydrochanges.htm).

3. Estimation of residence times of water masses

   The overall observational and modeling strategy should target the estimation of the residence time of the various water masses in the Mediterranean Sea. Simpler and more elaborate techniques (box models, GCMs, etc.) can both be applied. In this way, the effects of residence time and renewal processes on the physical and biogeochemical cycles can be better understood. Regular measurement of transient tracers in the Mediterranean Sea is an important and valuable tool for monitoring ventilation processes and any spatio–temporal changes in ventilation.

4. Attribution of renewal processes

   All the above should be related to the investigation of the relative importance of external forcing and internal modes of variability. The imprint of these on the deep-water masses is very significant in understanding the dynamical and biogeochemical functioning of the Mediterranean Sea.

3.3 Paleo-climate and past physical/biogeochemical changes in the Mediterranean

3.3.1 Present knowledge

The investigation of marine archives (like sediment cores) provides new and relevant evidence of physical and biogeochemical processes, which drove the dynamics of the Mediterranean basin, in terms of response to regional/global climate, exchanges with the Atlantic Ocean and internal processes on the sub-basin scale. This backward glance could extend our potential for understanding the deeper physical forces that presently drive the 3-D circulation system of the basin and regulate its relationships with the climate system.

So far the existing paleoclimate records provide data sets for different environmental indices, including microfaunal abundances (planktonic and benthic foraminifera), microfloral abundances (coccolithophores, dinocysts), and pollen. The analysis of stable isotopes and biomarkers/alkenones indicates rich and interesting dynamics in the eastern and western Mediterranean sub-basins in the past.

Despite the large amount of existing data, comparison between different records is problematic as archives are characterized mainly by different time resolutions and spatial distributions, while data collection is characterized by different methodological approaches. This lack of a systematic comparison between the paleoceanographic records inhibits an overall climatic consideration, which is an important step towards improving our understanding of the physical climate and its variability in response to natural and anthropogenic forcing.

A detailed study of the last 20 kyrs of Mediterranean sea dynamics was focused on selected time intervals like the transition from the last glacial maximum (LGM, ~20 kyr BP) to the Holocene, a distinctive cooling between 13 and 11.7 kyr BP (Younger Dryas), a rather mild climatic period between 10 and 6 kyr BP, a warm and wet period within the Mid Holocene, followed by the abrupt climatic deterioration observed all over the Northern Hemisphere at 4.2 kyr BP and several cooling and warming events during the last 2 kyrs.

As an example of climatic reconstruction, the Eastern Mediterranean circulation experienced a major phase of reduced thermohaline ventilation, causing anoxic sediment (sapropel) deposition between ~ 10 and 6 ka BP (Holocene climatic optimum and S1 deposition). A remarkable interruption, centred at 8.2 ka BP and reflecting an invigorated thermohaline circulation, occurred during this event in both the Adriatic and Aegean seas, with repopulation of deep-sea sediments by benthic foraminifera that had been absent before and after because of the anoxia (De Rijk et al., 1999). The 8.2 ka BP cooling event was found to be part of a repetitive sequence of rapid climate shifts (Bond et al., 1997) that can also be recognized in the Mediterranean Sea (Incarbona et al., 2008). The close association of these climatic shifts with abrupt changes in bottom water oxygenation and organic-rich matter (sapropel) deposition makes them uniquely suitable for the investigation of the changes in climatic forcing that caused thermohaline shutdowns/restarts.

Climatic reconstruction of the time interval during the last 20 kyr, however, is not simple. This is because regional factors, associated with the hydrologic, chemical–physical and climatic features of the Mediterranean, largely determine climatic variability (Cacho et al., 2002; Rohling et al., 2002). Moreover, despite the fact that the deep waters of the Mediterranean are currently well ventilated (Bethoux et al., 1999), the presence of sapropels in the eastern Mediterranean indicates that this mechanism has been absent several times in the past. Over the past million years, relatively small changes in the Mediterranean water budget have had a profound impact on the thermohaline circulation of the basin. Many studies have addressed the relationship between enhanced freshwater input into the basin at times of summer insolation maxima in the Northern Hemisphere and the formation of sapropels (Fig. 16); enhanced burial of organic carbon in sediments was likely initiated by the influx of low-salinity waters that slowed or halted the convective overturning in the eastern Mediterranean and reduced deep-water oxygenation (Rohling et al., 2002).

In the western Mediterranean, there is no record of sapropel deposits, but organic-rich, nonlaminated layer (ORL) deposit records are available. Although these cannot be considered to be true sapropels, some paleoceanographic studies dated and correlated eastern Mediterranean sapropels (Rohling, 1994) with ORL in the western Mediterranean (Perez-Folgado et al., 2004).

As the Mediterranean enhances the paleoceanographic and paleoclimatic signals, the basin can be used as a laboratory for climate change. This is the time to establish a paleoceanographic comparison between the eastern and western sides of the Mediterranean and evaluate paleoproductivity trends related to paleoclimatic events.

Although the analysis of proxy records to study the climate of the past is important, it provides only partial information regarding the state of the past ocean and the processes that regulated it. To complete our knowledge, model simulations are necessary. The modeling experiments have first to be validated with observations and can then be used to study the fully realistic, three-dimensional ocean dynamics.

The use of simple box models can clarify some mechanisms underlying the variability in the circulation of the Mediterranean. These simple models, which include Stommel's pioneering box model (Stommel, 1961), are useful for studying the multiple states of the Mediterranean Thermohaline Circulation (THC) in the different basins as well as the oscillations between them (e.g., the Bimodal Oscillating System, Gačić et al., 2010). An example of such a box model for the Eastern Mediterranean THC showing the shift in the deep water formation cell from the Adriatic to the Aegean seas has recently been proposed by Ashkenazy et al. (2012). These models, although not realistic, provide a conceptual view of the dynamics and the mechanisms underlying its variability. On the other end, hypotheses made by using simple models can be verified by using more complex ocean global circulation models and proxy data.

3.3.2 Unresolved issues

1. Impact of the Atlantic circulation on the Mediterranean THC

   No investigation exists about how the changes in the Atlantic waters inflowing at Gibraltar affect the Mediterranean thermohaline circulation as opposed to changes in surface heat fluxes.

2. Analysis of the limiting factors of the past interaction between North Atlantic and Mediterranean THC

   A proto-modern circulation was established in the Mediterranean when the eastern connection with the Indian Ocean was interrupted ~ 18 million years ago (Sprovieri et al., 2007). Thereafter, the Gibraltar sill regulated the Mediterranean circulation. In a broad sense, the North Atlantic and the Mediterranean can be seen as one connected system, whose internal dynamics is modulated by the exchanges at Gibraltar (Artale et al., 2006).

3. Investigation of the nonlinear behavior of the Mediterranean circulation

   The present state-of-the-art numerical climate models are able to provide reasonable simulations of the present climate but are unable to reproduce the rapid transitions from one climate state, such as glacial climate, to another such as the present state. These transitions are inherently nonlinear and involve the competition and interactions between external forcing, such as

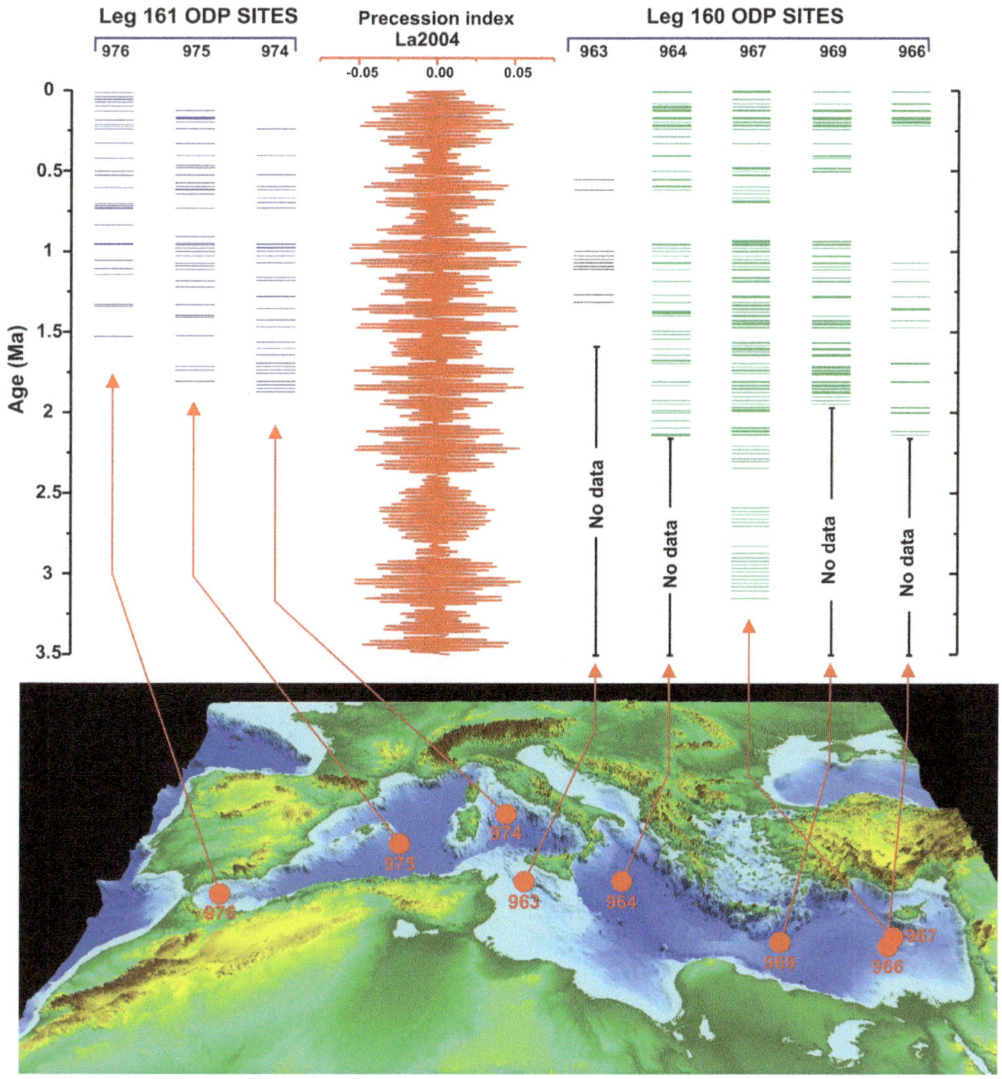

Fig. 16. Plot of sapropels and organic-rich layers (ORLs) collected in the Mediterranean Sea over the last 3.5 million years by the Ocean Drilling Program (ODP), legs 160 and 161. The chronology of sapropel layers from the central–eastern Mediterranean Sea (ODP sites 963, 964, 966, 967 and 969) follows Howell et al. (1998) and Emeis et al. (2000). The chronology of ORLs from the western Mediterranean Sea (ODP sites 974, 975 and 976) follows Murat (1999). The precession index of the same time slice is also shown (Laskar et al., 2004). Inset below: the bathymetric map of the Mediterranean Sea and the location of the ODP sites.

wind stress and heat–moisture fluxes, and internal dynamical mechanisms such as BIOS (Gačić et al., 2010; see also Sect. 3.4).

4. Study of the patterns of climate change of the Northern Hemisphere influencing Mediterranean climates

   The available evidence suggests that forced changes in dynamical modes of variability in the global ocean, such as the North Atlantic Oscillation (NAO), El Nino–Southern Oscillation (ENSO) and the Atlantic Multi-decadal Oscillation (AMO) play a key role in the patterns of climate variability in the Mediterranean region, and investigations are needed for remote times (see also Sect. 3.4).

3.4 Short-term and climatic variability in the SST and mixed layer heat budget over the Mediterranean and effects on the circulation and biota surface concentrations

3.4.1 Present knowledge

Instrumental records of increasing duration and spatial coverage as well as modeling efforts have documented substantial Mediterranean variability on timescales ranging from one day or less to decades and longer. Part of this variability can be related to known forcing mechanisms, but in many cases the relationship between observed variability and forcing has not been fully understood. Variability in the strength

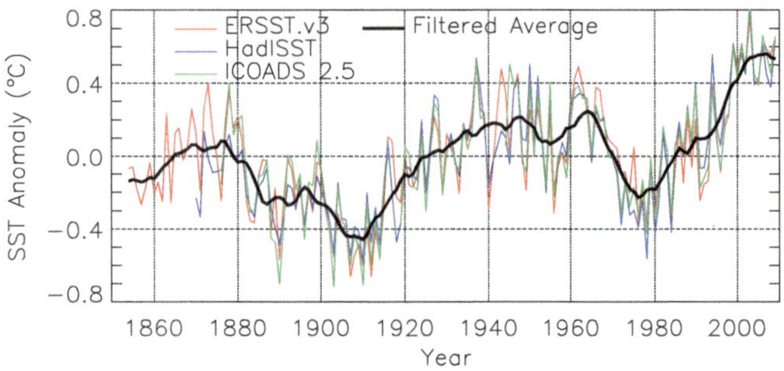

Fig. 17. Mediterranean annual SST anomaly (with respect to the 1971–2000 average) from 1854 to 2008. The red line shows ERSST.v3, the blue line HadISST, and the green line ICOADS 2.5. The solid black line represents the filtered average of the three data sets (from Marullo et al., 2011, American Meteorological Society; used with permission).

and location of local fluxes of heat, moisture and momentum are reflected in changes in the surface fields. Among these fields, the sea surface temperature (SST) is the more extensively measured surface parameter and, since it responds directly to the atmosphere–ocean interactions, represents a candidate for investigating the space–time surface variability of the Mediterranean system. Satellite SST data have been available, at least twice a day, since 1982. They have been used in a variety of studies to investigate the surface variability of the Mediterranean from the annual and interannual timescales (Borzelli and Ligi, 1999a, b; Marullo et al., 2007; Notarstefano et al., 2008; Borzelli, 2008; Borzelli et al., 2009) to the daily SST cycle (Marullo et al., 2011). The Group for High Resolution SST (GHRSST) supports these research activities by providing high-quality sea surface temperature data for applications in short, medium and decadal/climate timescales in an efficient manner through international collaboration and scientific innovation.

Marullo et al. (2011) used lagged-correlation analysis, the multitaper method (MTM) and singular spectral analysis (SSA) to reveal the presence of a significant oscillation with a period of about 70 years (Fig. 17), which is close to the Atlantic Multi-decadal Oscillation (AMO) period.

They found that, during winter, the Mediterranean SST and the North Atlantic Oscillation (NAO) vary coherently over periods longer than about 40 years, with a confidence limit between 90 % and 95 %. Over periods longer than 85–100 years, Mediterranean SST and AMO vary coherently with a confidence limit that exceeds 95–99 % (Fig. 18).

From the above analysis the question arises whether the Mediterranean Sea can be considered a component of the entire North Atlantic climate system, taking part in the deterministic mechanism proposed by Dima and Lohamnn (2007) and based on the interaction between atmosphere, ocean and ice. Furthermore, the system composed of the North Atlantic, the Mediterranean Sea/Gibraltar Strait (Artale et al., 2006) and the Arctic Sea/Fram Strait might work as a unique oceanographic entity, with the physical processes within the straits determining the exchange of the fresh and salty waters between the marginal seas and the open ocean (Sannino et al., 2009). The analysis of the Mediterranean SST variability alone, however, cannot provide a full answer to whether the forcing of the observed multi-decadal signal has an atmospheric origin or whether it is determined by changes in the Mediterranean Thermohaline Cell (THC).

Coupled ocean–atmosphere models could contribute to answering this question and to investigating the origin of the Mediterranean multi-decadal oscillation, separating the contributions of the atmosphere and the Mediterranean THC.

A very important link exists between SST and the heat content of the surface mixed layer with the phytoplankton distribution. The variability of the mixed layer temperature is governed by the heat equation that includes horizontal advection of the mean and eddy components and the entrainment at the base of the mixed layer. The heat equation is forced by the heat budget at the air–sea interface minus the short-wave radiation (SWR) not absorbed within the mixed layer. The portion of SWR absorbed within the mixed layer depends on the value of a diffuse attenuation coefficient for solar light. Several formulations are available to estimate the latter one, such as the one proposed by Foltz et al. (2003). The diffuse attenuation coefficient depends on the environmental conditions, being a measure of the water turbidity caused by the presence of suspended sediments and/or biological components. For the open sea, phytoplankton is the main factor determining this coefficient. Morel et al. (2007) proposed an empirical equation to determine its value from chlorophyll concentrations, obtaining a relationship that directly links the chlorophyill concentration to the percentage of absorbed solar radiation in the mixed layer. As an example of such a calculation, consider a typical April condition in the northwest Mediterranean when the mean mixed-layer depth is typically between 15 and 30 m, with a mean incoming SWR of 200 W m^{-2} and chlorophyll concentrations ranging from 1 to 0.1 mg m^{-3} inside and outside the most productive area. The SWR contribution to the mixed-layer heating would be 180 and 160 W m^{-2} for the two regions, respectively (D'Ortenzio et al., 2005). This example emphasizes the importance of evaluating the effect of biota concentrations in determining the mixed-layer thermodynamics.

3.4.2 Unresolved issues

1. Understanding the relation between multi-decadal SST oscillations and Mediterranean thermohaline cell variability

   The SST being the more extensively measured surface parameter of the ocean, many works have investigated its decadal or multi-decadal variability based solely on this parameter, supposing that some complex relation exists between the SST variability and the changes in

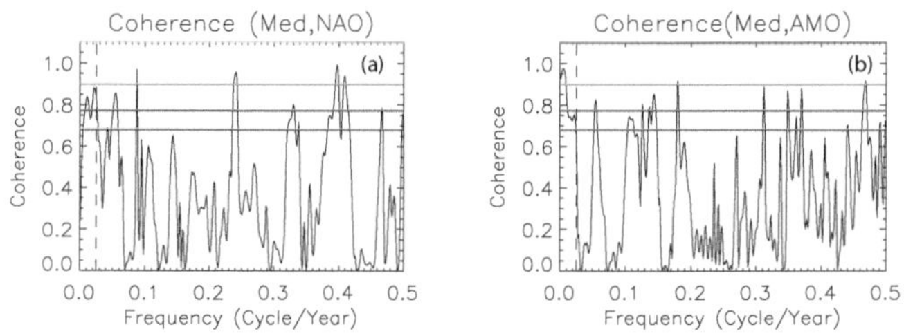

Fig. 18. Magnitude squared coherence estimation between **(a)** Mediterranean SST and NAO, and **(b)** Mediterranean SST and AMO. Green, blue, and yellow lines represent the 90 %, 95 %, and 99 % confidence limits, respectively (modified from Marullo et al., 2011, American Meteorological Society; used with permission).

the thermohaline cell or the heat content of the mixed layer over the same timescales.

The physical relation between SST and ocean interior dynamics exists, but unfortunately that relation is not as simple as replacing one with the other in the time series analysis obtaining the same trends or oscillations. We believe that future investigations should be devoted to understanding the extent to which an advanced use of combined surface data with numerical simulations will be able to describe the 3-D upper ocean with sufficient space and time resolution. A better understanding of the relation between the surface field and the 3-D structure of the ocean in the Mediterranean Sea will contribute to investigating to which extent feedback between changes in the Mediterranean thermohaline and multi-decadal variability of the SST exist.

2. Relation between multi-decadal Atlantic indices variability and Mediterranean SST oscillations

   Since Atlantic indices have been found to vary coherently over multi-decadal timescales with oscillations of the Mediterranean SST field, it would be very stimulating to investigate whether this statistical evidence can be related to some physical, likely nonlinear, process that drives the interaction between the two dynamical systems. Recently, Francombe et al. (2010) proposed that the multi-decadal variability of the North Atlantic is dominated by two main time scales (20–30 and 50–70 years). The first oscillation has an ocean-internal origin, while the second should be more related to atmospheric forcing. Considering the absence of lag between the AMO and the 70-year Mediterranean SST oscillation, the hypothesis of an atmospheric origin also for the Mediterranean 70-year signal can be considered reasonable. Actually, available information does not allow a definitive answer to the question of whether the forcing of the observed multi-decadal signal has an atmospheric origin or whether there is some contribution due to the Mediterranean THC or to the interaction with the Atlantic Ocean.

3. Bio-feedbacks in the Mediterranean Sea

   Bio-optical feedbacks among phytoplankton and upper ocean dynamics represent a promising field of investigation also for the Mediterranean Sea. The presence of phytoplankton in the ocean modifies the penetration of light and impacts the physical properties of the upper water column, modulating the spatial and temporal variability of the heat content in the mixed layer. In this context the use of coupled physical–biogeochemistry models gives us the opportunity to study the loop of effects between biology and physics. The direct inclusion of the optical properties of the upper ocean, modulated by the phytoplankton concentration, can contribute to understanding the way these bio-feedbacks work in the oligotrophic Mediterranean conditions and variation induced by spring blooms and transition between coastal case II and open ocean case I waters.

3.5 The carbonate system in the Mediterranean

3.5.1 Present knowledge

The characteristics of the Mediterranean Sea are such that it has the potential to sequester large amounts of anthropogenic CO_2 and C_{ant} (e.g., Álvarez et al., 2013). The buffer capacity of the Mediterranean Sea is particularly high due to the high alkalinity and temperatures throughout the water column. Furthermore, the active deep overturning circulation is effective in transporting the atmospheric imprint on the carbon cycle to the interior of the Mediterranean Sea. In fact, the column inventories of C_{ant} are higher in the Mediterranean than anywhere else in the world ocean (Schneider et al., 2010), and the C_{ant} storage in the Mediterranean is a significant portion of the global anthropogenic emissions of CO_2. However, the carbon observations in the Mediterranean are so scarce that it is difficult to quantify the sink of anthropogenic carbon in the Mediterranean accurately, and to quantify changes in the carbon cycle (Fig. 19).

This is manifested among other features as a shift in deep water formation from the Adriatic to the Aegean seas, and back again, the Adriatic being more important for deep water formation. The deep water formed from these two sources has different properties of salinity and the last few decades have seen dramatic changes in the circulation of the Mediterranean Sea's temperature and different biogeochemical signatures. Very little is known about how the recent changes in the Mediterranean overturning circulation have affected the storage rate of carbon and particularly C_{ant}.

Lateral exchanges of carbon to surrounding seas are significant (e.g., Ait-Ameur and Goyet, 2006; Huertas et al., 2009; Schneider et al., 2010; Flecha et al., 2012), although they are subject to large temporal variability and thus not

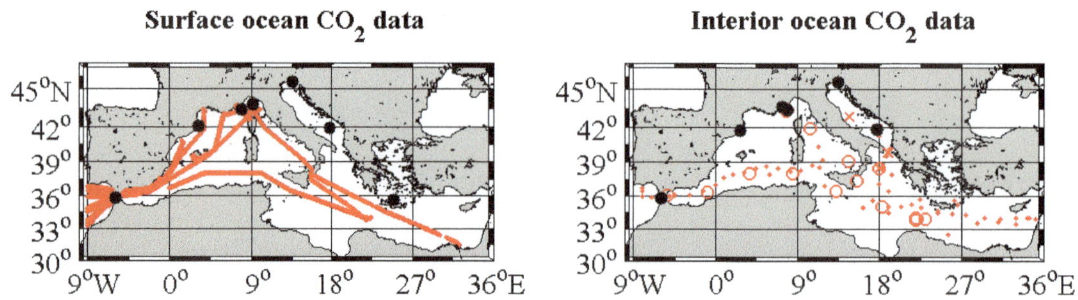

Fig. 19. Mediterranean Sea CO_2 data available in public databases. Left panel: pCO_2 data (surface ocean) from the SOCATv2 database (red dots), and time-series stations with pCO_2 data (black dots). Right panel: DIC data from the interior ocean; red dots Meteor cruises M51 and M84/3 in 2001 and 2011, red crosses MATER cruises from 1997 and 1998, red circles the PROSOPE cruise, black dots the time-series stations. The data from all time-series stations are not publicly available at this stage.

well constrained, similar to different methods for calculating C_{ant} in the Mediterranean Overflow Water (MOW), which lead to different magnitudes and even direction of the net flux through the Strait of Gibraltar. In particular, it seems that the TrOCA method overestimates the C_{ant} concentration in the MOW, leading to too large an export of C_{ant} from the Mediterranean to the Atlantic.

The increasing inorganic carbon content of the Mediterranean Sea leads to changes in the carbonate system, such that the concentration of the carbonate ion decreases and the pH decreases. In a recent study the sensitivity of the Mediterranean seawater to increasing pCO_2 concentrations is discussed in detail (Álvarez et al., 2013). Even though the pH of the Mediterranean is high in comparison to the world ocean, it is possible that these changes in the carbonate system can impact the ability of certain well-adapted groups of marine organisms (e.g., coccolithophores) to thrive. Decreasing pH leads to changes in chemical equilibrium in, for instance, trace metals and nutrients. The effects of these changes on the Mediterranean ecosystem are as yet poorly understood and an active field of research. Changes in the community structure as a direct effect of the changing carbonate system are conceivable.

3.5.2 Unresolved issues

1. Characterization of carbon distributions

   It is important to characterize the distributions and controls of natural and anthropogenic carbon (both organic and inorganic) distributions in the interior of the Mediterranean Sea better in order to understand any trends and variability. This includes (a) characterization of key areas for CO_2 penetration in the Mediterranean Sea, (b) evaluation of the role of the intermediate and deep water formation areas, and (c) evaluation of the impact of shelf events (flood, storms, shelf-water formation).

2. Temporal and spatial variability of the interior carbonate system

   The carbonate system is evidently not in a steady state due to increased influx across the air–sea interface, for instance. This motivates monitoring of the temporal variability of the interior carbonate system and determination of how this relates to increasing atmospheric CO_2 concentrations and changing climate, i.e., to the relation between anthropogenic and natural forcing. Observations of the interior ocean carbonate system are necessary on a regular basis in selected hydrographic sections and hotspots. This concept has been presented in some detail in the latest CIESM monograph (CIESM, 2012), which proposes a program called MED-SHIP (MEDiterranean Ship-based Hydrographic Investigation Program) for repeat hydrography in the Mediterranean Sea. Currently, ship-based observations are the only means of obtaining reliable carbonate data of the interior ocean of sufficiently high accuracy for determining temporal variability. This is particularly true for the regions of the Mediterranean Sea deeper than the Gulf of Lion, as 2000 m is the maximum depth of the present-day Argo floats. The MED-SHIP concept relevant for interior carbon data is based on zonal and meridional sections in the Mediterranean Sea repeated at regular intervals. This could be the cornerstone in an observational program for interior ocean carbon in the Mediterranean Sea. Furthermore, information from sensors mounted on gliders, floats and moorings is invaluable for the understanding of the temporal and spatial variability. Development of other sensors is rapid and any observational program needs to take advantage of additional sensors for a better understanding of the total carbonate system.

3. Air–sea fluxes of CO_2

   The variability of air–sea flux of CO_2 in the Mediterranean Sea is poorly constrained and needs to be understood better on various temporal (seasonal to annual to decadal) and spatial (local to sub-basin to basin) scales. Similarly, variability of surface ocean carbon will need to be connected to variability of the interior ocean carbon. Regular observations of surface pCO_2 (partial pressure of CO_2) on commercial vessels (cargo ships and ferries) that regularly cross the oceans is a well proven and useful concept. However, only a limited number of surface pCO_2 observations are available for the Mediterranean Sea, although time series like the DYFAMED site south of France provide important long-term information. Installation and operation of pCO_2 systems on a number of commercial vessels would be very useful for understanding the air–sea flux of CO_2 and its spatial and temporal variability. Ideally, these measurements should be coupled with measurements of additional biogeochemical variables concurrently, to help establish the forcings in the CO_2 variability. It is also important that an additional variable of the carbonate system, such as alkalinity, DIC or pH, is monitored simultaneously in order to understand the whole carbonate system.

4. Lateral carbon fluxes

   The poorly constrained lateral flux of carbon (natural and anthropogenic) to and from the Mediterranean Sea through the Strait of Gibraltar and the Bosphorus Strait system is highly variable and an integral component of the Mediterranean carbonate system. It is important to monitor the variability of the CO_2 system and ancillary data by means of repeat fixed stations or moorings (using biogeochemical sensors) in hotspot regions such as water mass formation areas (such as the Gulf of Lion and the northern Adriatic Sea), the Strait of Gibraltar and the Bosphorus Strait.

5. Monitoring the change and impact of decreasing pH

   The effect of increasing DIC concentration, leading to decreasing carbonate ion concentration and decreasing pH in Mediterranean ecosystems, is largely unresolved and needs to be monitored, for instance with current pH sensors for surface pH changes, in addition to monitoring of changes in the carbonate system itself in the Mediterranean Sea. The understanding of processes responsible for observed trends and variability is important and is probably best undertaken with the help of modeling schemes, data assimilations and/or inversion methods. Observations and modeling efforts need to go hand in hand and will benefit from close cooperation between the modeling and observational communities.

6. Anthropogenic CO_2 inference methods tailored to the Mediterranean Sea

   Calculations of the anthropogenic CO_2 penetration in the Mediterranean Sea using the existing, but limited, historical data record and the existing calculation techniques need to be performed. This should be coupled to the investigation of a new scheme for the calculation of anthropogenic CO_2, appropriate for the Mediterranean Sea.

4 Shelf–deep sea interactions and exchanges of physical/biogeochemical properties and how they affect the sub-basin circulation and property distribution

4.1 Filament and eddy formation and its effects on biogeochemical processes

4.1.1 Present knowledge

Frontal zones, filaments and eddies (FZFE hereafter, ranging from mesoscale (10–100 km) to submesoscale (1–10 km) dimensions) are dynamical features with sharp gradients having a large effect on circulation and on the distribution of heat, salt and matter in the ocean (Robinson, 1983; McGillicuddy et al., 1998). It has been shown that they are important routes of the energy cascade and dissipation in the ocean, and the transport of mass, energy, chemical compounds and biota among water masses by turbulent advection and mixing. Their characteristics and distributions play a major role on the basin scale and in the local energetics and circulation. They contribute to shaping the spatio–temporal distribution of biogeochemical variables by creating physical boundaries (separating distinct areas of completely different properties from the surroundings in the open sea) and by modulating the seasonal evolution (inducing, for example, sporadic events). In the coastal areas, they can transport land-based coastal and continental shelf natural and anthropogenic components to the open sea and vice versa. Often these dynamic features, created or destroyed by atmospheric interactions as well as internal instability mechanisms, interact with each other, going down even to the small scales of sub-mesoscale filaments and streamers only captured in high-resolution simulations and by satellite observations. Through energy cascading mechanisms of geophysical turbulence some features disintegrate or dissipate, while other small-scale features can coalesce to become coherent structures that persist for extended periods.

The challenges of characterizing the FZFE processes imply precise and high-resolution observations in addition to a multi-sensor approach on several spatial/temporal scales. Thermal NOAA/AVHRR satellite images were used to identify sites of highest frequency in cold filaments in the Mediterranean Sea (Bignami et al., 2008) and transport patterns were studied using satelite altimetry and Lagrangian numerical analysis in the open sea (d'Ovidio et al., 2004). This methodology was adapted to the coastal area using in situ observations to correct for the uncertainties connected to satelite altimetry data close to shore (Nencioli et al., 2011). Ruiz et al. (2009a, b) reported the first attempt to combine high-resolution (~ 0.5 km) hydrographic observations using glider technology and altimetry measurements to quantify vertical exchanges in an area with intense horizontal density gradients. The study was designed to be coincident with an OSTM/Jason-2 passage in the eastern Alboran Sea. Using the quasi geostrophic dynamics (Hoskins et al., 1978), they reported vertical velocities of about $1\,\mathrm{m\,day}^{-1}$, partially explaining an observed subduction of chlorophyll. Niewiadomska et al. (2009) used gliders equipped with optical biogeochemical sensors to detect upwards and downward motions not easily detected by direct measurements. More recently, Navarro et al. (2011) analyzed the coupled patterns of variability between satellite altimetry and chlorophyll data in the Alboran Sea. They demonstrated that the pelagic ecosystem in the Alboran Sea is controlled by the inverse barometer effect (first mode). They also showed that the distribution of chlorophyll-rich and poor areas can be explained by the second mode.

In the Balearic Basin, mesoscale eddies have been found to modify not only the local dynamics but also the large-scale patterns, as shown by Pascual et al. (2002) in a detailed study of the blocking effect of a large anti-cyclonic eddy, as well as a clear influence of the basin circulation on the phytoplankton biomass (Jordi et al., 2009). The (sub)meso-scale activity of boundary currents around the shelf break leads to strong patchiness of the nutrient and primary production distributions in the Gulf of Lion (Diaz et al., 2000). Bouffard et al. (2010) developed innovative strategies to characterize horizontal ocean flows, specifically in terms of current velocity associated with filaments, eddies or shelf–slope flow modifications close to the coast. These methodologies were applied to a series of glider missions carried out almost simultaneously and well co-localized along the satellite tracks, as part of a pilot initiative lead by IMEDEA (CSIC-UIB). In this context, Pascual et al. (2010) showed that the high-resolution hydrographic fields from the gliders combined with coastal altimetry revealed the presence of permanent and non-permanent signals, such as relatively intense eddies. Moreover, the almost synoptic view from altimetry and SST images during the glider missions provided a more detailed picture of regional small-scale features. Since January 2011, a new sustained observational program in the Balearic channels has been conducted by IMEDEA (CSIC-UIB) and SOCIB, the new Balearic Islands Coastal Observing and Forecasting System (Tintoré et al., 2012). This monitoring program consists of repeated transects between Mallorca, Ibiza and Denia. During 2011, 7 glider missions were successfully carried out in the Ibiza Channel, reporting an unprecedented spatial and temporal variability in transports (Heslop et al., 2011) compared to the literature values (Pinot et al., 2002). These new findings will have relevant consequences for improving our understanding of local ecosystem changes, as it is known that the variability of Atlantic Water through the Balearic channels is critical for the understanding of bluefin tuna spawning south of the Balearic Islands (Alemany et al., 2010). Another important observational/modeling program is MOOSE, the Mediterranean Ocean Observing System for the Environment, developed during recent years and now fully operational in the Liguro–Provencal basin (mio.pytheas.univ-amu.fr/moose/). MOOSE addresses not only continental inputs, mesoscale processes, biogeochemical cycles, biodiversity, and air–sea interactions in the northwestern gyre, but also has a modeling/data assimilation component.

Multi-scale interactions are especially relevant for the eastern Mediterranean, where coherent eddies either reinforce or block seasonal circulations, divert or entrain water masses of contrasting properties, and indirectly contribute to inter-basin transports (Feliks and Itzikowitz, 1987; Malanotte-Rizzoli et al., 1999; Zodiatis et al., 2005). In the Levantine Basin fast current systems known as the Mid Mediterranean Jet and the Asia Minor Current (Robinson et al., 1991, 2001) bear the energy that can sustain unstable FZFEs or evolve into long-lived features such as the Iera-Petra, Antalya and Shikmona eddies. Interaction between various eddies along the Asia Minor Current has often been observed in great detail (Onken and Yüce, 2000; Hamad et al., 2005, 2006). Modified Atlantic Water propagating from the west and the Levantine Intermediate Water created through convection events are entrained and transported by these eddies and jets and exchanged by FZFE interactions (Özsoy et al., 1989, 1991, 1993; Sur et al., 1992; Lascaratos and Nittis, 1998). Analysis of drifter data and satellite altimetry (Gerin et al., 2009; Menna et al., 2012) has expanded the knowledge of the mesoscale features. They showed the branching of the surface water entering through the Sicily Channel into the Ionian and the Levantine, the eastward flow along the Libyan and Egyptian slopes, and a cyclonic circuit along the Middle East and Turkish slopes. This flow is perturbed locally and temporally by instabilities of the current, wind and topography. The along-slope and offshore currents generate recurrently (western Egyptian coast, Ierapetra, Mersa–Matruh, southwest of Cyprus, Israel–Lebanon coast, Latakia) or persistent (Rhodes Gyre) eddies. Highly variable and energetic currents are observed between the Ierapetra and Mersa–Matruh regions, as a result of the interaction of the Mid Mediterranean Jet meandering in between and interacting with the eddies generated by the instability of the coastal current.

These interactions also affect the horizontal and vertical distributions of chemical parameters and plankton composition, as shown by conventional in situ sampling and analysis (for example, Özsoy et al., 1993; Ediger et al., 1999, 2005; Salihoğlu et al., 1990; Bingel et al., 1993; Yılmaz

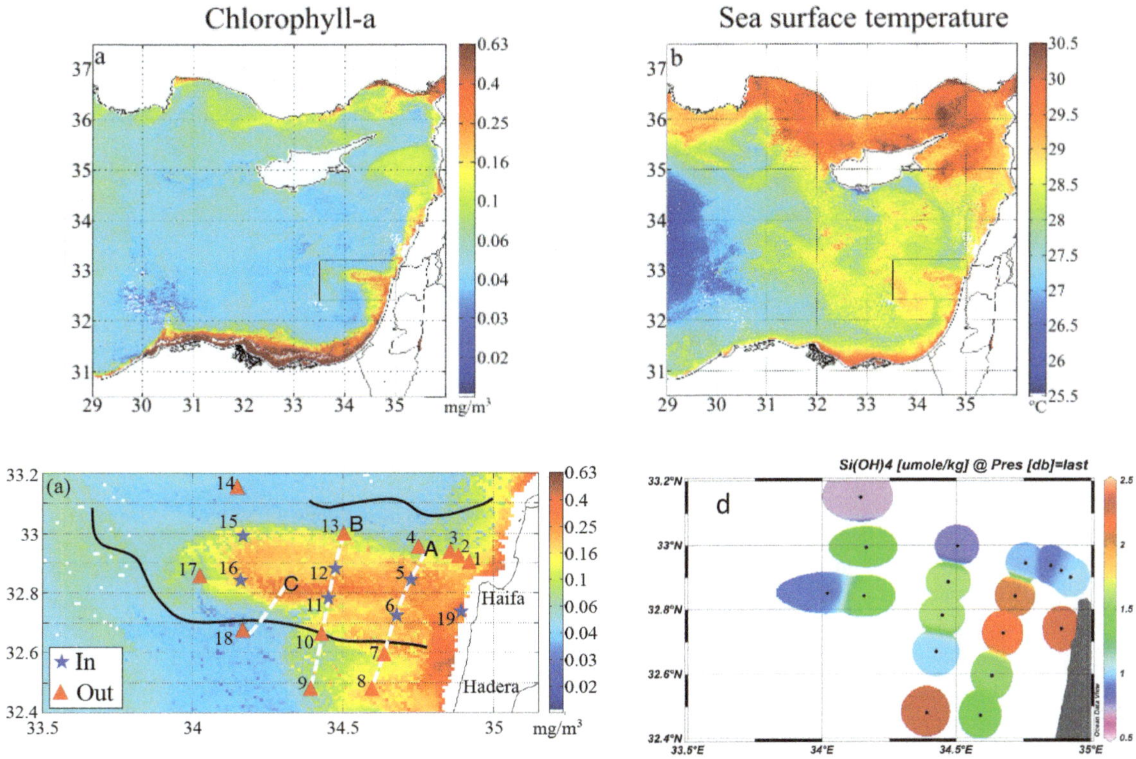

Fig. 20. Satellite images of chlorophyll *a* (upper panel) and sea surface temperature (middle panel) showing a filament protruding from the Israeli coast, easternmost Levantine Basin, in August 2011. The concentration of silicic acid measured at the surface (lower panel) was higher at the stations inside the filament than outside it. Modified from Efrati et al. (2013).

and Tuğrul, 1998; Krom et al., 1993). The surface and deep chlorophyll maxima formed near the base of the euphotic zone depend on light and biochemical variability imposed by FZFEs (Yılmaz et al., 1994; Yacobi et al., 1995; Ediger et al., 1996, 2005). In some Mediterranean areas FZFEs affect the distribution of plankton biomass and community composition either directly, due to the physical characteristics (e.g., salinity and temperature boundaries, convergence or divergence processes), or indirectly, through the influence on the chemical parameters (Fiala et al., 1994; Van Wambeke et al., 2004; Riandey et al., 2005; Siokou-Frangou et al., 2009, among others). Recently, relatively high concentrations of biogenic silica were found in some areas, perhaps related to the presence of eddies (Crombet et al., 2011). High silica in a filament stretching from the coast to the open sea in the Levantine Basin were reported as well (Efrati et al., 2013) (Fig. 20). A combination of near-real-time multi-satellite observations, a circulation model and in situ high-resolution sampling within the COSEM project (http://isramar.ocean.org.il/isramar2009/cosem/) showed that the filament was a semi-enclosed system, bounded within the mixed layer and separated from ambient waters by transport barriers induced by horizontal stirring. The physical–biogeochemical properties significantly differed from those of the waters surrounding it, with a volume flux on the order of 0.1 Sv.

4.1.2 Unresolved issues

1. Importance of frontal zones, filaments and eddies (FZFE) for the Mediterranean Sea circulation

   Although the existence of FZFE in the Mediterranean Sea has been documented, it is still unclear how important they are to the local and overall circulations. The unresolved issues that need to be addressed are (a) FZFE formation and decay and what the relative contributions to these processes of atmospheric forcing, topography and interaction between different features are; (b) investigation of the formation/decay timescales; (c) examination of the differences in these processes between the coastal and open sea regions; (d) FZFE role in the basin-wide and local energy budgets and how the multi-scale interactions transfer and partition energy; (e) FZFE areas are usually identified from their surface signature, which does not provide information about how deep they are. The penetration depth may be seasonal and area dependent and a function of the strength and character of the physical forcing causing the FZFEs.

2. Importance of FZFEs for the chemical and biological distribution and the biogeochemistry

   It has been shown that mesoscale and FZFEs affect the chemical and biological distribution in the water body as well as biogeochemical processes. However, it is still unclear if FZFEs are important features in the Mediterranean as a whole or are only of localized importance. The relative contributions of the FZFEs and the large-scale forcing mechanisms to the spatio–temporal distribution of chemical/biological parameters are not known. The unresolved issues that need to be addressed are (a) how the FZFEs affect the chemical and biological parameter distribution in the entire Mediterranean Sea; (b) determination of properties other than chlorophyll *a* that can be used to trace FZFEs remotely in the surface layer; (c) the extent to which chemical compounds can be used as passive tracers to track processes associated with FZFE; (d) the FZFEs may cause patchiness in the biomass distribution and, if so, areas need to be identified that are more likely to have this patchiness; (e) the FTFEs may change the estimated primary production and bridge the apparently high yield of fisheries compared to the low primary production (Estrada, 1996); (f) the FTFE contribution may affect the calculation of nutrient and carbon budgets in the Mediterranean; (g) thin layers of plankton in the Mediterranean may be found similar to those found elsewhere in frontal areas (Sullivan et al., 2010).

3. Integration of measurements from different platforms

   The study of these highly dynamic and space-/time-dependent FTFEs needs to integrate a multi-platform, multi-sensor approach, including a range of novel instrumentation for in situ and remote sensing observations, together with modeling efforts. As an example, gliders allow high-resolution oceanographic monitoring by autonomously collecting high-quality observations in three dimensions, and provide useful contributions to the understanding of mesoscale dynamics and multidisciplinary interactions (e.g., Hodges and Fratantoni, 2009). However, isolated measurements from fleets of gliders are not sufficient, as glider measurements remain scarce, both in space and time, for many small-scale processes. The combination of glider data with information from other platforms (satellite, ships cruises, high-frequency radar and buoys) will support the investigation of unresolved issues stated above. In addition, observational data should be integrated with satellite images and numerical analysis. Investigation of the transport patterns needs to be expanded using satelite altimetry and Lagrangian numerical analysis. Studies were performed in the open sea, also in the Mediterranean (d'Ovidio et al., 2004). Recently, this methodology was adapted to the coastal area using in situ observations to correct for the uncertainties connected to satellite altimetry data close to shore (Nencioli et al., 2011). A semi-operational tool is being operated along the southeastern coast of the Levantine Basin (Fig. 21). These patterns are important for predicting the distribution and transfer of characteristics and matter between areas.

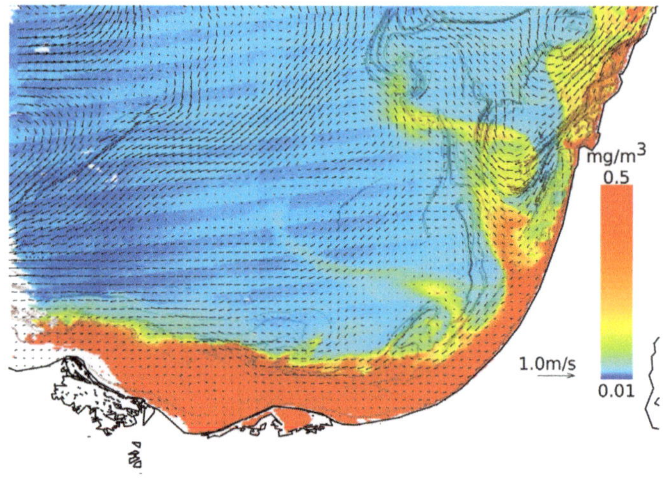

Fig. 21. Satellite image of chlorophyll (29 April 2013) showing filaments protruding from the coast at the southeastern Levantine Basin. Superimposed on the image are the currents and the Lagrangian analysis of passive tracer circulation. The satellite image was obtained from the MyOcean project and the currents and Lagrangian analysis from Israel's Marine Data Center (http://isramar.ocean.org.il/isramar2009/cosem/fsle.aspx).

4.2 On the role of surface forcing in the shelf–slope circulation

4.2.1 Present knowledge

The Mediterranean Sea is a region in which air–sea interaction plays a crucial role in shaping both the atmospheric and oceanic circulation. For the sea, the main components of the surface forcing are the wind stress, the surface heat flux, the direct surface fresh water flux (evaporation minus precipitation), and the terrestrial-based water flux in the form of runoff and river discharges. While early modeling studies of the Mediterranean used relatively coarse horizontal grid resolutions of ~ 25 km (e.g., Roussenov et al., 1995; Zavatarelli and Mellor, 1995), more recent studies have used higher resolution grids that are nearly eddy resolving (e.g., ~ 10 km in Beuvier et al., 2010) and even eddy resolving (~ 7 km in Tonani et al., 2008). Other studies of sub-basins (e.g., Lascaratos and Nittis, 1998) have also used eddy-resolving grids (~ 5 km) for process studies such as LIW formation. These studies show that a model with a reasonable spatial resolution is able to reproduce circulation features of the basin scale, sub-basin scale and mesoscale, in agreement with the small

Rossby radius of deformation for this sea. However, the common feature of these and other studies is that the surface forcing fields have been extracted from large-scale climatological or reanalysis data sets in which the horizontal resolution is at best on the order of 50–100 km. In most of these studies, the temporal resolution of the forcing was also relatively coarse (daily or longer means), with the exception of Tonani et al. (2008), who used 6-hourly data. While these studies have led to significant progress in understanding the circulation of the Mediterranean, it is clear that in most cases the models and the associated forcing are not able to resolve the fine details of the shelf and slope regions adequately, which are relatively narrow (typical width of a few tens of kilometers or less). One promising direction to improve this is the use of high-resolution shelf models (1–2 km resolution) for local circulation and process studies. Such models have been implemented within the framework of the operational ocean forecasting systems developed over the past 5–8 years for the Mediterranean (MFS, MOON, MyOcean, and MONGOOS; www.moon-oceanforecasting.eu) for daily forecasting. Some of these models have already been used to investigate the local circulation (e.g., Jordi et al., 2006).

Chelton et al. (2004) have discussed the importance of small-scale features in ocean winds, as well as the influences of the coastal and island orography on the wind stress curl and divergence as observed from the 25-km horizontal resolution of the wind stress from the Quickscatt radiometer. Such features, even on smaller scales (of about 1 km), are of great importance for the Mediterranean Sea and its adjacent basins, where orography and coastal geometry influence the wind pattern. For example, the effects of these small-scale variations have long been recognized in controlling the wind-driven circulation in the Adriatic Sea. Recent studies of the role of Bora wind events forcing the circulation in the northern Adriatic have demonstrated the importance of sufficiently high resolution in the observations (Lee et al., 2005) and models (Cushman-Roisin and Korotenko, 2007) for understanding the sea's response to the fine-scale atmospheric forcing. These studies also demonstrate the complex interaction that can occur in the shelf zone due to the combined effects of wind forcing and river discharge. Development of the operational ocean forecasting systems for the Mediterranean has led to the implementation of high spatial resolution (~ 10 km) and high-frequency (3 h or less) surface forcing to drive the short-term forecasts. This could provide the basis for producing a high-resolution surface forcing reanalysis for further detailed investigations of the shelf and slope circulation and process studies.

Other processes previously discussed in terms of the interaction of the shelf–slope circulation with the open sea are also highly dependent upon the surface forcing. The formation of dense water by cooling and/or evaporation on the broad, shallow northern shelf (generally shallower than 50 m) occurs when cold, dry air flows down from the Dinaric Alps during the winter. The resulting intense cooling (as much as $1000\,\mathrm{W\,m^{-2}}$) produces dense water (11 °C, 38.5 psu), which spills over the shelf edge and then flows southward along the Italian coast as a gravity current. In the South Adriatic Pit it mixes with the Levantine Intermediate Water (LIW) and eventually exits into the Ionian Sea through the Otranto Strait, both as a shelf vein or a slope vein mixed within the Adriatic Deep Water (Zoccolotti and Salusti, 1987; Manca et al., 2002). Bignami et al. (1990) observed the cascading dense waters at the Bari canyon in October 1987 and estimated the transit time from its formation in the north to its arrival in the Bari canyon to be 4–5 months. Such dense-water formation events are episodic and their small spatial scales make them difficult to observe. Based on data from 1981, 1987 and 1999, Vilibić and Supić (2005) found that the reduction in river discharges in the months preceding dense-water generation is a precondition for NAdDW formation. The reduced river discharge may also permit LIW to penetrate the northern Adriatic and bring highly saline water masses that when cooled sufficiently can form very dense water that is capable of cascading into the interior. Sustained cooling by severe Bora outbreaks from November to January are necessary to cool the water column and form the dense NAdDW. In addition to the convective mixing driven by air–sea heat loss and evaporation, the tidal mixing at the bottom as well as mixing from wind-driven waves may also play a role in triggering the cascading process.

In other basins of the Mediterranean a strong rim current usually flows along the bathymetry near the shelf break or over the slope. In the western Mediterranean this is manifested by the Algerian Current, which meanders and often generates coastal eddies due to baroclinic instability (Millot, 1985), while the Balearic Basin in the northwestern Mediterranean is characterized by frontal dynamics and shelf–slope exchanges in terms of mesoscale eddies and filaments (e.g., Bouffard et al., 2010). In the eastern Mediterranean the Libyo–Egyptian Current may generate mesoscale eddies that modify the pathway of Atlantic Water, dispatching it offshore (Gerin et al., 2009), or it may meander and form long-lived sub-basin scale features (Menna et al., 2012). Along the eastern boundary of the Levantine Basin, the northward flowing shelf-break current (Rosentraub and Brenner, 2007) also meanders, generating coastal eddies that drift out to the open sea, and occasionally it separates from the shelf break (Brenner, 2003). All of these current systems are an integral part of the general circulation of the Mediterranean Sea and play an important role in shelf–open sea interaction. The contribution of the high-resolution and high-frequency surface forcing to these processes is not well understood and warrants further investigation.

4.2.2 Unresolved issues

1. Remote-sensing and shelf–slope in situ measurements of the surface forcing

While it is clear that high spatial and temporal resolution surface forcing is important for properly simulating the ocean response in terms of mesoscale dynamics in the various basins and sub-basins of the Mediterranean Sea, it is also crucial for understanding the fundamental processes of the shelf and slope zone circulation. Remotely sensed winds on a continuous basis (with no interruptions due to the termination of the satellite mission) with at least a daily time step would be a valuable source of information on the synoptic wind patterns over the Mediterranean. This kind of data should be complemented with data from a coastal network that should fill the gap near land (15–30 km, Bourassa et al., 2010) where the satellite remotely sensed winds are not available. Moreover, there is a need for atmospheric forcing with adequate temporal resolution from in situ observations at platforms such as buoys, moorings, or drifters, located in the offshore regions, especially where wind-driven eddy/gyre systems are observed. Ideally, the remotely sensed wind data should be integrated with the data from autonomous self-recording and transmitting platforms with meteorological sensors moored in the regions of the shelf–slope interaction. Such instruments can provide the local meteorological conditions that cannot be remotely sensed with a sufficient temporal detail and duration. Close cooperation should be established with the operational ocean forecasting community.

2. Dedicated observations for specific process studies

   A similar approach should also be applied to regions where other important oceanic phenomena occur as a response to the local forcing, or dense-water formation (the Gulf of Lion, the southern Adriatic, and the southern Aegean). For example, Vilibić and Supić (2005) suggest that the importance of the preconditioning phase in NAdDW formation needs to be studied by examining the specific years in which NAdDW formation has been observed and documented. They also suggest investigating the precise location(s) of NAdDW formation and spreading in the northern Adriatic using more realistic, high-resolution surface forcing based on observational data and model simulations. The same conclusion was reached by Signell et al. (2005) when assessing the relevance of high-resolution meteorological models in the Adriatic.

3. High-resolution surface forcing reanalysis

   The use of high-resolution regional atmospheric models for downscaling of the surface forcing is a promising approach (e.g., Cushman-Roisin and Korotenko, 2007; Beuvier et al., 2010). High-resolution atmospheric fields (~ 10 km) are routinely provided to the operational ocean forecasting models of MONGOOS to drive the daily, short-term forecasts. The logical next step would be to develop a high-resolution, multi-decadal, regional surface forcing reanalysis data set. In this respect, close cooperation with the operational ocean forecasting community is paramount, since the atmospheric analyses and forecasts are produced daily. However, in order to make long-term surface forcing data sets useful, consistency of the data must be ensured using a reanalysis approach.

4.3 Interactions between the shelf–slope circulation and open sea in the Mediterranean

4.3.1 Present knowledge

Most of the Mediterranean Sea is characterized by relatively narrow shelf–slope zones so that throughout much of the basin the open sea is in close proximity to the coastal region. Notable exceptions are the northern Adriatic and Aegean seas, where the shelf is relatively wide but laterally confined. In many locations a prominent jet flows along the bathymetry near the shelf break or over the slope (Poulain et al., 2012). Often it meanders and generates eddies or filaments, thereby leading to a net cross shelf flow. Another process that leads to cross shelf flow is dense shelf water cascading (DSWC), in which dense water formed by cooling and evaporation over the relatively shallow continental shelf spills over the shelf edge and sinks as a bottom-trapped gravity current. Such processes are important for the shelf–open sea exchange of dissolved and suspended material and therefore may affect the ecosystem.

The circulation in the Alboran Basin is characterized by the intense inflow/outflow regime due to the exchange of water between the Atlantic and the Mediterranean through the Strait of Gibraltar. Atlantic Water entering the Mediterranean in the upper layer forms the intense Atlantic Jet. This jet meanders and forms the quasi-permanent West Alboran Gyre (e.g., Baldacci et al., 2001) and an intermittent Eastern Alboran Gyre (EAG). The eastern boundary of the EAG is defined by the Almería–Oran front (e.g., Ruiz et al., 2009a) and marks the start of the Algerian Current (AC) (Millot, 1985). As the AC progresses eastward, it forms baroclinically unstable meanders that can evolve into coastal eddies. Only anticyclonic eddies are long lived (Puillat et al., 2002; Isern-Fontanet et al., 2006). They usually propagate along the bathymetry, can detach from the current and become open-sea eddies (Millot and Taupier-Letage, 2005, and references therein). Some eddies are quasi-permanent and can divert the AW flow offshore as far as south of the Balearic Islands (Millot and Taupier-Letage, 2005).

The northern current in the Ligurian–Provençal Basin in the northwestern Mediterranean is the dominant feature. The flow is strongest during winter, with significant mesoscale variability, and weakens during summer (Millot and Taupier-Letage, 2005). In the Balearic Basin it splits into two branches, one recirculating into the Balearic Current (e.g., Ruiz et al., 2009a) and the other continuing south through the Ibiza Channel (Pinot et al., 2002). The Balearic Basin is also characterized by shelf–slope exchange due to mesoscale eddies (Pinot et al., 2002; Bouffard et al., 2010), filaments, and shelf–slope flow modifications (Wang et al., 1988; La Violette et al., 1990). The bathymetry plays a key role in controlling the transport between the northern and southern regions (Astraldi et al., 1999) and may enhance sub-mesoscale activity (Bouffard et al., 2012).

The Atlantic Ionian Stream (AIS) enters the eastern Mediterranean through the Strait of Sicily and usually follows the southern coast of Italy and the western coast of Greece (Robinson and Golnaraghi, 1994). Recent studies (e.g., Borzelli et al., 2009; Gačić et al., 2010) have shown that during certain periods the AIS can follow a more southerly pathway when the overall northwestern Ionian circulation switches from anticylonic to cyclonic. According to the POEM results (e.g., Robinson and Golnaraghi, 1994; Malanotte-Rizzoli et al., 1999) the AIS continues primarily as the eastward flowing Mid Mediterranean Jet that meanders through the center of the Levantine Basin. The POEM results also hinted that it may feed into an eastward flowing shelf-break jet that is part of the general, basinwide rim current. Studies based on SST and surface drifters suggest that this Libyo–Egyptian Current, LEC (Alhammoud et al., 2005; Hamad et al., 2005; Millot and Taupier-Letage, 2005), generates baroclinically unstable mesoscale eddies that transport AW offshore (Gerin et al., 2009). However, other recent studies based on subsurface drifters and satellite altimetry (Poulain et al., 2012; Menna et al., 2012) have shown that the LEC is persistent and stable and forms several stable, sub-basin scale eddies. This current continues eastward, turns northward following the coasts of Israel and Lebanon, and then turns westward following the coast of Turkey as the Asia Minor Current (Öszoy et al., 1993). Based on extensive in situ current measurements, Rosentraub and Brenner (2007) found that over the shelf and slope of Israel, the flow throughout most of the year is directed northward with strong currents that may reach $50\,\mathrm{cm\,s^{-1}}$ or more in both winter and summer. Further to the north, this current can form long-lived eddies through interaction with the MMJ (Menna et al., 2012), which may also lead to shelf open sea exchange. In the near-bottom layer at the shelf break, Rosentraub and Brenner (2007) found evidence of a net seaward, cross-shelf flow.

Understanding DSWC's interaction with the bottom morphology is also of paramount relevance. These energetic cascades, lasting up to a few weeks, are considered one of the main drivers at the oceanic margins. During DSWC events, cold, dense shelf waters spill over the shelf edge and flow along topographic features as a bottom-trapped gravity current. Upon reaching the level of neutral density they spread laterally. Of the 61 confirmed cases of dense-water cascades around the globe (e.g., Ivanov et al., 2004), three occurred in the Mediterranean. Over the broad northern Adriatic shelf the North Adriatic Dense Water (NAdDW) forms in the area exposed to the cold, dry Bora winds during the winter. One branch of NAdDW flows southward with a vein-like shape reaching the shelf break, while a second branch enters the Jabuka pit, from where the spillover can reach the Bari Canyon and the South Adriatic Pit. The cascading water is diluted by entrainment of ambient water masses. While DSWC in the northern Adriatic or at the northern side of the Cretan arc does not play a direct role in the exchange with the open sea, it is nevertheless important, since this process is crucial for the formation of the dense deep water that eventually fills the Eastern Mediterranean. Over the Gulf of Lions shelf DSWC does not occur every year, but when it does, it can reach the bottom (Dufau-Julliand et al., 2004; Palanques et al., 2009; Pascaual et al., 2010, and references therein) and can be an important source of deep water along with that formed by open sea convection.

4.3.2 Unresolved issues

1. Meandering coastal currents, fronts, and eddies

   The establishment of ocean observing networks, especially over the shelf, is being adopted as an important component of marine strategy by many countries with economically significant coastal areas. These new facilities are delivering new insights into coastal and open ocean variability. They also contribute to a more science-based and sustainable management of the coastal area. In the western Mediterranean, MOOSE (Mediterranean Ocean Observing Site for Environment) and SOCIB (Balearic Islands Coastal Observing and Forecasting System) are two examples of such networks. SOCIB addresses multidisciplinary research into mesoscale and frontal dynamics as a key element in the physical and ecosystem variability (Tintoré et al., 2012). It includes multi-platform observations and modeling services distributed through an integrated system. Such information, when used in synergy with satellite observations, will be particularly helpful for societal benefits, and for a better understanding of 3-D biogeochemical (Lévy et al., 2009) and energy transfers (Lapeyre and Klein, 2006) occurring at meso- and submeso-scales. Similar networks should be established at other select locations in the eastern Mediterranean in order to provide continuous, long-term measurements. Field campaigns to investigate specific aspects of the jet and frontal dynamics should be planned, and very high-resolution models for specific shelf regions should be developed or refined as necessary.

2. Dense shelf water cascading

   Dense-water formation is primarily a local phenomenon and therefore amenable to study by local observations and models. Present models do not have sufficient resolution to capture the small-scale processes responsible for dense-water formation and entrainment. Water mass preconditioning is crucial to dense-water formation and should be studied. In particular, turbulent mixing in the water column, driven by surface cooling and by tidal and storm-driven mixing near the bottom, is the key to homogenizing the shallow shelf waters so that they can be subsequently cooled. These must be further clarified through process-oriented studies combining ad hoc field campaigns to sample the state of the shelf water and the implementation of very high-resolution models that include more accurate parameterization of turbulent mixing.

Implementation of very high-resolution shelf models and data assimilation systems

Present high-resolution shelf models used for the operational ocean forecasting systems within the framework of MONGOOS typically have a horizontal resolution of ~ 1–2 km (e.g., Brenner et al., 2007; Zodiatis et al., 2008; Guarnieri et al., 2010), which is barely adequate for resolving the narrow shelf circulation regime. Similarly, the data typically collected as part of the operational systems provide reasonable coverage of the open sea but are more limited in the shelf zone. In order to investigate shelf–open sea interaction further, the models' horizontal and vertical resolutions must be refined further and the domains must extend far enough into the open sea. Non-hydrostatic models may also be necessary. An example of the potential advantage of higher resolution models is shown in Fig. 22, which compares the daily mean currents in the southeastern corner of the Mediterranean based on the MyOcean reanalysis (resolution of ~ 6.9 km), in the left panel, with the downscaled field produced by a model that has a resolution of ~ 1.7 km, in the right panel. The shelf break current in the high-resolution field is more sharply defined and stronger than in the coarser resolution field. Similarly, smaller-scale eddies appearing over the shelf in the downscaled field are absent in the coarse resolution field. Data routinely collected within the framework of the operational forecasting systems should be used for baseline studies and should be combined with data collected during dedicated field campaigns and the very high-resolution models through a data assimilation system. The experience and systems developed within the framework of operational forecasting in the Mediterranean will be invaluable here.

3. Additional advances in model development

In addition to refining the models' resolutions, several new parameterizations should be developed and implemented. Specifically, second moment closure (SMC) models of turbulent mixing (e.g., Kantha and Clayson, 2007) have enabled more accurate simulations of oceanic mixing processes, although there is still room for improvement. Recently, Kantha and Carniel (2009) refined turbulence models of stably stratified flows to allow mixing at all values of the gradient Richardson number. Non-local mixing models are the key to accurate depiction of mixing processes dominated by convection, prevalent during dense-water formation. Furthermore, parameterization of dense gravity currents for inclusion in larger-scale models is of interest, as are the effects of wind waves on mixing in shallow areas of dense-water formation, using a coupled hydrodynamic model–wave model. To the extent possible these model advances should draw on the experience of the operational forecasting community.

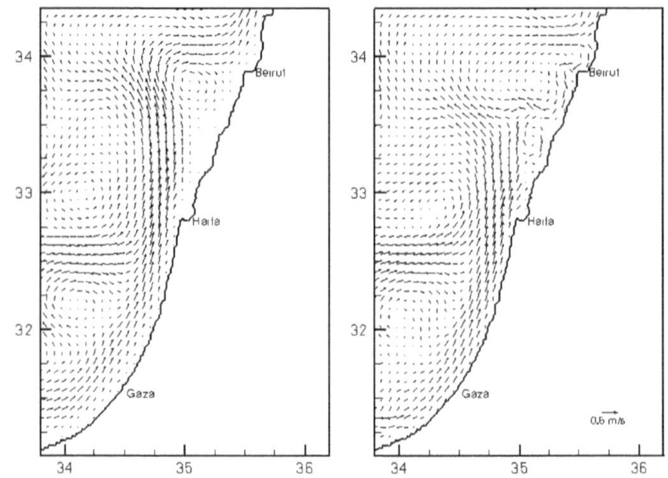

Fig. 22. Comparison of the daily mean near-surface (10 m) currents on 20 July 2008 from the MyOcean reanalysis, resolution ~ 6.9 km (left panel), with the downscaled field produced by a high-resolution, ~ 1.7 km, model (right panel).

5 Conclusions

The most important novel results that have emerged in the investigation of the Mediterranean Sea during recent decades can be summarized in a few focal points. First and foremost, the appearance and discovery in the early 90s of the Eastern Mediterranean Transient (EMT) has marked a remarkable transition in the understanding of the large-scale thermohaline circulation of the eastern basin. The EMT is manifested in the shift in deep water formation through deep winter convection processes from the southern Adriatic Sea to the Aegean and, later, back again to the Adriatic. This phenomenon led to a greater in-depth outlook for the western basin, where a similar phenomenon, evidenced in the early 2000s, was called the Western Mediterranean Transition (WMT). The WMT involved events of high production of anomalously warm and salty new deep water in the winters of 2004/2005 and 2005/2006. The intriguing possibility is that events of such a type had multiple occurrences in the past and might be the manifestation of multiple equilibria of the closed thermohaline circulations of the two basins. Such events may have very profound consequences for biogeochemical properties such as, as an important example, the carbonate system of the Mediterranean. The characteristics of the sea are such that it has the potential to sequester large amounts of anthropogenic CO_2 and C_{ant}. This high buffer capacity is due to the high alkalinity and temperatures throughout the water column.

Still, very little is known about how the recent changes in the thermohaline overturning circulations have affected the storage rate of C_{ant}.

A second major novel result has been the recent finding of multiannual oscillations again in the eastern basin, which manifest themselves as reversals in the Ionian upper-layer circulation. The first event occurred in the late 80s, with a reversal from a cyclonic to an anticyclonic circulation. In 1997 an opposite inversion occurred, with the return to a cyclonic pattern. These inversions of the Ionian circulation may have profound effects on the dynamics of the eastern basin ecosystems. In fact, the merging of the whole Mediterranean Sea into a single bio-province, as proposed in some studies, is rather questionable. The western basin, even though smaller, displays a larger number of trophic regimes than the eastern one, with the exception of the limited Adriatic sea area. The link between differences in multiannual through pluridecadal variability in the internal dynamics of the two basins and the biotic dynamics is still unexplored, and this issue deserves even more attention in view of the newly discovered multiannual oscillations.

Finally, historical hydrographic observations as well as recent observations prove that the Mediterrenean Sea is getting saltier, and this is particularly true for the western basin. The crucial question is whether this increase is due to gradual changes or to singular, shorter events. Furthermore, it is not clear if this increased saltiness of the western basin is due to a progressive increase in surface evaporation with downward diffusion of saltier water or through advection of saltier waters, for instance from the eastern basin. Again, such changes in the water mass properties have profound consequences for biogeochemical properties and processes.

Modeling techniques and computer power have been greatly improved during the last decades, enabling an ever increasing accuracy of results that can be applied on the Mediterranean and sub-basin scales for the investigation of long-term variability. The contribution of operational oceanography in the Mediterranean Sea is important, developing and implementing new and advanced modeling tools and techniques in the Mediterranean Sea in the framework of research projects and initiatives at the Mediterranean and European levels (e.g., MFSPP, MFSTEP, ECOOP, MERSEA, MyOcean, MOON/MONGOOS). Blending the two approaches through data assimilation procedures can produce useful results for filling spatio–temporal gaps

These are just a few highlights from the richness of novel, recent results presented in the various sections that have emerged from the investigations carried out during recent decades. More importantly, each section of this paper poses new questions and addresses unresolved issues. The hope is that these questions will produce new challenges and excite new discussions in the interdisciplinary Mediterranean community, which will lead to even more exciting collaborative research and subsequent discoveries.

Acknowledgements. We wish to thank Mario Hoppema for his excellent handling of the manuscript as topic editor throughout the successive phases of the review process and for his excellent comments and suggestions at the various stages of the revisions.

Edited by: M. Hoppema

References

Adani, M., Dobricic, S., and Pinardi, N.: Quality assessment of a 1985–2007 Mediterranean Sea reanalysis, J. Atmos. Ocean. Technol., 28, 569–589, 2011.

Ait-Ameur, N. and Goyet, C.: Distribution and transport of natural and anthropogenic CO_2 in the Gulf of Cadiz, Deep-Sea Res. Pt. I, 53, 1329–1343, 2006.

Alemany, F., Quintanilla, L., Velez-Belchí, P., García, A., Cortés, D., Rodríguez, J. M., Fernández de Puelles, M. L., González-Pola, C., and López-Jurado, J. L.: Characterization of the spawning habitat of Atlantic bluefin tuna and related species in the Balearic Sea (western Mediterranean), Prog. Oceanogr., 86, 21–38, 2010.

Alhammoud, B., Beranger, K., Mortier, L., Crepon, M., and Dekeyser, I.: Surface circulation of the Levantine Basin: comparison of model results with observations, Prog. Oceanogr., 66, 299–320, 2005.

Allen, J. I., Blackford, J. C., and Radford, P. J.: A 1-D vertically resolved modelling study of the ecosystem dynamics of the middle and southern Adriatic sea, J. Marine Syst., 18, 265–286, 1998.

Allen, J. T., Smeed, D. A., Tintoré, J., and Ruiz, S.: Mesoscale subduction at the Almeria–Oran front, Part 1: Agesotrophic flow, J. Marine Syst., 30, 263–285, doi:10.1016/S0924-7963(01)00062-8, 2001.

Anderson, T. R. and Turley, C. M.: Low bacterial growth efficiency in the oligotrophic eastern Mediterranean Sea: a modelling analysis, J. Plankton Res., 25, 1011–1019, 2003.

Arhonditsis, G. B. and Brett, M. T.: Evaluation of the current state of mechanistic aquatic biogeochemical modeling, Marine Ecol. Prog. Ser., 271, 13–26, 2004.

Artale, V., Calmanti, S., Malanotte-Rizzoli, P., Pisacane, G., Rupolo, V., and Tsimplis, M.: The Atlantic and Mediterranean Sea as connected systems, in: Mediterranean Climate Variability, edited by: Lionello, P., Malanotte-Rizzoli, P., and Boscolo, R., Elsevier, 283–323, 2006.

Ashkenazy, Y., Stone, P. H., and Malanotte-Rizzoli, P.: Box modelling of the Eastern Mediterranean Sea, Physica A, 391, 1519–1531, 2012.

Astraldi, M., Balopoulos, E., Candela, J., Font, J., Gacic, M., Gasparini, G. P., Manca, B., Theocharis, A., and Tintore, J.: The role of straits and channels in understanding the characteristics of Mediterranean circulation, Prog. Oceanogr., 44, 65–108, 1999.

Baldacci, A., Corsini, G., Grasso, R., Manzella, G., Allen, J. T., Cipollini, P., Guymer, T. H., and Snaith, H. M.: A study of the Alboran Sea mesoscale system by means of empirical orthogonal function decomposition of satellite data, J. Marine Syst., 29, 293–311, doi:10.1016/S0924-7963(01)00021-5, 2001.

Barausse, A. and Palmierim L.: A Comparative Analysis of Trophic Structure and Functioning in Large- Scale Mediterranean Marine Ecosystems, in: The Mediterranean Sea, its history and present challenges, edited by: Goffredo, S. and Dubinsky, Z., 678 pp., Springer, 2013.

Barron, C. N. and Kara, A. B.: Satellite based daily SSTs over the global ocean, Geophys. Res. Lett., 33, L15603, doi:10.1029/2006GL026356, 2006.

Bethoux, J. P., Gentili, B., Morin, P., Nicolas, E., Pierre, C., and Ruiz-Pino, D.: The Mediterranean Sea: a miniature ocean for climatic and environmental studies and a key for the climatic functioning of the North Atlantic, Prog. Oceanogr., 44, 131–146, 1999.

Beuvier, J., Sevault, F., Hermann, M., Kontoyiannis, H., Ludwig, W., Rixen, M., Stanev, E., Béranger, K., and Somot, S.: Modelling the Mediterranean Sea inter annual variability during 1961–2000: focus on the Eastern Mediterranean Transient (EMT), J. Geophys. Res., 115, C08017, doi:10.1029/2009JC005950, 2010.

Bianchi, C. N. and Morri, C.: Marine biodiversity of the Mediterranean Sea: situation, problems and prospects for future research, Mar. Pollut. Bull., 40, 367–376, 2000.

Bignami, F., Mattietti, G., Rotundi, A., and Salusti, E.: On a Suigimoto-Whitehead effect in the Mediterranean Sea: sinking and mixing of a bottom current in the Bari Canyon, southern Adriatic Sea, Deep-Sea Res., 37, 657–665, 1990.

Bignami, F., Böhm, E., D'Acunzo, E., D'Archino, R., and Salusti, E.: On the dynamics of surface cold filaments in the Mediterranean Sea, J. Marine Syst., 74, 429–442, 2008.

Bingel, F., Özsoy, E., and Ünluata, Ü.: A review of the state of the fisheries and the environment of the northeastern Mediterranean (Northern levantine Basin), General Fisheries Council for the Mediterranean,No 65, 74 pp,Rome, FAO, 1993.

Boero, F., Bouillon, J., Gravili, C., Miglietta, M. P., Parsons, T., and Piraino, S.: Gelatinous plankton: irregularities rule the world (sometimes), Mar. Ecol.-Prog. Ser., 356, 299–312, 2008.

Bond, G., Showers, W., Cheseby, M., Lotti, R., Almasi, P., de-Menocal, P., Priore, P., Cullen, H., Hajdas, I., and Bonani, G.: A pervasive millennial-scale cycle in North Atlantic Holocene and glacial climates, Science, 278, 1257–1266, 1997.

Borzelli, G. L. E.: Scales and variability of the sea surface temperature distribution in the Adriatic Sea, J. Geophys. Res., 113, C11S01, doi:10.1029/2007JC004396, 2008.

Borzelli, G. and Ligi, R.: Empirical orthogonal function analysis of sst image series: a physical interpretation, J. Atmos. Ocean. Tech., 16, 682–690, 1999a.

Borzelli, G. and Ligi, R.: Autocorrelation scales of the SST distribution and water masses stratification in the channel of Sicily, J. Atmos. Ocean. Tech., 16, 776–781, 1999b.

Borzelli, G. L. E., Gačić, M., Cardin, V., and Civitarese, G.: Eastern Mediterranean Transient and reversal of the Ionian Sea circulation, Geophys. Res. Lett., 36, L15108, doi:10.1029/2009GL039261, 2009.

Bouffard, J., Pascual, A., Ruiz, S., Faugère, Y., and Tintoré, J.: Coastal and mesoscale dynamics characterization using altimetry and gliders: a case study in the Balearic Sea, J. Geophys. Res., 115, C10029, doi:10.1029/2009JC006087, 2010.

Bouffard, J., Renault, L., Ruiz, S., Pascual, A., Dufau, C., and Tintoré, J.: Sub-surface small scale eddy dynamics from multi-sensor observations and modelling, Progr. Oceanogr., 106, 62–79, 2012.

Bourassa, M., Stoffelen, A., Bonekamp, H., Chang, P., Chelton, D. B., Courtney, J., Edson, R., Figa, J., He, Y., Hersbach, H.,Hilburn, K., Jelenak, Z., Kelly, K. A., Knabb, R., Lee, T., Lindstrom, E. J., Liu, W. T., Long, D. G., Perrie, W., Portabella, M., Powell, M., Rodriguez, E., Smith, D. K., Swail, V., and Wentz, F. J.: Remotely sensed winds and wind stresses for marine forecasting and ocean modeling, in: Proceedings of OceanObs'09: Sustained Ocean Observations and Information for Society, vol. 2, edited by: Hall, J., Harrison, D. E., and Stammer, D., ESA Publication WPP-306, doi:10.5270/OceanObs09.cwp.08, Venice, Italy, 21–25 September 2009, 2010.

Brenner, S.: Long-term evolution and dynamics of a persistent warm core eddy in the Eastern Mediterranean Sea, Deep-Sea Res. Pt. II, 40, 1193–1206, 1993.

Brenner, S.: High-resolution nested model simulations of the climatological circulation in the southeastern Mediterranean Sea, Ann. Geophys., 21, 267–280, doi:10.5194/angeo-21-267-2003, 2003.

Brenner, S., Gertman, I., and Murashkovsky, A.: Pre-operational ocean forecasting in the southeastern Mediterranean: model implementation, evaluation, and the selection of atmospheric forcing, J. Marine Syst., 65, 268–287, doi:10.1016/j.jmarsys.2005.11.018, 2007.

Bruno, M., Chioua, J., Romero, J., Vazquez, A., Macias, D., Dastis, C., Ramirez-Romero, E., Echevarria, F., Reyes, J., and Garcia, C. M.: The importance of submesoscale processes for the exchange of properties through the Strait of Gibraltar, Progr. Oceanogr., 116, 66–79, 2013.

Cacho, I., Grimalt, J. O., and Canals, M.: Response of the Western Mediterranean Sea to rapid climatic variability during the last 50 000 years: a molecular biomarker approach, J. Marine Syst., 33, 253–272, 2002.

Canals, M., Puig, P., de Madron, X. D., Heussner, S., Palanques, A., and Fabres, J.: Flushing submarine canyons, Nature, 444, 355–357, doi:10.1038/nature05271, 2006.

Candela, J.: Mediterranean water and the global circulation, in: Ocean Circulation and Climate: Observing and Modeling the Global Ocean, edited by: Siedler, G., Church, J., and Gould, J., Academic Press, 419–429, 2001.

Chelton, D. B., Schlax, M. G., Freilich, M. H., and Millif, R. H.: Satellite measurements reveal persistent small-scale features in ocean winds, Science, 303, 978–983, 2004.

CIESM: Dynamics of Mediterranean deep waters, No 38, CIESM Workshop Monographs, edited by: Briand, F., Monaco, CIESM, 16 Boulevard de Suisse, MC-98000, Monaco, 132 pp., 2009.

CIESM: Designing Med-SHIP: a program for repeated oceanographic surveys, No 43, CIESM Workshop Monographs, edited by: Briand, F., Monaco, CIESM, 16 Boulevard de Suisse, MC-98000, Monaco, 164 pp., 2012.

Civitarese, G., Gačić, M., Lipizer, M., and Eusebi Borzelli, G. L.: On the impact of the Bimodal Oscillating System (BiOS) on the biogeochemistry and biology of the Adriatic and Ionian Seas (Eastern Mediterranean), Biogeosciences, 7, 3987–3997, doi:10.5194/bg-7-3987-2010, 2010.

Coll, M. and Libralato, S.: Contributions of food webmodelling to the ecosystem approach to marine resource management in the Mediterranean Sea, Fish and fisheries, 13, 60–88, 2012.

Coll, M., Lotze, H. K., and Romanuk, T. N.: Structural degradation in Mediterranean Sea food webs: testing ecological hypotheses using stochastic and mass-balance modelling, Ecosystems, 11, 939–960, 2008.

Cossarini, G., Solidoro, C., and Fonda Umani, S.: Dynamics of biogeochemical properties in temperate coastal areas of freshwater influence: Lessons from the Northern Adriatic Sea (Gulf of Trieste), Estuarine, Coast. Shelf Sci., 115, 63–74, 2012.

Crise, A., Crispi, G., and Mauri, E.: Seasonal three-dimensional study of the nitrogen cycle in the Mediterranean Sea, Part 1: Model implementation and numerical results, J. Marine Syst., 18, 287–312, 1998.

Crispi, G., Crise, A., and Solidoro, C.: Three-dimensional oligotrophic ecosystem models driven by physical forcings: the Mediterranean Sea case, Environ. Model. Softw., 13, 483–490, 1998.

Crombet, Y., Leblanc, K., Quéguiner, B., Moutin, T., Rimmelin, P., Ras, J., Claustre, H., Leblond, N., Oriol, L., and Pujo-Pay, M.: Deep silicon maxima in the stratified oligotrophic Mediterranean Sea, Biogeosciences, 8, 459–475, doi:10.5194/bg-8-459-2011, 2011.

Cushman-Roisin, B. and Korotenko, K. A.: Mesoscale-resolving simulations of summer and winter bora events in the Adriatic Sea, J. Geophys. Res., 112, C11S91, doi:10.1029/2006JC003516, 2007.

Demirov, E. and Pinardi, N.: Simulation of the Mediterranean Sea circulation from 1979 to 1993: Part 1. The interannual variability, J. Marine Syst., 33, 23–50, 2002.

De Rijk, S., Hayes, A., and Rohling, E. J.: Eastern Mediterranean sapropel S1 interruption: an expression of the onset of climatic deterioration around 7 kyr BP, Mar. Geol., 153, 337–343, 1999.

Diaz, F., Raimbault, P., Conan, P.: Small-scale study of primary productivity during spring in a Mediterranean coastal area (Gulf of Lions), Cont. Shelf Res., 20, 975–996, 2000.

Dima, M. and Lohmann, G.: A hemispheric mechanism for the Atlantic multidecadal oscillation, J. Climate, 20, 2706–2719, 2007.

Dobricic, S., Pinardi, N., Adani, M., Tonani, M., Fratianni, C., Bonazzi, A., and Fernandez, V.: Daily oceanographic analyses by Mediterranean Forecasting System at the basin scale, Ocean Sci., 3, 149–157, doi:10.5194/os-3-149-2007, 2007.

Dobricic, S., Pinardi, N., Testor, P., and Send, U.: Impact of data assimilation of glider observations in the Ionian Sea (Eastern Mediterranean), Dynam. Atmos. Oceans, 50, 78–92, 2010.

D'Ortenzio, F. and Ribera d'Alcalà, M.: On the trophic regimes of the Mediterranean Sea: a satellite analysis, Biogeosciences, 6, 139–148, doi:10.5194/bg-6-139-2009, 2009.

D'Ortenzio, F., Iudicone, D., Montegut, C. B., Testor, P., Antoine, D., Marullo, S., Santoleri, R., and Madec, G.: Seasonal variability of the mixed layer depth in the Mediterranean Sea as derived from in situ profiles, Geophys. Res. Lett., 32, L12605, doi:10.1029/2005GL022463, 2005.

D'Ovidio, F., Fernández, V., Hernández-García, E., and López, C.: Mixing structures in the Mediterranean Sea from finite-size Lyapunov exponents, Geophys. Res. Lett., 31, L17203, doi:10.1029/2004GL020328, 2004.

Dufau-Julliand, C., Marsaleix, P., Petrenko, A., and Dekeyser, I.: Three-dimensional modeling of the Gulf of Lions hydrodynamics (northwest Mediterranean during January 1999 and late winter 1999: Western Mediterranean Intermediate Water's (WIW's) formation and its cascading over the shelf break, J. Geophys. Res., 109, C11002, doi:10.1029/2003JC002019, 2004.

Ediger, D. and Yılmaz, A.: Characteristics of deep chlorophyll maximum in the North-eastern Mediterranean with respect to environmental conditions, J. Marine Syst., 9, 291–303, 1996.

Ediger, D., Tuğrul, S., and Yılmaz, A.: Vertical profiles of particulate organic matter and its relationship with chlorophyll *a* in upper layer of the NE Mediterranean Sea, J. Marine Syst., 55, 311–326, 2005.

Efrati, S., Lehahn, Y., Rahav, E., Kress, N., Herut, B., Gertman, I., Heifetz, E.: Intrusion of coastalWaters into the pelagic eastern Mediterranean: in situ and satellite-based characterization, Biogeosciences, 10, 3349–3357, doi:10.5194/bg-10-3349-2013, 2013.

Emeis, K.-C., Sakamoto, T., Wehausen, R., and Brumsack, H. J.: The sapropel record of the eastern Mediterranean Sea – results of Ocean Drilling Program Leg 160, Palaeogeogr. Palaeoclim. Palaeoecol., 158, 371–395, 2000.

Estrada, M.: Primary production in the northwestern Mediterranean, Sci. Mar., 60, 55–64, 1996.

Feliks, Y. and Itzikowitz, S.: Movement and geographical distribution of anticyclonic eddies in the Eastern Levantine Basin, Deep-Sea Res., 34, 1499–1508, 1987.

Fiala, M., Sournia, A., Claustre, H., Marty, J. C., Prieur, L., and Vétion, G.: Gradients of phytoplankton abundance, composition and photosynthesis pigments across the Almeria–Oran front (SW Mediterranean Sea), J. Marine Syst., 5, 223–233, 1994.

Finkel, Z. V., Beardall, J., Flynn, K. J., Quigg, A., Rees, T. A. V., and Raven, J. A.: Phytoplankton in a changing world: cell size and elemental stoichiometry, J. Plankton Res., 32, 119–137, 2010.

Flecha, S., Pérez, F. F., Navarro, G., Ruiz, J., Olivé, I., Rodríguez-Gálvez, S., Costas, E., and Huertas, I. E.: Anthropogenic carbon inventory in the Gulf of Cádiz, J. Mar. Systems, 92, 67–75, doi:10.1016/j.jmarsys.2011.10.010, 2012.

Foltz, G. R., Grodsky, S. A., and Carton, J. A.: Seasonal mixed layer heat budget of the tropical Atlantic Ocean. J. Geophys. Res., 108, 3146, doi:10.1029/2002JC001584, 2003.

Francombe, L. M., von der Heydt, A., and Dijkstra, H. A.: North Atlantic Multidecadal Climate Variability: an Investigation of

Dominant Time Scales and Processes, J. Climate, 23, 3626–3638, doi:10.1175/2010JCLI3471.1, 2010.

Fraschetti, S., Guarnieri, G., Bevilacqua, S., Terlizzi, A., Claudet, J., Russo, G. F., and Boero, F.: Conservation of Mediterranean habitats and biodiversity countdowns: what information do we really need?, Aquatic Conservation: Marine and Freshwater Ecosystems, 21, 299–306, 2011.

Gačić, M., Eusebi Borzelli, G. L., Civitarese, G., Cardin, V., and Yari, S.: Can internal processes sustain reversals of the ocean upper circulation? The Ionian Sea example, Geophys. Res. Lett., 37, L09608, doi:10.1029/2010GL043216, 2010.

Gačić, M., Civitarese, G., Eusebi Borzelli, G. L., Kovačević, V., Poulain, P.-M., Theocharis, A., Menna, M., Catucci, A., and Zarokanellos, N.: On the relationship between the decadal oscillations of the Northern Ionian Sea and the salinity distributions in the Eastern Mediterranean, J. Geophys. Res., 116, C12002, doi:10.1029/2011JC007280, 2011.

Gačić, M., Schroeder, K., Civitarese, G., Cosoli, S., Vetrano, A., and Eusebi Borzelli, G. L.: Salinity in the Sicily Channel corroborates the role of the Adriatic-Ionian Bimodal Oscillating System (BiOS) in shaping the decadal variability of the Mediterranean overturning circulation, Ocean Sci., 9, 83–90, doi:10.5194/os-9-83-2013, 2013.

Galil, B. S.: A sea under siege-alien species in the Mediterranean, Biol. Invasions, 2, 177–186, 2000.

Galil, B. S., Nehring, S., and Panov, V.: Waterways as invasion highways-impact of climate change and globalization, in: Biological invasions, 59–74, Springer Berlin Heidelberg, 2007.

Garcia-Lafuente, J., Delgado, J., Sanchez Roman, A., Soto, J., Carracedo, L., and Dýaz del Rio, G.: Interannual variability of the Mediterranean outflow observed in Espartel sill, western Strait of Gibraltar, J. Geophys. Res., 114, C10018, doi:10.1029/2009JC005496, 2009.

Gasparini, J. P., Ortona, A., Budillon, G., Astraldi, M., and Sansone, E.: The effect of the Eastern Mediterranean Transient on the hydrographic characteristics in the strait of Sicily and in the Tyrrhenian Sea, Deep-Sea Res. Pt. I, 52, 915–935, 2005.

Gerin, R., Poulain, P. M., Taupier-Letage, I., Millot, C., Ben Ismail, S., and Sammari, C.: Surface circulation in the Eastern Mediterranean using drifters (2005–2007), Ocean Sci., 5, 559–574, 2009,
http://www.ocean-sci.net/5/559/2009/.

Gertman, I., Pinardi, N., Popov, Y., and Hecht, A.: Aegean Sea Water Masses during the Early Stages of the Eastern Mediterranean Climatic Transient (1988–1990), J. Phys. Oceanogr., 36, 1841–1859, 2006.

Gomez Gesteira, J. L., Dauvin, J. C., and Salvande Fraga, M.: Taxonomic level for assessing oil spill effects on soft-bottom sublittoral benthic communities, Mar. Pollut. Bull., 46, 562–572, 2003.

Guarnieri, A., Oddo, P., Bortoluzzi, G., Pinardi, N., and Ravaiolo, M.: The Adriatic Basin Forecasting System: new model and system development, in: Proceedings of the 5th International Conference on EuroGOOS, edited by: Dhalin, H., Fleming, N., and Petersson, S. E., 184–190, 2010.

Guerzoni, S., Chester, R., Dulac, F., Herut, B., Migon, C., Molinaroli, E., Moulin, C., Rossini, P., Saydam, C., Soudine, A., and Ziveri, P.: The role of atmospheric deposition in the biogeochemistry of the Mediterranean Sea, Prog. Oceanogr., 44, 147–190, 1999.

Hamad, N., Millot, C., and Taupier-Letage, I.: A new hypothesis about the surface circulation in the eastern basin of the Mediterranean Sea, Prog. Oceanogr., 66, 287–298, 2005.

Hamad, N., Millot, C., and Taupier-Letage, I.: The surface circulation in the eastern basin of the Mediterranean Sea, Sci. Mar., 70, 457–503, 2006.

Herrmann, M., Somot, S., Sevault, F., Estournel, C., and Déqué, M.: Modeling the deep water convection in the northwestern Mediterranean Sea using an eddy-permitting and an eddy-resolving model: Case study of winter 1986–1987, J. Geophys. Res., 113, C04011, doi:10.1029/2006JC003991, 2008.

Herrmann, M., Diaz, F., Estournel, C., Marsaleix, C., Pulses, C.: Impact of atmospheric and oceanic interannual variability on the Northwestern Mediterranean Sea pelagic planktonic ecosystem and associated carbon cycle, J. Geophys. Res., 118, 5792–5813, doi:10.1002/jgrc.20405, 2013.

Heslop, E., Ruiz, S., Garau, B., Allen, J., Tintoré, J., Lopez-Jurado, J.-L., and Schroeder, K.: Variability in upper layer transports in the Balearic Sea, using new data from glider missions, 5th Everyone's Gliding Observatories Workshop, Gran Canaria, Spain, 14–18 March, 2011.

Hodges, B. A. and Fratantoni, D. M.: A thin layer of phytoplankton observed in the Philippine Sea with a synthetic moored array of autonomous gliders, J. Geophys. Res., 114, C10020, doi:10.1029/2009JC005317, 2009.

Hoskins, B. J., Draghici, I., and Davies, H. C.: A new look at the omega-equation, Q. J. Roy. Meteorol. Soc., 104, 31–38, 1978.

Howell, M. W., Thunell, R. C., Di Stefano, E., Sprovieri, R., Tappa, E. J., and Sakamoto, T.: Stable isotope chronology and paleoceanographic history of sites 963 and 964, Eastern Mediterranean Sea, in: Proc. ODP, Sci. Results, edited by: Robertson, A. H. F., Emeis, K.-C., Richter, C., and Camerlenghi, A., 160, 167–180, 1998.

Huertas, I. E., Ríos, A. F., García-Lafuente, J., Makaoui, A., Rodríguez-Gálvez, S., Sánchez-Román, A., Orbi, A., Ruíz, J., and Pérez, F. F.: Anthropogenic and natural CO_2 exchange through the Strait of Gibraltar, Biogeosciences, 6, 647–662, doi:10.5194/bg-6-647-2009, 2009.

Huertas, I. E., Rios, A. F., Garcia-Lafuente, J., Navarro, G., Makaoui, A., Sanchez-Roman, A., Rodriguez-Galvez, S., Orbi, A., Ruiz, J., and Perez, F. F.: Atlantic forcing of the Mediterranean oligotrophy, Global Biogeochem. Cy., 26, GB2022, doi:10.1029/2011GB004167, 2012.

Incarbona, A., Di Stefani, E., Patti, B., Pelosi, N., Bonomo, S., Mazzola, S., Sprovieri, R., Tranchida, G., Zgozi, S., and Bonanno, A.: Holocene millennial-scale productivity variations I the Sicily channel (Mediterranean Sea), Paleoceanography, 23, PA3204, doi:10.1029/2007PA001581, 2008.

Ioannone, A., Catucci, A., Grasso, M., and Eusebi Borzelli, G. L.: Decadal variability and scales of the sea surface structure in the northern Ionian, Cont. Shelf Res., 31, 37–46, 2011.

Isern-Fontanet, J., Garcia-Ladona, E., and Font, J.: Vortices of the Mediterranean Sea: an altimetric perspective, J. Phys. Oceanogr., 36, 87–103, 2006.

Ivanov, V. V., Shapiro, G. I., Huthnance, J. M., Aleynik, D. L., and Golovin, P. N.: Cascades of dense water around the world ocean, Prog. Oceanogr., 60, 47–98, 2004.

Jordi, A., Basterretxea, G., Orfila, A., and Tintoré, J.: Analysis of the circulation and shelf-slope exchanges in the continental margin of the northwestern Mediterranean, Ocean Sci., 2, 173–181, doi:10.5194/os-2-173-2006, 2006.

Jordi, A., Basterretxea, G., and Angléss, S.: Influence of ocean circulation on phytoplankton biomass distribution in the Balearic Sea: study based on sea-viewing wide field-of-view sensor and altimetry satellite data, J. Geophys. Res., 114, C11005, doi:10.1029/2009JC005301, 2009.

Josey, S. A., Somot, S., and Tsimplis, M.: Impacts of atmospheric modes of variability on Mediterranean Sea surface heat exchange, J. Geophys. Res., 116, C02032, doi:10.1029/2010JC006685, 2010.

Kantha, L. H. and Carniel, S.: A note on mixing in stably stratified flows, J. Atmos. Sci., 66, 2501–2505, doi:10.1175/2009JAS3041.1, 2009.

Kantha, L. H. and Clayson, C. A.: On leakage of energy from turbulence to internal waves in the oceanic mixed layer, Ocean Dynam., 57, 151–156, doi:10.1007/s10236-006-0100-3, 2007.

Kontoyiannis, H., Theocharis, A., and Nittis, K.: Structures and characteristics of newly formed water masses in the NW Levantine during 1986, 1992, 1995, in: The Eastern Mediterranean as a Laboratory Basin for the Assessment of Contrasting Ecosystems, edited by: Malanotte-Rizzoli, P. and Eremeev, V. N., Kluwer Academic Publishers, 1999.

Kontoyiannis, H., Papadopoulos, V., Kazmin, A., Zatsepin, A., and Georgopoulos, D.: Climatic variability of the sub-surface sea-temperatures in the Aegean-Black Sea system and relation to meteorological forcing, Clim. Dynam., 39, 1507–1525, doi:10.1007/s00382-012-1370-8, 2012.

Koppelmann, R., Bottger-Schnack, R., Mobius, J., and Weikert, H.: Trophic relationships of zooplankton in the eastern Mediterranean based on stable isotope measurements, J. Plankton Res., 31, 669–686, 2009.

Krichak, S. O. and Alpert, P.: Decadal trends in the East Atlantic/West Russia pattern and the Mediterranean precipitation, Int. J. Climatol., 25, 183–192, 2005.

Krom, M. D., Brenner, S., Kress, N., Neori, A. and Gordon, L. I.: Nutrient distributions during an annual cycle across a warmcore eddy from the E. Mediterranean Sea, Deep Sea Res. Pt. 1, 40, 805–825, 1993.

Krom, M. D., Herut, B., and Mantoura, R. F. C.: Nutrient budget for the Eastern Mediterranean: implications for phosphorus limitation, Limnol. Oceanogr., 49, 1582–1592, 2004.

Lapeyre, G. and Klein, P.: Impact of the small-scale elongated filaments on the oceanic vertical pump, J. Mar. Res., 64, 835–851, 2006.

Lascaratos, A. and Nittis, K.: A high-resolution three-dimensional numerical study of intermediate water formation in the Levantine Sea, J. Geophys. Res., 103, 18497–18511, doi:10.1029/98JC01196, 1998.

Lascaratos, A., Roether, W., Nittis, K., and Klein, B.: Recent changes in deep water formation and spreading in the eastern Mediterranean Sea: a review, Prog. Oceanogr., 44, 5–36, 1999.

Laskar, J., Robutel, P., Joutel, F., Gastineau, M., Correia, A. C. M., and Levrard, B.: A long-term numerical solution for the insolation quantities of the Earth, Astronomy Astrophys., 428, 261–285, 2004.

Lauzier, M. S. and Sindlinger, L.: On the source of Mediterranean overflow water property changes, J. Phys. Oceanogr., 39, 1800–1817, 2009.

La Violette, P. E., Tintoré, J., and Font, J.: The surface circulation of the Balearic Sea, J. Geophys. Res., 95, 1559–1568, 1990.

Lazzari, P., Solidoro, C., Ibello, V., Salon, S., Teruzzi, A., Béranger, K., Colella, S., and Crise, A.: Seasonal and interannual variability of plankton chlorophyll and primary production in the Mediterranean Sea: a modelling approach, Biogeosciences, 9, 217–233, doi:10.5194/bg-9-217-2012, 2012.

Lee, C. M., Askari, F., Book, J., Carniel, S., Cushman-Roisin, B., Dorman, C., Doyle, J., Flament, P., Harris, C. K., Jones, B. H., Kuzmic, M., Martin, A., Ogston, M., Orlic, H. Perkins, P.-M., Poulain, J., Pullen, A., Russo, C. Sherwood, P., Signell, R. P., and Thaler, D.: Northern Adriatic response to a wintertime Bora wind event, Eos T. AGU, 86, 157–165, doi:10.1029/2005EO16001, 2005.

Lermusiaux, P. F. J. and Robinson, A. R.: Features of dominant mesoscale variability, circulation patterns and dynamics in the Strait of Sicily, Deep Sea Res., 48, 1953–1997, 2001.

Lévy, M., Klein, P., and Ben Jelloul, M.: New production stimulated by high-frequency winds in a turbulent mesoscale eddy field, Geophys. Res. Lett., 36, L16603, doi:10.1029/2009GL039490, 2009.

Libralato, L. and Solidoro, C.: Bridging biogeochemical and food web models for an End-to-End representation of marine ecosystem dynamics: The Venice lagoon case study, Ecol. Model., 220, 2960–2971, 2009.

Longhurst, A. R.: Ecological Geography of the Sea, Elsevier Science, New York, 552 pp., 1998.

Malanotte-Rizzoli, P. and Bergamasco, A.: The wind and thermally driven circulation of the Eastern Mediterranean Sea, Part 2: The baroclinic case, Dynam. Atmos. Oceans, 15, 355–419, 1991.

Malanotte-Rizzoli, P., Manca, B., Ribera d'Alcalà, M., Theocharis, A., Bergamasco, A., Bregant, D., Budillon, G., Civitarese, G., Georgopoulos, D., Michelato, A., Sansone, E., Sarazzato, P., Sovermezoglou, E.: A synthesis of the Ionian Sea hydrography, circulation and water mass pathways during POEM-Phase I, Progr. Oceanogr., 39, 153–204, 1997.

Malanotte-Rizzoli, P., Manca, B. B., Ribera d'Alcalà, M., Theocharis, A., Brenner, S., Budillon, G., and Özsoy, E.: The Eastern Mediterranean in the 80s and in the 90s: the big transition in the intermediate and deep circulations, Dynam. Atmos. Oceans, 29, 365–395, 1999.

Malanotte-Rizzoli, P., Manca, B. B., Marullo, S., Ribera d'Alcalà, M., Roether, W., Theocharis, A., Bergamasco, A., Budillon, G., Sansone, E., Civitarese, G., Conversano, F., Gertman, I., Hernt, B., Kress, N., Kioroglou, S., Kontoyannis, H., Nittis, K., Klein, B., Lascaratos, A., Latif, M. A., Özsoy, E., Robinson, A. R., Santoleri, R., Viezzoli, V., and Kovacevic, D., The LIWEX group: the Levantine Intermediate Water Experiment, the Levantine basin as a laboratory for multiple water mass formation processes, J. Geophys. Res., 9, 8101, doi:10.1029/2002JC001643, 2003.

Manca, B. B., Kovačević, V., Gačić, M., and Viezzoli, D.: Dense water formation in the Southern Adriatic Sea and spreading into the Ionian Sea in the period 1997–1999, J. Marine Syst., 33, 133–154, 2002.

Markaki, Z., Loye-Pilot, M.-D., Violaki, K., Benyahya, L., and Mihalopoulos, N.: Variability of atmospheric deposition of dissolved nitrogen and phosphorus in the Mediterranean and possible link to the anomalous seawater N/P ratio, Mar. Chem., 120, 187–194, 2010.

Marullo, S., Buongiorno Nardelli, B., Guarracino, M., and Santoleri, R.: Observing the Mediterranean Sea from space: 21 years of Pathfinder-AVHRR sea surface temperatures (1985 to 2005): re-analysis and validation, Ocean Sci., 3, 299–310, doi:10.5194/os-3-299-2007, 2007.

Marullo, S., Artale, V., and Santoleri, R.: The SST multidecadal variability in the Atlantic–Mediterranean region and its relation to AMO, J. Climate, 24, 4385–4401, doi:10.1175/2011JCLI3884.1, 2011.

McGill, D. A.: A preliminary study of the oxygen and phosphate distribution in the Mediterranean Sea, Deep Sea Res. (1953), 8, 259–268, 1961.

McGillicuddy Jr., D. J., Robinson, A. R., Siegel, D. A., Jannasch, H. W., Johnson, R., Dickey, T. D., McNeil, J., Michaels, A. F., and Knap, A. H.: Influence of mesoscale eddies on new production in the Sargasso Sea, Nature, 394, 263–265, 1998.

Menna, M., Poulain, P.-M., Zodiatis, G., and Gertman, I.: On the surface circulation of theLevantinesub-basin derived from Lagrangian drifters and satellite altimetry data, Deep Sea Res. Pt. I, 65, 46–58, 2012.

MerMex group: Marine Ecosystems Responses to climatic and anthropogenic forcings in the Mediterranean, Progr. Oceanogr., 91, 97–166, 2011.

Miller, A. R.: Physical Oceanography of the Mediterranean Sea: a discourse, Rapp. Comm. Int. Mer Med., 17, 857–871, 1963.

Millot, C.: Some features of the Algerian Current, J. Geophys. Res., 90, 7169–7176, 1985.

Millot, C.: Circulation in the Western Mediterranean Sea, J. Marine Syst., 20, 423–442, 1999.

Millot, C.: Interannual salinification of the Mediterranean inflow, Geophys. Res. Lett., 34, L21609, doi:10.1029/2007GL031179, 2007.

Millot, C.: Levantine Intermediate Water characteristics: an astounding general misunderstanding! Scientia Marina, 77, 0214-8358, doi:10.3989/scimar.03518.13A, 2013.

Millot, C. and Taupier-Letage, I.: Circulation in the Mediterranean Sea, in: The Handbook of Environmental Chemistry, vol. 5, Part K, edited by: Saliot, A., Springer-Verlag, 29–66, 2005.

Millot, C., Candela, J., Fuda, J.-L., and Tber, Y.: Large warming and salinification of the Mediterranean outflow due to changes in its composition, Deep-Sea Res., 53, 656–666, doi:10.1016/j.dsr.2005.12.017, 2006.

Morel, A., Huot, Y., Gentili, B., Werdell, P. J., Hooker, S. B., and Franz, B. A.: Examining the consistency of products derived from various ocean color sensors in open ocean (Case 1) waters in the perspective of a multi-sensor approach, Remote. Sens. Environ., 111, 69–88, 2007.

Moutin, T., Van Wambeke, F., and Prieur, L.: Introduction to the Biogeochemistry from the Oligotrophic to the Ultraoligotrophic Mediterranean (BOUM) experiment, Biogeosciences, 9, 3817–3825, doi:10.5194/bg-9-3817-2012, 2012.

Murat, A.: Pliocene-Pleistocene occurrence of sapropels in the Western Mediterranean Sea and their relation to Eastern Mediterranean sapropels, in: Proc. ODP, Sci. Results, edited by: Zahn, R., Comas, M. C., and Klaus, A., 161, 519–527, 1999.

Navarro, G., Vázquez, Á., Macías, D., Bruno, M., and Ruiz, J.: Understanding the patterns of biological response to physical forcing in the Alborán Sea (western Mediterranean), Geophys. Res. Lett., 38, L23606, doi:10.1029/2011GL049708, 2011.

Nencioli, F., d'Ovidio, F., Doglioli, A. M., and Petrenko, A. A.: Surface coastal circulation patterns by in-situ detection of Lagrangian coherent structures, Geophys. Res. Lett., 38, L17604, doi:10.1029/2011GL048815, 2011.

Niewiadomska, K., Claustre, H., Prieur, L., and d'Ortenzio, F.: Submesoscale physical–biogeochemical coupling across the Ligurian current (northwestern Mediterranean) using a bio-optical glider, Limnol. Oceanogr., 53, 2210–2225, 2009.

Nilsson, J. A., Dobricic, S., Pinardi, N., Taillandier, V., and Poulain, P. M.: On the assessment of Argo float trajectory assimilation in the Mediterranean Forecasting System, Ocean Dynam., 61, 1475–1490, 2011.

Onken, R. and Yüce, H.: Winter circulation and convection in the Antalya basin (Eastern Mediterranean), J. Phys. Oceanogr., 30, 1099–1110, 2000.

Özsoy, E., Hecht, A., and Ünlüata, Ü.: Circulation and hydrography of the Levantine Basin, Results of POEM coordinated experiments 1985/1986, Prog. Oceanogr., 22, 125–170, 1989.

Özsoy, E., Hecht, A., Ünlüata, Ü., Brenner, S., Oğuz, T., Bishop, J., Latif, M. A., and Rosentroub, Z.: A review of the Levantine Basin circulation and its variability during 1985–88, Dynam. Atmos. Oceans, 15, 421–456, 1991.

Özsoy, E., Hecht, A., Ünlüata, Ü., Brenner, S., Sur, H. I., Bishop, J., Latif, M. A., Rozentraub, Z., and Oğuz, T.: A synthesis of the Levantine circulation and hydrography, 1985–1990, Deep Sea Res. Part II, 40, 1075–1119, 1993.

Palanques, A., Puig, P., Latasa, M., and Scharek, R.: Deep sediment transport induced by storms and dense shelf-water cascading in the northwestern Mediterranean basin, Deep-Sea Res. Pt. I, 56, 425–434, 2009.

Parsons, T. R. and Lalli, C. M.: Jellyfish population explosions: revisiting a hypothesis of possible causes, La mer, 40, 111–121, 2002.

Pascual, A., Buongiorno Nardelli, B., Larnicol, G., Emelianov, M., and Gomis, D.: A case of an intense anticyclonic eddy in the Balearic Sea (western Mediterranean), J. Geophys. Res.-Oceans, 107, 4–1, 2002.

Pascual, A., Ruiz, S., and Tintoré, J.: Combining new and conventional sensors to study the Balearic Current, Sea Technol., 51, 32–36, 2010.

Pasqual, A., Sanchez-Vidal, A., Zúñiga, D., Calafat, A., Canals, M., Durrieu de Madron, X., Puig, P., Heussner, S., Palanques, A., and Delsaut, N.: Flux and composition of settling particles across the continental margin of the Gulf of Lion: the role of dense shelf water cascading, Biogeosciences, 7, 217–231, doi:10.5194/bg-7-217-2010, 2010.

Pérez-Folgado, M., Sierro, F. J., Flores, J.-A., Grimalt, J. O., and Zahn, R.: Paleoclimatic variations in foraminifer assemblages from the Alboran Sea (Western Mediterranean) during the last 150 kyr in ODP Site 977, Mar. Geol., 212, 113–131, 2004.

Pierini, S. and Rubino, A.: Modeling the oceanic circulation in the area of the Strait of Sicily: the remotely forced dynamics, J. Phys. Ocean, 31, 1397–1412, 2001.

Pinardi, N. and Coppini, G.: Preface "Operational oceanography in the Mediterranean Sea: the second stage of development;;, Ocean Sci., 6, 263–267, doi:10.5194/os-6-263-2010, 2010.

Pinardi, N. and Masetti, E.: Variability of the large scale general circulation of the Mediterranean Sea from observations and modelling: a review, Palaeoecology, 158, 153–173, 2000.

Pinot, J. M., Lopez-Jurado, J. L., and Riera, M.: The Canales experiment (1996–1998): interannual, seasonal, and mesoscale variability of the circulation in the Balearic channels, Prog. Oceanogr., 55, 335–370, doi:10.1016/S0079-6611(02)00139-8, 2002.

Pisacane, G., Artale, V., Calmanti, S., and Rupolo, V.: Decadal oscillations in the Mediterranean Sea: a result of the overturning circulation variability in the Eastern Basin?, Clim. Res., 31, 257–271, 2006.

POEM group: General circulation of the Eastern Mediterranean, Earth Sci. Rev., 32, 285–309, doi:10.1016/0012-8252(92)90002-B, 1992.

Polimene, L., Pinardi, N., Zavatarelli, M. and Colella, S.: The Adriatic Sea ecosystem seasonal cycle: validation of a three-dimensional numerical model, J. Geophys. Res, 112, C03S19, doi:10.1029/2005JC003260, 2007.

Poulain, P.-M., Menna, M., and Mauri, E.: Surface geostrophic circulation of the Mediterranean Sea derived from drifter and satellite altimeter data, J. Phys. Oceanogr., 42, 973–990, 2012.

Puillat, L., Taupier-Letage, I., and Millot, C.: Algerian eddies lifetime can near 3 years, J. Marine Syst., 31, 245–259, 2002.

Riandey, V., Champalbert, G., Carlotti, F., Taupier-Letage, I., and Thibault-Bothac, D.: Zooplankton distribution related to the hydrodynamic features in the Algerian Basin (western Mediterranean Sea) in summer 1997, Deep-Sea Res. Pt. I, 52, 2029–2048, 2005.

Ribera d'Alcalà, M., Civitarese, G., Conversano, F., and Lavezza, R.: Nutrient ratios and fluxes hint at overlooked processes in the Mediterranean Sea, J. Geophys. Res., 108, 8106, doi:10.1029/2002JC001650, 2003.

Ridame, C., Moutin, T., and Guieu, C.: Does phosphate adsorption onto Saharan dust explain the unusual N/P ratio in the Mediterranean Sea?, Ocean. Acta, 32, L12608, 629–634, 2003.

Rixen, M., Beckers, J., Levitus, S., Antonov, J., Boyer, T., Maillard, C., Fichaut, M., Balopoulos, E., Iona, S., Dooley, H., Garcia, M., Manca, B., Giorgetti, A., Manzella, G., Michailov, N., Pinardi, N., and Zavatarelli, M.: The Western Mediterranean Deep Water: a proxy for climate change, Geophys. Res. Lett., 32, L12608, doi:10.1029/2005GL022702, 2005.

Robinson, A.: Eddies in Marine Science (Topics in Atmospheric and Oceanographic Sciences), Springer-Verlag, 609 pp., 1983.

Robinson, A., Leslie, W., Theocharis, A., and Lascaratos, A.: Mediterranean Sea circulation, in: Encyclopedia of Ocean Sciences, Academic Press Ltd., London, 1689–1706, 2001.

Robinson, A. R., Golnaraghi, M., Leslie, W. G., Artegiani, A., Hecht, A., Lazzoni, E., Michelato, A., Sansone, E., Theocharis, A., and Ünlüata, Ü.: The Eastern Mediterranean general circulation: features, structure and variability, Dynam. Atmos. Oceans, 15, 215–240, 1991.

Robinson, A. T. and Golnaraghi, M.: The physical and dynamical oceanography of the Mediterranean Sea, in: Ocean Processes in Climate Dynamics: Global and Mediterranean Examples, edited by: Malanotte-Rizzoli, P. and Robinson, A. R., Kluwer Academic Publishers, 255–306, 1994.

Roether, W. and Lupton, J. E.: Tracers confirm downward mixing of Tyrrhenian Sea upper waters associated with the Eastern Mediterranean Transient, Ocean Sci., 7, 91–99, doi:10.5194/os-7-91-2011, 2011.

Roether, W., Manca, B. B., Klein, B., Bregant, D., Georgopoulos, D., Beitzel, V., Kovacević, V., and Luchetta, A.: Recent changes in Eastern Mediterranean deep waters, Science, 271, 333–335, 1996.

Roether, W., Klein, B., Manca, B. B., Theocharis, A., and Kioroglou, S.: Transient Eastern Mediterranean deep waters in response to the massive dense-water output of the Aegean Sea in the 1990s, Prog. Oceanogr., 74, 540–571, 2007.

Rohling, E. J.: Review and new aspects concerning the formation of Mediterranean sapropels, Mar. Geol., 122, 1–28, 1994.

Rohling, E. J., Cane, T. R., Cooke, S., Sprovieri, M., Bouloubassi, I., Emeis, K. C., Schiebel, R., Kroon, D., Jorissen, F. J., Lorre, A., and Kemp, A. E. S.: African monsoon variability during the previous interglacial maximum, Earth Planet. Sc. Lett., 202, 61–75, 2002.

Rosentraub, Z. and Brenner, S.: Circulation over the southeastern continental shelf and slope of the Mediterranean Sea: direct current measurements, winds, and numerical model simulations, J. Geophys. Res., 112, C11001, doi:10.1029/2006JC003775, 2007.

Roussenov, V., Stanev, E., Artale, V., and Pinardi, N.: A seasonal model of the Mediterranean Sea general circulation, J. Geophys. Res., 100, 13515–13538, 1995.

Ruiz, S., Pascual, A., Garau, B., Faugère, Y., Alvarez, A., and Tintoré, J.: Mesoscale dynamics of the Balearic Front, integrating glider, ship and satellite data, J. Marine Syst., 78, 3–16, 2009a.

Ruiz, S., Pascual, A., Garau, B., Pujol, I., and Tintoré, J.: Vertical VELOCITIES in the upper ocean from glider and altimetry data, Geophys. Res. Lett., 36, p. 72, doi:10.1029/2009GL038569, 2009b.

Salihoğlu, İ., Saydam, C., Baştürk, Ö., Goemen, D., Hatipoğlu, E., and Yılmaz, A.: Transport and distribution of nutrients and chlorophyll *a* by mesoscale eddies in the Northeastern Mediterranean, Mar. Chem., 29, 375–390, 1990.

Sannino, G., Herrmann, M., Carillo, A., Rupolo, V., Ruggiero, V., Artale, V., and Heimbach, P.: An eddy-permitting model of the Mediterranean Sea with a two-way grid refinement at the Strait of Gibraltar, Ocean Model., 30, 56–73, doi:10.1016/j.ocemod.2009.06.002, 2009.

Santinelli, C., Manca, B. B., Gasparini, G. P., Nannicini, L., and Seritti, A.: Vertical distribution of dissolved organic carbon (DOC) in the Mediterranean Sea, Clim. Res., 31, 205–216, 2006.

Santinelli, C., Nannicini, L., and Seritti, A.: DOC dynamics in the meso and bathypelagic layers of the Mediterranean Sea, Deep-Sea Res. Pt. II, 57, 1446–1459, 2010.

Saunders, P. M.: The instability of a baroclinic vortex, J. Phys. Oceanogr., 3, 61–65, 1973.

Schneider, A., Tanhua, T., Körtzinger, A., and Wallace, D. W. R.: High anthropogenic carbon content in the eastern Mediterranean, J. Geophys. Res., 115, C12050, doi:10.1029/2010JC006171, 2010.

Schroeder, K., Ribotti, A., Borghini, M., Sorgente, R., Perilli, A., and Gasparini, G. P.: An extensive western Mediterranean deep water renewal between 2004 and 2006, Geophys. Res. Lett., 35, L18605, doi:10.1029/2008GL035146, 2008.

Schroeder, K., Josey, S. A., Herrmann, M., Grignon, L., Gasparini, G. P., and Bryden, H. L.: Abrupt warming and salting of the Western Mediterranean Deep Water after 2005: atmospheric forcings and lateral advection, J. Geophys. Res., C08029, doi:10.1029/2009JC005749, 2010.

Schroeder, K., Garcia-Lafuente, J., Josey, S. A., Artale, V., Buongiorno Nardelli, B., Carrillo, A., Gacic, M., Gasparini, G. P., Herrmann, M., Lionello, P., Ludwig, W., Millot, C., Ozsoy, E., Pisacane, G., Sanchez-Garrido, J. C., Sannino, G., Santoleri, R., Somot, S., Struglia, M., Stanev, E., Taupier-Letage, I., Tsimplis, M. N., Vargas-Yanez, M., Zervakis, V., and Zodiatis, G.: Climate of the Mediterranean Region – From the past the future, edited by: Lionello, P., Elsevier, 187–256, 2012.

Send, U., Font, J., Krahmann, G., Millot, C., Rhein, M., and Tintore, J.: Recent advances in observing the physical oceanography of the western Mediterranean Sea, Prog. Oceangr., 44, 37–64, 1999.

Shea, K. and Chesson, P.: Community ecology theory as a framework for biological invasions, TRENDS in Ecology Evolution, 17, 170–176, 2002.

Signell, R. P., Carniel, S., Cavaleri, L., Chiggiato, J., Doyle, J., Pullen, J., and Sclavo, M.: Assessment of wind quality for oceanographic modeling in semi-enclosed basins, J. Marine Syst., 53, 217–233, doi:10.1016/j.jmarsys.2004.03.006, 2005.

Siokou-Frangou, I., Zervoudaki, S., Christou, E. D., Zervakis, V., and Georgopoulos, D.: Variability of mesozooplankton spatial distribution in the North Aegean Sea, as influenced by the Black Sea waters outflow, J. Marine Syst., 78, 557–575, 2009.

Sprovieri, M., Pelosi, N., Sprovieri, R., Incarbona, A, and Ribera d'Alcalà, M.: L'evoluazione del clima nell'area mediterranea durante l'intervallo 20 000–70 000 anni, in Clima e Cambiamenti Climatic: le Attivitá di Ricerca del CNR, edited by: Carli, B., Cavarretta, G., Colacino, M., and Fuzzi, S., Consiglio Nazionale delle Ricerche, Rome, 177–180, 2007.

Stammer, D.: Global characteristics of ocean variability estimated from regional TOPEX/POSEIDON altimeter measurements, J. Phys. Oceanogr., 27, 1743–1769, 1997.

Stommel, H.: Thermohaline convection with two stable regimes of flow, Tellus, 2, 224–230, 1961.

Sullivan, J. M., McManus, M. A., Cheriton, O. M., Benoit-Bird, K. J., Goodman, L., Wang, Z., Ryan, J. P., Stacey, M., Van Holliday, D., Greenlaw, C., Moline, M. A., and McFarland, M.: Layered organization in the coastal ocean: an introduction to planktonic thin layers and the LOCO project, Cont. Shelf Res., 30, 1–6, 2010.

Sur, H., Özsoy, E., and Ünlüata, Ü.: Simultaneous deep and intermediate depth convection in the Northern Levantine Sea, Winter 1992, Oceanol. Acta, 16, 33–43, 1993.

Taupier-Letage, I., Puillat, I., Millot, C., and Raimbault, P.: Biological response to mesoscale eddies in the Algerian Basin, J. Geophys. Res., 108, 3245, doi:10.1029/1999JC000117, 2003.

Theocharis, A., Balopolos, E., Kioroglou, S., Kontoyannis, H., and Iona, A.: A synthesis of the circulation and hydrography of the South Aegean Sea and the Straits of the Cretan Arc (March 1994–January 1995), Prog. Oceanogr., 44, 469–509, 1999a.

Theocharis, A., Nittis, K., Kontoyiannis, H., Papageorgiou, E., and Balopoulos, E.: Climatic changes in the Aegean Sea influence the Eastern Mediterranean thermohaline circulation (1986–1997), Geophys. Res. Lett., 26, 1617–1620, 1999b.

Theocharis, A., Klein, B., Nittis, K., and Roether, W.: Evolution and status of the Eastern Mediterranean Transient (1997–1999), J. Mar. Syst., 33–34, 91–116, 2002a.

Theocharis, A., Lascaratos, A., and Sofianos, S.: Variability of sea water properties in the Ionian, Cretan and Levantine seas during the last century, in: Tracking Long-Term Hydrological Change in the Mediterranean Sea, CIESM Workshop Series, 16, edited by: Briand, F. and Millot, C., Monaco, CIESM, 16 boulevard de Suisse, MC-98000, Monaco, 71–78, 2002b.

Tintoré, J., Vizoso, G., Casas, B., Renault, L., Ruiz, S., Garau, B., Pascual, A., Martínez-Ledesma, M., Gomez-Pujol, L. L., and Orfila, A.: Designing Med-SHIP: a Program for Repeated Oceanographic Surveys, CIESM Workshop Monographs, 43, edited by: Briand, F., CIESM, 16 boulevard de Suisse, MC-98000, Monaco, 164 pp., 2012.

Tonani, M., Pinardi, N., Dobricic, S., Pujol, I., and Fratianni, C.: A high-resolution free-surface model of the Mediterranean Sea, Ocean Sci., 4, 1–14, doi:10.5194/os-4-1-2008, 2008.

Triantafyllou, G., Hoteit, I., Korres, G. and Petihakis, G.: Ecosystem modeling and data assimilation of physical-biogeochemical processes in shelf and regional areas of the Mediterranean Sea, App. Num. Anal. Comp. Math., 2, 262–280, 2005.

Triantafyllou, G., Korres, G., Hoteit, I., Petihakis, G., and Banks, A. C.: Assimilation of ocean colour data into a Biogeochemical Flux Model of the Eastern Mediterranean Sea, Ocean Sci., 3, 397–410, doi:10.5194/os-3-397-2007, 2007.

UNEP-MAP-RAC/SPA: Classification of Benthic Marine Habitat Types for the Mediterranean Region, UNEP RAC SPA, Tunis, 2006.

Van Wambeke, F., Lefèvre, D., Prieur, L., Sempéré, R., Bianchi, M., Oubelkheir, K., and Bruyant, F.: Distribution of microbial biomass, production, respiration, dissolved organic carbon and factors controlling bacterial production across a geostrophic front (Almeria-Oran, SW Mediterranean Sea), Mar. Ecol. Prog. Ser., 269, 1–15, 2004.

Vervatis, V. D., Sofianos, S. S., and Theocharis, A.: Distribution of the thermohaline characteristics in the Aegean Sea related to water mass formation processes (2005.2006 winter surveys), J. Geophys. Res., 116, C09034, doi:10.1029/2010JC006868, 2011.

Vilibić, I. and Supić, N.: Dense water generation on a shelf: the case of the Adriatic Sea, Ocean Dynam., 55, 403–415, 2005.

Volpe, G., Banzon, V. F., Evans, R. H., Santoleri, R., Mariano, A. J., and Sciarra, R.: Satellite observations of the impact of dust in a low-nutrient, low-chlorophyll region: Fertilization or artifact?, Global Biogeochem. Cy., 23, GB3007, doi:10.1029/2008GB003216, 2009.

Wang, F., Vieira, M., Salat, J., Tintoré, J., and La Violette, P. E.: A shelf/slope filament off the Northeast Spanish Coast, J. Mar. Res., 46, 321–332, 1988.

Winters, K. B. and Young, W. R.: Available potential energy and buoyancy variance in horizontal convection, J. Fluid Mech., 629, 221–231, 2009.

Winters, K. B., Lombard, P. N., Riley, J. J., and D'Asaro, E. A.: Available potential energy and mixing in density stratified fluids, J. Fluid Mech., 289, 115–128, 1995.

Wunsch, C.: The vertical partition of oceanic horizontal kinetic energy, J. Phys. Oceanogr., 27, 1770–1794, 1997.

Wunsch, C. and Stammer, D.: The global frequency-wave number spectrum of oceanic variability estimated from TOPEX/POSEIDON altimetric measurements, J. Geophys. Res., 100, 24895–24910, 1995.

Xoplaki, E., Gonzalez-Rouco, F. J., Luterbacher, J., and Wanner, H.: Mediterranean summer air temperature variability and its connection to the large-scale atmospheric circulation and SSTs, Clim. Dynam., 20, 723–739, 2003.

Xoplaki, E., Gonzalez-Rouco, F. J., Luterbacher, J., and Wanner, H.: Wet season Mediterranean precipitation variability: influence of large-scale dynamics, Clim. Dynam., 23, 63–78, doi:10.1007/s00382-004-0422-0, 2004.

Yacobi, Y. Z., Zohary, T., Kress, N., Hecht, A., Robarts, R. D., Waiser, M., Wood, A. M. and Li, W. K. W.: Chlorophyll distribution throughout the southeastern Mediterranean in relation to physical structure of the water mass, J. Marine Syst., 6, 179–190, 1995.

Yılmaz, A. and Tuğrul, S.: The effect of cold and warm-core eddies and the distribution and stoichiometry of dissolved nutrients in the Northeastern Mediterranean, J. Marine Syst., 16, 253–268, 1998.

Yılmaz, A., Ediger, D., Baştürk, Ö., and Tuğrul, S.: Phytoplankton fluorescence and deep chlorophyll maxima in the northeastern Mediterranean, Oceanol. Acta, 17, 69–77, 1994.

Zavatarelli, M. and Mellor, G. L.: A numerical study of the Mediterranean Sea circulation, J. Phys. Oceanogr., 25, 1384–1414, 1995.

Zervakis, V., Geogopoulos, D., and Deakopoulos, P. G.: The role of the North Aegean in triggering the recent Eastern Mediterranean climatic changes, J. Geophys. Res., 105, 26103–26116, doi:10.1029/2000JC900131, 2000.

Zoccolotti, L. and Salusti, E.: Observation of vein of very dense water in the southern Adriatic Sea, Cont. Shelf Res., 7, 535–551, 1987.

Zodiatis, G., Drakopoulos, P., Brenner, S., and Groom, A. S.: Variability of the Cyprus warm core eddy during the CYCLOPS project, Deep-Sea Res. Part II, 52, 2897–2910, 2005.

Zodiatis, G., Hayes, R. D., Lardner, R., and Georgio, G.: Subregional forecasting and observing system in the Eastrn Mediterranean Levantine Basin: the Cyprus Coastal Ocean Forecasting and Observing System (CYCOFS), in: Towards an integrated system of Mediterranean marine observatories, edited by: Briand, F. CIESM Workshop Monographs, 34, 101–106, 2008.

8

Simulated melt rates for the Totten and Dalton ice shelves

D. E. Gwyther[1,2], B. K. Galton-Fenzi[3,2], J. R. Hunter[2], and J. L. Roberts[3,2]

[1]Institute for Marine and Antarctic Studies, University of Tasmania, Private Bag 129, Hobart, Tasmania 7001, Australia
[2]Antarctic Climate & Ecosystems Cooperative Research Centre, University of Tasmania, Private Bag 80, Hobart, Tasmania 7001, Australia
[3]Australian Antarctic Division, Channel Highway, Kingston, Tasmania 7050, Australia

Correspondence to: D. E. Gwyther (david.gwyther@gmail.com)

Abstract. The Totten Glacier is rapidly losing mass. It has been suggested that this mass loss is driven by changes in oceanic forcing; however, the details of the ice–ocean interaction are unknown. Here we present results from an ice shelf–ocean model of the region that includes the Totten, Dalton and Moscow University ice shelves, based on the Regional Oceanic Modeling System for the period 1992–2007. Simulated area-averaged basal melt rates (net basal mass loss) for the Totten and Dalton ice shelves are 9.1 m ice yr^{-1} (44.5 Gt ice yr^{-1}) and 10.1 m ice yr^{-1} (46.6 Gt ice yr^{-1}), respectively. The melting of the ice shelves varies strongly on seasonal and interannual timescales. Basal melting (mass loss) from the Totten ice shelf spans a range of 5.7 m ice yr^{-1} (28 Gt ice yr^{-1}) on interannual timescales and 3.4 m ice yr^{-1} (17 Gt ice yr^{-1}) on seasonal timescales.

This study links basal melt of the Totten and Dalton ice shelves to warm water intrusions across the continental shelf break and atmosphere–ocean heat exchange. Totten ice shelf melting is high when the nearby Dalton polynya interannual strength is below average, and vice versa. Melting of the Dalton ice shelf is primarily controlled by the strength of warm water intrusions across the Dalton rise and into the ice shelf cavity. During periods of strong westward coastal current flow, Dalton melt water flows directly under the Totten ice shelf further reducing melting. This is the first such modelling study of this region to provide a valuable framework for directing future observational and modelling efforts.

1 Introduction

Understanding how changing ocean circulation and properties are causing increased basal melt of Antarctic ice shelves is crucial for predicting future sea level rise. The Totten Glacier (see Fig. 1) drains over 570 000 km^2 of East Antarctica (Rignot et al., 2008). Most of the ice sheet that drains through the Totten Glacier is from the Aurora Subglacial Basin (Young et al., 2011) and is marine based (i.e. the ice base is grounded below sea level; Roberts et al., 2011) making the region potentially vulnerable to rapid ice sheet collapse. Weertman (1974) showed that ice grounded below sea level is inherently unstable, particularly where the bedrock slopes downwards, away from the ocean, as is the case in the Aurora Subglacial Basin (Young et al., 2011). More recent studies have indicated lateral stresses may act to stabilise ice retreat on a reverse bed slope (Jamieson et al., 2012), but the applicability of this on a broad drainage basin, where lateral stresses are lower, is uncertain.

The Antarctic ice sheet contains enough ice to raise sea level by over 58 metres (Fretwell et al., 2013). Observations of global mean sea level rise over the period 1993–2008, indicate an average rate of 2.61 ± 0.55 mm yr^{-1}, with an Antarctic contribution of 0.43 ± 0.2 mm yr^{-1} (Church et al., 2011). Projections of sea level rise have a large uncertainty in the contribution of Antarctica, due to lack of observations of ice discharge rates and surface mass balance (Gregory et al., 2013).

The ice sheet flows towards the ocean through both broad, slow ice-sheet flow and many fast-flowing glaciers. Ice shelves form in coastal regions where the ice thickness of the ice sheet or glacier is insufficient to maintain contact

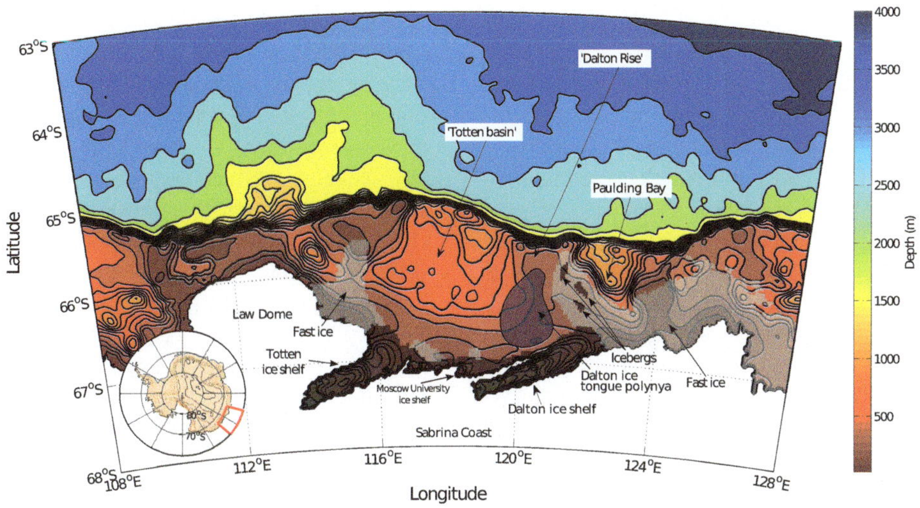

Fig. 1. Model bathymetry and domain. The contour interval is 100 m up to 1500 m depth and 500 m up to 4000 m depth. Ice shelves and icebergs (dark shading), fast ice (bright shading), polynya (mauve shading with dark outline) and major bathymetric features are indicated. Inset shows model domain location in Antarctica.

with the bedrock and the ice begins to float on the ocean. An ocean-filled cavity, insulated from the atmosphere, forms beneath. The sub-ice shelf seabed can be found at depths of over 2800 m below mean sea level (e.g. the deepest part of the Amery ice shelf cavity; Galton-Fenzi et al., 2008). The freezing point of seawater decreases with increasing pressure. Therefore, at the back of deep ice-shelf cavities, the decreased freezing point of seawater provides a large potential for thermal driving of melting, leading to melting at depth and a buoyant melt water plume (Lewis and Perkin, 1986).

Increased melting leads to the localised thinning of an ice shelf, and potentially to the acceleration of the glacier behind it, as the buttressing effect of the shelf is decreased (Dupont and Alley, 2005). An acceleration in flow rate of a glacier causes a mass budget imbalance in the surrounding ice sheet due to the longer response time of the interior ice sheet to flow change; hence the grounded ice sheet thins. For example, observations show surging, acceleration and retreat of tributary glaciers after the collapse of the Larsen A and B ice shelves (De Angelis and Skvarca, 2003; Scambos et al., 2004). It is this dynamic coupling between the ice shelf and the glacier which makes the grounded ice sheet susceptible to oceanic changes (Dupont and Alley, 2005). The ice shelf–ocean interaction is therefore an important control on the discharge of grounded ice into the oceans and subsequently on sea level. Recent satellite observations have shown rapid and accelerated thinning of glaciers along the coastal margins of Antarctica (Pritchard et al., 2009). For example, Pine Island Glacier thinned at rates up to 6 m yr^{-1} while the Totten Glacier displayed thinning rates of 1.9 m yr^{-1} for the period 2003–2007 (Pritchard et al., 2009), three times the rate previously reported (Rignot, 2006). The high thinning rates of marine terminating glaciers suggests a common oceanic driving, for example through increased basal melting (Pritchard et al., 2012).

Glaciologically derived estimates of area-averaged basal melt rate for the Totten ice shelf are 20 ± 9 m yr^{-1} (Rignot, 2002); 26 ± 8 m yr^{-1} (Rignot and Jacobs, 2002); and 10.5 ± 0.7 m yr^{-1} (Rignot et al., 2013). These estimates are derived from satellite interferometric synthetic aperture radar (InSAR) observations of ice surface velocity and the first two are calculated as the difference in flux of ice from the Totten Glacier across the grounding line and across a flux gate 10–30 km downstream, divided by the area of enclosed ice shelf. These measurements should thus be treated as an average melt rate for the grounding line region. The latter estimate was also derived from InSAR observations, but was calculated using the volume flux divergence for the whole ice shelf area. The last estimate is equivalent to a mass loss rate of 63.2 ± 4 Gt yr^{-1} (Rignot et al., 2013). Comparatively, the InSAR estimate for the Moscow University Ice Shelf shows basal melt of 4.7 ± 0.8 m yr^{-1} and a mass loss rate of 27.4 ± 4 Gt yr^{-1} (Rignot et al., 2013). It should be noted that this work defines the Dalton ice shelf as being separate from the Moscow University Ice Shelf. We define the Dalton ice shelf as the ice shelf separated from the ocean by the grounded peninsula which runs north-east from 67.2° S, 118.5° E. This is done since the grounded peninsula is likely to protect the Dalton ice shelf from oceanic influences, and will change the mass loss characteristics of the Dalton as compared to the Moscow University Ice Shelf.

These estimates show the Totten and Dalton basal melting and mass loss to be exceptionally high compared to other ice shelves in East Antarctica (Rignot et al., 2013). Other East Antarctic ice shelves that are experiencing similar basal melt rates are typically smaller ice shelves, and thus mass loss rates are smaller than the mass loss of the Totten. For example, the Wilma/Robert/Downer ice shelves have a basal melt rate of $11.7 \pm 0.7\,\mathrm{m\,yr^{-1}}$, but with an area of only $858\,\mathrm{km^2}$, mass loss ($10.0 \pm 0.6\,\mathrm{Gt\,yr^{-1}}$) is only 16 % as large as the Totten mass loss. Furthermore, unlike many other high melt locations, such as the Pine Island ice shelf which is estimated to have an area-averaged mass loss of $101.2 \pm 8\,\mathrm{Gt\,yr^{-1}}$ (Rignot et al., 2013), no definite causal mechanisms for basal melting of the Totten and Dalton ice shelves are known.

Here we examine the temporal variability and driving mechanisms controlling the basal melt rates of the Totten and Dalton ice shelves using a numerical model. The Totten and Dalton ice shelves are prime candidates for investigation, due to their rapid thinning rates, high estimated basal melt and the limited existing research. Section 2 provides an overview of the Totten and Dalton ice shelf–ocean region. Section 3 presents different hypotheses for drivers of basal melting and in Sect. 4, the regional circulation and heat transport model is described. Sections 5 and 6 present the results of the modelling and a discussion on cross shelf flow and atmosphere exchange processes.

2 Description of region

The Totten ice shelf (centred at 67° S, 116° E) is located on the Sabrina coast in East Antarctica (see Fig. 1). The Sabrina coast is characterised by marine terminating glaciers, ice shelves (such as the Totten, Moscow University and Dalton ice shelves), large tracts of year-round ice adjoining the land (fast ice), extensive seasonal sea ice growth and a region of strong sea ice formation (the Dalton ice tongue polynya). The continental shelf seas are approximately 1000 m deep, with the shelf break occurring between 65° S to 65.5° S in this region. The Antarctic Circumpolar Current is located in the deep abyssal waters to the north of the continental shelf break (Orsi et al., 1995). The Antarctic Circumpolar Current is composed of several water masses, with the Antarctic Surface Water overlying relatively warm and salty Circumpolar Deep Water (CDW). Poleward mixing of CDW with cold and dense Antarctic waters produces "modified" CDW or MCDW (Orsi et al., 1995), at the southern boundary of the Antarctic Circumpolar Current.

The embayed position of the Totten ice shelf, located on the eastern flank of Law Dome and with a broad continental shelf, makes it relatively isolated from the eastward flowing Antarctic Circumpolar Current. The two main oceanic currents near to the coastal ice shelves flow westwards. The Antarctic Coastal Current (ACoC) flows within 50 km of the coast and is associated with the east wind drift. The Antarctic Slope Current (ASC) is topographically controlled and is an important influence on transport across the continental shelf break (Jacobs, 1991).

A large polynya forms west of the Dalton ice shelf tongue (Massom et al., 1998). Polynyas form when strong gravity-driven katabatic winds flow down the Antarctic ice sheet and turn westwards under Coriolis acceleration, removing freshly formed sea ice. Heat is then lost from the uncovered ocean, leading to more sea ice formation and brine rejection as a latent-heat polynya. These are important as they produce large quantities of high-salinity shelf water (HSSW), which contribute to bottom water formation (Rintoul et al., 2001), an important driver of the global climate system (Jacobs, 2004). The easterly katabatic winds are an important controlling factor in polynya strength (Massom et al., 1998). As the ACoC is also controlled by the strength of the easterly winds, a correlation between the strength of both the polynya and the ACoC is to be expected.

3 Overview of drivers of basal melting

The largest process responsible for causing ice shelf thinning is thought to be changes in the magnitude of basal melting (Pritchard et al., 2012; Rignot et al., 2013; Depoorter et al., 2013). Ice shelf thinning by increased basal melt suggests increased oceanic heat available for melting at the ice shelf cavity (Holland et al., 2008; Jacobs et al., 2013). Proposed hypotheses explaining increased heat supply to ice shelf cavities include; redirected coastal ocean currents leading to increased heat content beneath ice shelves (Hellmer et al., 2012; Jacobs et al., 2011); regional upper-ocean hydrography (Hattermann et al., 2014); dynamic eddy-scale activity (Moffat et al., 2009); a link with sea ice conditions (Holland et al., 2010); polynya activity (Cougnon et al., 2013); wind-driven currents (Dinniman and Klinck, 2004); and changes in the Antarctic Circumpolar Current (Gille, 2008; Böning et al., 2008). This paper investigates the mechanisms that influence the oceanic heat supply that drives basal melting.

3.1 Circumpolar Deep Water intrusions

Increased heat supply for ice shelf melting has been suggested to be due to periodic incursions of modified Circumpolar Deep Water (MCDW) onto the continental shelf which can then flow along isopycnals and be delivered to the ice shelf base (Jacobs et al., 1996; Thoma et al., 2008). Observations near and beneath several Antarctic ice shelves show the signature of MCDW. In the case of Pine Island ice shelf, which is showing high melt rates, water measurements taken from an autonomous underwater vehicle show MCDW mixing with cold melt water. This indicates that MCDW is responsible for the high melt rates of Pine Island ice shelf (Jenkins et al., 2010; Jacobs et al., 2011).

In the Totten Glacier region, Conductivity, Temperature and Depth (CTD) measurements have indicated the presence of MCDW on the south side of the continental shelf break, ~ 200 km from the Totten ice shelf (Williams et al., 2011). However, no observations yet exist to determine the temporal and spatial variability or causal mechanism of the MCDW incursion. Consequently, it is not yet clear whether MCDW is the water source driving melting of the Totten ice shelf.

3.2 Interaction with bathymetric features

Observations suggest that intrusions of warm water onto the continental shelf are highly dependent on bathymetric features (e.g. Martinson and McKee, 2012). Geostrophic eddy interaction with small scale bathymetry features can lead to enhanced oceanic mixing via internal wave breaking (Nikurashin and Ferrari, 2010). Broad scale bathymetry features such as the shelf break (as opposed to small scale sills and ridges) can also lead to net onshore intrusion. Inertia of the along-slope water flow can carry the water up the slope and onto the shelf, depending on the curvature and orientation of the continental shelf (Dinniman and Klinck, 2004; Klinck and Dinniman, 2010). In the case of Pine Island Glacier, modelling has shown warm waters being channelled through irregular bathymetric features and submarine troughs (Thoma et al., 2008). The bathymetry in front of the Totten ice shelf is suggestive of possible pathways of MCDW intrusion onto the continental shelf. For example, Paulding Bay (see Fig. 1) is bordered by a low continental shelf break, possibly allowing bathymetric intrusion. The western side of the Dalton rise is another location where the deepening of the continental shelf break may allow intrusion of MCDW. Current bathymetry products of the continental shelf break and shelf proper in this region are informed by satellite gravity inversions, and have limited ship-based groundtruthing. Consequently, there is a strong need to acquire accurate ship-based and airborne bathymetry data.

3.3 Eddies

Eddies may also transport and mix heat across the shelf break. Observations in the western Antarctic Peninsula have characterised eddies as transporting Circumpolar Deep Water onto the continental shelf (see Moffat et al., 2009). The Antarctic Circumpolar Current is recognised as an important generator of eddies, which carry heat polewards (Rintoul et al., 2001). Closer to the continental shelf, interaction of the westward flowing coastal current with bathymetric features, such as ridges along the shelf break may lead to eddy generation. Numerical models have shown topographic Rossby wave interaction with bathymetry is an important factor for generating eddies, which carry a significant amount of heat across the continental shelf (St-Laurent et al., 2013).

Direct modelling of eddy generation and evolution is limited by model grid resolution and bathymetry detail. Eddy-permitting flow can be simulated with horizontal grid resolutions of approximately 2–3 km (Klinck and Dinniman, 2010), however, a horizontal grid resolution of 1 km is required in order to resolve individual eddies (St-Laurent et al., 2013). The modelling work presented here (with a horizontal grid resolution of 2.5–3.5 km) is eddy-permitting, where the mean flow and associated interactions are resolved.

3.4 Atmospheric forcing

In analysing the source of increased heat supply to the Totten ice shelf region, we must also consider the effect of the atmospheric forcing on the ocean. Atmospheric forcing consists of transfers of momentum (surface wind stress) and buoyancy (heat and/or salt fluxes), such brine rejection in sea ice formation and polynya regions.

In some cases, local meteorological conditions favour wind stress patterns that drive currents over the continental shelf break and onto the shelf. For example, modelling of Pine Island ice shelf showed a strong correlation between flow of CDW onto the continental shelf and local temporal variations in synoptic wind stress conditions (Thoma et al., 2008).

The polynya located to the east of the Totten ice shelf produces dense HSSW. HSSW flow may interact in different ways with the local environment, and until observations are taken, modelling presents the only method to understand the implications of the polynya for the oceanography of the region. Kämpf (2005) suggested that dense water cascading down submarine canyons and off the shelf break could induce upwelling of deeper water onto the shelf, which was confirmed in numerical simulations and laboratory experiments (Kämpf, 2007). Alternatively, HSSW produced by polynya activity could be guided by the westward flowing coastal current, directly towards the mouth of the Totten ice shelf cavity, potentially cooling sub-ice shelf waters and quenching melting.

Ice concentration observations taken by the Special Sensor Microwave Imager are used to infer heat and salt flux out of the ocean (Tamura et al., 2008). The heat flux out of the Dalton ice tongue polynya using the method of Tamura et al. (2008) is shown in Fig. 2a, where a positive value indicates loss of heat to the atmosphere and formation of HSSW, and is considered as a proxy for polynya activity. A peak in polynya activity coincides with the Austral winter, due to strong wind over the region (see Fig. 2a, green line) and low air temperatures. Between approximately mid-February and mid-November the polynya removes heat from the ocean. The persistence and duration of this polynya before the satellite era is unknown.

Long-term polynya activity can be analysed by considering the polynya heat flux anomaly – calculated as the difference between the heat flux and a climatological value for each month, smoothed with a Gaussian filter (the filter standard deviation is $\sigma_{\mathrm{filter}} = 4$), as shown in Fig. 2b. The climatology is calculated over the 1992–2007 period. The heat flux anomaly for the Dalton ice tongue polynya region is derived from the buoyancy flux data (see Tamura et al., 2008) used to force the open ocean surface of the model. A positive heat flux anomaly indicates above average heat loss from

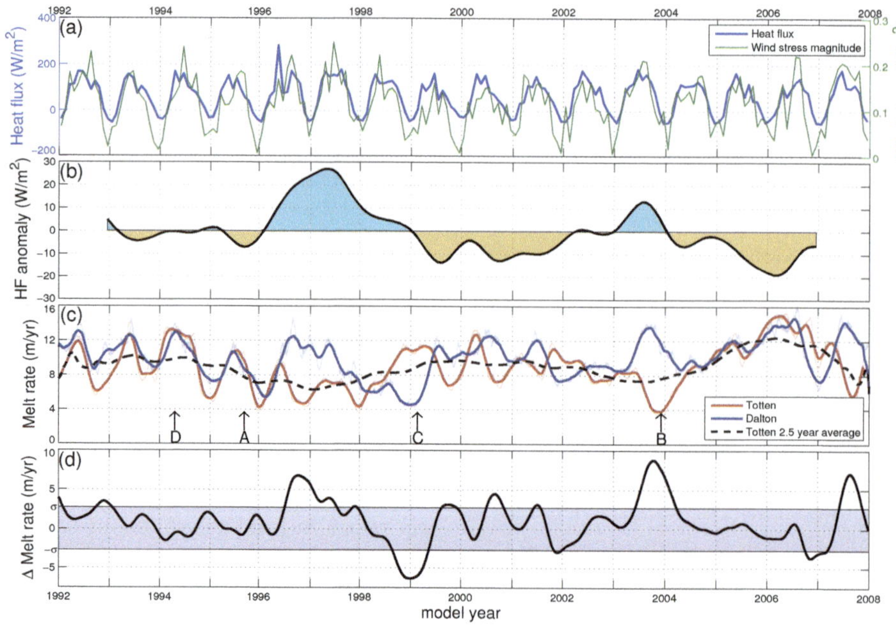

Fig. 2. **(a)** Heat flux out of ocean from SSM/I observations (Tamura et al., 2008) and the magnitude of wind stress, for the Dalton ice tongue polynya. **(b)** Difference between monthly heat flux and climatological heat flux (calculated from 1992–2007 data) for respective months, smoothed with a Gaussian filter ($\sigma_{\text{filter}} = 4$), and referred to as the "heat flux anomaly". Heat flux is defined out of the ocean, so positive anomaly periods indicate stronger than average heat loss from the ocean. **(c)** Model melt rates for the Totten and Dalton ice shelves. The times of the snapshots in Fig. 5 are marked A to D. **(d)** Difference between melt rates of the Totten and Dalton ice shelves. A positive difference indicates Dalton melting that is stronger than Totten melting. Raw melt rates are smoothed with a 2.5-year moving average filter, ± 1 standard deviation ($2.3\,\text{m}\,\text{yr}^{-1}$) of difference is shaded and years are marked for the 1st of January. In **(c)**, light blue and light red lines are the raw Dalton and Totten melt rates, blue and red lines are the raw melt rates smoothed with a 3-month moving average filter and the dashed black line is the Totten melt rate smoothed with a 2.5-year moving average filter.

the ocean and stronger than average polynya activity and thus, the resulting signal is a proxy for interannual polynya strength.

4 Model details

A modified version of Regional Oceanic Modeling System (ROMS; Shchepetkin and McWilliams, 2005), a 3-D primitive equation finite difference model, is used to simulate ice shelf–ocean interaction. ROMS was modified to include realistic tidal forcing, thermodynamic interaction between ocean and the ice shelf (see Galton-Fenzi et al., 2012; Dinniman et al., 2007) and frazil thermodynamics (see Galton-Fenzi et al., 2012). A three-equation formulation parameterises ice–ocean thermodynamics, following Holland and Jenkins (1999).

The model domain (part of which is shown in Fig. 1) covers 104.5° E to 130° E and 60° S to 68° S, with a grid resolution of 2.5 km to 3.5 km and 31 terrain-following layers in the vertical. Model bathymetry in the open ocean is provided by RTOPO-1 (Timmermann et al., 2010), which in turn is based on S-2004 (Marks and Smith, 2006), a combination of the gravity map of Smith and Sandwell (1997) and the ship-based echo soundings of the General Bathymetry Chart of the Oceans (GEBCO; British Oceanographic Data Centre, 2003). Ice shelf thickness over the Totten, Dalton and Moscow University ice shelves is inferred by hydrostatic inversion of ICESat ice surface elevation measurements. The bathymetry under the ice shelves is unknown, so cavity thickness, T_C, is set to a nominal depth, $T_C = 300$ m, representative of the typical thickness of an ice shelf cavity along the cavity centreline (Galton-Fenzi et al., 2008), while along the grounding line, $T_C = 0$. Within the cavity, T_C is linearly interpolated between the cavity centreline and grounding line, except where the centreline approaches the grounding line. Here, T_C is not set, allowing a smooth interpolation from grounding line to centreline. The choice of T_C along the centreline is justifiable, as uncertainty in this depth should only affect the spatial distribution of melt/freeze regions (Holland et al., 2008) and not open ocean circulation and heat transport into the cavity. Furthermore, the agreement between modelled basal melt rates and glaciological estimates (see Sect. 5.1) attests to the scale of the ocean cavity being of the correct order. Fast ice is included from the maps of Fraser et al. (2012) with a 5 m draft (Massom et al., 2001).

There are several sea ice models available for integration into ROMS (Budgell, 2005). However, sea ice production from coupled sea ice models is sensitive to the representation of grounded icebergs (Kusahara et al., 2010), leading to the possibility of poor representation of polynyas forming in the lee of grounded icebergs (such as the Dalton ice tongue polynya). Consequently, we parameterise sea ice formation using the heat and salt ux algorithms of Tamura et al. (2008), which incorporate observed ice concentration for the period 1992–2007 from the Special Sensor Microwave Imager (SSM/I).

Currents, heat and salt fluxes on the lateral boundaries are relaxed to monthly values from the ECCO2 reanalysis product (Menemenlis et al., 2008). The surface is forced with wind stress from COREv2 (Large and Yeager, 2009) and the buoyancy fluxes of Tamura et al. (2008).

The model was run for by 32 years; comprised of 16 years of repeated 1992 forcing, followed by observed forcing for the period 1992–2007. After the 16-year spinup phase, the model was observed to have attained a pseudo-steady state as shown by the total ocean heat content for the spinup phase in Fig. 3.

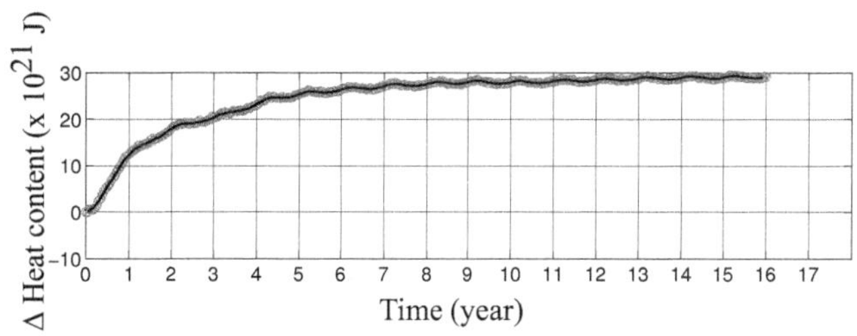

Fig. 3. The total ocean heat content within the model for the spinup period. In the spinup, the model is run with 1992 forcing repeated for 16 years.

5 Results

5.1 Computed ice shelf melt rate

Melting and freezing at the ice–ocean interface is calculated for the ice in the domain. Area-averaged melt rates for the Totten and Dalton ice shelves are shown in Fig. 2c, while the spatial melt rate distributions are shown in Fig. 4. These are at the same times as the corresponding circulation patterns in Fig. 5. In Fig. 2c, the smoothed interannual Totten melt rate is shown by a dashed black line, and is calculated by smoothing the area-averaged Totten melt rate with a 2.5-year moving average filter. The area of the Totten and Dalton ice shelves, which are used for calculating the area-averaged melt rates and mass loss, are defined as $5402\,km^2$ and $5113\,km^2$, respectively.

For both the Totten and Dalton ice shelves, simulated whole ice shelf area-averaged melt rates range from $4\,m\ ice\,yr^{-1}$ to $16\,m\ ice\,yr^{-1}$. The Totten ice shelf area-averaged melt rate (mass loss) averaged over the period 1992–2007 is modelled as $9.1\,m\ ice\,yr^{-1}$ ($44.5\,Gt\ ice\,yr^{-1}$), which compares well to the most recent estimate of area-averaged melt rate of $10.5 \pm 0.7\,m\,yr^{-1}$ (Rignot et al., 2013). The interannual variability in the Totten melt rate (shown as the dashed black line in Fig. 2c) has a mean range of $5.7\,m\ ice\,yr^{-1}$ (mass loss of $26\,Gt\ ice\,yr^{-1}$), while the seasonal variability has a mean range of $3.4\,m\ ice\,yr^{-1}$ (mass loss of $17\,Gt\ ice\,yr^{-1}$). The area-averaged melt rate in the grounding line region is likely to be among the highest rates found under the Totten ice shelf. The model area-averaged melt rate in the Totten grounding line region ranges from $16\,m\ ice\,yr^{-1}$ to $45\,m\ ice\,yr^{-1}$ over the 1992 to 2007 period, with an average value of $28\,m\ ice\,yr^{-1}$, in good agreement with the estimates from InSAR ($20 \pm 9\,m\,yr^{-1}$; Rignot, 2002; $26 \pm 8\,m\,yr^{-1}$; Rignot and Jacobs, 2002).

While there are no glaciological estimates of melt rate for the Dalton ice shelf, the model area-averaged Dalton melt rate (mass loss) for the 1992–2007 period is $10.1\,m\ ice\,yr^{-1}$ ($46.6\,Gt\ ice\,yr^{-1}$). The strength of basal melt of the Dalton ice shelf, similar in magnitude to the Totten ice shelf, suggests that this region is also important for the Antarctic mass balance budget.

5.2 Ocean circulation features

Simulated model currents are shown in Fig. 5a–d. We find four different atmospheric and ocean circulation regimes, and subfigures in Fig. 5 are chosen at times representing these regimes. The model adequately reproduces the westward flowing ACoC with maximum transport during periods of strong easterly winds (see Fig. 5a), and very low transport during weak easterly winds (see Fig. 5d).

The ASC is also reproduced, flowing westwards along the continental shelf break (see Fig. 5b). The current is channelled by bathymetric features, such as where the shelf break curves north-west at 116° E, 64.5° S and approximately flows between the 2000 m and 2500 m isobaths. Consequently, the ASC is differentiated from coastal flow associated with the ACoC by both the zonal location and temperature of water within the flow. Currents for the ASC can be as high as $20\,cm\,s^{-1}$, but have a mean flow of approximately $5\,cm\,s^{-1}$. These model simulated currents agree well with the sparse nearby observations of Bindoff et al. (2000) and Williams et al. (2011), which reported observed currents of approximately $2\text{–}15\,cm\,s^{-1}$ and $2.3\text{–}6.4\,cm\,s^{-1}$, respectively. Modelling studies of circum-Antarctic ASC transport predict similar velocities (Mathiot et al., 2011).

The bathymetry of the continental shelf break acts to guide flow onto the shelf. The ASC interacts with several barriers as it flows westwards through the model domain. The eastern edge of the rise located at 122° E, 66° S, henceforth named "Dalton rise" can act to guide water southwards towards the mouth of the Dalton ice shelf. West of the Dalton rise, a large basin with an average depth of 750 m exists, henceforth named the "Totten basin". Interaction of the westward

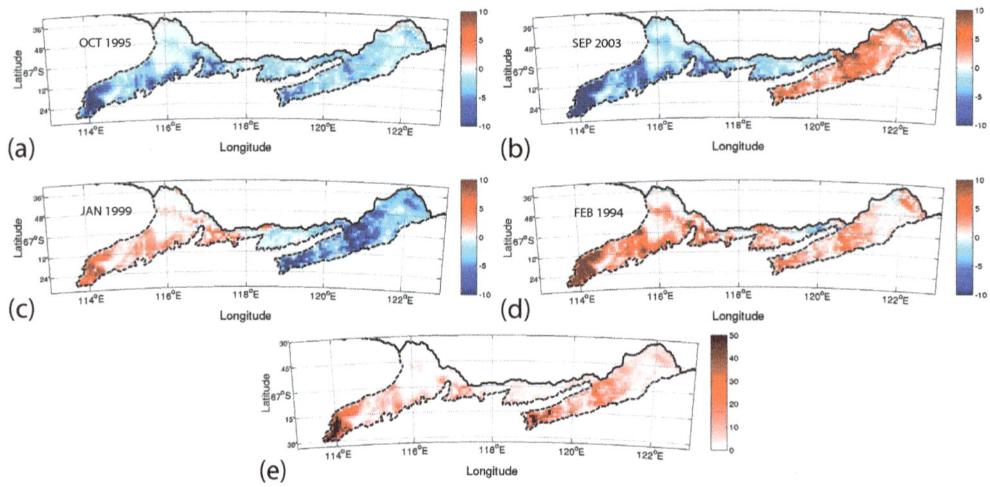

Fig. 4. The anomaly from the mean melt rate over the period 1992–2007 is shown at **(a)** October 1995 **(b)** September 2003 **(c)** January 1999 **(d)** February 1994. **(e)** The mean melt rate over the period 1992–2007. Larger positive numbers indicate higher melting.

flowing ASC with the the western edge of the Dalton rise generates a southward flowing jet clockwise around the Totten basin. This interaction, which is dependent on the meridional location of the ASC front, floods the Totten basin with warmer water, supplying heat that is available to drive the melting of the Totten ice shelf.

Observations to support these model results do not exist, as the region is covered by sea ice for much of the year. This makes long term estimation of sea surface currents by satellite observation impossible, and ship-borne observations sparse and difficult to acquire.

5.3 Heat

The temperature at each layer of the model generally decreases polewards. The water that makes up the ASC is typically between 0 °C to 0.4 °C, though at several points during the 1992–2007 time period, the temperature rises to approximately 0.5 °C. The water that makes up the ACoC (−1 °C to −0.5 °C) is cooler than the ASC, as a result of mixing with cold meltwater from glaciers and ice shelves, and heat lost to the atmosphere.

Model output shows vertical water structure at the shelf break generally composed of relatively warm and salty CDW (with temperatures warmer than 1 °C and salinity approximately 34.7) overlayed with fresher and cooler Antarctic Surface Water (temperatures of ∼ −2 °C and salinity of 34.2). Flow is mostly barotropic along the shelf break except for baroclinic bottom-intensified flows at areas of changing bathymetry (such as the Dalton rise).

On the Totten region continental shelf, the water masses include slightly cooler but salty CDW (MCDW formed through mixing with shelf waters); this is overlayed by cold and fresh Antarctic Surface Water; while along the coastline, a seasonally varying cold and salty HSSW (with temperatures of ∼ −2 °C or less and salinity of 34.7 or greater) flows westwards. On the shelf, model output shows the westward flow of HSSW is bottom-intensified and can entrain surface waters with it. The flow of the ACoC is generally surface-intensified and carries meltwater from the Dalton ice shelf and ice tongue westwards, however, these two flows can combine to form approximately-barotropic westward flow. A large clockwise recirculation feature forms in the Totten basin. This feature, which is surface intensified, carries heat clockwise from the shelf break towards both major ice shelves, and is composed of water with temperature between −0.2 to 0.4 °C and salinity of 34.5 to 34.6.

6 Discussion

Model results show variability in melt rate over the course of each year. However, the interannual Totten melt rate, the dashed black line in Fig. 2c, shows variation from high to low melt rate and vice versa, with a ∼2–3 year period. The ∼2–3 year modulation of melt rate signal is noticeable as an increasing annual average melt between 1992 to 1994 (most visible in the 3-month Totten melt rate, shown as the red line in Fig. 2c), 1998 to 2002, 2004 to 2007 and decreasing annual average melt between 1994 to 1998 and between 2002 and 2004. This pattern of increasing and decreasing melt repeats over the 1992–2007 period. However, there is no overall trend (over the 1992–2007 period) in Totten ice shelf melt discernible above the ∼2–3 year melt variation. Thus, we focus the analysis on understanding the processes governing simulated interannual variability in melt rates.

Satellite-derived estimates of ice shelf thinning indicate that for the period 2003–2008 thinning of the Totten ice shelf was the fourth highest in Antarctica (Pritchard et al., 2012). Over the period 2003–2007, the model area-averaged melt

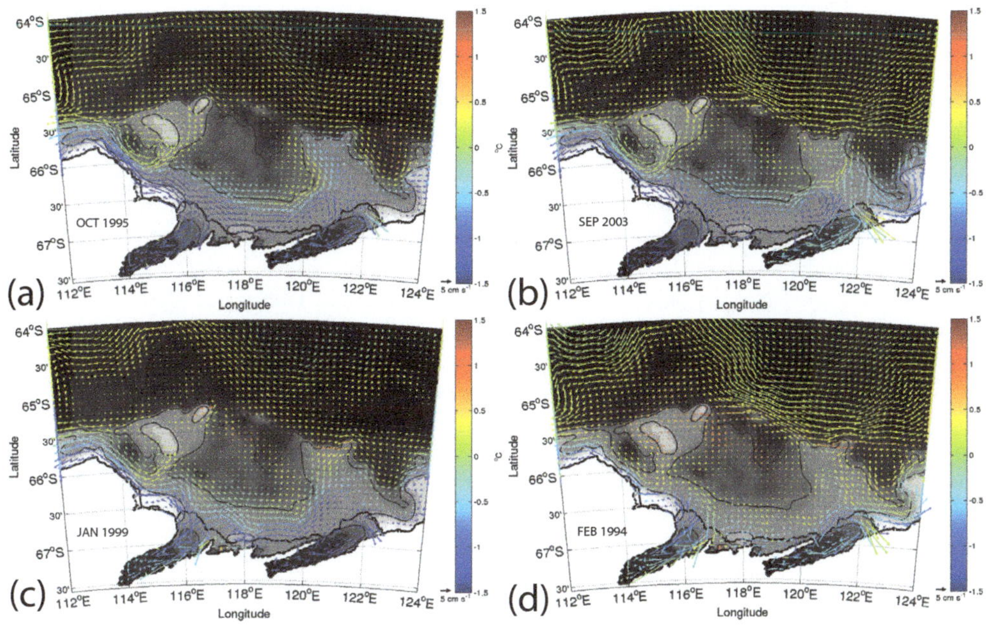

Fig. 5. Simulated column-averaged currents (excluding the surface layer), where the vectors are coloured for mean column temperature, and scaled according to the reference vector. **(a)** October 1995. Strong winds drive polynya activity and the ACoC, inhibiting heat flow to the Totten ice shelf. There is no flow over the Dalton rise. **(b)** September 2003. Strong polynya activity increases HSSW production, preventing warm water entry to the Totten ice shelf cavity and decreasing melt rates. Flow over the Dalton rise supplies heat for melting to the Dalton ice shelf. **(c)** January 1999. Polynya activity and ACoC are weak, allowing heat supply to cause melting of the Totten ice shelf. There is no flow over the Dalton rise, so Dalton ice shelf melt rates drop. **(d)** February 1994. Polynya activity and ACoC are very weak, allowing warm inflow to Totten ice shelf. Flow over the Dalton rise can increase melting of Dalton ice shelf.

rate increased to its highest value, in agreement with an ice shelf that is rapidly thinning.

Oceanic heat is supplied to both ice shelves across the continental slope. However, variability in heat supply at the ice shelves is also dependent on the nearby Dalton ice tongue polynya. HSSW produced by the polynya can flow westwards, reducing supply of heat into the Totten ice shelf cavity. We proceed by examining model currents and atmosphere–ocean exchange together with variability in ice shelf melt rate.

6.1 Atmosphere–ocean exchange

Strong katabatic winds that drive polynya activity, are also partly responsible for driving the ACoC. Generally, as polynya activity increases from March to April, large quantities of HSSW are formed in the model. When the polynya is strongly active, the ACoC is also strong and the resulting westward flow directs most of the HSSW towards the Totten ice shelf. As polynya activity decreases in August, production of HSSW decreases, allowing warm slope front current (which flows across the continental shelf break and into the Totten basin), access to the ice shelf cavity to drive high melt of the Totten ice shelf.

Times when HSSW production and ACoC are strong lead to cold, dense HSSW entering the ice shelf cavity and decreased melting. However, the timing of the minimum of Totten ice shelf melt (generally in December) lags the peak in polynya activity and HSSW production in June (compare Fig. 2c and a). Increasing polynya activity causes the rate of change of melt rate to become negative and so the melt rate passes through a local maximum. The melt rate continues to decrease until polynya activity decreases, at which point the rate of change of melt rate becomes positive. The melt rate then passes through a local minimum and increases. The lag between polynya maximum and Totten melt minimum is created because the bulk volume of water below the ice shelf requires time to cool as HSSW is added and time to heat up as warm water is added. The Dalton ice shelf exhibits a similar temporal pattern of melt rate as the Totten ice shelf, suggesting that polynya activity is an influential factor on Dalton melt rate as well.

The heat flux anomaly for the Dalton ice tongue polynya region is shown in Fig. 2b. Here, a positive signal indicates stronger than average heat loss to the atmosphere and stronger HSSW production.

From 1993 to 1995, the heat flux anomaly was negative during the middle of the year, suggesting weaker peak polynya strength. Consequently, both ice shelves display strong melting. From 1996 to 1998, the heat flux anomaly was strongly positive, leading to strong HSSW production and the Totten progressively decreasing in melt rate during this period. From mid-1999 to mid-2002, the heat flux anomaly was negative and consequently, HSSW production dropped and oceanic heat supply increased to both ice shelves. Annual average melt was thus higher for this period than for the 1996–1998 period when the heat flux anomaly was strongly positive. From late 2002 to 2004, the heat flux anomaly was weakly positive, coinciding with weakened melt rates. For mid-2004 and from 2005 to 2007, the heat flux anomaly was again strongly negative, which led to reduced production of HSSW and increased melt as heat supply increased. The years with a strong seasonal variation in melt rate (see 1992 through to 1996) are years when atmosphere–ocean exchange (in the form of heat loss to the atmosphere) was the dominant mechanism modulating seasonal variation in the melt rate. Other years, such as 1998 and 2003, displayed a more complex melt rate evolution that was likely a combination of atmospheric forcing and melting caused by a time-dependent warm water source.

The seasonal changes in atmosphere–ocean fluxes explains the seasonal evolution of melt of both ice shelves, while the heat flux anomaly explains the interannual trend in melt rate of both ice shelves. However, there were several periods when the melt rate of the Totten and Dalton ice shelves fluctuated quite differently from each other, indicating melting not driven solely by polynya activity.

6.2 Shelf exchange processes and heat supply for melting

The primary source of heat for melting is supplied across the continental shelf. Bathymetry is important for controlling intrusion of warm water by guiding the slope current. Model results show that regional topographic features (the Dalton rise and Totten basin) act to guide water onto the continental shelf. These features are regions of intermittent warm inflow. The mechanism by which the flow of the ASC can be diverted by bathymetry is probably different for each topographic feature.

Geostrophic flows tend to conserve potential vorticity f/H, where f is the Coriolis parameter (which varies with latitude) and H represents the water depth, leading to the tendency of geostrophic flows primarily following isobaths. If a change in bathymetry is encountered, resulting flow will tend to divert to a different latitude to counter the change in depth. In the model results, we see this by the poleward current generated as the ASC interacts with the edge of the Totten basin. The narrow current formation is supported by idealised modelling studies, which suggest a significant onshore flow is generated by interaction with flow along the shelf break and a trough (St-Laurent et al., 2013).

Melting of the Dalton ice shelf is strengthened by occasional flow of warm water into the basin of Paulding Bay, east of the Dalton rise. Model currents suggest counter-clockwise circulation forming in Paulding Bay when the ASC impacts with the north-eastern tip of the Dalton rise. The resulting circulation is onto the Dalton rise and into the ice shelf cavity (see Fig. 5b).

The difference between the simulated Totten and Dalton melt rates is shown in Fig. 2d. A positive value indicates the Totten melting is lower than Dalton, while a negative value indicates Dalton melting is weaker than Totten. The signal is smoothed by a 6-month moving average and ± 1 standard deviation is shaded. A difference greater in magnitude than 1 standard deviation indicates periods when the melt rates of the Totten and Dalton ice shelves are significantly uncorrelated. Since polynya activity drives the melt rate signal on a seasonal timescale, significantly different melt rates indicate times when the melt rate of at least one of the ice shelves is not purely driven by polynya activity.

Late 1996, late 2000, late 2003 and late 2007 were times when there was a strong supply of heat to the Dalton ice shelf, but not to the Totten ice shelf (see Fig. 2d). During these times, oceanic currents supplied heat over the Dalton rise only to the Dalton ice shelf, but polynya activity and formation of cold, dense HSSW was strong enough to limit Totten melting (see Fig. 5b). Circulation also shows cold Dalton melt water leaving the cavity and flowing westwards to the Totten ice shelf cavity.

Late 1998, early 1999 and late 2007 were times when the Totten ice shelf experienced strong melting but the Dalton ice shelf did not. Polynya activity at these times was weaker than average, which led to lower rates of HSSW production. Model circulation at these times shows weak flow of HSSW to the Totten ice shelf (see Fig. 5c). Warm water that intrudes into the Totten basin can consequently enter the ice shelf cavity and drive increased melting. Weaker westward ACoC flow results in HSSW flowing into the nearby Dalton ice shelf cavity, decreasing Dalton melting.

6.3 Melt rate forcing by a combination of polynya activity and circulation

The melt rates displayed by both the Totten and Dalton ice shelves are a complex combination of a seasonal signal with occasional periods of very high and very low melt. Comparing melt rate and polynya activity, as shown in Fig. 2c and b, we see that in general, periods of lower melt are correlated with periods of stronger polynya activity. However, there were several short periods when ocean circulation features caused the melt rate of the Dalton and Totten ice shelves to be out of phase. This suggests that a combination of atmosphere–ocean exchange and shelf break flow mechanisms is required to fully explain the melt rate of the Dalton and Totten ice shelves.

We propose a set of four main states of circulation and polynya activity, that drive melt of the Dalton and Totten ice shelves. These four states describe the configuration of on-shelf flow and polynya activity, and explain the seasonal and interannual variation of the melt rate of the Totten and Dalton ice shelves. These states are described as follows:

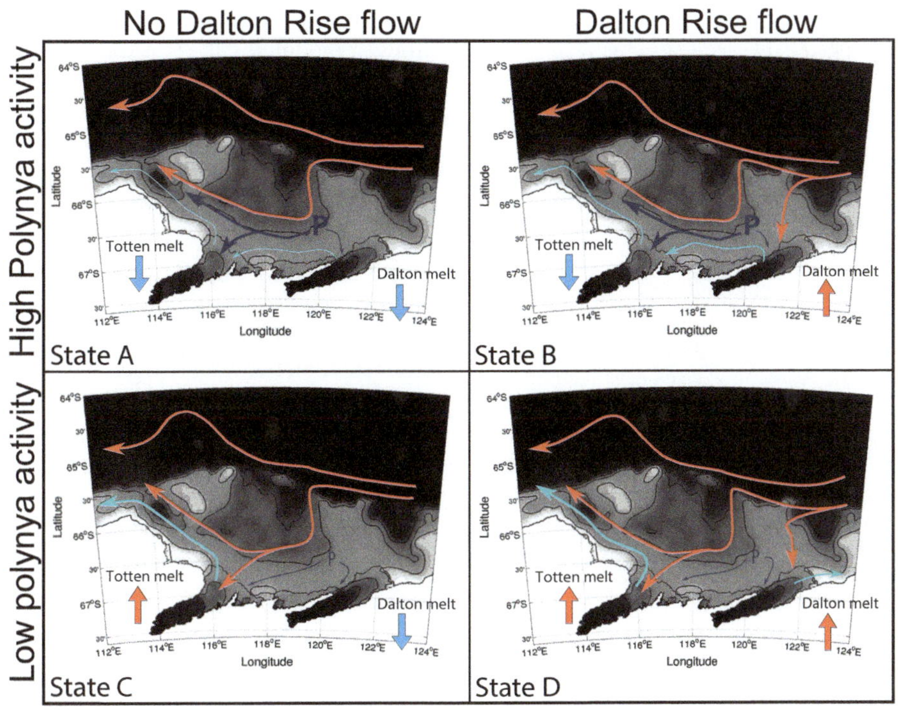

Fig. 6. ASC (light red line), HSSW (dark blue line) and ice shelf meltwater (light blue line) are shown for; States **(A)**: here, polynya activity is strong (as shown by a bold "P"), but the ASC only enters the Totten basin. Consequently, both ice shelves show lowered melt rates. **(B)**: polynya activity is strong, however the ASC comes onshore on both sides of the Dalton rise, leading to increased Dalton ice shelf melt rates. HSSW produced by the polynya and Dalton ice shelf melt water combine to lower Totten melt rates. **(C)**: polynya activity is weak (as shown by a thinner "P") and warm water only flows into the Totten basin. Consequently, the Totten ice shelf displays strong melting, while the Dalton ice shelf displays weak melting. **(D)**: polynya activity is weak and warm water onflow over the Dalton rise and into the Totten basin lead to high melt rates for both ice shelves.

- State A: The meridional position of the ASC means that it interacts only with the western edge of the Dalton rise. The resulting jet flows southward into the Totten basin, following the isobaths clockwise until it rejoins the ASC north of Law Dome. The polynya and ACoC are strongly active, causing HSSW to flow into the Dalton ice shelf cavity and westwards to the Totten ice shelf cavity inhibiting melting of both ice shelves. This state is illustrated in Fig. 6a and shown in model currents in Fig. 5a and melt rate anomaly distribution is shown in Fig. 4a.

- State B: Polynya activity is strong, but now the ASC interacts with the bathymetry offshore of the Dalton rise, which gives rise to a warm current over the Dalton rise, directly to the mouth of the Dalton ice shelf. The increased melting leads to a strong meltwater outflow westwards towards the Totten ice shelf, inhibiting melting of the Totten. The Totten ice shelf shows very low melt rates, while the Dalton ice shelf displays strong melting, as illustrated in Fig. 6b and shown in model currents in Fig. 5b and melt rate anomaly distribution in Fig. 4b.

- State C: Weak polynya activity and ACoC strength leads to reduced HSSW flow to both the Totten and Dalton ice shelves. The ASC only interacts with the western edge of the Dalton rise, generating a warm current into the Totten basin, which drives strong Totten ice shelf melting. Since the Dalton ice shelf has very little warm inflow, the lowered HSSW production is enough to inhibit melting. This state is shown in Fig. 6c and shown in model currents in Fig. 5c and in melt rate anomaly distribution in Fig. 4c.

- State D: Polynya activity and ACoC strength is much weaker than in States A and B. As a result, HSSW production is lowered and transport of dense, cold water to both the Dalton and Totten ice shelves decreases. The ASC interacts with the Dalton rise, leading to a warm currents flooding the Totten and the Dalton ice shelves. Both ice shelves thus display high melting, but the weak ACoC means that Dalton melt water (which now flows eastwards) does not limit Totten melting as in State B. This state is illustrated in Fig. 6d and shown in model currents in Fig. 5d and in melt rate anomaly distribution in Fig. 4d.

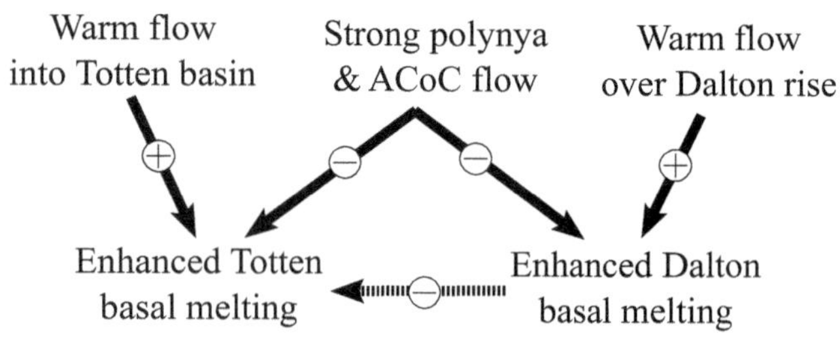

Fig. 7. The interaction between the Totten and Dalton ice shelves, the Dalton ice tongue polynya and on shelf flow is summarised. A negative interaction is shown by ⊖, while a positive interaction is shown by ⊕. Strong polynya and ACoC flow has a negative impact on warm water supply to both ice shelves, while warm water flow into the Totten basin and over the Dalton rise has a positive impact on basal melting of the Totten and Dalton ice shelves, respectively. Increased basal melting of the Dalton ice shelf can have a strong negative affect on the Totten ice shelf. The dependence of this last interaction on strong ACoC flow is shown by the dashed line.

The interactions described by these four states, represented in Fig. 7, summarise the melt rate controls of the Dalton and Totten ice shelves. This interaction chart shows the strong negative impact of polynya activity and ACoC strength on heat supply to both ice shelves. Warm water intrusions that flow over the shelf break can interact with the shelf bathymetry and supply heat to either ice shelf, which will result in enhanced basal melting. If heat is supplied to the Dalton ice shelf, increased melting may lead to a decrease in heat supply for the Totten ice shelf (via outflow of cold Dalton melt water).

These interactions provide a hypothesis describing likely causes of melting of these under-researched ice shelves. A similar mechanism has been proposed linking basal melt of the Mertz Glacier Tongue and a nearby polynya (Cougnon et al., 2013).

This is also the first evidence of the interaction between two ice shelves, where the melt rate of one ice shelf affects the melt rate of the other. This indicates that the melt rate and glaciology of the Totten ice shelf cannot be studied in isolation from the Dalton ice shelf.

7 Conclusions

Remote sensing observations show thinning of the Totten Glacier and suggest strong basal melting of the Totten ice shelf, driven by changing ocean conditions as the most likely explanation. Previous studies have indicated that enhanced exchange of heat across the shelf break is a major cause of increased basal melting (see for example Jacobs et al., 2013). This study shows that, along with exchange of heat across the shelf break, atmosphere–ocean interaction processes on the continental shelf can modify oceanic heat supply to ice shelves, and modulate basal melting.

Dense water formation in the Dalton ice tongue polynya strongly modulates the seasonality of melting for both ice shelves. Melting of the Totten ice shelf is increased when the interannual strength of the Dalton polynya is below average, and vice versa. The Dalton ice shelf melt rate is primarily controlled by intrusions of warm water across the Dalton rise and into the ice shelf cavity. Simulated area-averaged melt rates for the Totten ice shelf agree well with recent glaciological estimates, and suggest the Dalton ice shelf, like the Totten, as being a region of high basal melting.

Furthermore, this study suggests important evidence for the interaction between two ice shelves, where the melting of one directly influences the melting of the other. Lastly, strong interannual variability in heat supply to the ice shelves and subsequent melt rates suggest that observations will need to be long to correctly determine attributable mechanisms.

This is the first such modelling study of this region, and provides valuable information for directing future observations.

Acknowledgements. This work was supported by the Australian Government's Cooperative Research Centre Programme through the Antarctic Climate & Ecosystems Cooperative Research Centre. David Gwyther is supported by the CSIRO and UTAS through the Quantitative Marine Science PhD program.

David Gwyther wishes to acknowledge and thank the helpful comments and suggestions given by Ben Caspani. Early guidance with analysis was provided by Guy Williams. We also thank the two reviewers for their thorough review of the manuscript. Computing resources were provided by both the Tasmanian Partnership for Advanced Computing and the National Computational Infrastructure under grant m68.

Edited by: M. Hoppema

References

Bindoff, N. L., Rosenberg, M. A., and Warner, M. J.: On the circulation and water masses over the Antarctic continental slope and rise between 80 and 150° E, Deep-Sea Res.-Pt. II, 47, 2299–2326, doi:10.1016/S0967-0645(00)00038-2, 2000.

Böning, C. W., Dispert, A., Visbeck, M., Rintoul, S. R., and Schwarzkopf, F. U.: The response of the Antarctic Circumpolar Current to recent climate change, Nat. Geosci., 1, 864–869, doi:10.1038/ngeo362, 2008.

British Oceanographic Data Centre: Centenary Edition of the GEBCO Digital Atlas, [CD-ROM]: Published on behalf of the Intergovernmental Oceanographic Commission and the International Hydrographic Organization, Liverpool, UK, 2003.

Budgell, W.: Numerical simulation of ice-ocean variability in the Barents Sea region, Ocean Dynam., 55, 370–387, doi:10.1007/s10236-005-0008-3, 2005.

Church, J. A., White, N. J., Konikow, L. F., Domingues, C. M., Cogley, J. G., Rignot, E., Gregory, J. M., van den Broeke, M. R., Monaghan, A. J., and Velicogna, I.: Revisiting the Earth's sea-level and energy budgets from 1961 to 2008, Geophys. Res. Lett., 38, 1–8, doi:10.1029/2011GL048794, 2011.

Cougnon, E. A., Galton-Fenzi, B. K., Meijers, A. J. S., and Legrésy, B.: Modelling inter-annual dense shelf water export in the region of the Mertz Glacier Tongue (1992–2007), J. Geophys. Res.-Oceans, 118, 5858–5872, doi:10.1002/2013JC008790, 2013.

De Angelis, H. and Skvarca, P.: Glacier surge after ice shelf collapse, Science, 299, 1560–1562, doi:10.1126/science.1077987, 2003.

Depoorter, M. A., Bamber, J. L., Griggs, J. A., Lenaerts, J. T. M., Ligtenberg, S. R. M., van den Broeke, M. R., and Moholdt, G.: Calving fluxes and basal melt rates of Antarctic ice shelves, Nature, 502, 89–92, doi:10.1038/nature12567, 2013.

Dinniman, M. S. and Klinck, J. M.: A model study of circulation and cross-shelf exchange on the west Antarctic Peninsula continental shelf, Deep-Sea Res.-Pt. II, 51, 2003–2022, doi:10.1016/j.dsr2.2004.07.030, 2004.

Dinniman, M. S., Klinck, J. M., and Smith, W. O.: Influence of sea ice cover and icebergs on circulation and water mass formation in a numerical circulation model of the Ross Sea, Antarctica, J. Geophys. Res., 112, C11013, doi:10.1029/2006JC004036, 2007.

Dupont, T. K. and Alley, R. B.: Assessment of the importance of ice-shelf buttressing to ice-sheet flow, Geophys. Res. Lett., 32, 1–4, doi:10.1029/2004GL022024, 2005.

Fraser, A. D., Massom, R. a., Michael, K. J., Galton-Fenzi, B. K., and Lieser, J. L.: East Antarctic Landfast Sea Ice Distribution and Variability, 2000–08, J. Climate, 25, 1137–1156, doi:10.1175/JCLI-D-10-05032.1, 2012.

Fretwell, P., Pritchard, H. D., Vaughan, D. G., Bamber, J. L., Barrand, N. E., Bell, R., Bianchi, C., Bingham, R. G., Blankenship, D. D., Casassa, G., Catania, G., Callens, D., Conway, H., Cook, A. J., Corr, H. F. J., Damaske, D., Damm, V., Ferraccioli, F., Forsberg, R., Fujita, S., Gim, Y., Gogineni, P., Griggs, J. A., Hindmarsh, R. C. A., Holmlund, P., Holt, J. W., Jacobel, R. W., Jenkins, A., Jokat, W., Jordan, T., King, E. C., Kohler, J., Krabill, W., Riger-Kusk, M., Langley, K. A., Leitchenkov, G., Leuschen, C., Luyendyk, B. P., Matsuoka, K., Mouginot, J., Nitsche, F. O., Nogi, Y., Nost, O. A., Popov, S. V., Rignot, E., Rippin, D. M., Rivera, A., Roberts, J., Ross, N., Siegert, M. J., Smith, A. M., Steinhage, D., Studinger, M., Sun, B., Tinto, B. K., Welch, B. C., Wilson, D., Young, D. A., Xiangbin, C., and Zirizzotti, A.: Bedmap2: improved ice bed, surface and thickness datasets for Antarctica, The Cryosphere, 7, 375–393, doi:10.5194/tc-7-375-2013, 2013.

Galton-Fenzi, B., Maraldi, C., Coleman, R., and Hunter, J.: The cavity under the Amery Ice Shelf, East Antarctica, J. Glaciol., 54, 881–887, doi:10.3189/002214308787779898, 2008.

Galton-Fenzi, B. K., Hunter, J. R., Coleman, R., Marsland, S. J., and Warner, R. C.: Modeling the basal melting and marine ice accretion of the Amery Ice Shelf, J. Geophys. Res., 117, C09031, doi:10.1029/2012JC008214, 2012.

Gille, S. T.: Decadal-Scale Temperature Trends in the Southern Hemisphere Ocean, J. Climate, 21, 4749–4765, doi:10.1175/2008JCLI2131.1, 2008.

Gregory, J. M., White, N. J., Church, J. A., Bierkens, M. F. P., Box, J. E., Van den Broeke, M. R., Cogley, J. G., Fettweis, X., Hanna, E., Huybrechts, P., Konikow, L. F., Leclercq, P. W., Marzeion, B., Oerlemans, J., Tamisiea, M. E., Wada, Y., Wake, L. M., and Van de Wal, R. S. W.: Twentieth-century global-mean sea-level rise: Is the Whole Greater than the sum of the parts?, J. Climate, 26, 4476–4499, doi:10.1175/JCLI-D-12-00319.1, 2013.

Hattermann, T., Smedsrud, L. H., Nøst, O. A., Lilly, J. M., and Galton-fenzi, B. K.: Eddy-resolving model reveals two states of basal melting below Fimbul Ice Shelf, Antarctica, in review, 2014.

Hellmer, H. H., Kauker, F., Timmermann, R., Determann, J., and Rae, J.: Twenty-first-century warming of a large Antarctic ice-shelf cavity by a redirected coastal current, Nature, 485, 225–228, doi:10.1038/nature11064, 2012.

Holland, D. M. and Jenkins, A.: Modeling Thermodynamic Ice-Ocean Interactions at the Base of an Ice Shelf, J. Phys. Oceanogr., 29, 1787–1800, doi:10.1175/1520-0485(1999)029<1787:MTIOIA>2.0.CO;2, 1999.

Holland, P. R., Jenkins, A., and Holland, D. M.: The Response of Ice Shelf Basal Melting to Variations in Ocean Temperature, J. Climate, 21, 2558–2572, doi:10.1175/2007JCLI1909.1, 2008.

Holland, P. R., Jenkins, A., and Holland, D. M.: Ice and ocean processes in the Bellingshausen Sea, Antarctica, J. Geophys. Res., 115, 1–16, doi:10.1029/2008JC005219, 2010.

Jacobs, S. S.: On the nature and significance of the Antarctic Slope Front, Mar. Chem., 35, 9–24, doi:10.1016/S0304-4203(09)90005-6, 1991.

Jacobs, S. S.: Bottom water production and its links with the thermohaline circulation, Antarct. Sci., 16, 427–437, doi:10.1017/S095410200400224X, 2004.

Jacobs, S. S., Hellmer, H. H., and Jenkins, A.: Antarctic ice sheet melting in the Southeast Pacific, Geophys. Res. Lett., 23, 957–960, 1996.

Jacobs, S. S., Jenkins, A., Giulivi, C. F., and Dutrieux, P.: Stronger ocean circulation and increased melting under Pine Island Glacier ice shelf, Nat. Geosci., 4, 519–523, doi:10.1038/ngeo1188, 2011.

Jacobs, S. S., Giulivi, C., Dutrieux, P., Rignot, E., Nitsche, F., and Mouginot, J.: Getz Ice Shelf melting response to changes in ocean forcing, J. Geophys. Res.-Oceans, 118, 4152–4168, doi:10.1002/jgrc.20298, 2013.

Jamieson, S. S. R., Vieli, A., Livingstone, S. J., Cofaigh, C. Ó., Stokes, C., Hillenbrand, C.-D., and Dowdeswell, J. A.: Ice-stream stability on a reverse bed slope, Nat. Geosci., 5, 799–802, doi:10.1038/ngeo1600, 2012.

Jenkins, A., Dutrieux, P., Jacobs, S. S., McPhail, S. D., Perrett, J. R., Webb, A. T., and White, D.: Observations beneath Pine Island Glacier in West Antarctica and implications for its retreat, Nat. Geosci., 3, 468–472, doi:10.1038/ngeo890, 2010.

Kämpf, J.: Cascading-driven upwelling in submarine canyons at high latitudes, J. Geophys. Res., 110, C02007, doi:10.1029/2004JC002554, 2005.

Kämpf, J.: On the magnitude of upwelling fluxes in shelf-break canyons, Cont. Shelf Res., 27, 2211–2223, doi:10.1016/j.csr.2007.05.010, 2007.

Klinck, J. M. and Dinniman, M. S.: Exchange across the shelf break at high southern latitudes, Ocean Sci., 6, 513–524, doi:10.5194/os-6-513-2010, 2010.

Kusahara, K., Hasumi, H., and Tamura, T.: Modeling sea ice production and dense shelf water formation in coastal polynyas around East Antarctica, J. Geophys. Res., 115, C10006, doi:10.1029/2010JC006133, 2010.

Large, W. G. and Yeager, S. G.: The global climatology of an interannually varying air–sea flux data set, Clim. Dynam., 33, 341–364, doi:10.1007/s00382-008-0441-3, 2009.

Lewis, E. L. and Perkin, R. G.: Ice Pumps and Their Rates, J. Geophys. Res., 91, 11756–11762, doi:10.1029/JC091iC10p11756, 1986.

Marks, K. M. and Smith, W. H. F.: An Evaluation of Publicly Available Global Bathymetry Grids, Mar. Geophys. Res., 27, 19–34, doi:10.1007/s11001-005-2095-4, 2006.

Martinson, D. G. and McKee, D. C.: Transport of warm Upper Circumpolar Deep Water onto the western Antarctic Peninsula continental shelf, Ocean Sci., 8, 433–442, doi:10.5194/os-8-433-2012, 2012.

Massom, R. A., Harris, P. T., Michael, K. J., and Potter, M. J.: The distribution and formative processes of latent-heat polynyas in East Antarctica, Ann. Glaciol., 27, 420–426, 1998.

Massom, R. A., Hill, K. L., Lytle, V. I., Worby, A. P., Paget, M. J., and Allison, I.: Effects of regional fast-ice and iceberg distributions on the behaviour of the Mertz Glacier polynya, East Antarctica, Ann. Glaciol., 33, 391–398, 2001.

Mathiot, P., Goosse, H., Fichefet, T., Barnier, B., and Gallée, H.: Modelling the seasonal variability of the Antarctic Slope Current, Ocean Sci., 7, 455–470, doi:10.5194/os-7-455-2011, 2011.

Menemenlis, D., Campin, J.-M., Heimbach, P., Hill, C., Lee, T., Nguyen, A., Schodlok, M., and Zhang, H.: ECCO2 : High Resolution Global Ocean and Sea Ice Data Synthesis, Mercator Ocean Quarterly Newsletter, 31, 13–21, 2008.

Moffat, C., Owens, B., and Beardsley, R. C.: On the characteristics of Circumpolar Deep Water intrusions to the west Antarctic Peninsula Continental Shelf, J. Geophys. Res., 114, C05017, doi:10.1029/2008JC004955, 2009.

Nikurashin, M. and Ferrari, R.: Radiation and Dissipation of Internal Waves Generated by Geostrophic Motions Impinging on Small-Scale Topography: Application to the Southern Ocean, J. Phys. Oceanogr., 40, 2025–2042, doi:10.1175/2010JPO4315.1, 2010.

Orsi, A. H., Whitworth III, T., and Nowlin Jr., W. D.: On the meridional extent and fronts of the Antarctic Circumpolar Current, Deep-Sea Res.-Pt. I, 42, 641–673, 1995.

Pritchard, H. D., Arthern, R. J., Vaughan, D. G., and Edwards, L. A.: Extensive dynamic thinning on the margins of the Greenland and Antarctic ice sheets, Nature, 461, 971–975, doi:10.1038/nature08471, 2009.

Pritchard, H. D., Ligtenberg, S. R. M., Fricker, H. A., Vaughan, D. G., van den Broeke, M. R., and Padman, L.: Antarctic ice-sheet loss driven by basal melting of ice shelves, Nature, 484, 502–505, doi:10.1038/nature10968, 2012.

Rignot, E. J.: Mass balance of East Antarctic glaciers and ice shelves from satellite data, Ann. Glaciol., 34, 217–227, doi:10.3189/172756402781817419, 2002.

Rignot, E.: Changes in ice dynamics and mass balance of the Antarctic ice sheet, Philos. T. Roy. Soc. A, 364, 1637–1655, doi:10.1098/rsta.2006.1793, 2006.

Rignot, E. J. and Jacobs, S. S.: Rapid bottom melting widespread near Antarctic Ice Sheet grounding lines, Science, 296, 2020–2023, doi:10.1126/science.1070942, 2002.

Rignot, E. J., Bamber, J. L., van den Broeke, M. R., Davis, C., Li, Y., van de Berg, W. J., and van Meijgaard, E.: Recent Antarctic ice mass loss from radar interferometry and regional climate modelling, Nat. Geosci., 1, 106–110, doi:10.1038/ngeo102, 2008.

Rignot, E., Jacobs, S., Mouginot, J., and Scheuchl, B.: Ice Shelf Melting Around Antarctica, Science, 341, 266–270, doi:10.1126/science.1235798, 2013.

Rintoul, S. R., Hughes, C. W., and Olbers, D.: Chapter 4.6 The antarctic circumpolar current system, in: Ocean Circulation and Climate Observing and Modelling the Global Ocean, edited by: Gerold Siedler, J. C. and Gould, J., Vol. 77 of International Geophysics, 271–302, Academic Press, doi:10.1016/S0074-6142(01)80124-8, 2001.

Roberts, J. L., Warner, R. C., Young, D., Wright, A., van Ommen, T. D., Blankenship, D. D., Siegert, M., Young, N. W., Tabacco, I. E., Forieri, A., Passerini, A., Zirizzotti, A., and Frezzotti, M.: Refined broad-scale sub-glacial morphology of Aurora Subglacial Basin, East Antarctica derived by an ice-dynamics-based interpolation scheme, The Cryosphere, 5, 551–560, doi:10.5194/tc-5-551-2011, 2011.

Scambos, T. A., Bohlander, J. A., Shuman, C. A., and Skvarca, P.: Glacier acceleration and thinning after ice shelf collapse in the Larsen B embayment, Antarctica, Geophys. Res. Lett., 31, L18402, doi:10.1029/2004GL020670, 2004.

Shchepetkin, A. F. and McWilliams, J. C.: The regional oceanic modeling system (ROMS): a split-explicit, free-surface, topography-following-coordinate oceanic model, Ocean Model., 9, 347–404, doi:10.1016/j.ocemod.2004.08.002, 2005.

Smith, W. and Sandwell, D.: Global sea floor topography from satellite altimetry and ship depth soundings, Science, 2, 209–215, 1997.

St-Laurent, P., Klinck, J. M., and Dinniman, M. S.: On the Role of Coastal Troughs in the Circulation of Warm Circumpolar Deep Water on Antarctic Shelves, J. Phys. Oceanogr., 43, 51–64, doi:10.1175/JPO-D-11-0237.1, 2013.

Tamura, T., Ohshima, K. I., and Nihashi, S.: Mapping of sea ice production for Antarctic coastal polynyas, Geophys. Res. Lett., 35, L07606, doi:10.1029/2007GL032903, 2008.

Thoma, M., Jenkins, A., Holland, D., and Jacobs, S.: Modelling Circumpolar Deep Water intrusions on the Amundsen Sea continental shelf, Antarctica, Geophys. Res. Lett., 35, L18602, doi:10.1029/2008GL034939, 2008.

Timmermann, R., Le Brocq, A., Deen, T., Domack, E., Dutrieux, P., Galton-Fenzi, B., Hellmer, H., Humbert, A., Jansen, D., Jenkins, A., Lambrecht, A., Makinson, K., Niederjasper, F., Nitsche, F., Nøst, O. A., Smedsrud, L. H., and Smith, W. H. F.: A consistent data set of Antarctic ice sheet topography, cavity geometry, and global bathymetry, Earth Syst. Sci. Data, 2, 261–273, doi:10.5194/essd-2-261-2010, 2010.

Weertman, J.: Stability of the junction of an ice sheet and an ice shelf, J. Glaciol., 13, 3–11, 1974.

Williams, G. D., Meijers, A. J. S., Poole, A., Mathiot, P., Tamura, T., and Klocker, A.: Late winter oceanography off the Sabrina and BANZARE coast (117–128° E), East Antarctica, Deep Sea Res.-Pt. II, 58, 1194–1210, doi:10.1016/j.dsr2.2010.10.035, 2011.

Young, D. A., Wright, A. P., Roberts, J. L., Warner, R. C., Young, N. W., Greenbaum, J. S., Schroeder, D. M., Holt, J. W., Sugden, D. E., Blankenship, D. D., van Ommen, T. D., and Siegert, M. J.: A dynamic early East Antarctic Ice Sheet suggested by ice-covered fjord landscapes, Nature, 474, 72–75, doi:10.1038/nature10114, 2011.

9

The effects of global climate change on the cycling and processes of persistent organic pollutants (POPs) in the North Sea

K. O'Driscoll[1], B. Mayer[2], J. Su[2], and M. Mathis[2]

[1]Marine Research Group, School of Planning, Architecture & Civil Engineering, Queen's University Belfast, UK
[2]Institute of Oceanography, University of Hamburg, Bundesstrasse 53, Hamburg 20146, Germany

Correspondence to: K. O'Driscoll (kieran.odriscoll@qub.ac.uk)

Abstract. The fate and cycling of two selected legacy persistent organic pollutants (POPs), PCB 153 and γ-HCH, in the North Sea in the 21st century have been modelled with combined hydrodynamic and fate and transport ocean models (HAMSOM and FANTOM, respectively). To investigate the impact of climate variability on POPs in the North Sea in the 21st century, future scenario model runs for three 10-year periods to the year 2100 using plausible levels of both in situ concentrations and atmospheric, river and open boundary inputs are performed. This slice mode under a moderate scenario (A1B) is sufficient to provide a basis for further analysis. For the HAMSOM and atmospheric forcing, results of the IPCC A1B (SRES) 21st century scenario are utilized, where surface forcing is provided by the REMO downscaling of the ECHAM5 global atmospheric model, and open boundary conditions are provided by the MPIOM global ocean model.

Dry gas deposition and volatilization of γ-HCH increase in the future relative to the present by up to 20 % (in the spring and summer months for deposition and in summer for volatilization). In the water column, total mass of γ-HCH and PCB 153 remain fairly steady in all three runs. In sediment, γ-HCH increases in the future runs, relative to the present, while PCB 153 in sediment decreases exponentially in all three runs, but even faster in the future, due to the increased number of storms, increased duration of gale wind conditions and increased water and air temperatures, all of which are the result of climate change. Annual net sinks exceed sources at the ends of all periods.

Overall, the model results indicate that the climate change scenarios considered here generally have a negligible influence on the simulated fate and transport of the two POPs in the North Sea, although the increased number and magnitude of storms in the 21st century will result in POP resuspension and ensuing revolatilization events. Trends in emissions from primary and secondary sources will remain the key driver of levels of these contaminants over time.

1 Introduction

Persistent organic pollutants (POPs) are highly toxic chemicals. As their name suggests, they degrade slowly and, thus, have long half lives, generally on the order of years to tens of years. In productive ocean waters, which are often shallow and coastal, POPs enter the food web, bioaccumulate in different kinds of tissues and become hazardous to living organisms including humans (AMAP, 1998). Even though primary emissions of legacy POPs have been declining, concentrations of some POPs in the atmosphere over open ocean areas have not reduced accordingly which can be the result of emissions from capacitors such as oceans, soil and sediments, e.g. Nizzetto et al. (2010), Gioia et al. (2008). Hence, the study of POPs under a changing climate scenario is important and interesting.

POPs in the ocean are subject to a wide range of processes including mechanical, chemical, physical and biological processes. For the North Sea, large amounts of POPs enter the system from the surrounding countries. Major entrance pathways of POPs to the North Sea are through atmospheric deposition, river inputs and adjacent seas, i.e. the North Atlantic Ocean, the English Channel and the Baltic Sea (see Fig. 1). Ilyina et al. (2006), who developed the fate and transport ocean model (FANTOM), were the first to investigate the fate of POPs in the southern North Sea with a high-resolution combined hydrodynamic and fate and transport model. O'Driscoll et al. (2013) extended the model domain to the entire North Sea system while improving several POPs processes in the model and the quality of input and forcing data.

In this study, we have selected two very different legacy POPs for the simulations. Lindane, γ-HCH, is a legacy POP that was widely used as an insecticide but was banned under the Stockholm Convention. Although one of the lesser hydrophobic POPs, it mostly dissolves in water and has half-lives of about 1.2 and 1.6 years in water and sediment, respectively (Ngabe et al., 1993; Hornsby et al., 1996). PCB 153, on the other hand, is very hydrophobic, so it does not readily dissolve in water but mostly sorbs to organic material in aquatic environments. It is also subject to the terms of the Stockholm Convention. It causes a host of medical problems in humans, see e.g. Ghosh et al. (2011) and references therein. It has half-lives of about 13.7 and 18.8 years in water and sediment, respectively (Sinkkonen and Paasivirta, 2000). For a further discussion of properties influencing the fate and transport of γ-HCH and PCB 153 in the marine environment, see O'Driscoll et al. (2013). A complete list of physical–chemical property values used with the simulations is provided in the Supplement (Table S1).

A number of recent modelling studies have investigated the effects of global climate change on POPs. Lamon et al. (2009) used a multimedia chemical fate model to evaluate the influence of climate change on the global distribution of PCB 28 and PCB 153. They found that higher temperatures increased primary and secondary volatilization emissions of PCBs, thereby enhancing transport from temperate regions to the Arctic, while the largest relative increase in concentrations of both PCB congeners in air occurs in the high Arctic and the remote Pacific Ocean. Their modelling results indicate that climate change will increase volatilization emissions and increase mobility of POPs with properties similar to those of PCBs. Reviewing the roles of multimedia environmental fate and food-web bioaccumulation models to investigate potential POP changes due to climate change, Gouin et al. (2013) found that alterations to exposure concentrations are typically within a factor of 2 of the baseline output, while recommending caution when interpreting the relative importance of global climate change on how these changes influence chemical fate and bioavailability. In their overview, Armitage et al. (2011) provide context for northern residents of the relative importance of different aspects of exposure to organic contaminants related to global climate change. Important processes include the influence of temperature on contaminant amplification and distribution, particularly in the Arctic where accumulation is expected to be sensitive to changing ice cover and precipitation events. Using a global-scale multimedia chemical fate model, Wöhrnschimmel et al. (2013) analyse and quantify the impact of climate change on emissions and fate of two legacy POPs, PCB153, and α-HCH, and their transport to the Arctic. Given a phase of ongoing primary emissions, they found that modelled concentrations in Arctic contamination increase by up to a factor of 2 in air and water, and are driven mostly by changes in emission patterns. Following a phase-out, they found increases of up to a factor of 2 in air and 4 in water, which are mostly attributable to changes in transport and fate of chemicals. These changes are also relative to the 20th century climate baseline model output since their model runs were for the period 1900–2100.

However, there have to date been no studies investigating the fate of POPs in coastal or shelf seas with high-resolution simulations under climate change scenarios. Therefore, we feel that the present study is an important contribution to the literature, by complementing previous studies and giving us insight into the effect of climate change on important POP processes in shelf seas, thus helping policy makers to make well-informed decisions.

To investigate the impact of climate variability on POPs in the North Sea in the 21st century, future scenario model runs for three 10-year periods to the year 2100 (2006–2015, 2046–2055 and 2090–2099) using plausible levels of both in situ concentrations and atmospheric, river and open boundary inputs are performed.

A 3-D hydrodynamic model is a necessary prerequisite tool for modelling their fate in the ocean. For this purpose, we have applied a very high resolution version of the Hamburg Shelf Ocean Model (HAMSOM) for the North Sea developed at the Institute of Oceanography, University of Hamburg, see O'Driscoll et al. (2013) for details.

2 Model description and approach

2.1 Model description

Here, we present a brief overview of both the HAMSOM and FANTOM models; for a more detailed description see O'Driscoll et al. (2013). Comprehensive descriptions of both models are available in the literature for the FANTOM (Ilyina et al., 2006) and HAMSOM (Backhaus, 1985; Pohlmann, 2006).

Model bathymetry is obtained from Larsen and She (2001) and presented in Fig. 1. Time steps (5 min) are the same in both models.

The HAMSOM is a baroclinic shallow water equation ocean circulation model, which has been successfully applied to different regional seas, see e.g. Mayer et al. (2010), Su and Pohlmann (2009), and O'Driscoll et al. (2013). Meteorological forcing is provided by the REMO downscaling of the ECHAM5 global atmospheric model IPCC A1B (SRES) 21st century scenario, see Jacob et al. (2001) and Roeckner et al. (2006). Open boundary conditions (temperature, salinity and sea surface height) are provided by the global ocean model MPIOM, see Jungclaus et al. (2006). Temperature and

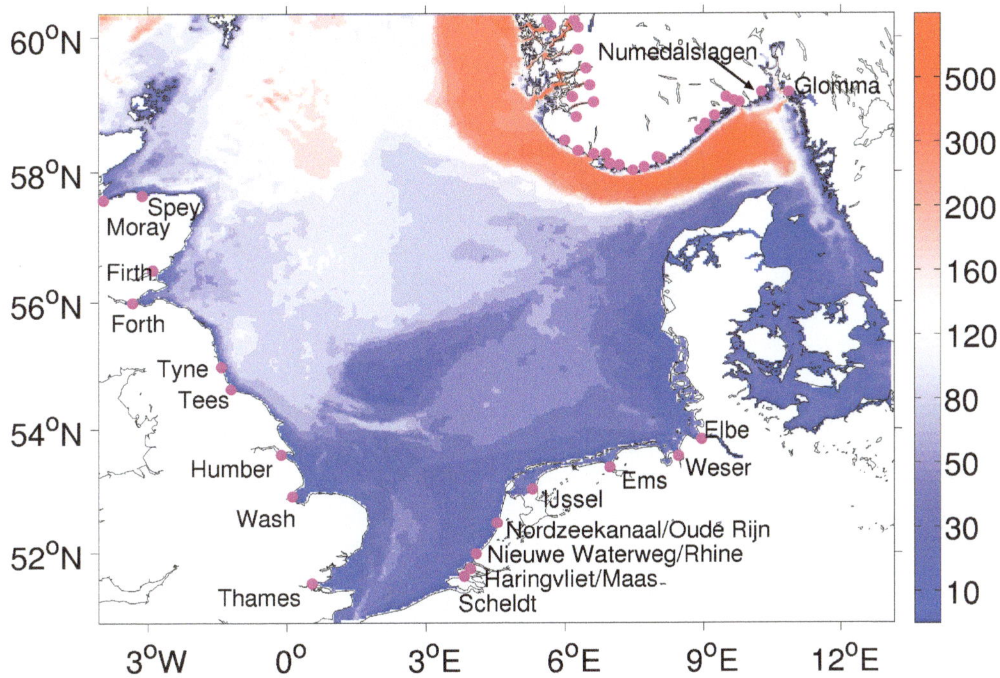

Figure 1. The model domain with topography (metres). River sources of freshwater are shown by magenta dots.

salinity are prescribed under inflow conditions, while a radiation condition by Orlanski (1976) is utilized when the flow is outward. Locations of freshwater runoff into the model domain are shown in Fig. 1. Monthly climatological river runoff data are used for the entire run, see O'Driscoll et al. (2013).

The FANTOM is a 3-D Eulerian model for simulating the fate of POPs and other pollutants in coastal regions and shelf seas. Model processes fall into five main categories: transport due to advection and diffusion; air–sea exchange; phase (re-)distribution; degradation; exchange between the water column and sediment (deposition and resuspension). A summary of model processes is provided in the Supplement (Sect. S2). See O'Driscoll et al. (2013) and Ilyina et al. (2006) for full details. For reasonable simulation of POPs exchange between water column and sediment, the FANTOM has 20 sediment layers representing the upper 2 cm of the sediment bed.

Both models (HAMSOM and FANTOM) have been evaluated for the period 1996–2005 (O'Driscoll et al., 2013) and the simulations were shown to be of a very satisfactory quality relative to observations.

2.2 Model approach

Since estimates of future concentration levels of POPs in the atmosphere, oceans and rivers are not available, our approach was to reutilize 2005 values in the atmosphere, rivers and at the open ocean boundaries for every year of the simulations – that is seasonally varying concentrations are reused each year – while using the (final) result for POP concentrations in water and sediment from the 1996–2005 runs of O'Driscoll et al. (2013) as the initial condition for all three future runs. In this way, we attribute differences between the three 10-year simulations to climate change only. Atmospheric concentrations of POPs are provided by output from the atmospheric model of the MSC-E, EMEP (Gusev et al., 2009) for the year 2005, while concentrations for riverine input have been calculated from available data sets for the year 2005. We think this is a reasonable approach since concentrations in the atmosphere, rivers and oceans (primary sources) were greatly reduced in 2005 relative to 1996 (O'Driscoll et al., 2013) but capacitors (secondary sources) will continue to play an important role in the coming years due to their long-term storage ability, see discussions in Nizzetto et al. (2010) and Lohmann et al. (2007). Concerning external pollutant conditions, since, to date, we have only vague ideas about future input changes of POPs through the atmosphere or rivers (net North Atlantic input into the North Sea is negligible), we cannot perform a projection for the POPs as we do for the climate itself. When considering changes in both climate and POP inputs, we would not be able to distinguish which of these is responsible for changes in POPs. We can make some estimates on the likely direction and magnitude of change based on emission, fate and transport rates. For example, using a global mass balance model approach, Breivik et al. (2007) found that emissions of PCBs (ΣPCB_{22}) will decrease exponentially in the 21st century, while Wöhrnschimmel et al. (2013), using a global model, found that PCB 153 and α-HCH concentrations in the Arctic will decrease by 2–3 orders of magnitude in 2100 relative to 2005 mostly due to

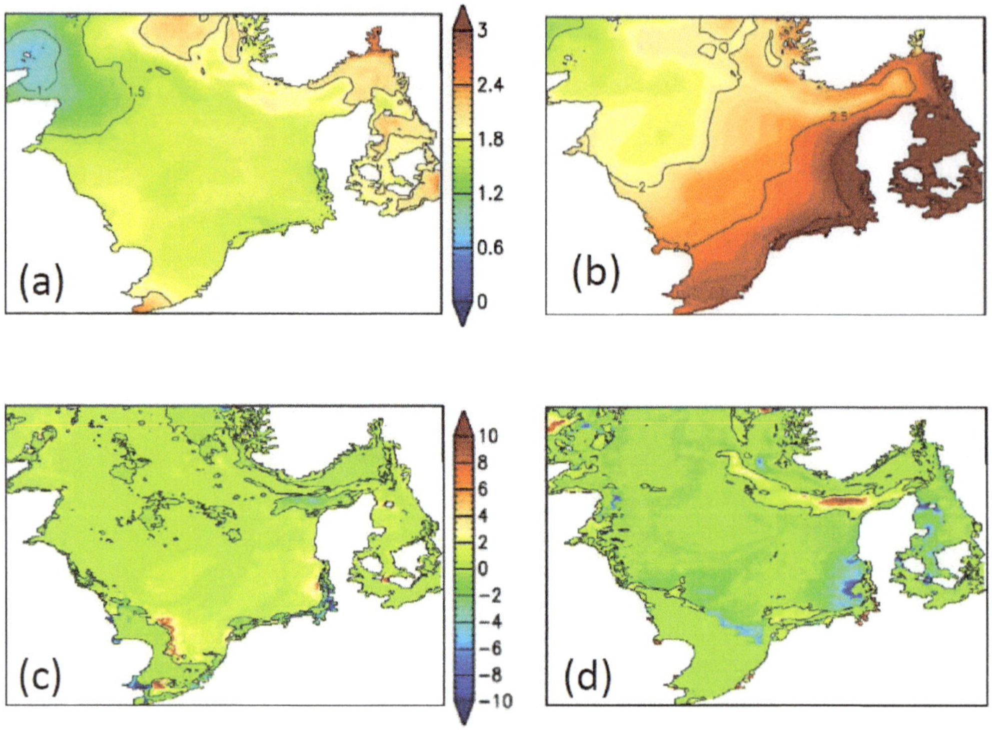

Figure 2. Distributions of 21st century trends. **(a)** SST in summer (June, July, August); **(b)** SST in winter (December, January, February), both in °C per 100 years; **(c)** relative occurrence of erosion in percentage points per 100 years; **(d)** relative occurrence of deposition in percentage points per 100 years. The upper scale is for **(a)** and **(b)**, the lower scale for **(c)** and **(d)**.

phase-out. So we can expect to find similar decreases in atmospheric and open ocean concentrations at lower latitudes. However, since the area of interest in this work is the regional, semi-enclosed, shallow North Sea, we again stress that concentrations of some POPs in the atmosphere over open ocean areas have not reduced accordingly, which can be the result of emissions from capacitors such as oceans, soil and sediments (Nizzetto et al., 2010; Gioia et al., 2008) and so we can expect capacitors to be important in the coming years due to their long-term storage ability (Nizzetto et al., 2010; Lohmann et al., 2007).

We chose the A1B scenario because it is a moderate choice. A more extreme scenario would introduce more variance into the solution, resulting in more biased conclusions. Also, since more computer power is required to simulate other scenarios, which will not lead to any new conclusion in addition to those of the present study, it is not worth the effort involved.

Monthly climatological river runoff data for major continental and British rivers are used for the entire period in both models (O'Driscoll et al., 2013).

3 Results and discussion

3.1 Hydrodynamics

The basic hydrodynamic results for the model domain are as follows. Analysis of the atmospheric forcing data for the North Sea reveals a continuous increase of the spatially averaged 2 m air temperature of 2.38 °C within the scenario period 2001–2100, where warming in winter is 25 % higher than the mean, but is 22 % lower in summer. Similarly, the volume-averaged water temperature of the North Sea increases by 2.0 °C (to 2100) with warming 4 % weaker in summer and 4 % stronger in winter in 2100 relative to 2000, whereas sea-surface temperature (SST) trends show average increases of 1.66 and 2.31 °C in summer and winter, respectively (see Fig. 2a and b). The analysis of 10 m wind speed does not show a significant change of the mean wind speeds throughout the year (in 2100). Local maximum wind speeds, however, indeed show an increase of 1.38 $m\,s^{-1}$ in winter and a decrease of 0.75 $m\,s^{-1}$ in summer. The duration of gale wind conditions per year, indicated by mean wind speeds greater than 17 $m\,s^{-1}$, increases by about 65 h or by 30 %. Since mean wind speeds hardly change, both more gale and more calm wind conditions tend to occur, particularly in winter.

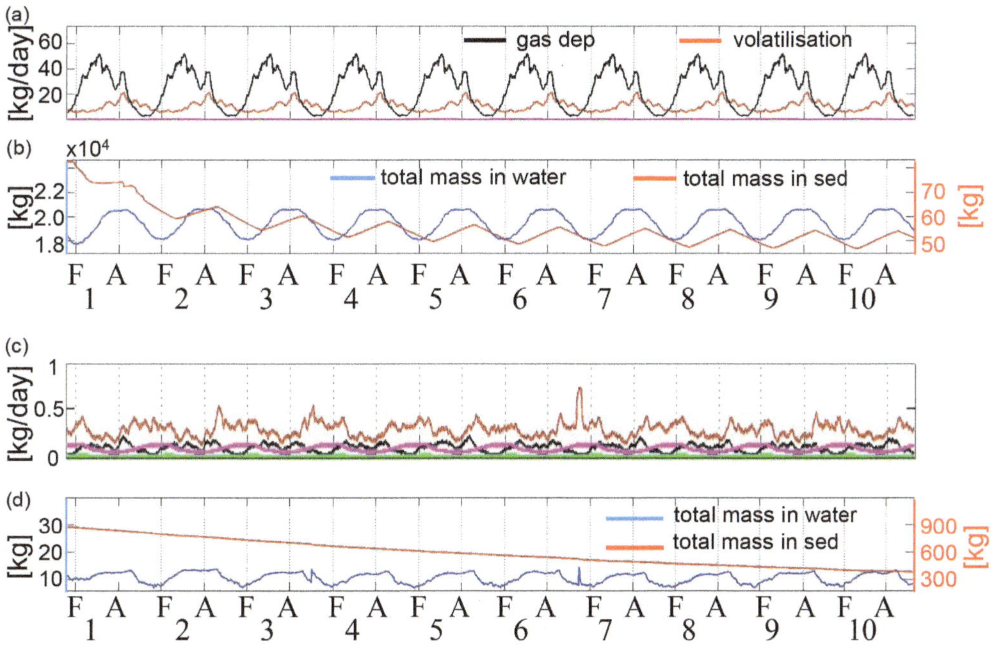

Figure 3. Time series of surface fluxes and total mass in water and sediment of γ-HCH, top two panels, respectively, and the same for PCB 153, bottom two panels, respectively, for the period 2006–2015. February and August (F and A) and the year number (1–10) are shown on the x axis. River input, wet deposition and dry deposition are plotted in magenta, green and blue (panels 1 and 3), respectively. Note the different scales (on the left and right) in second and bottom panels showing the different amounts of the substance in water and sediment.

The effect of the change in wind conditions during the 21st century on bottom erosion and deposition of suspended particulate matter (SPM), which are important for our POP dynamics, is shown in Fig. 3c and d, respectively. Plots show trends of relative frequency of erosion and deposition periods in percentage points per 100 years, based on annual mean percentages of erosion and deposition for the 21st century. A slight increase/decrease in annual periods of erosion/deposition are found.

It is the effect of these changes on the studied POPs (previous paragraph) that are analysed. The full hydrodynamic analysis of the A1B scenario for the North Sea is to be discussed elsewhere.

3.2 Time series of surface fluxes of γ-HCH and PCB 153 and total masses in water and sediment

Time series of surface fluxes are given in Fig. 3 (2006–2015), Fig. 4 (2046–2055) and Fig. 5 (2090–2099). Air–sea exchange of γ-HCH, top panels, is mostly due to dry gas deposition and volatilization, which are much greater than the other processes (wet deposition, particle deposition and river input). In general, dry gas deposition exhibits an annual cycle with greatest values in the spring and summer with peak values around May, and least values in autumn/winter, bottoming out in November–December when atmospheric concentrations are least (generally an order of magnitude or more less than in the summer months; Gusev et al., 2009). We remind the reader that partitioning of the POPs between air and water is dependent upon temperature, wind speed and the physical–chemical properties of the compound and its abundance in the environment. Thus, when dry gas deposition increases, concentration and total mass in water increases. This subsequently leads to increased volatilization, as described in the following paragraph.

Volatilization also exhibits a clear annual cycle, with greatest values in late summer and the early autumn months (August–September) and least values in late winter and early spring (February–March). Dry gas deposition is generally 2–4 times greater than volatilization, except between about September and December, when deposition is lowest. Volatilization is still high at this time but decreases with the total mass in water curve, second panel, which in turn is due to the decrease in dry gas deposition.

Comparing the different time series, it is seen that in 2006–2015 there is very little inter-annual variability in dry gas deposition and volatilization of γ-HCH (Fig. 3) with dry gas deposition fluxes peaking at around 50 kg day^{-1} in May, and volatilization at about 20 kg day^{-1} in August–September. Contrast this with the 2046–2055 and 2090–2099 runs, respectively. Maximum gas deposition fluxes increases in most years in both runs, generally approaching 60 kg day^{-1} in May and clearly exceeding it in four and three years in the 2050 and 2090 runs, respectively. Volatilization increases in the 2046–2055 run, relative to the 2006 run, and is even greater in the 2090 run.

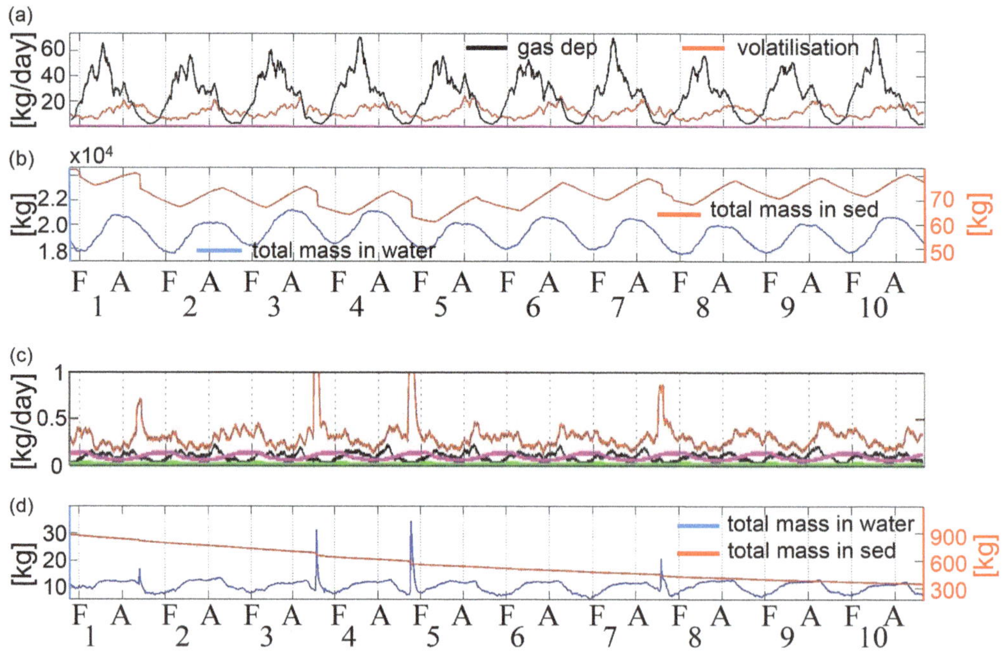

Figure 4. As Fig. 3, but for 2046–2055. Note the different scales (on the left and right) in second and bottom panels showing the different amounts of the substance in water and sediment.

So we conclude that climate change in the 21st century results in increased dry gas deposition and volatilization of γ-HCH. This is further discussed and clarified in Sect. 3.3.

Clear annual cycles in total mass of γ-HCH in water (Figs. 3–5, second panels) is seen in the three 10-year runs. Values are greatest around August and least around February, due to the accumulated net flux of dry gas deposition over volatilization in the spring and summer and the net deficit (reverse) in winter, respectively. Clear annual cycles in total mass of γ-HCH in sediment, second panels, are also seen for all three runs with maximum values observed around September–October and minimum values around March–April. The 1–2-month lag in these extrema, relative to the water column, is due to settling in summer and resuspension due to storm events and weather conditions in general in winter. Because of plankton blooms, more particulate organic carbon (POC) is contained in the water column in the spring and summer months. POPs sorb to this POC before ultimately sinking out of the water column and into sediment. In winter months, winds and stormy weather lead to resuspension of POPs (see discussion in O'Driscoll et al., 2013). Total mass is diminished somewhat in the first year of all three runs, which is consistent with the trend in 1996–2005 (O'Driscoll et al., 2013), where decreasing atmospheric concentrations and river input resulted in a downward trend of total mass in sediment. Note also that the total mass in water exceeds that in sediment by more than two orders of magnitude.

For PCB 153 (Figs. 3–5, third panels) volatilization (red) is greater than any of the surface sources, dry gas deposition (black) and river input (magenta) are generally 2–4 times less than volatilization, while particle (blue) and wet deposition (green) are much smaller. Volatilization and river input are in phase with largest values in winter and lowest in summer, while dry gas deposition is highest in spring and summer and least in winter. Volatilization generally exceeds the sum of the sources (deposition and river input), so that the North Sea is net volatilizational to the atmosphere for PCB 153, which was already shown to be the case for 1996–2005 (O'Driscoll et al., 2013), but can be net depositional in the summer months when the sum of dry gas, wet and aerosol deposition and river input can exceed volatilization. The other main feature is the peaks, or rapid increases, in volatilization due to storms.

A seasonal cycle in total mass of PCB 153 in water (Figs. 3–5, bottom panels) is seen with highest (lowest) values found in the summer (winter) months. Total mass is generally lowest in winter and starts to increase during the spring bloom (primary production) when depositional PCB 153 sorbs to POC in the water column. Values remain high through the spring and summer, when the North Sea can be net depositional for PCB 153, before decreasing in September–October, when the primary production decreases, reaching lowest values in January–February, thus completing the annual cycle. The spikes observed are the result of winter storms, which cause erosion of fine fraction of sediment and resuspension of the POP into the water column, resulting in rapid increases in total mass in water. The recovery time for these storms is about 1–2 months in the water column when concentrations return to pre-storm values. These events

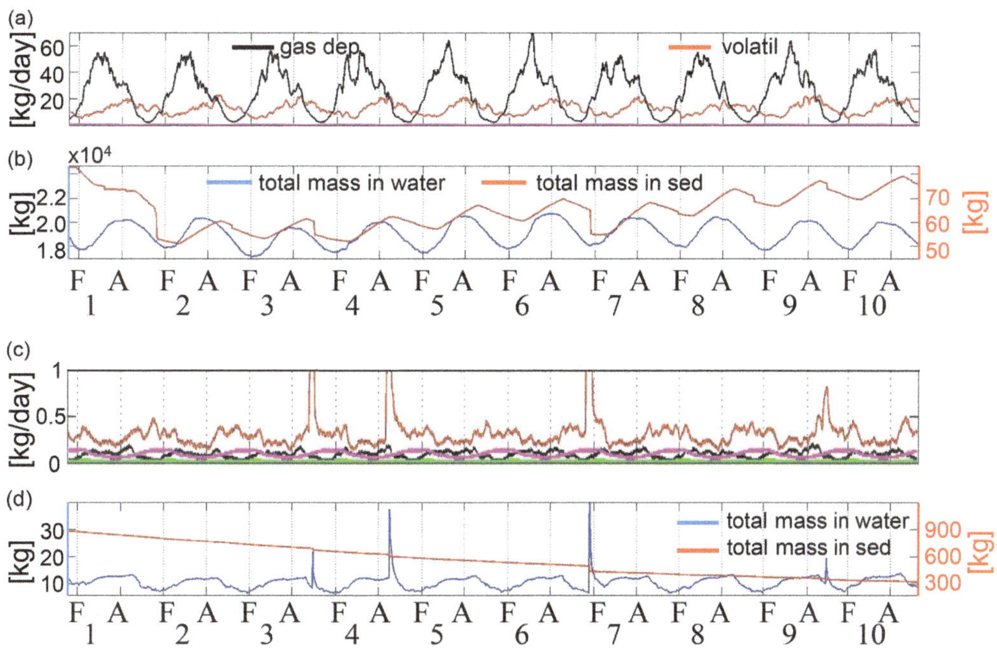

Figure 5. As Fig. 3. but for 2090–2099. Note the different scales (on the left and right) in second and bottom panels showing the different amounts of the substance in water and sediment.

are also found in the sediment and γ-HCH time series (total mass of PCB 153 in sediment decreases simultaneously, as does that of γ-HCH but so little of it is contained there, relative to water, that it makes no impression on the total mass in water time series).

Total mass of PCB 153 in sediment (Figs. 3–5, bottom panels) decays exponentially, with values reduced to less than half by the end of the three 10-year runs. An annual cycle, though not obvious, occurs with maximum values in spring and summer, due to primary productivity and sinking on POC, and minimum values in winter due to erosion resulting from storms and strong winds. Since PCB 153 is hydrophobic (or lipophilic) most of it sorbs to POC before settling in the sediment, and we see that total mass in sediment is more than an order of magnitude greater than that in water. Therefore, continuous resuspension of PCB 153 is enough to keep a constant small amount of it dissolved in water but the total mass in sediment will decay exponentially unless it is supplemented through atmospheric or river input.

During storm events, erosion causes mobilization and resuspension of PCB 153 into the water column, increasing its concentration there, thus resulting in a consequent increase in volatilization. Storms will generally resuspend sediment at depths of tens of metres, while very strong storms can cause resuspension down to 50 m or more. But most of the pollutants are found in shallow water, close to polluted rivers, in the southern North Sea (see O'Driscoll et al., 2013, Figs. 6 and 7). Because the total mass of PCB 153 in water is so much less than that in sediment, and because of increased volatilization during and following a storm, and net volatilization in general, the total mass in sediment cannot recover following a storm event. This is in contrast to γ-HCH, which is generally net depositional from atmosphere to ocean, has total mass in water more than two orders of magnitude greater than that in sediment, and thus pre-storm mass in sediment can be recovered. From this, we conclude that the change in number and strength of storms, and with this erosion events, has only a small impact on the concentration of γ-HCH in sediment and water.

Concerning the predictability of North Sea storms, Weisse et al. (2005) found relatively good agreement between the NCEP–NCAR global reanalysis (Kalnay et al., 1996) and actual events, where storms events were generally reasonably reproduced for all parameters. Using a limited ensemble of 21st century scenarios, Weisse et al. (2009) have been able to estimate potential future met-ocean conditions while also making a first guess about the underlying uncertainties. The increased number of storms they found (Weisse et al., 2009, Fig. 4), makes us even more confident about the increase in storm size and number found in our scenario runs, particularly in the 2090–2099 decade.

3.3 Comparison of time series of total mass of γ-HCH and PCB 153 in water and sediment

Time series of total mass of γ-HCH and PCB 153 in water and sediment for the entire model domain for the three 10-year runs are plotted against each other for comparison in Fig. 6. For γ-HCH in water (Fig. 6a) there is little inter-annual variability, the annual cycle generally faithfully

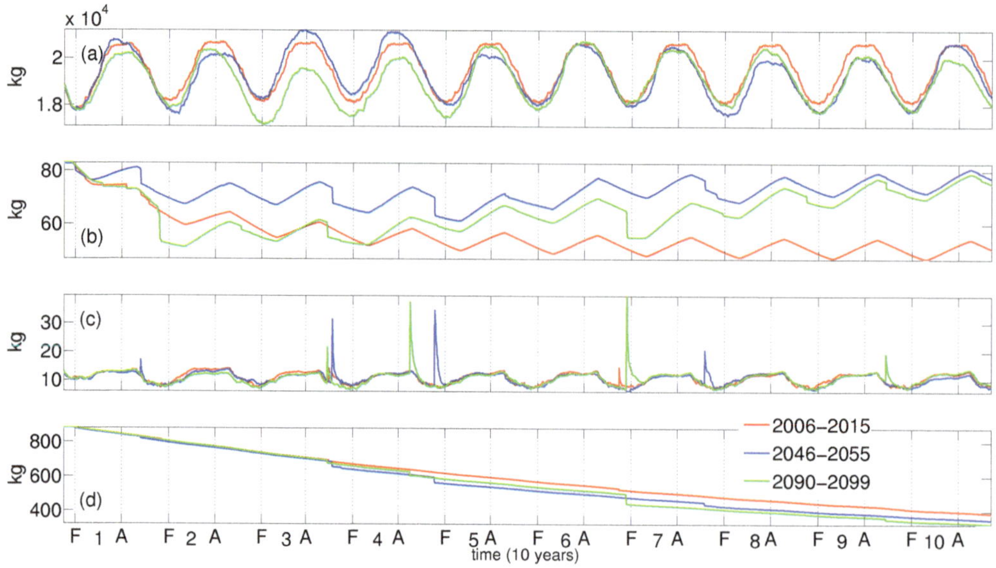

Figure 6. Time series of total mass of γ-HCH in water **(a)**, and sediment **(b)**, and PCB 153 in water **(c)**, and sediment **(d)**, for the three 10-year periods 2006–2015, 2046–2055 and 2090–2099. February and August (F and A) and the year number (1–10) are shown on the x axis. The time series presented in this paper are the sum values within the model domain.

repeating itself. The almost identical repetition of the annual cycle of dry gas deposition and volatilization in 2006–2015 (Fig. 3, top panel) results in the repeated corresponding annual cycle for total mass in water. Inter-annual variability in both dry gas deposition and volatilization for both future runs (Figs. 4 and 5, top panels) are responsible for the same variability in total mass in water. For example, the increase in dry gas deposition in the third and fourth years of the 2046 run, relative to the 2006 run, results in the increase in mass in water, while decreasing deposition and increasing volatilization are responsible for the decrease in mass in water in the eighth and ninth years. In the 2090 run the increase in volatilization, relative to that in dry gas deposition (Fig. 5) in the second to fifth and the last three years is responsible for the decrease in total mass in water.

Comparing total mass of γ-HCH in sediment for the three runs (Fig. 6b) shows that while the mass diminishes somewhat in the first year of all three runs (by $\sim 20\,\%$ in 2006, 15 % in 2046 and 40 % in 2090), it continues to decrease between 2006–2011 (to $\sim 60\,\%$ of 2006 mass in 2011) but remains steady thereafter; it diminishes a little further in the period 2047–2049 (to $\sim 80\,\%$ of 2046 values) but increases thereafter to 2055 (back to initial values); and following a decrease resulting from a storm in 2091 it increases from 2092–2099, also back to initial levels. So there is an upward trend of mass in sediment in the future runs relative to the present (2006–2015). This is due to increased dry gas deposition in the future runs, which causes an increase of the amount of γ-HCH sorbing to POC in the water column and, subsequently, resulting in sinking of an increased mass of the POP into the sediment. This small increase of total mass in sediment is not observed in the water column in Fig. 6a because it is small relative to total mass in the water column.

Comparing total masses of PCB 153 in water (Fig. 6c), it is seen the mass remains steady for all three cases. In sediment (Fig. 6d), total mass in the 2090–2099 run decreases faster than the 2046–2055 run, which decreases faster again than the 2006–2015 run. This is due to the increased number and intensity of storms observed in the 2050 and 2090 runs. These storms cause resuspension of PCB 153 into the water column, part of which becomes available for volatilization due to the equilibrium partitioning of the POP.

3.4 Annual fluxes of γ-HCH and PCB 153

Total annual mass fluxes of γ-HCH and PCB 153 for 2015, and the ratio of the fluxes for the years 2055/2015 and 2099/2015 are presented in Fig. 7. The patterns are fairly similar for each of the two POPs but there are important differences. The sum of the sources is always less than the sum of the sinks, with some variability in the different years and chemicals.

For γ-HCH, dry gas deposition and volatilization both increase somewhat in the future runs (2055, 2099), relative to the present (2015). Gas deposition is up by 7.5 % and 4.0 %, while volatilization is up 15.9 % and 16.1 % in 2055 and 2099, respectively, relative to 2015. These changes are due to increased air and water temperature and wind speed, see supplementary information for details. Degradation flux (the total mass that is degraded) in water decreases, down by 1 % and 3 % in 2055 and 2099, respectively, but remains higher than dry gas deposition. Exchange at the open boundaries is fairly negligible. Total sum of sources increase by 7.3 % and

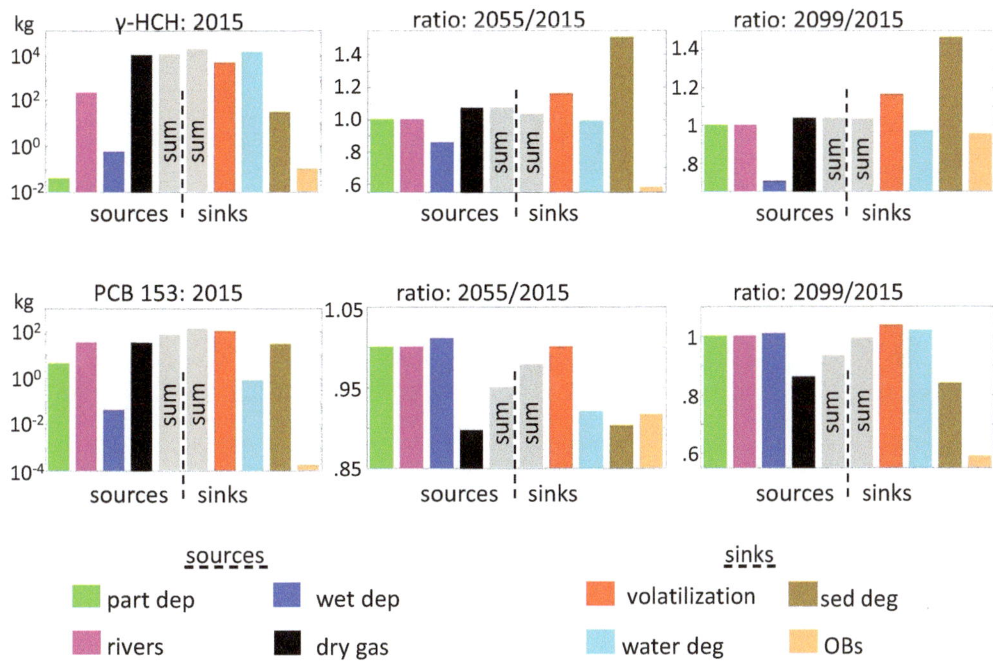

Figure 7. Total annual mass fluxes, [kg] log10 scale, of North Sea sources and sinks for 2015, left panels. Ratios of total annual mass fluxes: 2015/2055 (centre panels) and 2099/2015 (right panels). Sources are deposition with particles, input through rivers, wet deposition and gas deposition. Sinks are volatilization, degradation in water, degradation in sediment and net outward flux through the open ocean boundaries (OBs). The sums of sources and sinks are given by the grey panels.

3.9 %, and sinks by 3.2 % and 2.6 % in 2055 and 2099, respectively.

For PCB 153, dry gas deposition decreases in the future runs, by 10.3 % and 13.7 % in 2055 and 2099, respectively, while volatilization increases, by less than 1 % in 2055 and 3 % in 2099. This is in agreement with Lamon et al. (2009) who found that higher temperatures increased primary and secondary volatilization emissions of PCBs, thereby enhancing transport from temperate regions to the Arctic. Degradation flux in sediment decreases due to decreased mass there, to 90 % and 84 % of 2015 values in 2055 and 2099, respectively. Exchange at the open boundaries is fairly negligible. Sum of sources decreases by 5 and 7 % in 2055 and 2099, respectively, while the sinks decrease by 2 and 1 %, respectively.

Wöhrnschimmel et al. (2013) also found that modelled concentrations of PCB153 and α-HCH in Arctic contamination increase in air and water by a factor of 2, given a phase of ongoing primary emissions, increases of up to a factor of 2 in air and 4 in water following a phase-out, mostly driven by emissions, and attributed to transport and fate changes of the substances. This also agrees with our results and those of Lamon et al. (2009), whereby higher concentrations would be partially due to transport of the chemicals into the Arctic from lower latitudes.

So the general result is that there are small changes in annual mass fluxes due to the changing climate.

3.5 Spatial changes of concentrations in water and sediment

Distributions of vertically integrated concentration ratios for the summer season (JJA) and the winter season (DJF) of γ-HCH and PCB 153 between 2005 and the last year of the three model runs are plotted in Fig. 8 (2015/2005, 2055/2005 and 2099/2005). Clear seasonal trends and patterns are apparent in both cases. PCB 153 values are mostly less than 1 (decreased relative to 2005) while γ-HCH ratios are often greater than 1 (increased relative to 2005).

Starting with γ-HCH, the first feature that stands out is the region of decreased concentration in the future runs (blue) in the very south of the model domain. The reasons for this are twofold. First, concentration in gas in the atmosphere is greatest along the British coast, particularly around southern England and the Thames estuary. Concentrations in 2005 are substantially reduced relative to earlier years (e.g. about 2–4 fold less than in 2000 and around 20–30 % less than in 2004, see Gusev et al., 2009). Dry gas depositions are also greatest adjacent to the British coast and have also been shown to drop substantially between 1997 and 2004 (see Fig. 11 in O'Driscoll et al., 2013). Secondly, the Thames and Rhine are two major river point sources. Although the total flux of river sources into the North Sea is more than an order of magnitude less than dry gas deposition, they are substantial local sources. These concentrations have also dropped substantially annually since 1995 (see O'Driscoll et al., 2013,

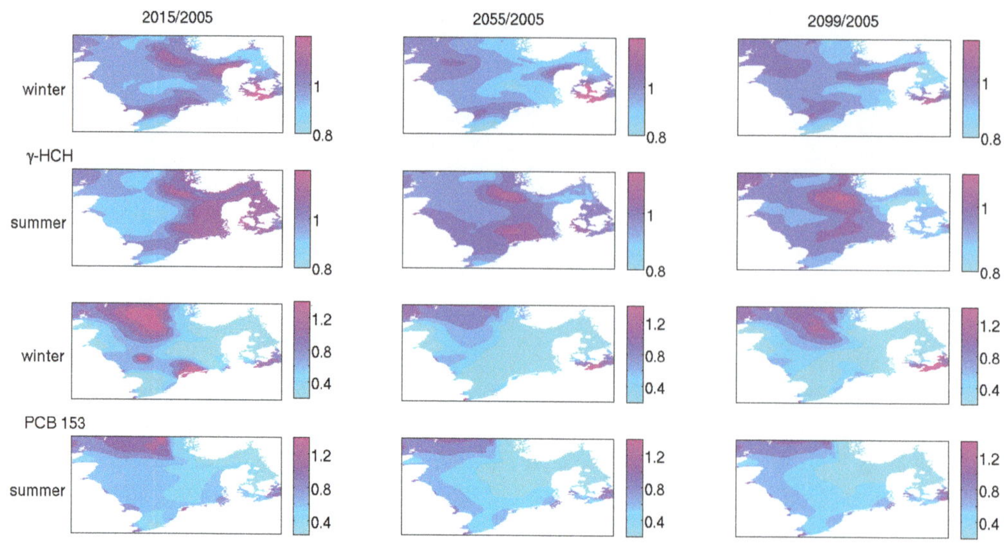

Figure 8. Distributions of concentration ratios integrated through the entire water column, 2015/2005, 2055/2005 and 2099/2005, from left to right, respectively, for γ-HCH in winter (top) and summer (second row), and for PCB 153 in winter (third row) and summer (bottom). Units are ng L^{-1}.

for details). Together, these changes combine and result in the reduced concentration band found in the very south of the model domain. Outside of this region, a number of regions of increased and decreased change are seen. The two regions of increased concentration in the future (pink) between East Anglia and the IJsselmeer and to the west of Jutland and the Norwegian coast (winter) diminish in the future runs because of increased water temperature, leading to increased volatilization. In summer, increases (pink) are greatest away from the British coast where gas concentrations have not been substantially reduced. This is modulated by the inflow of lower concentration Atlantic water.

For PCB 153, winter distributions are fairly similar for the three years, although somewhat higher values are found out to the open North Atlantic and in the IJsselmeer region, which can be put down to stronger inflow at the open boundary and stronger bottom currents in the southern North Sea, respectively. Lesser future decreases in concentration are also found near British and Dutch rivers, between southern Britain and Holland, around German river sources, out into the German Bight and Wadden Sea, and into the Kattegat. This is due to the decreased resuspendable content of PCB 153 in the sediment, compared to 2005, as a result of previously ongoing higher erosion activities, in combination with the anti-clockwise circulation that carries these polluted waters into the Kattegat. In summer, concentrations are somewhat reduced in 2055 and 2099 relative to 2015, which is due to the decrease in dry gas deposition in the future runs (2055 and 2099) and reduced concentrations in sediment. In summer, ratios are generally less than 1, with values of 1 or slightly greater found at the open North Atlantic boundaries.

Distributions of ratios in summer and winter in sediment between the last year of each of the three model runs and 2005 are shown in Fig. 9 (2015/2005, 2055/2005 and 2099/2005). For γ-HCH, concentrations are generally reduced from 2005 values. The greatest losses (blue) are found in 2015 when dry gas deposition is less than in 2055 and 2099, with high values found particularly around the southern British and continental coasts and southernmost regions of the North Sea. Patterns are similar in winter and summer but values are lower in summer (less blue) due to increased dry gas deposition, particularly in 2055 and 2099. Increased concentrations are seen off the western and northern parts of Jutland in all three runs but are greater in 2055 and 2099, and, also, in the future runs, off parts of the British and continental coasts and small areas of the southern North Sea, which can be due to the increased dry gas deposition but also erosion and redeposition in coastal areas, in particular close to river sources (see O'Driscoll et al., 2013, for distributions in 2004).

For PCB 153, there is little notable change in distributions and concentrations between 2005 and the last year of each of the three runs. This is due to the large masses in sediment that have been either degraded or eroded (Figs. 3–7) and account for most of the difference. However, it is interesting to plot the ratios at the end of the three 10-year runs against themselves (Fig. 10). Beginning with the winter plots, higher values in 2015 relative to 2055 (dark pink) are found in the coastal regions of the southern North Sea, especially around the Wash, Thames, offshore of the Rhine and southern Dutch coast, and offshore of the Elbe and the Wadden Sea region, which is due to increased gale conditions in the 2055 run. Areas of decrease are also found in the southern

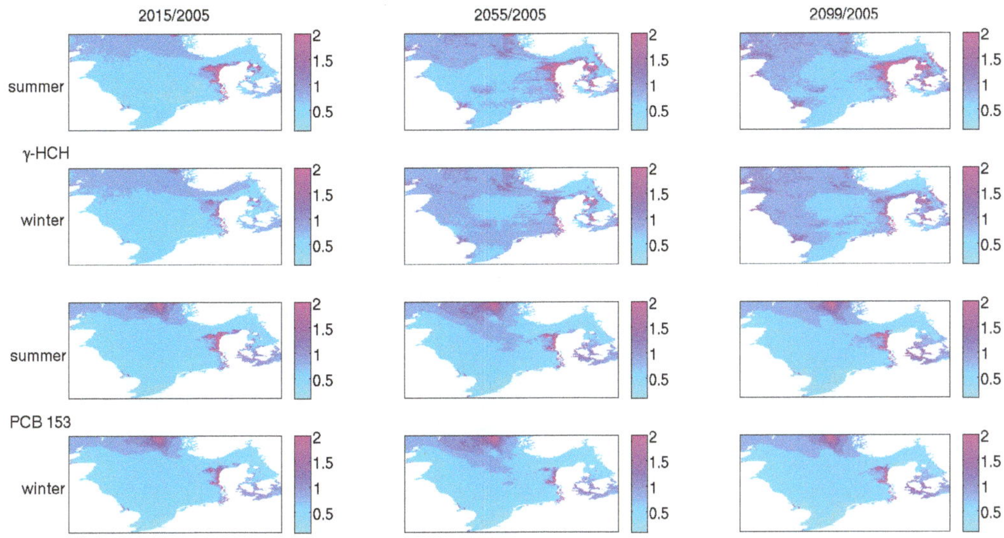

Figure 9. Distributions of concentration ratios in sediment, 2015/2005, 2055/2005 and 2099/2005, from left to right, respectively, for γ-HCH in winter (top) and summer (second row), and PCB 153 in winter (third row) and summer (bottom). Units are ng g^{-1} dry weight.

and northern North Sea but their magnitude is small, though there is a region of higher increase around the Firth of Forth and Moray Firth which may be due to localized storms in the 2015 run. Increases in 2015 relative to 2099 are again found in the aforementioned areas of the southern North Sea but this time extend along the entire British coast and an extended area seaward of the Wash. This is to be expected due to the increased stormy conditions. For the 2055/2099 case, large positive values are seen from East Anglia up to the northern Scottish coast, indicating stronger storm activity here in the 2090s, which is to be expected from examining the previous two plots. Large negative values are found around the Thames, central Dutch coast, Friesian Islands, German Bight, and Wadden Sea, and are also as expected from the previous two plots. Summer patterns are similar to those in winter, since most of the erosion takes place in winter, but some differences are found which can be the result of both deposition and erosion.

4 Summary and conclusions

The impact of climate variability on the fate and cycling of two selected legacy POPs, γ-HCH and PCB 153, in the North Sea in the 21st century, is investigated with a combined hydrodynamic (HAMSOM) and Fate and Transport Ocean Model (FANTOM). Future scenario model runs for three 10-year periods to the year 2100 (2006–2015, 2046–2055 and 2090–2099) using plausible levels of both in situ concentrations and atmospheric, river and open boundary inputs are performed. Our approach was to reutilize 2005 POP values in the atmosphere, rivers and at the open boundaries for every year of the simulations, while using the (final) result of our 1996–2005 runs, see O'Driscoll et al. (2013), as the initial condition for all three future runs. In this way, differences between the three 10-year simulations are attributed to climate change only.

Dry gas deposition and volatilization of γ-HCH increase in the future, relative to the present, and volatilization in 2090–2099 is even greater than 2046–2055, as the direct result of hydrodynamic forcing due to climate change.

Total mass of PCB 153 in sediment decreases exponentially in all three runs due to degradation in sediment, but also due to erosion and subsequent volatilization, particularly during storms, and concentrations in rivers and the atmosphere are not high enough to sustain sediment concentration levels. Total mass decreases even faster in the future runs, relative to the present, because of the increased number and intensity of storms. γ-HCH is also eroded during storms, but concentrations in sediment are recovered because so much of it is dissolved in water. Increased dry gas deposition of γ-HCH in the future results in more of it sorbing to POC in the water column, thus the increase of it in sediment in the future runs.

Annual net sinks exceed annual net sources at the end of all periods, leading to a loss of POPs in the system. Annual fluxes into and out of the North Sea system for both POPs show a slight future decrease during the simulations and also the second and third periods relative to the first period. The North Sea is net volatilizational for PCB 153, while γ-HCH is net depositional at the North Sea surface.

For the last year of each of the three runs, concentrations of γ-HCH in the water column are decreased relative to 2005 between the Thames and Rhine estuary regions due to decreased river and atmospheric concentrations. For PCB 153 there are no notable differences between the three runs in winter, but concentrations are somewhat reduced in summer in the future runs because of reduced dry gas deposition

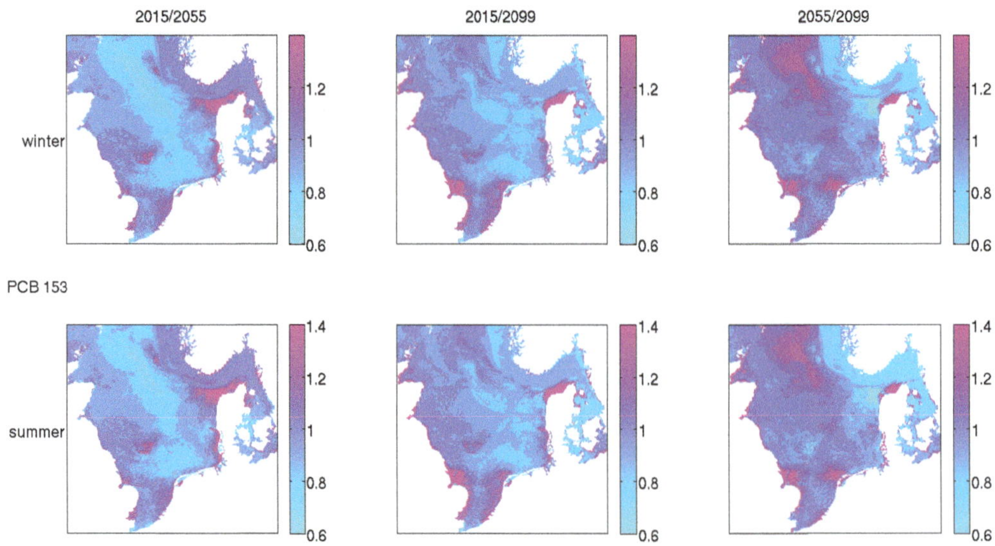

Figure 10. Distributions of concentration ratios of PCB 153 in sediment at the end of the three runs, 2015/2055, 2015/2099 and 2055/2099, from left to right, respectively, in winter (top) and summer (bottom). Units are ng g^{-1} dry weight.

and reduced concentrations in sediment. In winter, future decreases are found near British and Dutch rivers and the southern North Sea generally.

In sediment, concentrations of γ-HCH are generally reduced in the future runs relative to 2005, with greatest losses found in 2015 because of less dry gas deposition relative to 2055 and 2099. For PCB 153, there is little notable change in distributions and concentrations between 2005 and the last of each of the three 10-year runs, which is due to the large amount of sediment that has been either degraded or eroded.

The results agree with other studies that found increased volatilization of PCBs due to rising temperatures, thus contributing to increased transport from temperate regions to the Arctic (Lamon et al., 2009; Gouin et al., 2013; Armitage et al., 2011; Wöhrnschimmel et al., 2013). There is some uncertainty though, since investigators using global multimedia models found that concentrations of PCB153 and α-HCH in Arctic contamination increase in air and water by factors of 2 to 4 (Wöhrnschimmel et al., 2013), respectively, whereas volatilization rates of PCB 153 in our model increase by only a few percent. We remind the reader there are differences between our modelling approach and assumptions and those of Wöhrnschimmel et al. (2013). Wöhrnschimmel et al. (2013) present 150-year simulations driven by global-scale emission estimates over time. Accordingly, the modelled responses to climate change are not directly comparable in terms of absolute magnitude. Wöhrnschimmel et al. (2013) found that the key influences on surface ocean concentrations in the Arctic for PCB 153 are temperature-driven increases in primary emissions and reductions in sea ice cover. Although the increased volatilization of North Sea POPs would lead to increased concentrations in the Arctic, it appears that other sources would also have to contribute to the increase, even though the increased rates in marginal seas would contribute substantially to increased concentrations found in the Arctic.

Finally, although the influence of climate change on the two POPs simulated here is generally small, the increased number and magnitude of storms in the 21st century will result in large amounts of hydrophobic POPs being resuspended in shallow seas with ensuing revolatilization events. This can have important consequences including movement of POPs up through the marine food chain and delivery of POPs into Europe.

Acknowledgements. The authors thank the German Science Foundation(DFG) for funding the project under grant no. PO 614/8-1. The authors thank Tatiana Ilyina and Thomas Pohlmann for valuable discussions. We thank four reviewers for their thoughtful comments.

Edited by: M. Hoppema

References

AMAP Assessment Report: Arctic Pollution Issues: Arctic Monitoring and Assessment Programme, Oslo, Norway, 1998.

Armitage, J. M., Quinn, C. L., and Wania, F.: Global climate change and contaminants – an overview of opportunities and priorities for modelling the potential implications for long-term human exposure to organic compounds in the Arctic, J. Environ. Monitor., 13, 1532–1546, 2011.

Backhaus, J. O.: A three-dimensional model for the simulation of shelf sea dynamics, Deutsche Hydrographische Zeitschrift, 38, 165–187, 1985.

Breivik, K., Sweetman, A., Pacyna, J. M., and Jones, K. C.: Towards a global historical emission inventory for selected PCB congeners – A mass balance approach: 3. An update, Sci. Total Environ., 377, 296–307, 2007.

Ghosh, S., Zang, S., Mitra, P. S., Ghimbovschib, S., Hoffman, E. P., and Dutta, S. K.: Global gene expression and Ingenuity biological functions analysis on PCBs 153 and 138 induced human PBMC in vitro reveals differential mode(s) of action in developing toxicities, Environ. Int., 37, 838–857, doi:10.1016/j.envint.2011.02.010, 2011.

Gioia, R., Nizzetto, L., Lohmann, R., and Dachs, J.: Polychlorinated biphenyls (PCBs) in air and seawater of the Atlantic Ocean: sources, trends and processes, Environ. Sci. Technol., 42, 1416–1422, 2008.

Gouin, T., Armitage, J., Cousins, I., Muir, D., Ng, C. A., Reid, L., and Tao, S.: Influence of global climate change on chemical fate and bioaccumulation: The role of multimedia models, Environ. Toxicol. Chem., 32, 20–31, 2013.

Gusev, A., Rozovskaya, O., Shatalov, V., Sokovyh, V., Aas, W., and Breivik, K.: Persistent Organic Pollutants in the Environment, EMEP Meteorological Synthesizing Centre-East, Status Report, 3, 77 pp., 2009.

Hornsby, A. G., Wauchope, R. D., and Herner, A. E.: Pesticide Properties in the Environment, Springer-Verlag, New York, 1996.

Ilyina, T., Pohlmann, T., Lammel, G., and Sündermann, J.: A fate and transport ocean model for persistent organic pollutants and its application to the North Sea, J. Marine Syst., 63, 1–19, 2006.

Jacob, D., Van den Hurk, B. J. J. M., Andrae, U., Elgered, G., Fortelius, C., Graham, L. P., Jackson, S. D., Karstens, U., Käopken, C., Lindau, R., Podzun, R., Rockel, B., Rubel, F., Sass, B. H., Smith, R. N. B., and Yang, X.: A comprehensive model inter-comparison study investigating the water budget during the BALTEX-PIDCAP period, Meteorol. Atmos. Phys., 77, 19–43, 2001.

Jungclaus, J. H., Botzet, M., Haak, H., Keenlyside, N., Luo, J. J., Latif, M., Marotzke, J., Mikolajewicz, U., and Roeckner, E.: Ocean circulation and tropical variability in the coupled model ECHAM5/MPI-OM, J. Climate, 19, 3952–3972, doi:10.1175/JCLI3827.1, 2006.

Kalnay, E., Kanamitsu, M., Kistler, R., Collins, W., Deaven, D., Gandin, L., Iredell, M., Saha, S., White, G., Woollen, J., Zhu, Y., Chelliah, M., Ebisuzaki, W., Higgins, W., Janowiak, J., Mo, K. C., Ropelewski, C., Wang, J., Leetmaa, A., Reynolds, R., Jenne, R., and Joseph, D.: The NCEP/NCAR 40-year reanalysis project, B. Am. Meteorol. Soc., 77, 437–471, 1996.

Lamon, L., von Waldow, H., MacLeod, M., Scheringer, M., Marcomini, A., and Hungerbuhler, K.: Modeling the global levels and distribution of Polychlorinated Biphenyls in air under a climate change scenario, Environ. Sci. Technol., 43, 5818–5824, 2009.

Larsen, J. and She, J.: Optimisation of a bathymetry database for the North European Shelf Seas, Technical Report 01-21, Danish Meteorological Institute, 2001.

Lohmann, R., Breivik, K., Dachs, J., and Muir, D.: Global fate of POPs: Current and future research directions, Environ. Pollut., 150, 150–165, 2007.

Mayer, B., Damm, P. E., Pohlmann, T., and Rizal, S.: What is driving the ITF? An illumination of the Indonesian throughflow with a numerical nested model system, Dynam. Atmos. Oceans, 50, 301–312, doi:10.1016/j.dynatmoce.2010.03.002, 2010.

Ngabe, B., Bidleman, T. F., and Falconer, R. L.: Base Hydrolysis of alpha- and gamma- Hexachlorocyclohexanes, Environ. Sci. Technol., 27, 1930–1933, 1993.

Nizzetto, L., Lohmann, R., Gioia, R., Dachs, J., and Jones, K. C.: Atlantic ocean surface waters buffer declining atmospheric concentrations of persistent organic pollutants, Environ. Sci. Technol., 44, 6978–6984, doi:10.1021/es101293v, 2010.

O'Driscoll, K., Mayer, B., Ilyina, T., and Pohlmann, T.: Modelling the cycling of persistent organic pollutants (POPs) in the North Sea system: fluxes, loading, seasonality, trends, J. Marine Syst., 111–112, 69–82, doi:10.1016/j.jmarsys.2012.09.011, 2013.

Orlanski, I.: A simple boundary condition for unbounded hyperbolic flows, J. Comput. Phys., 21, 251–269, 1976.

Pohlmann, T.: A meso-scale model of the central and southern North Sea: Consequences of an improved resolution, Cont. Shelf Res., 26, 2367–2385, 2006.

Roeckner, E., Brokopf, R., Esch, M., Giorgetta, M., Hagemann, S., Kornblueh, L., Manzini, E., Schlese, U., and Schulzweida, U.: Sensitivity of simulated climate to horizontal and vertical resolution in the ECHAM5 atmosphere model, J. Climate, 19, 3771–3791, 2006.

Sinkkonen, S. and Paasivirta, J.: Degradation half-life times of PCDDs, PCDFs and PCBs for environmental fate modeling, Chemosphere, 40, 943–949, 2000.

Su, J. and Pohlmann, T.: Wind-topography interaction in an upwelling system in the eastern Hainan coast, J. Geophys. Res., 114, C06017, doi:10.1029/2008JC005018, 2009.

Weisse, R., von Storch, H., and Feser, F.: Northeast Atlantic and North Sea storminess as simulated by a regional climate model 1958–2001 and comparison with observations, J. Climate, 18, 465–479, 2005.

Weisse, R., von Storch, H., Callies, U., Chrastansky, A., Feser, F., Grabbemann, I., Günther, H., Pluess, A., Stoye, T., Tellkamp, J., Winterfeldt, J., and Woth, K.: Regional meteorological-marine reanalyses and climate change projections: Results for Northern Europe and potential for coastal and offshore applications, B. Am. Meteorol. Soc., 90, 849–860, 2009.

Wöhrnschimmel, H., MacLeod, M., and Hungerbühler, K.: Emissions, fate and transport of persistent organic pollutants to the Arctic in a changing global climate, Environ. Sci. Technol., 47, 2323–2330, 2013.

Numerical modelling of physical processes governing larval transport in the southern North Sea

M. C. H. Tiessen[1], L. Fernard[2], T. Gerkema[1], J. van der Molen[2], P. Ruardij[1], and H. W. van der Veer[1]

[1]Royal Netherlands Institute for Sea Research, P.O. Box 59, 1790 AB Den Burg, Texel, the Netherlands
[2]Centre for Environment, Fisheries and Aquaculture Science (CEFAS), Pakefield Road, Lowestoft, Suffolk NR33 0HT, UK

Correspondence to: M. C. H. Tiessen (meinard.tiessen@nioz.nl)

Abstract. A three-dimensional hydrodynamic model (GETM) was coupled with a particle tracking routine (GITM) to study the inter-annual variability in transport paths of particles in the North Sea and English Channel. For validation, a comparison with observed drifter trajectories is also presented here. This research investigated to what extent variability in the hydrodynamic conditions alone (reflecting passive particle transport) contributed to inter-annual variability in the transport of eggs and larvae. In this idealised study, no a priori selection of specific spawning grounds or periods was made and no active behaviour (vertical migration) or mortality was included. In this study, egg and larval development towards coastal nursery areas was based solely on sea water temperature, while settlement areas were defined by a threshold water depth. Results showed strong inter-annual variability in drift direction and distance, caused by a combination of wind speed and direction. Strong inter-annual variability was observed both in absolute amount of settlement in several coastal areas, and in the relative importance of the different areas. The effects of wind and temperature variability are minor for settlement along the western shores of the North Sea and in the English Channel, but have a very significant impact on settlement along the eastern shores of the North Sea. Years with strong south-westerly winds across the Dover Straight resulted in higher settlement figures along its eastern shores of the North Sea (standard deviation 37 % of the mean annual settlement value). Settlement in the western Dutch Wadden Sea did not only show inter-annual variability, but patterns were also variable within each year and revealed seasonal changes in the origin of particles: during winter, stronger currents along with colder temperatures generally result in particles originating from further away.

1 Introduction

The pelagic phase is a characteristic component of the life cycle of the majority of marine species. Within the range from holoplankton to holobenthos species a variety of different life strategies occur, including benthic resting stages in plankton species (such as, among others, dinoflagellates (Dale, 1985) and cnidaria (Boero et al., 1991)) and pelagic egg and larval stages in benthic bivalves and demersal fish species. Marine species can only exist in geographic locations within which there can be continuity in the life cycle; i.e. "in a geographical setting within which retention (membership) of the population exceeds losses (vagrancy) in some integrated sense for the life cycle as a whole" (cf. Sinclair, 1988). The importance of life cycle closure is continuously topical since at any stage of the life cycle an individual can become separated from its population either by spatial (losses from distributional areas) or by energetic (predation, starvation, disease) processes. The variety of all these complex life stages can be viewed as species-specific and/or local adaptations to ultimately achieve life cycle closure.

During the pelagic phase physical factors such as advection and temperature are acting on and potentially resulting in the dispersal of individuals, depending on strength and duration of the physical forcing. Especially for species with a pelagic egg and larval stage, this period of egg and larval drift is considered to be an important factor in year-class strength regulation (bivalves: Thorson, 1966; fishes: Leggett and Deblois, 1994; flatfishes: Van der Veer, 1986; Van der Veer et al., 2000; Van der Veer and Leggett, 2005). As a consequence, the impact of inter-annual variability in hydrodynamic conditions on egg and larval transport has been studied since large-scale two-dimensional and later three-dimensional hydrodynamic transport models have become available. For the North Sea area this has resulted in a number of detailed studies on egg and larval drift and transport of fish larvae, especially flatfishes (Van der Veer et al., 1998; de Graaf et al., 2004; Bolle et al., 2009; Dickey-Collas et al., 2009; Savina et al., 2010; Hufnagl et al., 2013; Lacroix et al., 2013) and all these studies stress the importance of drift and transport during the pelagic phase.

Passive transport is the only agent during egg development but from hatching onwards also larval behaviour such as active vertical migration may come into play (Rijnsdorp et al., 1985). In the recent three-dimensional hydrodynamic modelling studies by van der Molen et al. (2007), Dickey-Collas et al. (2009), Bolle et al. (2009), Savina et al. (2010), Hufnagl et al. (2013), and Lacroix et al. (2013) both passive transport and active behaviour components were included thereby preventing an analysis of their relative importance. In particular, it remains unclear to what extent physical factors alone can be responsible for inter-annual variability.

The aim of the present paper was to determine the importance of passive transport by advection and the impact of physical factors alone on the dispersal of particles in the southern North Sea (Fig. 1), whereby we used plaice *Pleuronectes platessa* as model species. By way of validation, we start with a comparison of simulated model trajectories with observed drifter paths. Particle tracking was introduced by means of a three-dimensional Lagrangian tracer model in combination with a hydrodynamic model. Instead of selecting specific release areas (reflecting spawning grounds) and time periods (reflecting spawning periods), as was done by de Graaf et al. (2004) and Bolle et al. (2009), no a priori selection was made; the entire southern side of the North Sea and the English Channel were selected as potential starting points (cf. Hufnagl et al., 2013) for particles over a 4-month period in winter and spring, reflecting the most important period for egg and larval drift in plaice larvae (Talbot, 1977; Harding et al., 1978). Particles were released at random over latitude, longitude, depth and time. With regards to the transport of larvae (such as of plaice *Pleuronectes platessa*), we used sea water temperature to define a pelagic life span (Bolle et al., 2009). When development into a juvenile was completed, settling was assumed to take place at the first opportunity when a particle would encounter an area with a certain minimum threshold depth. In this way, the success of settling could be defined and mapped.

2 Material and methods

2.1 Hydrodynamic model

The hydrodynamic model used in this research is the three-dimensional General Estuarine Transport Model (GETM). GETM incorporates the General Ocean Turbulence Model (GOTM), which is a one-dimensional model over the vertical

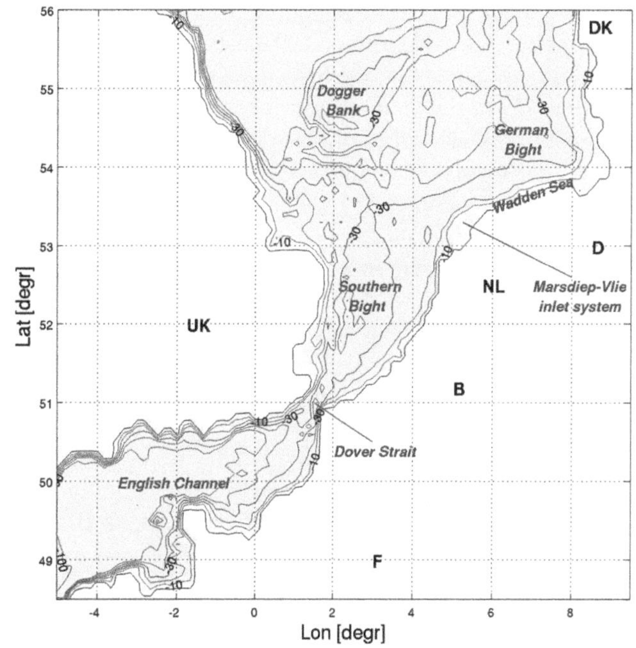

Figure 1. Topography of the research area, along with area and location names. Note that the area shown does not correspond with the entire domain of the hydrodynamic simulations, which spans 48.2–60.1° N, −7–16.9° E, and includes part of the Baltic sea.

that provides a variety of turbulence closure schemes. The combined 3-D hydrodynamic model suite was developed by Burchard and Bolding (2002). It has been applied to coastal seas such as the North Sea and Baltic Sea (Ruardij et al., 2005; Lenhart et al., 2010; Gräwe and Burchard, 2011; Van Leeuwen et al., 2012; Aldridge et al., 2012). The model can implement the inflow of fresh water and describes the development of stratification in the domain, both as a result of salinity and temperature gradients. The vertical tides and meteo-forcing are prescribed at the open boundaries.

In the present study, the model runs were based on GETM-ERSEM (European Regional Seas Ecosystem Model; Van Leeuwen et al., 2012; Ruardij et al., 2005) and were set up to describe the North Sea and part of the Baltic (48.2–60.1° N, −7–16.9° E). The model grid was 6 nautical miles (approximately 10 km) in the horizontal, and 26 sigma-layers were used over the vertical. Open model boundaries were located at the English Channel and between the North Sea and North Atlantic. At these boundaries the surface elevation and currents were forced by data from a barotropic shelf-wide model, which in turn was forced with the Topex Poseidon altimetry data (Topex Poseidon, http://podaac.jpl.nasa.gov/). Monthly sea water temperature and salinity values according to (Janssen et al., 1999) were introduced at each grid-point as initial conditions. Both the volume and salinity effects of the various rivers flowing into the North Sea were provided by local agencies (see Acknowledgements). The ECMWF-reanalysis provided the meteorological forcing (ECMWF, http://www.ecmwf.int/). The hydrodynamic model was run for the entire duration between 1993 and 2005. From December to July of each year, the hydrodynamic conditions were stored at 45 min intervals to provide a high enough temporal resolution to describe currents during the M2 tidal cycle, which is the main tidal component in the central and southern North Sea.

2.2 Particle tracking model

The three dimensional Lagrangian tracer model (GITM) used in combination with the GETM hydrodynamic model was originally developed by Wolk (2003). A re-programmed version has recently been expanded upon, by Van der Molen (unpublished data), with modules to allow for egg and larval development and vertical migration behaviour in similarity to the work presented in Van der Molen et al. (2007). The model describes the motion of a number of discrete particles through a 3-D velocity field. This velocity field is provided by the hydrodynamic model. The particle tracking model divides the motion into an advective and a diffusive component. The advection equation is solved analytically. Turbulence is accounted for by the random walk model (Hunter et al., 1993) where the vertical diffusivity is calculated by the turbulence model in GETM using the Kolmogorov–Prandtl relation (Burchard and Bolding, 2002). Horizontal diffusion is here excluded from the particle drift, and the displacement of particles is therefore governed by the output of current velocities from the model. The internal time step of 10 s is chosen to ensure accuracy of the tracer model.

2.3 Set-up of plaice model runs

Instead of selecting specific spawning grounds and periods (as was done by: Van der Veer et al., 1998; de Graaf et al., 2004; Van der Molen et al., 2007; Bolle et al., 2009), here no a priori selection has been made. This allows for more flexibility in the interpretation. The entire southern side of the North Sea and the English Channel were selected as starting points for particles. Particles were released at random over latitude, longitude, depth and time. A total of 1 000 000 particles were released between 48.5–56° N and −5–9.5° E, between 1 December (of the preceding year) and 31 March (Fig. 2). For instance, for the Southern Bight of the North Sea, this translates in approximately 5 800 particles being released every week. Development rates of particles (simulating plaice larvae) were derived from Bolle et al. (2009), but no biological behaviour or active vertical migration was included in the later stages of the particle evolution. After the final larval stage, a settlement stage, lasting up to 30 days, followed. During this stage, a particle would settle if it encountered favourable conditions. The only prerequisite for a suitable settlement location was a predefined threshold water depth: when particles would flow through a grid-cell whose minimum depth (at one of the corners) was less than 10 m,

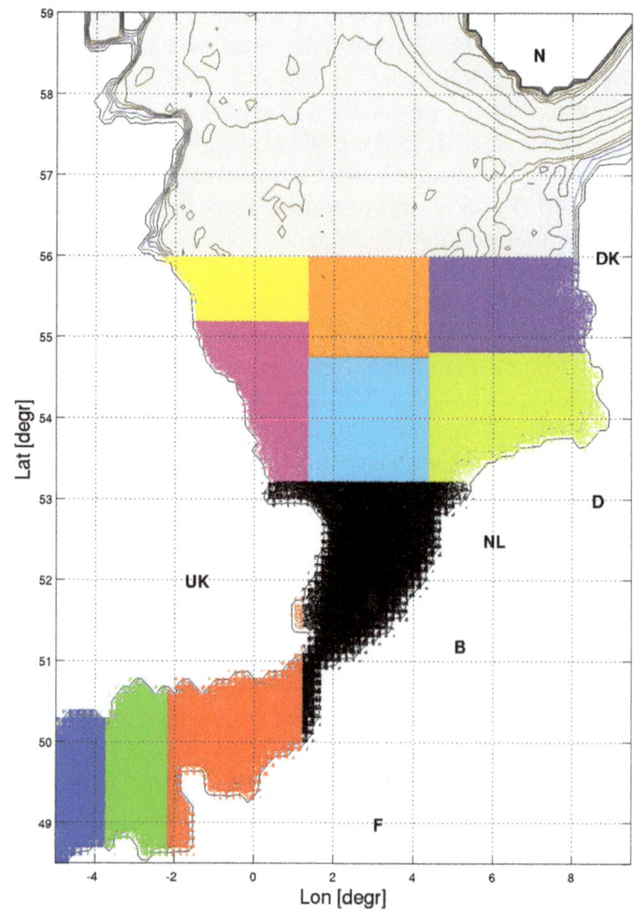

Figure 2. Release distribution of particles for the numerical runs. Colour of the dots solely indicates the release position.

then it was assumed this particle would settle. The duration of the pelagic life span of particles depended on the water temperature, as this determined the duration of the different life stages (as was the case in Bolle et al., 2009). The water temperature experienced by each particle was determined over time by interpolation of the 3-D temperature fields provided by the hydrodynamic model. A maximum pelagic life span (which included the settlement stage) of 120 days was imposed, at which point model runs were truncated. No mortality was incorporated.

2.4 Validation of the drift model

Validation of the model was done by a comparison with a number of trajectories made by drifter buoys. During the spring and summer of 2000, CEFAS (Centre for Environment, Fisheries and Aquaculture Science; UK) released over 30 drifters in the central and southern North Sea. These drifters were equipped with a drogue floating at a pre-prescribed depth, which was attached to a floating buoy providing the communication with a satellite. The drogue consisted of a large cylindrical shape with a diameter of 1.5 m and a length of approximately 7 m. The drogue was kept at a depth of 10 m. Drifter paths lasted up to 60 days, but were all eventually truncated. This was generally due to fishing activities accidentally cutting the wire between the drogue and the surface buoy. The subsequent drift of the surface buoy was (although recorded) not included in the comparison here, as the ensuing (rapid) drift was driven by surface waves and direct wind effects, which were not included in the numerical simulations.

The present validation comparison focussed on six drifters released in the Southern Bight, since this formed the target area for plaice larvae drift. Additionally, only drift durations of more than 20 days were included. Certain drifter buoys were also excluded as they drifted shoreward into areas with a water depth of less than 15 m, which would mean that the drogue might be touching the bed, affecting the drift direction and speed.

The drift trajectories were reproduced by releasing particles in the numerical model at the same time and location and at a depth of 10 m. This depth was chosen to cover both the drag effect of the drogue and the minor contribution of the surface buoy (Niiler et al., 1995). For these runs vertical diffusion was excluded to ensure that the modelled particles remained at the same depth as the drifter buoys.

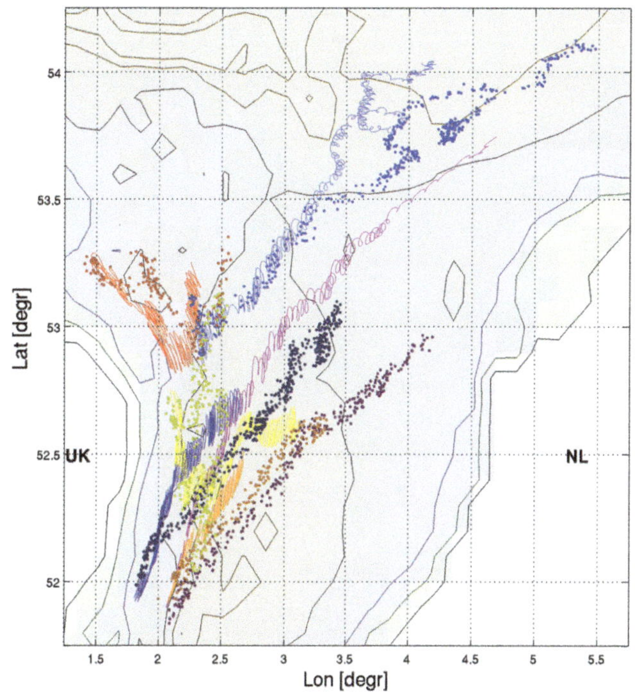

Figure 3. Comparison between six drifter paths (different coloured dots) and modelled trajectories (shown in differently coloured lines). The starting-point of each of the drifter-buoy pairs is shown as a thicker circle, and is generally located close to the UK coast.

3 Results

3.1 Validation

A comparison between the drift of the buoys and their modelled counterparts shows that the model reproduces the general trend of the drifters reasonably well (Fig. 3). At the end of the drift, buoy positions and modelled particle locations have diverged, as can be seen by the distance between the final dot for each colour and the end of the colour-corresponding line. However, there is no general trend in this divergence, suggesting that the model does not over- or under-predict the overall residual current patterns. The error that developed over time between each of the drifter buoys and the modelled particles was less than 10 % of the total drifted distance of each of the buoys, indicating a fairly good agreement between predicted and observed drift paths. To put the discrepancy between modelled and actual trajectories into perspective, this divergence was compared to the modelled spread of particles released in the proximity of the drifter release point (Fig. 4) Along with the single modelled particle being released at the same position and time as each of the drifter buoys, additional modelled particles were positioned in a 10 km-wide circle around the drifter release-point. The initial distance between the central particle and the particles in the circle was chosen to roughly correspond to the tidal excursion as was displayed by the drifter buoys. With time, the circle gets deformed, and the distance between particles in the circle and the central particle increases (Fig. 4). Significant differences between the different drifter buoys and their modelled equivalents exist. For the buoys that cover the longest distance (blue and purple, see Fig. 3), the particles in the circle show the smallest deformation to develop over time, with the circle of particles remaining close to 0. However, the buoys that show a more complicated drift pattern (in particular red, yellow and dark blue) show a much bigger spread of the circle of particles to develop over time. Overall, however, the accuracy of the model predictions is fairly reasonable, in the sense that the maximum spread of the particles originally released in a circle is of the same size as the difference between the drifter buoys and the corresponding modelled central particle.

3.2 Particle drift and settlement success

In the following sections, an in-depth investigation into the hydrodynamic causes of inter-annual variability in larval transport and settlement success is presented. The analysis will first be applied to the numerical runs of plaice eggs and larvae for a single year (2001), followed by a comparison between the different years. The year 2001 was chosen as this year seems to be a very moderate year, not only considering hydrodynamic conditions (for a comparison with SST (sea surface temperature) values for other years, see Van der Veer and Witte (1999), but also with respect to settlement success

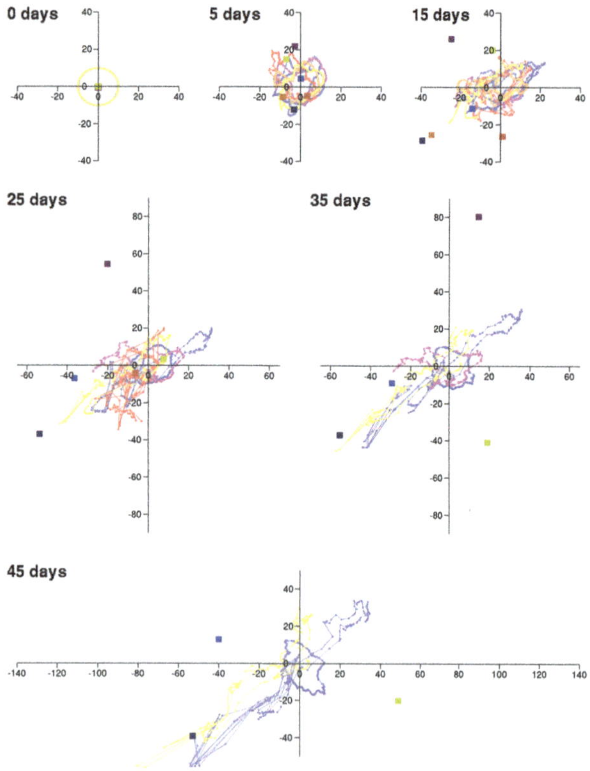

Figure 4. At different moments during the drifter-buoy track, the difference between the drifter path and the corresponding modelled trajectory is shown (as squares, with the different tracks shown in different colours) with, on top of that, the current position of particles released in a circle around the drifter position. The x and y axis represent kilometres of spread. With time, the distance between the actual drifter position and that of the modelled particles increases (as can also be seen in Fig. 3). The spread of particles originally released in a circle around the central particle deforms and expands over time.

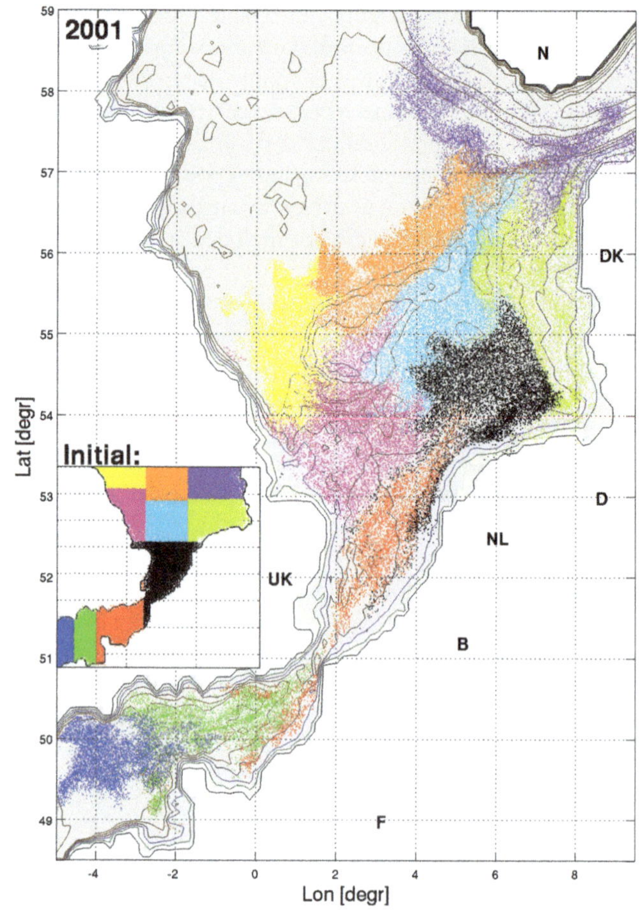

Figure 5. Distribution of particles at the end of their pelagic life span (up to 120 days). Only shown are particles that do not settle and to improve clarity only 100 000 particles that were released around 1 January are included. The colour of the dots corresponds to the release location of the particles, which is shown in the inset.

rates, as shown in data of the Balgzand area (Van der Veer et al., 2009).

In Fig. 5, the final position of particles is shown at the end of their pelagic life span (up to 120 days). Please note that in order to facilitate the interpretation of the figure, only a small section of the total number of particles released is shown in this figure. Only particles released from 25 December to 6 January (100 000 particles) are shown. Particles that "settled" (i.e. reached an area shallower than the threshold depth of 10 m in their final pelagic stage) are not included in this figure as their drift was truncated prematurely. The particles showed a significant displacement over this period of time. Especially particles released in the Southern Bight (black) and the English Channel (red) showed drift over a large distance. The particles drifted in an anticlockwise fashion, especially in the central North Sea. This conformed with the typical circulation pattern (Otto et al., 1990; Sündermann and Pohlmann, 2011). Just south-west of Dover Straight and along the eastern shores, particles showed the largest drift distance (towards NE), whereas drift distances around Dogger Bank were smallest.

Settlement success (of all of the 1 000 000 particles released) was not only dependent on the particle drift direction, but also on the particle development (Fig. 6). Particles mainly settled along the coastlines of the North Sea, but also in the central parts of the Southern Bight and the English Channel. Additionally, pie charts show the percentages of different outcomes of the numerical simulations: prolonged development inhibiting particles to reach settlement stage (red) only occurred in the German Bight. This red segment of the pie charts corresponds to particles that are still in their larval stage when simulations are truncated after 120 days. However, also large parts of the central North Sea (for at least a period of time) contained particles with reduced development rates, as these only reached the beginning of the settlement stage (orange). The orange segment represents particles that had spent fewer than 30 days in their settlement stage, before the simulation was truncated after 120 days. As a result,

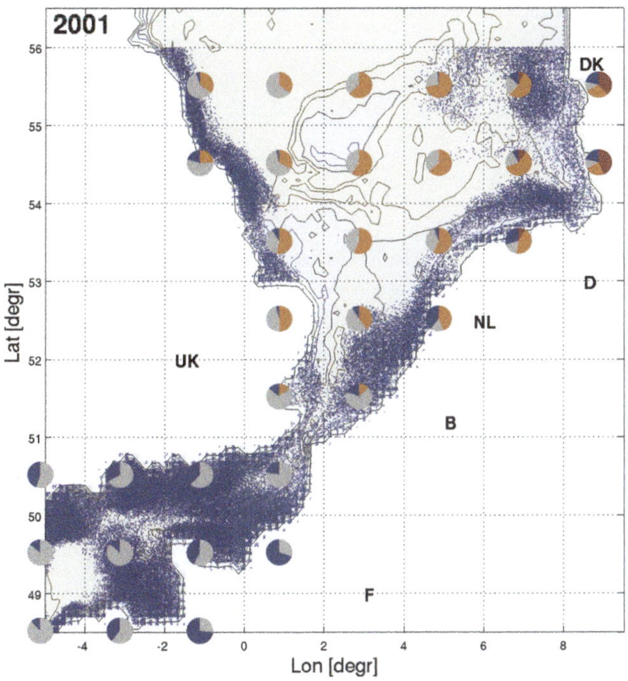

Figure 6. Distribution of the origin of particles that settle (shown in blue dots). Please note that all 1 000 000 particles are included in this figure. On top of that, different pie charts show the geographic variation of the different outcomes of particle development and drift: (blue) particles that settle; (grey) particles that do not settle; (orange) particles that do not settle but also do not reach end of the 3-day settlement stage before the numerical simulation is truncated after 120 days; (red) particles that do not settle as these have not yet reached the beginning of their 30-day settlement stage, before simulation is truncated. The red segment of the pie charts indicates particles that are still in their larval stage after 120 days.

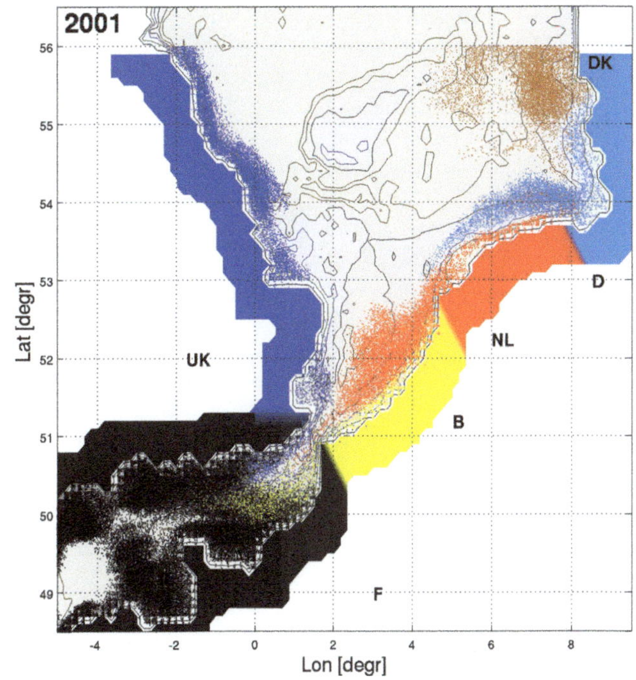

Figure 7. Release location of particles, with the settlement location shown in the colour of the dots. Additionally, the different settlement areas are indicated with a colour-corresponding band along the coastline. (dark blue) Settlement in the North Sea coastal zone of the UK; (black) settlement along the coast of the English Channel (both France and UK); (yellow) settlement in the French, Belgian and Dutch North Sea coast; (red) settlement in the western part of the Wadden Sea; (light blue) settlement in the northern part of the Wadden Sea; (orange) settlement outside of the shown domain (generally along the northern Danish and Norwegian coast). Please note that the coarse model grid does not include any Wadden Sea islands, and therefore there is no distinction between settlement on the North Sea side of these islands, and in the Wadden Sea itself.

these particles had only a reduced period during which a particle could drift into a suitable settlement location. The geographic variation in the success rate of a particle reaching the settlement stage closely corresponded to the water temperature (van der Veer and Witte, 1999). The lowest temperatures occurred in the German Bight and along the Danish Coast, the same area where particles did not reach the settling stage due to their prolonged development. The highest water temperatures occurred in the English Channel, where all released particles reached the end of the settling stage before the 120-day model-run duration ended. Note, however, that the particle development took place during the entire particle drift, and that therefore particles may have passed through a range of water temperatures in space and time.

The geographic origin of particles that settled in the different coastal zones can be seen in Fig. 7. The coastlines around the southern North Sea and the English Channel are divided into different zones. Particles settling in the western part of the Wadden Sea (the red part) largely originated from the Southern Bight. A clear divide can be seen between the origin of particles settling in the western Wadden Sea and the north-eastern Wadden Sea (light blue). Particles released in the Dutch coastal zone generally settled in the western Wadden Sea, whereas particles released only slightly further offshore, drifted into the German Bight before settling. This principle also applied to the other continental coastal areas.

3.3 Inter-annual variability

Numerical runs were carried out for the years 1994–2005. Some characteristic values for all the different years will be compared but, initially, a comparison between two extreme years (1996 and 1998) and the "moderate" year (2001) presented previously will be discussed.

The overall displacement of the section of particles released around 1 January of 1996 and 1998 show significant differences (Fig. 8). The particle drift direction of 1998 was roughly the same as that presented for 2001, mainly towards the north-east along the eastern shores of the North Sea and

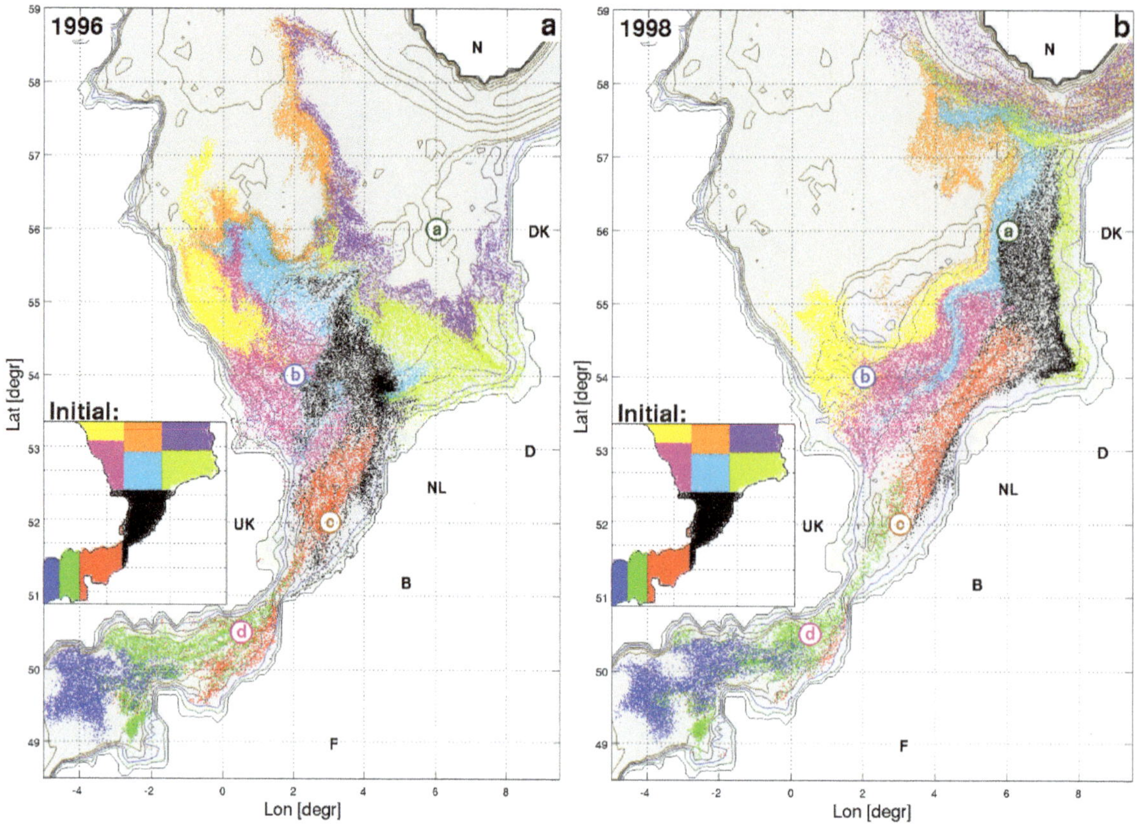

Figure 8. Distribution of particles at the end of their pelagic life span (up to 120 days) **(a)** 1996; **(b)** 1998. Only shown are particles that do not settle. Only 100 000 particles that were released around 1 January are included. The colour of the dots corresponds to the release location of the particles, shown at the inset. Locations "a" to "d" correspond to data shown in Figs. 9 and 14.

Table 1. Relationship between variability in forcing conditions and changes in settlement success. Settlement success is represented be the variability in the number of settling particles and by the number of particles with truncated development. The forcing conditions constitute the SST over February for each year at station d, and the percentages of wind intensity from the south-west at station d (the intensity is determined as the average wind speed from this direction times the percentage of time the wind is coming from this direction, this value is then divided over sum of intensities for all wind directions). CV stands for coefficient of variation (SD/mean × 100), SD for standard deviation, and Corr. for Correlation coefficient. Bold numbers indicate a high correlation.

	Mean	SD	CV (%)	Corr. Temp	Corr. Wind
Total number of settling particles	119 412	9827	8	0.61	0.50
Total number of particles with truncated development	235 441	147 904	63	**−0.77**	**−0.72**

in the English Channel. In 1996, no clear overall drift direction could be observed. There was only a very small portion that showed a similar trend to that of 1998 and 2001, located around the Dover Strait. Particles released in the central North Sea seemed to drift towards the west and north-west, instead of towards the north-east (as occurred for 1998 and 2001). Drift distance was also very different for both years, with 1998 showing bigger drift distances than 2001, while these were significantly reduced for 1996. Both the changes in drift direction and distance were the result of significantly different wind patterns for 1996 in comparison with 1998 (with 2001 being similar to 1998) (Fig. 9). In 1998 and 2001 the prevailing wind direction and the highest average wind speeds were south-westerly for stations c and d. For 1996, the exact opposite wind forcing can be observed.

The different drift directions in 1996 resulted in very different settlement success levels (of the 1 000 000 particles released) (Fig. 10). A very large portion of particles across the southern and eastern parts of the North Sea did not reach the settlement stage within 120 days (shown in the red segments of the pie-charts), and particles did not reach the end of the settling stage (shown in orange) in the entire North Sea. Much smaller numbers of particles released from the Southern and German bights settled compared to both 2001 and

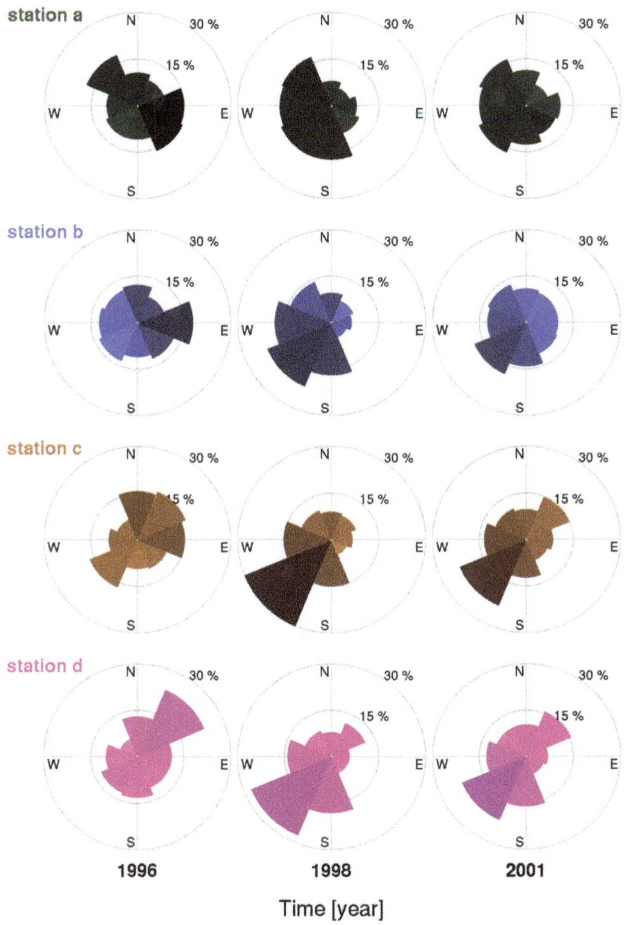

Figure 9. Wind speed and direction at the different stations (see Fig. 8) for 1996 (left), 1998 (middle) and 2001 (right) over the 8 months over which the particle drift was computed, starting with December of the previous year, while finishing in July. The size of the pie-pieces is defined by the percentage of time that the wind is coming from this direction. The colour of the pieces indicates the average wind speed.

1998. Not only were the currents in 1996 very different from 2001 and 1998, also the water temperature was much lower. The inflow of warm English Channel water was significantly reduced while the cold water patch in the German Bight and along the Danish coast was chilled and expanded across the central North Sea, resulting in the slow development experienced by particles released in these areas. The SST distribution in 1998 was similar to that of 2001, with only a slightly increased SST across the model domain. The result was that almost all (except in the most easterly parts of the German Bight) particles reached both the start and end of the settlement stage. In comparison with 2001, 1998 showed a very similar distribution of successful settlement of particles.

The inter-annual variability in currents and water temperature caused not only changes in settlement success but also in location of settlement (Fig. 11). Along the eastern coast of the UK and in the English Channel, the settlement locations of particles were similar for 1996, 1998 and 2001. However, the origin of particles settling in for instance the western part of the Wadden Sea (red) changed significantly over the years. In 1998, particles released from the Dover Strait reached this settlement area (similar to 2001), whereas in 1996 the origin of particles settling here was restricted to the settlement area itself and along the Dutch North Sea coast. Additionally, the stronger currents experienced in 1998 (compared to 2001), increased the number of particles settling along the northern shores of Denmark and along Norway: in 2001, a collection of particles released just north of the Dutch Wadden Isles would settle in the German Bight and along the Danish Wadden Sea coast; in 1998 such a plume could not be seen and particles released only slightly further offshore now reached the Norwegian Trench (shown in orange).

The final status of each particle was grouped in different outcomes for each year separately in Fig. 12. The total number of settling particles (shown in blue) showed only limited variability, with a coefficient of variability (CV) of only 8 %, while the number of particles with truncated development (depicted in orange and red) showed a strong variability over the years (CV = 63 %), which can be linked to hydrodynamic conditions. Table 1 shows the correlation coefficients between the final status of a particle (more specifically: settlement or truncated development) and two characteristic forcing conditions. The inter-annual variability in the monthly average sea surface temperature for February (as shown in Fig. 10c, d) at station d, shows a good correlation with the number of particles with truncated development, as increased water temperatures reduce development times. Additionally, in years with a higher percentage of south-westerly prevailing winds (at station d), there was a drift of warm English Channel water into the Southern Bight, increasing the water temperatures experienced by the majority of particles, resulting in a high correlation between this wind characteristic and development success. Additionally, with the small variability in the total number of settling particles, the correlation between this variability and the forcing conditions is smaller (temperature: 0.61; wind: 0.50).

Closer inspection of the number of particles that settled in the different zones (as shown in Fig. 7), showed that the largest number of settling particles did so in the English Channel (Fig. 13). As the particles were released at random over space (including depth), a significantly large portion of particles was released in the (relatively deep) English Channel. The inter-annual trends in settling numbers for each zone illustrated that the number of settling particles along the North Sea coast of the UK (dark blue) and the English Channel (black) was reasonably constant (CV is 8 and 6 %, respectively, see Table 2), with the majority of the variability originating from the particles settling along the eastern shores of the North Sea (yellow, red, light blue and orange). These settlement zone characteristics roughly coincided with

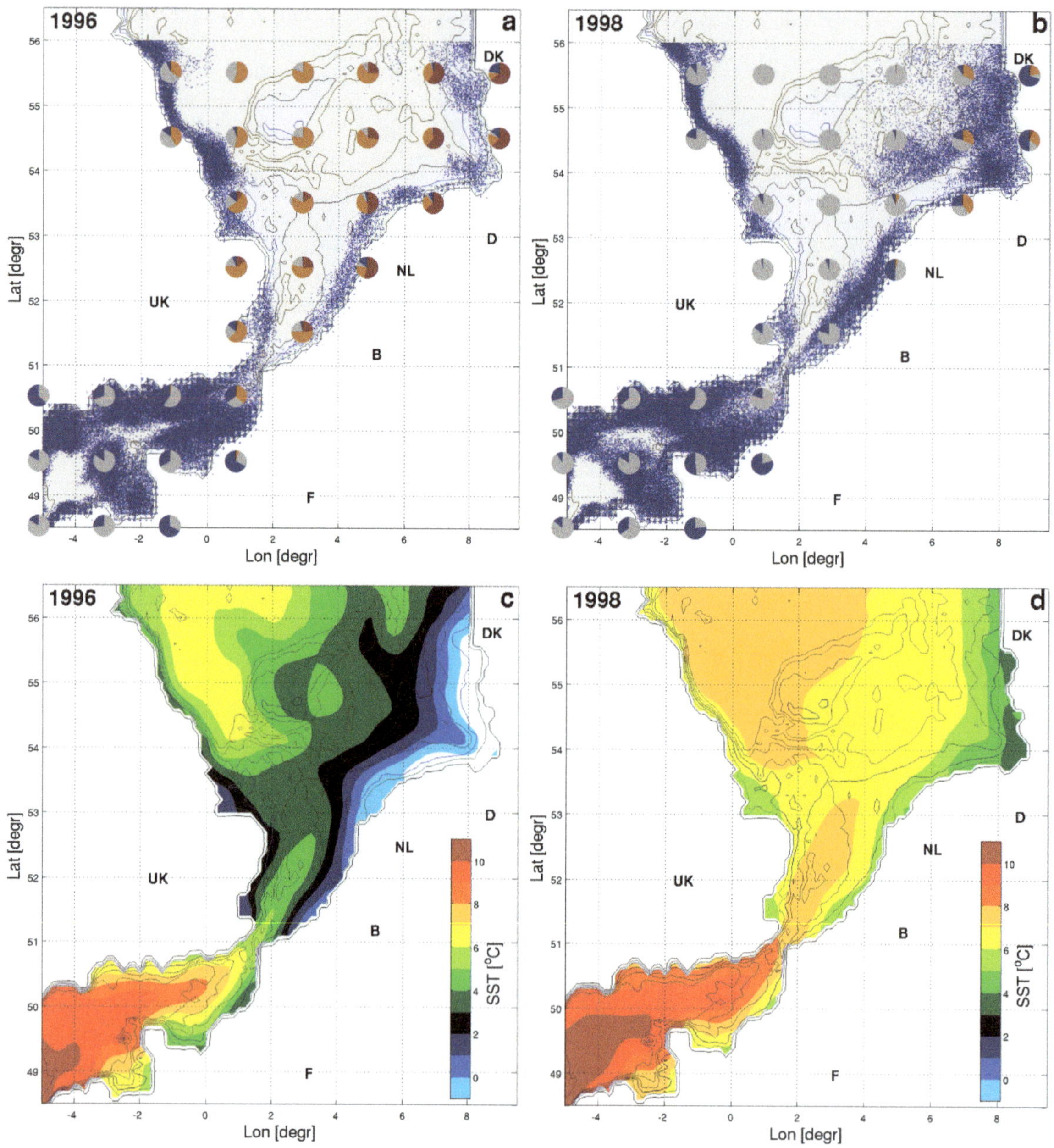

Figure 10. Inter-annual differences (**a, c** 1996; **b, d** 1998) in development rates give rise to variations in settlement success (for all 1 000 000 particles released). **(a, b)** Particle development characteristics (see Fig. 6 for an explanation of the colour coding); **(c, d)** sea surface temperature in February.

increased numbers of particles that show truncated development (correlation coefficient of 0.68), as shown in Fig. 12.

Water temperature in the western part of the North Sea (station b) showed less variability over the years (Fig. 14), as the residual current patterns (and therefore also variations in this) generally resulted in temperature effects along the eastern shores. This was also (partially) the reason for the more constant number of particles settling along the UK's North Sea coast. Similarly, the higher temperatures recorded in the English Channel (station d) resulted in all particles reaching the end of their settling stage in this area, therefore also (partially) explaining the reasonably constant settlement numbers in this area over the years. Wind conditions also influenced particle settlement success. In general, years with increased numbers of particles settling along the eastern North Sea shores (specifically 1998, 2000, 2002, 2004 and 2005 see Fig. 13) coincided with strong south-westerly winds in the English Channel (see Table 2, correlation coefficient of 0.79 for the eastern Southern Bight coast, and 0.71 for the western Wadden Sea) and to a limited extent in

Table 2. Relationship between forcing conditions (at station d) and settlement dynamics at the different coastal zones (see Fig. 13 for a description of the different coastal zones). Bold numbers indicate a high correlation, whereas numbers in italics correspond to low correlation coefficients.

	Mean	SD	CV (%)	Corr. Temp	Corr. Wind
UK North Sea	20 647	1708	8	*−0.13*	*−0.12*
English Channel	69 145	4101	6	−0.63	**−0.71**
Eastern Southern Bight	4153	931	22	0.43	**0.79**
Western Wadden Sea	8264	2535	31	*0.37*	**0.71**
Northern Wadden Sea	4646	1686	36	*0.18*	0.52
Denmark and Norway	12 557	8995	72	**0.79**	0.52

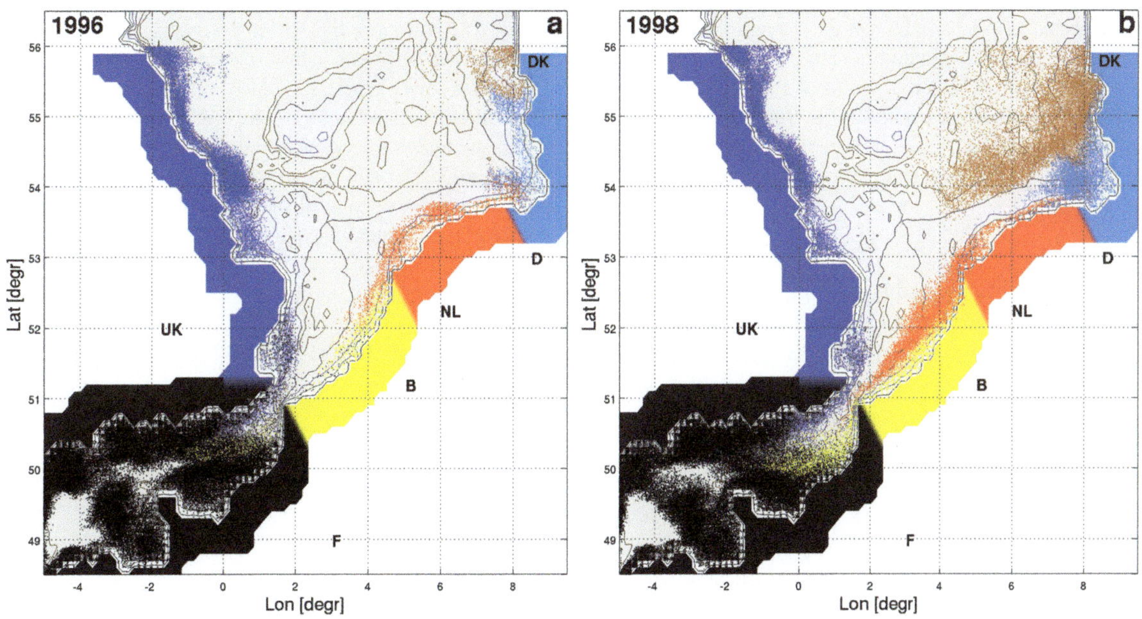

Figure 11. Settle locations of all particles shown in the colour of the dots, for different years (**a** 1996; **b** 1998). For explanation of colour coding see Fig. 7.

the Southern Bight (not shown), similar to those observed for 1998 (shown in Fig. 9). However, years with small settlement values for the most northerly settlement areas (shown in orange, in particular 1997, 2001 and 2003) generally coincided with smaller than normal wind speeds in the English Channel (correlation coefficient of 0.52).

3.4 Settlement in the western Wadden Sea

The western Wadden Sea forms one of the main nursery areas of plaice juveniles in the North Sea basin (Zijlstra, 1972). In the following section, settlement characteristics for this area have been studied in more detail, to highlight not only inter-annual variability in settlement dynamics, but also present some intra-annual processes that are caused by variability in the hydrodynamic conditions.

The characteristics and origin of particles settling in this domain are presented in Figs. 15 and 16. The settlement area that is considered to represent the western Wadden Sea, geographically roughly encompasses the Marsdiep–Vlie inlet system; although with the low model resolution used in these runs, neither can be distinguished here. To compensate for this low resolution the settlement domain was chosen to partially encompass the neighbouring North Sea as well (black box in Fig. 15a, c, e). Please note that, as was previously the case, the settlement domain is further limited solely to grid cells where the water depth of at least one grid-point is below 10 m.

In Fig. 15 a link between the origin of particles settling in the western Wadden Sea, and the relation between the release and arrival times in this area were investigated for the three years already discussed in Sect. 3.3. The variability in origin of particles settling in the western Wadden Sea was strongly dependent on wind patterns, and showed great inter-annual variability, as was discussed in the previous section. However, also the moment of settlement in this area was strongly variable. In 1996, low water temperatures prolonged particle

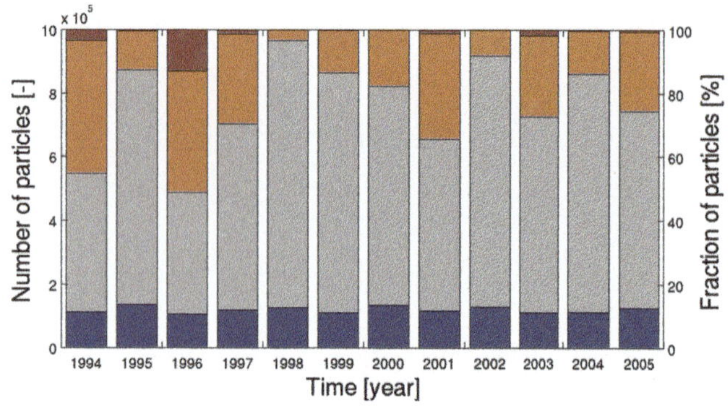

Figure 12. Final status of all particles for each year: (blue) particles that settle; (grey) particles that do not settle; (orange) particles that do not settle but also do not reach end of the 30-day settlement stage before the numerical simulation is truncated after 120 days; (red) particles that do not settle as these have not yet reached the beginning of their 30-day settlement stage, before simulation is truncated.

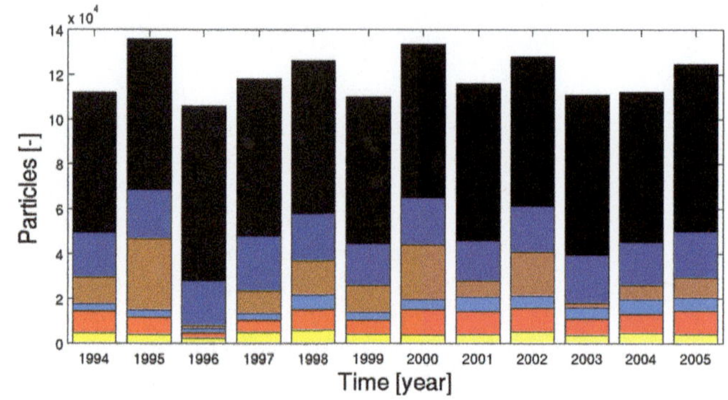

Figure 13. Numbers of particles settling in the different coastal zones for the different years: (dark blue) settlement in the North Sea coastal zone of the UK; (black) settlement in the English Channel; (yellow) settlement along the French, Belgian and Dutch North Sea coast; (red) settlement in the western part of the Wadden Sea; (light blue) settlement in the northern part of the Wadden Sea; (orange) settlement outside of the shown domain (generally along the northern Danish and Norwegian coast).

development, and only after yearday 145, particles started to settle in the domain. These particles were released around yearday 25, meaning that the particle drift duration lasted the 120-day simulation limit (an investigation into the consequences of this 120-day simulation limit can be found in Sect. 3.5). This was in stark contrast with 1998 and 2001, when particles that were released during the entire simulation period settle. The distributions of release times versus settling times for 1998 and 2001 (Fig. 15d, f) show a temperature-related, curved lower boundary of the settlement distribution. For 2001, initially a small effect of the 120-day simulation limit can also be observed at the upper-boundary of the distribution; instead of it being similarly curved as the lower boundary, it is straight and following the 120-day limit. During the early stages of the model simulation, water temperatures dropped, causing the particle development to be prolonged, resulting in a shorter duration over which particles could settle. With the gradual increase of water temperatures in spring, development accelerated, and particles could settle earlier. Under extremely cold conditions (as happened in 1996) the temperature related bottom-curve cut through the 120-day limit, inhibiting any particles to settle for the duration of this colder period. The minor differences in forcing conditions experienced for 1998 and 2001 seem to balance each other out with regards to the origin of the settling particles. The stronger south-westerly winds for 1998 caused the particles to propagate faster. However, the coinciding higher water temperatures resulted in a reduction in the drift duration (as can be seen in the slightly earlier settle times of particles in 1998). As a result, the maximum drift distance (the yellow dots) was fairly similar for 1998 and 2001. There was a rough sorting through the settlement time in the origin of particles: in 1998 in particular, particles from the Dover Straight (shown in yellow in Fig. 15) generally arrived during

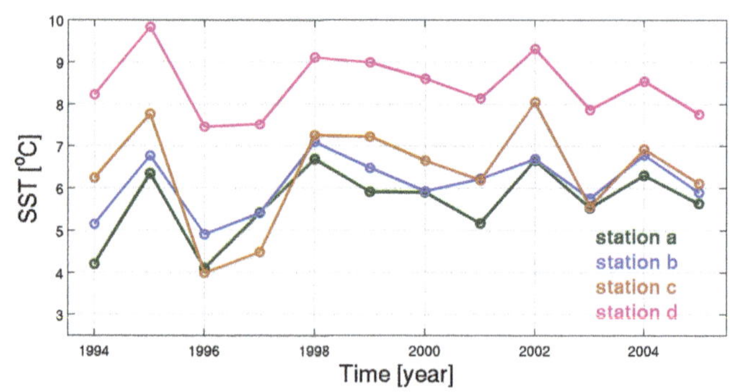

Figure 14. Sea surface temperature for the different years of particle distribution runs, averaged over February for each year. The colour of the lines represents the different stations as presented in Fig. 8.

the start of the year, whereas particles that originated from close to the settlement domain (purple and blue) mainly arrived in the second half of the modelled period. Another remarkable feature was that the particles released close to the settle area (blue) settled after the shortest drift duration (release time−settle time). Also of note was the grouping of settling particles at certain moments in time (right-hand side of Fig. 15). During periods of several days very small numbers of particles settled, whereas at other moments particles released from a broad range of release times all settled at once (for instance just before yearday 160 in 2001, the particles that settled were released from yearday 40 until yearday 90). Only occasionally did the wind conditions force particles into the western Wadden Sea, causing large numbers of particles to drift into suitable settling areas.

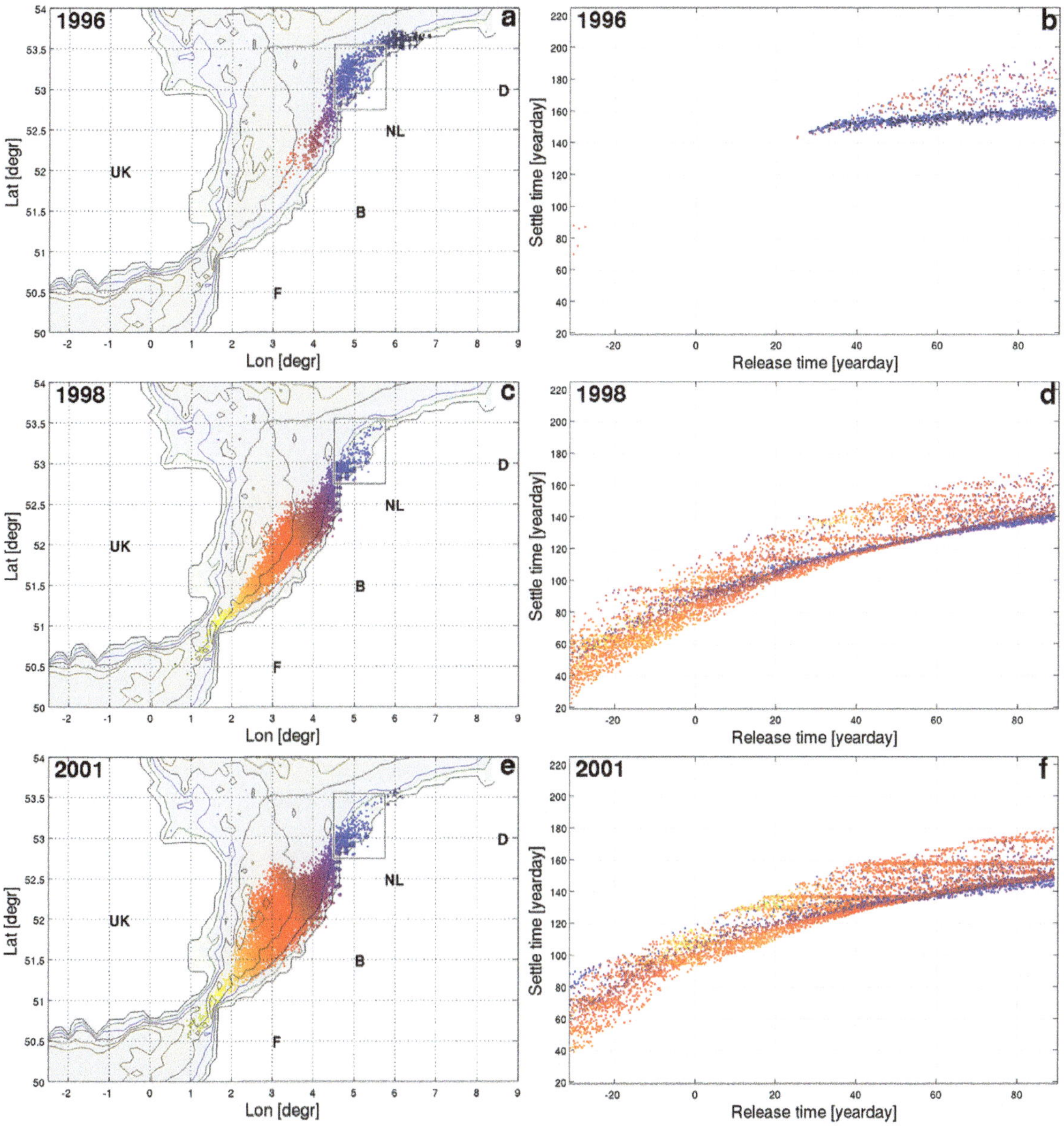

Figure 15. Particle settlement in the western Wadden Sea. **(a, c, e)** Release position of particles settling in the western Wadden Sea (shown as a black box). The colour of the dots represents the release position of the particles. **(b, d, f)** The drift duration, presented as the link between the release time and settle time of the particles. Colour coding is the same as in **(a, c, e)**. The curved bottom edge is the result of the development rate of the particles: colder temperatures in late winter (release time around yearday 30) result in prolonged development rates. The threshold at the top of the plots is primarily the result of the 30-day development threshold. **(a, b)** 1996; **(c, d)** 1998; **(e, f)** 2001.

A comparison over all the modelled years of the settle times and drift durations can be seen in Fig. 16 where the maximum drift duration of 120 days is presented as a vertical edge in the distribution, while the diagonal distribution edges at the bottom (clearly visible for instance in 1995 and 1998) and top (visible for all years) correspond with the start and end of the simulated spawning period. Using the same colour coding of the dots as in Fig. 15, the features observed for the 3 years studied above can be recognised for most other years. With regards to the temperature-related threshold in drift duration shift over the years, colder years see a prolonged development duration before settling commences. This shift corresponded well with the surface water temperatures observed in Fig. 14 (especially for stations c and d). The origin of particles also varied over the years, with particles in 1994 and 1995 (dark yellow) originating furthest away, well into the English Channel. Years with the largest drift distances generally coincided with stronger winds from

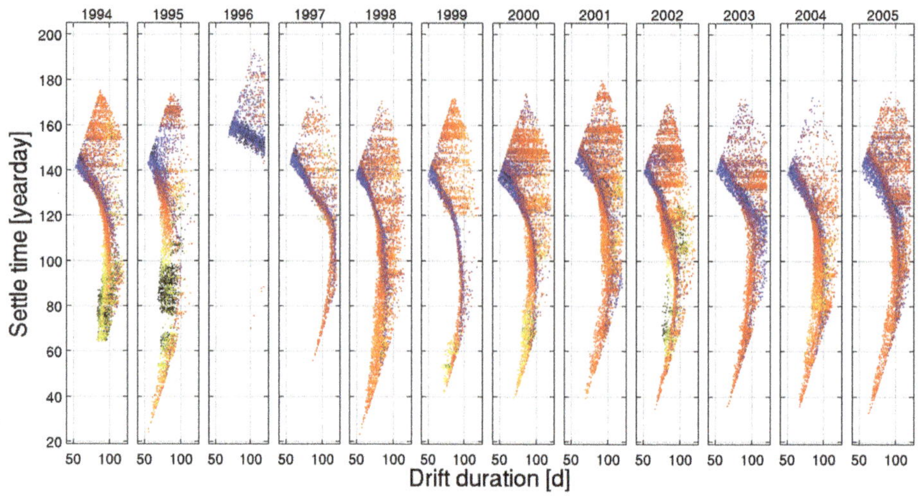

Figure 16. Drift duration and settle time of particle settling in the western Wadden Sea for the period 1994–2005. Colour of the dots represents again the distance from the settle location, and is the same as in Fig. 15.

the SW for the two southerly stations (c and d, with correlation coefficients of 0.70 and 0.66, respectively). The particles originating from the area closest to the settlement domain (in blue) seemed to settle after the shortest drift duration, but only when water temperatures had increased to enable the particles to reach their settle stage early (which generally occurs around yearday 140). Prior to that, particles released further south, experienced warmer temperatures promoting faster development and earlier settlement. The direct relationship between development rate and sea water temperature, results in high correlation coefficients between the SST for February for stations c and d and the average drift duration (−0.88 and −0.95 respectively).

3.5 Sensitivity analysis

In the previous numerical simulations, various settings were used whose impact on the particle success rate is investigated in the following section.

Firstly, a 10 m depth limit was chosen to define the boundaries of suitable settlement areas (for at least one corner of the grid cell). The effect of changing this threshold on settlement levels not only shows an increase in the total number of settling particles, but also indicates a shift in importance of different settlement areas (Fig. 17). Settlement in the English Channel (black) was relatively constant for different settlement depths. The depth in this region was in general significantly larger than in the other areas. In 1996, the settlement in the Southern and German bights of the North Sea was non-existent, as the particles were released in a very cold period, truncating their development (see Sect. 3.3). Additionally, the settlement along the English coast (dark blue) showed an (relatively) abrupt increase in settling particles, when the settle-depth threshold was increased from 15 to 20 m, which is caused by the Dogger Bank coming into play, and becoming a suitable settlement area. All other areas showed a more constant increase in the number of settling particles, with the increasing settlement depth. The increasing settlement depth forces the boundaries of the settlement areas gradually outwards, enabling particles that drift further offshore to settle.

In this research, the hydrodynamic conditions were updated every 45 min. This value was chosen to give a high enough temporal resolution to describe the M2 tidal cycle. However, in other particle tracking publications, larger hydrodynamic update intervals are generally used (Van der Veer et al., 1998; Hufnagl et al., 2013). The impact of larger hydrodynamical time steps, was evaluated by reproducing the drifter-buoy tracks discussed in Sect. 3.1, whereby the hydrodynamic update interval was varied from 45 min to 24 h (Fig. 18), with all the values above 45 min being averages over the specified periods. In general, tracks that showed large residual currents were reasonably well reproduced for bigger hydrodynamic update intervals (not shown). However, if the residual current was less important, the coarser temporal interval caused significant differences between the different modelled tracks to develop. A low frequency of the hydrodynamic updates could be very computationally advantageous. However, by substituting the tidal motion with tidally averaged motion, even the residual drift patterns could not be well reproduced.

Another aspect studied here is the impact of the temperature-dependent development rate of plaice eggs and larvae on their settlement success. To investigate this, results for a fixed development duration were compared to the temperature-dependent development. In the new set-up, the development duration was set at 80 days (similar to the average duration for settling particles in 2001), after which the particles would have 30 days to settle (the same as in the other simulations). Results presented in Fig. 19, show a change with regards to drift duration, which in the fixed

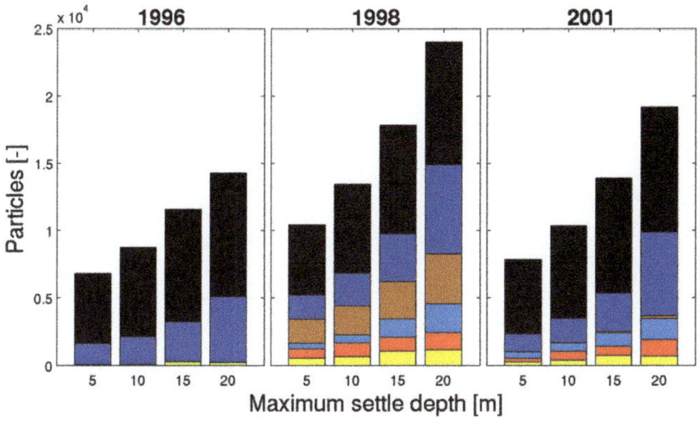

Figure 17. Settlement into the different areas as shown in Fig. 7, for different maximum settle depths for 1996, 1998 and 2001. Only 100 000 particles released around 1 January, are taken into consideration (similar to Fig. 5).

development duration case lacks the prolonged development rates during the colder periods. Additionally, the results for 1996 show a larger number of settling particles when this constant development duration is applied. However, the origin of the particles as well as the settlement time (in particular for 1998 and 2001) remain relatively unchanged, and seem to be the result mainly of the current patterns (as can also be seen in Table 2).

The impact of the 120-day simulation limit on the results presented here is considered by comparing the original results with those with a prolonged drift duration. In these new results, the numerical simulation limit was extended to 180 days. Figure 20 shows the release times against settlement times for particles settling in the western Wadden Sea for the 3 years already presented in Fig. 15. The main differences can be observed for 1996, where in the original result a release period existed when no particles would eventually settle. In the new simulations, this gap cannot be observed. The timing of settlement does not change significantly though, as the main settlement time has only shifted around 10 days (from yearday 160 to yearday 150). In the new simulations, settlement commences during a fairly brief period. This indicates that particles show a very wide range of drift durations before settling; for particles released in December, drift durations can last up to 170 days, whereas at the end of the simulations, this has been reduced to around 100 days. A large portion of settling particles that were released in December originated around the German Bight. To what extent these results would be observed in the field remains to be seen, as spawning is expected to occur only in February and March that far east. For 1998 and 2001, the increase in the maximum lifespan does not result in significant changes to the number, origin and distribution of the settling plaice. Furthermore, for all 3 years examined here, the extension of the simulation duration resulted only in a slight broadening in the origin of particles settling in this area, and the overall origin and timing of settlement was also not altered significantly.

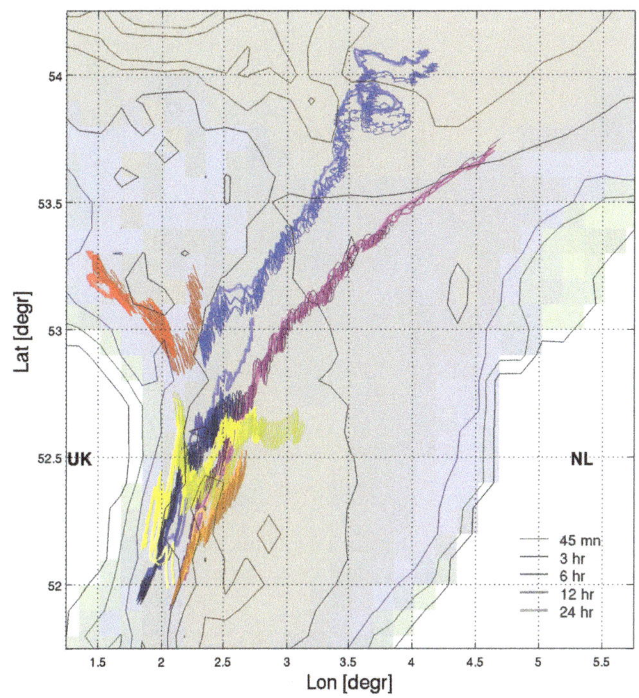

Figure 18. The effect of altering the hydrodynamic forcing time step on the particle drift paths, reproducing the drifting buoys (discussed in Sect. 3.1). In the particle tracking routine, the hydrodynamics were updated every 45 min, every 3 6 12 and 24 h. These update intervals are shown in the thickness and darkness of the lines, while the colour of the lines correspond to the different drifter buoys (as presented in Fig. 3).

4 Discussion

The aim of this research was to investigate whether changes in hydrodynamic conditions alone can potentially cause significant inter-annual settlement variability of North Sea plaice juveniles whereby the focus was to investigate the influence of hydrodynamic forcing on plaice egg and larval drift patterns and not to reproduce the actual settlement demographics. In the following sections, various aspects concerning the findings presented here are put into a broader perspective. Both a comparison with other simulations, as well as several steps towards a future comparison with observations of settlement data are discussed.

4.1 Overall findings

The simulations of plaice drift and settlement showed that the inter-annual changes to the environmental conditions (specifically water temperature and wind forcing) significantly influence drift direction and distance, as well as settlement dynamics along the eastern shores of the North Sea (CV 37 %). Strong south-westerly wind speeds in the English Channel and Southern Bight generally coincided with higher drift distances for particles settling in the western Wadden Sea

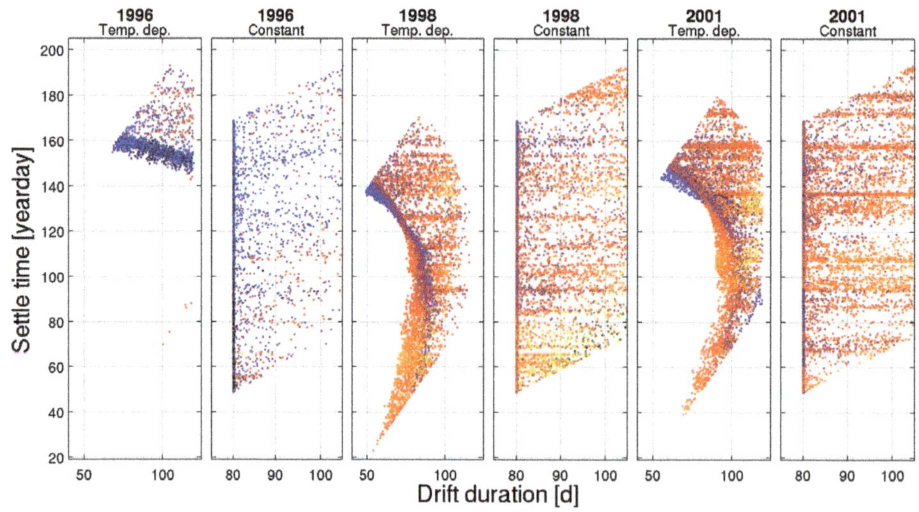

Figure 19. Comparison between the temperature-dependent development rate and a fixed development duration for 1996, 1998 and 2001. Shown is the drift duration against the settlement time for particles settling in the western Wadden Sea.

(correlation coefficient of 0.66 and 0.70 respectively). Effectively, these winds enhance the general anticlockwise circulation pattern that transports particles from the Southern Bight and English Channel along the eastern shores of the North Sea and past various nursery areas. Years with reduced (or reversed) wind direction and intensity (for instance 1996) result in significant changes to this general observed drift direction. This in turn affects settlement dynamics and changes the connectivity between spawning grounds and nursery areas. In 1996 the origin of settling plaice juveniles in the western Wadden Sea was significantly shifted towards the east, in comparison to most other years (see Fig. 16). Strong winds result in a large influx of warmer English Channel water into the Southern Bight, increasing water temperature, and reducing drift durations for particles settling in the western Wadden Sea (correlation coefficient of −0.88). As mortality has not been included in the present set-up, the impact of prolonged drift durations on survivability has not been studied. Results from field studies of settled plaice juveniles suggest that the accompanying lower temperatures would result in reduced mortality rates (Van der Veer et al., 2000; Van der Veer and Witte, 1999; Bolle et al., 2009). In the present research, years with reduced development rates generally coincide with reduced south-westerly winds and as a result fewer particles settling in the Wadden Sea nursery area.

The results presented here with respect to the drift of particles across the southern North Sea compare well with other papers regarding the general trend and inter-annual variability. The reproduction of the paths of the CEFAS drifter buoys showed that the coupled (GETM–GITM) model reproduced the hydrodynamic conditions reasonably well, with the error between the buoy and the particle for each drifter buoy being less than 10 % of the total drifted distance. Additionally, other studies have shown similar inter-annual variability in the drift paths of fish eggs and larvae: Hufnagl et al. (2013) using an approach similar to the one used here, showed that the connectivity of the different spawning areas to various coastal zones corresponds well with the general settlement dynamics observed here. The same authors indicated similar geographic distributions of development times, partially caused by the same temperature-dependent development rate. With regards to vertical migration of larvae, Savina et al. (2010) showed that for 2 examined years (1995 and 1996) the impact of inter-annual variability in the hydrodynamic conditions on the drift of sole exceeded the influence of various vertical migration regimes. Rochette et al. (2012) showed that for sole the settlement success in the English Channel area shows significant inter-annual variability. This is in contrast to the limited variability in settlement figures for the English Channel presented here. This difference might be due to the different set-up of spawning areas with regards to release area, numbers and timing. One aspect that differed from ours was that spawning figures were based on the number of adults, which varied over the years. However, they concluded that these variable initial conditions only have limited influence on the eventual settlement success. The inter-annual drift patterns of herring were investigated by Dickey-Collas et al. (2009). With slight variations in hatching (release) time around 1 January, similar drift patterns and inter-annual changes were presented as shown here, with the drift during 1996 being significantly shorter than in the majority of the examined years. Bolle et al. (2009) using a two-layer model, showed drift trends during the early life stages (without vertical migration) as those observed here, and corresponding inter-annual variability for Southern Bight spawning grounds. The inter-annual variability in the settlement success of sole is of a similar magnitude relative to the overall settlement as that observed here

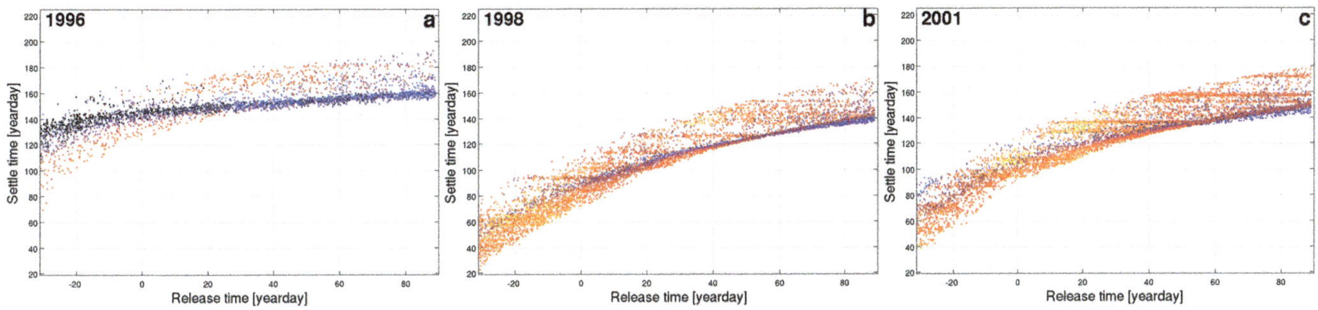

Figure 20. Settlement of particles in the western Wadden Sea, for prolonged numerical simulations (180 days). **(a)** 1996; **(b)** 1998; **(c)** 2001.

in the nursery areas along the eastern shores (Lacroix et al., 2013). Whilst our study showed a significant reduction in settlement in the western Wadden Sea during 1996, the work of Lacroix et al. (2013) suggests that settlement is enhanced for 1996 in comparison to 1998 and 2001. However, the timing of spawning and early life-stage mortality of sole are both temperature-related and, therefore, variable over the years. Additionally, plaice and sole have different spawning seasons and vertical migration is included in the work presented by Lacroix et al. (2013). As a result, the observed drift trends presented here are only partially matched with the results presented by Lacroix et al. (2013): The significant differences between drift distances for 1996, 1998, and to a lesser extent 2001, presented here, correspond to a limited extent to the results presented by Lacroix et al. (2013), where reduced drift distances were observed for 1996 while slightly enhanced drift distances occurred for 1998 and 2001 for sole spawning in the English Channel.

4.2 Comparison with field data

No direct comparison between plaice larvae drift patterns and numerical model results is presented here. In fact, such a comparison will face a number of hurdles, placing it beyond the scope of this paper. Comprehensive data sets combining the numbers of plaice eggs at spawning locations, the drift of eggs and larvae along the various transport paths, and of juveniles upon arrival at nursery areas are non-existent. Quantitative field data from spawning grounds is sparse (Taylor et al., 2007), while the drift patterns of plaice are even less well known (Talbot, 1977). Field data that can be obtained most easily is that of the arrival of plaice juveniles in the nursery areas (Van der Veer et al., 2009). Given the uncertainty about plaice spawning areas, numbers and times, and only very sparse data from the egg and larval drift, the numerical simulations can only be linked with settlement data (as was done by for instance Bolle et al., 2009). However, if structured cruise data of egg and larval abundance are available, presence–absence maps can be compiled and compared (Van der Molen et al., 2007). However, assumptions have then to be made about spawning and drift of plaice eggs and larvae.

4.3 Selection of spawning grounds and periods

In the present study, particles were released across the entire southern and central North Sea, and also encompassing the English Channel (similar to Hufnagl et al., 2013). Studies of egg abundance in the different parts of the study domain have shown a strong variability in egg abundance over the different areas, as well as over the years (Taylor et al., 2007; Loots et al., 2010a, b). There is some sorting in spawning areas within a year; in the English Channel spawning can already occur in December, whereas the maximum spawning in the German Bight occurs around February and lasts into March (Bergman et al., 1988). In the present set-up, limits could be placed with respect to specific spawning grounds and periods (cf. De Graaf et al., 2004; Bolle et al., 2009). However, the variability over the years as well as the uncertainty in the location and boundaries, start time, and the duration of the different spawning areas in the different parts of the North Sea mean that any limits on these cannot be prescribed based solely on hydrodynamic circumstances, which would inhibit the original aim of this research.

4.4 Vertical position of plaice eggs and larvae

To investigate the pre-juvenile drift of fish eggs and larvae, a particle tracking routine may be used that combines hydrodynamic forcing conditions and active behaviour, such as vertical migration, to describe the drift of particles from spawning grounds to nursery areas. With respect to plaice larvae, several studies have been carried out where, during the latter stages of pre-juvenile evolution, it is assumed that plaice show a tendency to migrate vertically over the water column, either in a daily frequency (Rijnsdorp et al., 1985), or over a tidal cycle (de Graaf et al., 2004; Fox et al., 2006, 2009; Van der Molen et al., 2007; Bolle et al., 2009; Hufnagl et al., 2013). The latter will cause plaice larvae to be able to drift faster when the tidal current is towards suitable nursery areas and reduce the backward drift, when the tidal flow is opposite. The overall impact of vertical migration on the particle drift might be limited as it only occurs during the brief later stages of particle development (Bolle et al., 2009). However, the impact on the numbers of juveniles successfully settling in a certain domain can be significant, as particles that would otherwise drift too far offshore of the settling area, will now be able to drift towards it (de Graaf et al., 2004; Fox et al., 2006). The presented results show a very strong short-term variability in the numbers of particles arriving in a nursery area. These intra-annual changes in settling figures are forced by local changes in (residual) current patterns. Residual current patterns are however significantly smaller than the tidal currents, and active vertical migration would most likely result in a reduction of these intra-annual fluctuations and cause a more constant flow of particles to settle in the settling domain. Additionally, plaice eggs are assumed to exhibit a positive buoyancy (Coombs et al., 1990), suggesting a higher abundance of plaice eggs in the upper layers of the sea. This has not been incorporated into the present study, but would affect the drift distance during this early development stage since currents are generally enhanced further up the water column.

4.5 Mortality based on predation and food abundance

In the present simulations, particles released from very different parts of the North Sea and English Channel drifted into suitable settlement areas. Without food abundance or predation included in the present runs, particle drift is only truncated when a particle settles, or because a particle reaches the end of its pelagic life span, either coming to the end of the 30-day settlement period without settling, or because of reaching the 120-day simulation limit (cf. de Graaf et al., 2004; Van der Molen et al., 2007; Bolle et al., 2009; Hufnagl et al., 2013). As a result (along with the lack of spawning-ground and -period selection), most years show continuous settlement of particles for almost the entire modelled duration. A constant mortality (Van der Veer et al., 1998) or mortality based on temperature (Fox et al., 2006, 2009) could give insights into the survival rates during egg and larval stages. However, an intra-annual variability in food abundance could form an extra threshold for particle development, and survival. Expressing the changing particle development rate with regards to the particle's dynamic energy budget (DEB) would form a promising way to model more realistic circumstances in which plaice eggs and larvae might evolve. Additionally, intra-annual population dynamics of predator species might give a second input towards seasonal changes in survivability rates of plaice eggs and larvae before settlement.

4.6 Drift duration

As was mentioned previously, no direct comparison between numerical simulations and field measurements of numbers of arriving plaice juveniles in the western Wadden Sea is presented, as this would involve unverifiable assumptions towards spawning grounds and periods. However, the total lack of arriving plaice juveniles in the early stages of 1996 in the numerical simulations (Fig. 15) is in stark contrast with the peak in plaice juveniles arriving in the western Wadden Sea according to field studies. However this can also partially be due to lower mortality rates in cold years (Van der Veer et al., 2000; Van der Veer and Witte, 1999; Bolle et al., 2009). In 1996, the development of many particles is truncated due to low water temperatures resulting in slow particle development. The resulting lack of settlement in the western Wadden Sea is caused by slow development rates in combination with the 120-day numerical simulation limit. This limit was chosen based on drift durations presented in previous studies (Bolle et al., 2009; Hufnagl et al., 2013). However the results presented in the sensitivity analysis, concerning the prolonged simulation limit, suggested that both the peak in settlement and the origin of the settling particles only show limited variability.

The duration of the settlement period (30 days), in which particles drift until reaching a suitable settlement area, is based on existing literature (Bolle et al., 2009). The duration of this period influences the settlement dynamics, as a longer (or shorter) period would enable more (or fewer) particles to drift into a suitable settlement area. However, we would like to argue that since the majority of settling particles do so at the onset of the settling period, a reduction in the duration will have smaller impact than the magnitude of the reduction itself would imply.

4.7 Suitability of settlement areas

As already discussed in the sensitivity analysis, settlement areas are probably defined by food abundance (cf. Creutzberg et al., 1978) and lack of predation of plaice juveniles, whereas the boundaries of settlement areas are presently solely defined by a threshold water depth. Similar to the possibility to make a priori a selection of spawning grounds and periods, also different nursery areas could be pre-defined based on field data. This would not only lead to an increase in settling particles in the (more limited) number of nursery areas, but also possibly to shifts in importance of different settlement areas. In this respect the Dogger Bank is a puzzling area: settlement does not seem to be successful, however transplantation experiments indicate that it could be a potential nursery area with good growth conditions for plaice (Borley, 1919).

Nursery grounds could be defined based around soil characteristics as well as water depth (similar to Lacroix et al., 2013). These areas could be limited to certain soil compositions, such as fine sands and mud for instance, thereby confining plaice juvenile areas to estuarine environments such as the Wadden Sea. The current assumption that a particle would settle at the first possible opportunity (during their juvenile stage) when a depth of less than 10 m is encountered is an oversimplification of this process. With better understanding of the ideal settlement conditions and better knowledge of the occurrence of such conditions, a different approach might be chosen whereby the drift of particles could be prolonged beyond this first settlement opportunity: if, for instance, environmental conditions would improve with longer drift distances, settlement could then be postponed. In this case, settlement along the Dutch North Sea coast would be avoided, as these particles would tend to propagate further towards the better conditions in the Wadden Sea. Additionally, prolonged drift during this settlement stage may result in juveniles reaching optimal sites within nursery areas with regards to food abundance and predation.

5 Conclusions

In this study, we demonstrate a substantial effect of physical factors alone on the inter-annual variability in the transport of eggs and larvae and their subsequent success rate in settling. These idealised studies are not intended to explain or reproduce observations (insofar as they exist, mainly from nursery grounds). This would require that assumptions of spawning grounds and periods, mortality and possibly active behaviour (such as vertical migration) be included in the model. The present study serves therefore to give an indication of the extent of variability that can be attributed to physical factors.

Inter-annual variability in wind speed and direction caused residual current patterns to show significant changes over the years, strongly influencing the direction and distance of the particle drift. As a result, different years had a significant variability in the origin of the particles settling in the different nursery areas. In general, an anticlockwise circulation in the central and southern North Sea could be observed, and a north-easterly drift direction when released in the English Channel. However, in years with strong easterly winds there was less drift along coastlines, and more into the central North Sea, limiting settlement chances and also giving rise to different connections between spawning areas and nursery grounds.

In combination with wind-driven current variability, the sea water temperature also varied over the years. Assuming that particle development was dependent on water temperature, differences in experienced water temperature significantly affected their success rate. With higher temperatures in early winter, late spring and summer, particle development was accelerated, and the settlement stage was reached earlier, thereby reducing the distance between spawning ground and settlement area. A maximum particle lifespan of 120 days in combination with temperature dependent trajectory, meant that for some years (1996 in particular) large numbers of particles did not reach the settling stage before the end of their pelagic lifespan.

Settlement in the western Wadden Sea showed strong inter-annual variability due to changes in the hydrodynamic conditions. Not only did the origin of particles settling in this area vary, but the general arrival time was also influenced by hydrodynamic conditions. Over each season, the origin and drift duration of particles settling in this nursery area changed: during winter, particles generally originated from more distant areas than particles settling in the western Wadden Sea in spring and summer. The low sea water temperatures during late winter and early spring resulted in slower development rates and longer drift durations. Additionally, stronger wind conditions during winter and early spring, generally forced particles to drift further. Settlement success also varied substantially as a result of temporary local conditions: for example, if the right wind conditions prevailed large numbers of particles, whose origin could be widely variable over space and time, would be forced into this settlement area over a short period.

Acknowledgements. The authors would like thank the GETM developers for the use of the numerical code and the two referees for their detailed and useful comments. The authors gratefully acknowledge the ECMWF (European Centre for Medium Range Weather Forecasting) for allowing use of the meteorological data. The authors thank all contributors to the daily riverine loads database. French water quality data were supplied by the Agence de l'eau Loire-Bretagne, Agence de l'eau Seine-Normandie and IFREMER. UK water quality data were processed from raw data provided by the Environment Agency, the Scottish Environment Protection Agency and the National River Flow Archive. The German river loads are based on data from the ARGE Elbe, the Niedersächsisches Landesamt für Ökologie and the Bundesanstalt für Gewässerkunde. The river load data for the Netherlands were supplied by the DONAR database. Drifter observations supplied by DEFRA project AE1225.

Edited by: E. J. M. Delhez

References

Aldridge, J., van der Molen, J., and Forster, R.: Wider ecological implications of Macroalgae cultivation, ISBN: 978-1-906410-38-4, 95 pp., The Crown Estate, Lowestoft, UK 2012.

Bergman, M. J. N., Veer, H. W., and Zulstra, J. J.: Plaice nurseries: effects on recruitment, J. Fish Biol., 33, 201–218, 1988.

Boero, F., Bouillon, J., and Gravili, C.: The life cycle of Hydrichthys mirus (Cnidaria: Hydrozoa: Anthomedusae: Pandeidae), Zool. J. Linn. Soc., 101, 189–199, 1991.

Bolle, L. J., Dickey-Collas, M., van Beek, J. K. L., Erftemeijer, P. L. A., Witte, J. I. J., van der Veer, H. W., and Rijnsdorp, A.: Variability in transport of fish eggs and larvae. III. Effects of hydrodynamics and larval behaviour on recruitment in plaice, Mar. Ecol. Prog. Ser., 390, 195–211, 2009.

Borley, J. O.: Report on the Experimental Transplantation of Plaice to the Dogger Bank Carried Out by the Marine Biological Association in the Years 1904–8, 79 pp. Lowestoft Laboratories, Lowestoft, UK, 1919.

Burchard, H. and Bolding, K.: GETM – A general estuarine transport model, Scientific documentation (Technical report No. EUR 20253 EN), European Commission, 2002.

Coombs, S. H., Nichols, J. H., and Fosh, C. A.: Plaice eggs (*Pleuronectes platessa* L.) in the southern North Sea: abundance, spawning area, vertical distribution, and buoyancy, J. du Cons., 47, 133–139, 1990.

Creutzberg, F., Eltink, A. T. G. W., and van Noort, G. J.: The migration of plaice larvae Pleuronectes platessa into the western Wadden Sea, in: Proc. 12th Europ. Mar. Biol. Symp. Presented at the Physiology and behaviour of marine organisms, edited by: McLusky, D. S. and Berry, A. J., Pergamon Press, New York, 243–251, 1978.

Dale, B.: Dinoflagellate cysr analysis of upper quaternary sediments in core GIK-15530-4 from the Skagerrak, Nor. Geol. Tidsskr., 65, 97–102, 1985.

De Graaf, M., Jager, Z., Vreugdenhil, C. B., and Elorche, M.: Numerical simulations of tidally cued vertical migrations of flatfish larvae in the North Sea, Estuar. Coast. Shelf S., 59, 295–305, 2004.

Dickey-Collas, M., Bolle, L. J., van Beek, J. K. L., and Erftemeijer, P. L. A.: Variability in transport of fish eggs and larvae. II. Effects of hydrodynamics on the transport of Downs herring larvae, Mar. Ecol. Prog. Ser., 390, 183–194, 2009.

Fox, C. J., Mccloghrie, P., Young, E. F., and Nash, R. D. M.: The importance of individual behaviour for successful settlement of juvenile plaice (*Pleuronectes platessa* L.): a modelling and field study in the eastern Irish Sea, Fish. Ocean., 15, 301–313, 2006.

Fox, C. J., McCloghrie, P., and Nash, R. D. M.: Potential transport of plaice eggs and larvae between two apparently self-contained populations in the Irish Sea, Estuar. Coast. Shelf S., 81, 381–389, 2009.

Gräwe, U. and Burchard, H.: Storm surges in the Western Baltic Sea: the present and a possible future, Clim. Dynam., 39, 165–183, 2011.

Harding, D., Nichols, J. H., and Tungate, D. S.: The spawning of plaice (Pleuronectes platessa) in the southern North Sea and English Channel, Rapp. Procès Verbaux Réunions Cons. Int. Pour Explor. Mer, 172, 102–113, 1978.

Hufnagl, M., Peck, M. A., Nash, R. D. M., Pohlmann, T., and Rijnsdorp, A. D.: Changes in potential North Sea spawning grounds of plaice (*Pleuronectes platessa* L.) based on early life stage connectivity to nursery habitats, J. Sea Res., 84, 26–39, 2013.

Hunter, J. R., Craig, P. D., and Philips, H. E.: On the use of random walk models with spatially variable diffusivity, J. Comput. Phys., 106, 366–376, 1993.

Janssen, F., Schrum, C., and Backhaus, J. O.: A climatological data set of temperature and salinity for the Baltic Sea and the North Sea, Dtsch. Hydrogr. Z., 51, 5–245, 1999.

Lacroix, G., Maes, G. E., Bolle, L. J., and Volckaert, F. A. M.: Modelling dispersal dynamics of the early life stages of a marine flatfish (*Solea solea* L.), J. Sea Res., 84, 13–25, 2013.

Leggett, W. C. and Deblois, E.: Recruitment in marine fishes: Is it regulated by starvation and predation in the egg and larval stages?, Neth. J. Sea Res., 32, 119–134, 1994.

Lenhart, H.-J., Mills, D. K., Baretta-Bekker, H., van Leeuwen, S. M., van der Molen, J., Baretta, J. W., Blaas, M., Desmit, X., Kühn, W., Lacroix, G., Los, H. J., Ménesguen, A., Neves, R., Proctor, R., Ruardij, P., Skogen, M. D., Vanhoutte-Brunier, A., Villars, M. T., and Wakelin, S. L.: Predicting the consequences of nutrient reduction on the eutrophication status of the North Sea, J. Mar. Syst., 81, 148–170, 2010.

Loots, C., Vaz, S., Koubbi, P., Planque, B., Coppin, F., and Verin, Y.: Inter-annual variability of North Sea plaice spawning habitat, J. Sea Res., 64, 427–435, 2010a.

Loots, C., Vaz, S., Koubbi, P., Planque, B., and Koubbi, P.: Spawning distribution of North Sea plaice and whiting from 1980 to 2007, J. Oceanogr., Research and Data, 3, 77–95, 2010b.

Niiler, P. P., Sybrandy, A. S., Bi, K., Poulain, P. M., and Bitterman, D.: Measurements of the water following capacity of holeysock and tristar drifters, Deep Sea Res., 42, 1951–1964, 1995.

Otto, L., Zimmerman, J. T. F., Furnes, G. K., Mork, M., Saetre, R., and Becker, G.: Review of the physical oceanography of the North Sea, Neth. J. Sea Res., 26, 161–238, 1990.

Rijnsdorp, A. D., van Stralen, M., and van der Veer, H. W.: Selective Tidal Transport of North Sea Plaice Larvae Pleuronectes platessa in Coastal Nursery Areas, T. Am. Fish. Soc., 114, 461–470, 1985.

Rochette, S., Huret, M., Rivot, E., and Le Pape, O.: Coupling hydrodynamic and individual-based models to simulate long-term larval supply to coastal nursery areas, Fish. Oceanogr., 21, 229–242, doi:10.1111/j.1365-2419.2012.00621.x, 2012.

Ruardij, P., Veldhuis, M. J. W., and Brussaard, C. P. D.: Modeling the bloom dynamics of the polymorphic phytoplankter Phaeocystis globosa: impact of grazers and viruses, Harmful Algae, 4, 941–963, 2005.

Savina, M., Lacroix, G., and Ruddick, K.: Modelling the transport of common sole larvae in the southern North Sea: Influence of hydrodynamics and larval vertical movements, J. Mar. Syst., 81, 86–98, 2010.

Sinclair, M.: Marine populations: an essay on population regulation and speciation, Books in recruitment fishery oceanography, Washington Sea Grant Program?: Distributed by University of Washington Press, Seattle [Wash.], 1988.

Sündermann, J. and Pohlmann, T.: A brief analysis of North Sea physics, Oceanologia, 53, 663–689, 2011.

Talbot, J. W.: The dispersal of plaice eggs and larvae in the Southern Bight of the North Sea, J. Cons., 37, 221–248, 1977.

Taylor, N., Fox, C. J., Bolle, L., Dickey-Collas, M., Fossum, P., Kraus, G., Munk, P., Rolf, N., van Damme, C., Vorbach, M., Taylor, N., Fox, C. J., Bolle, L., Dickey-Collas, M., Fossum, P., Kraus, G., Munk, P., Rolf, N., van Damme, C., and Vorbach, M.: Results of the Spring 2004 North Sea Ichthyoplankton Surveys. The distribution of fish eggs and larvae from the International Ichthyoplankton Survey, 2007.

Thorson, G.: Some factors influencing the recruitment and establishment of marine benthic communities, Neth. J. Sea Res., 3, 267–293, 1966.

Van der Molen, J., Rogers, S. I., Ellis, J. R., Fox, C. J., and McCloghrie, P.: Dispersal patterns of the eggs and larvae of spring-spawning fish in the Irish Sea, UK, J. Sea Res., 58, 313–330, 2007.

Van der Veer, H. W.: Immigration, settlement and dependent mortality of a larval and early post-larval 0-group plaice (Pleuronectes platessa) population in the western Dutch Wadden Sea, Mar. Ecol. Prog. Ser., 29, 223–236, 1986.

Van der Veer, H. W. and Leggett, W. C.: Recruitment, in: Flatfishes, edited by: Gibson, R. N., Blackwell Science Ltd, Oxford, UK, 120–137, 2005.

Van der Veer, H. W. and Witte, J. I. J.: Year-class strength of plaice Pleuronectes platessa in the Southern Bight of the North Sea:a validation and analysis of the inverse relationship with winter seawater temperature, Mar. Ecol. Prog. Ser., 184, 245–257, 1999.

Van der Veer, H. W., Ruardij, P., Van den Berg, A. J., and Ridderinkhof, H.: Impact of interannual variability in hydrodynamic circulation on egg and larval transport of plaice *Pleuronectes platessa* L. in the southern North Sea, J. Sea Res., 39, 29–40, 1998.

Van der Veer, H. W., Geffen, A. J., and Witte, J. I. J.: Exceptionally strong year classes in plaice Pleuronectes platessa:are they generated during the pelagic stage only, or also in the juvenile stage?, Mar. Ecol. Prog. Ser., 199, 255–262, 2000.

Van der Veer, H. W., Bolle, L. J., Geffen, A. J., and Witte, J. I. J.: Variability in transport of fish eggs and larvae. IV. Interannual variability in larval stage duration of immigrating plaice in the Dutch Wadden Sea, Mar. Ecol. Prog. Ser., 390, 213–223, 2009.

Van Leeuwen, S. M., van der Molen, J., Ruardij, P., Fernand, L., and Jickells, T.: Modelling the contribution of deep chlorophyll maxima to annual primary production in the North Sea, Biogeochemistry, 113, 137–152, 2012.

Wolk, F.: Three-dimensional Lagrangian Tracer Modelling in Wadden Sea Areas (Diploma thesis), Carl von Ossietzky University, Oldenburg, 2003.

Zijlstra, J. J.: On the importance of the Waddensea as a nursery area in relation to the conservation of the southern North Sea fishery resources, Symp. Zool. Soc. Lond., 19, 233–258, 1972.

Assessment of the structure and variability of Weddell Sea water masses in distinct ocean reanalysis products

T. S. Dotto[1], R. Kerr[1], M. M. Mata[1], M. Azaneu[1], I. Wainer[2], E. Fahrbach[3], and G. Rohardt[3]

[1]Laboratório de Estudos dos Oceanos e Clima, Instituto de Oceanografia, Universidade Federal do Rio Grande – FURG, Rio Grande, RS, 96203-900, Brazil

[2]Laboratório de Oceanografia Física, Clima e Criosfera, Instituto Oceanográfico, Universidade de São Paulo – USP, São Paulo, SP, 05508-120, Brazil

[3]Stiftung Alfred-Wegener-Institut für Polar- und Meeresforschung in der Helmholtz-Gemeinschaft, Fachbereich Klimawissenschaften, Sektion Messende Ozeanographie, Postfach 120121 Bremerhaven, Germany

Correspondence to: T. S. Dotto (tiagodotto@furg.br)

Abstract. We assessed and evaluated the performance of five ocean reanalysis products in reproducing essential hydrographic properties and their associated temporal variability for the Weddell Sea, Antarctica. The products used in this assessment were ECMWF ORAS4 (European Centre for Medium-Range Weather Forecasts Ocean Reanalysis System 4), CFSR (Climate Forecast System Reanalysis), MyOcean UR025.4 (University of Reading), ECCO2 (Estimating the Circulation and Climate of the Ocean, Phase II) and SODA (Simple Ocean Data Assimilation). The present study focuses on the Weddell Sea deep layer, which is composed of the following three main water masses: Warm Deep Water (WDW), Weddell Sea Deep Water (WSDW) and Weddell Sea Bottom Water (WSBW). The MyOcean UR025.4 product provided the most accurate representation of the structure and thermohaline properties of the Weddell Sea water masses when compared with observations. All the ocean reanalysis products analyzed exhibited limited capabilities in representing the surface water masses in the Weddell Sea. The CFSR and ECCO2 products were not able to represent deep water masses with a neutral density $\geq 28.40\,\mathrm{kg\,m^{-3}}$, which was considered the WSBW's upper limit throughout the simulation period. The expected WDW warming was only reproduced by the SODA product, whereas the ECCO2 product was able to represent the trends in the WSDW's hydrographic properties. All the assessed ocean reanalyses were able to represent the decrease in the WSBW's density, except the SODA product in the inner Weddell Sea. Improvements in parameterization may have as much impact on the reanalyses assessed as improvements in horizontal resolution primarily because the Southern Ocean lacks in situ data, and the data that are currently available are summer-biased. The choice of the reanalysis product should be made carefully, taking into account the performance, the parameters of interest, and the type of physical processes to be evaluated.

1 Introduction

The Southern Ocean is considered an important region for better understanding the global overturning circulation (GOC) because of the regional formation and export of bottom waters to the global ocean (e.g., Talley, 2013). The GOC's deeper branch starts with the formation of Antarctic Bottom Water (AABW), which occurs regionally around the Antarctic margins (Whitworth et al., 1998) as a result of the mixing of warm and salty intermediate waters with near-surface freezing-point shelf or ice-shelf waters. The AABW's properties are dependent on several complex physical processes coupled with atmosphere–ocean–cryosphere processes, including sea-ice formation, opening of coastal polynyas, melting under deep ice shelves, deep ocean convection, and entrainment of overlying or surrounding waters (e.g., Carmack and Foster, 1975; Foldvik et al., 1985; Nicholls et al., 2009; Ohshima et al., 2013).

The Weddell Sea is thought to be the major contributor to AABW's formation and export to the global ocean (e.g., Orsi et al., 1999; Huhn et al., 2008; Kerr et al., 2012a; van Sebille et al., 2013). Regionally, the bottom layer consists of Weddell Sea Bottom Water (WSBW), the densest AABW variety in the Weddell Sea. WSBW is produced by a mixture of Shelf Waters (SW) with Warm Deep Water (WDW) or modified WDW (MWDW) near the shelf–slope break (e.g., Foster and Carmack, 1976; Foldvik et al., 1985). This bottom water mass is primarily confined to the Weddell Basin (Orsi et al., 1993) and eventually can be exported from the source region due to mixture with overlying Weddell Sea Deep Water (WSDW) or flow through deep channels (Orsi et al., 1995). WSDW is the less dense Weddell Sea deep water variety that contributes to the AABW after leaving the source areas. It can be formed either directly or by a mixture of WSBW with WDW during the downslope flow (Orsi et al., 1993, 1999). Because WSDW is less dense than WSBW, it is easily exported from the Weddell Sea into the global ocean through the narrow passages of the South Scotia Ridge (e.g., Naveira Garabato et al., 2002; Franco et al., 2007). WDW is a branch of the Circumpolar Deep Water (CDW) that enters the Weddell Sea at ca. 30° E (Gouretski and Danilov, 1993). Thus, any change occurring during the AABW-formation process can be reflected in global circulation via the deep branch of the overturning cell (Lumpkin and Speer, 2007; Talley, 2013).

Over the past few decades, changes in the thermohaline properties of AABW source waters have been reported, such as freshening of the dense waters in the shelf regions (e.g., Hellmer et al., 2011; Azaneu et al., 2013) and long-term warming of WDW within the Weddell Sea (e.g., Robertson et al., 2002; Smedsrud, 2005). In turn, WSBW in the inner Weddell Sea also experienced warming during the second half of the 1990s (Fahrbach et al., 2004, 2011). Moreover, Huhn et al. (2013) found that all deep water masses in the Weddell Sea were continually growing older and becoming less ventilated from 1984 to 2011. In concordance with these findings, a decrease in the WSBW's contribution ($\sim 20\,\%$) to the total water mass mixture in the Weddell Basin occurred in the 1980s and 1990s near the Greenwich Meridian and at the tip of the Antarctic Peninsula (Kerr et al., 2009a). More recently, Azaneu et al. (2013) fully investigated the most complete Southern Ocean data set available and found a reduction in the volume of AABW in addition to warming and decreasing density from 1958 to 2011 in the deep and bottom layers south of 60° S. Despite the reported freshening of the AABW layer at the Drake Passage during the period 1993–2010 (Jullion et al., 2013), no sign of this freshening trend was found by Azaneu et al. (2013) in WSDW/WSBW layers in the last 50 years (1958–2011) in the inner Weddell Sea. In a global context, the AABW's layer in the global basins underwent a contraction from the 1980s to the 2000s (Purkey and Johnson, 2012).

In spite of the efforts made to understand the physical processes associated with those long-term changes, the regional seas of the Southern Ocean have limited and generally summer-biased sampling opportunities. The lack of consistent in situ observations precludes a better understanding of connections between those processes and their possible implications for the global climate. To overcome this limitation in data coverage, numerical ocean models powered by data-assimilation systems (i.e., reanalysis systems) are potentially valuable tools. Reanalysis provides a physical picture of the global climate over a period during which observational data are available, making it possible to minimize the information gaps in spatial and temporal coverage in those regions. However, ocean reanalysis systems can produce spurious trends and inhomogeneity (Carton and Santorelli, 2008) caused by the limited and summer-biased sampling, especially at high southern latitudes (Bromwich et al., 2011). Moreover, a good representation of the physical processes occurring in ocean and climate models together with accurate hydrographic data observed in the Weddell Sea, which can be investigated through ocean reanalysis products, should significantly influence the hydrography of the Southern Ocean and South Atlantic (Hellmer et al., 2005). In this way, validation of ocean reanalysis products is needed to evaluate the suitability, consistency, and applicability of these products for long-term investigations in the Southern Ocean.

The present study aims to assess and compare the representation and variability of the hydrographic properties of Weddell Sea deep water masses using five recent ocean reanalysis products to identify which reanalysis product best reproduces the main regional oceanographic features. The paper is organized as follows: Sect. 2 provides a description of the five ocean reanalysis products investigated here. The observational data set used for the structure and variability assessments of the Weddell Sea water masses is described in Sect. 3. A comparison of the results of each ocean reanalysis product is described in Sect. 4. Finally, Sect. 5 summarizes and addresses the study's main results and conclusions.

2 Ocean reanalysis data sets

We assessed the capabilities of the five ocean reanalysis products, which are briefly described below, in terms of the potential temperature (θ), salinity (S) and neutral density (γ^{n}; Jackett and McDougall, 1997) of seawater. The main characteristics of the ocean reanalysis products are reported in Table 1. To assess the robustness of those products for modeling the Weddell Sea, we compared the ocean reanalysis data sets against an observational data set for the period spanning from the 1980s to the 2000s.

The European Centre for Medium-Range Weather Forecasts Ocean Reanalysis System 4 (ECMWF ORAS4) is a global reanalysis system based on the ocean model Nucleus

Table 1. Summary of the main characteristics of the ocean reanalyses evaluated in this study. See the text for acronyms. For more information about the reanalyses, see the references indicated.

System	Ocean model and resolution	Sea-ice model	Atmospheric forcing	Assimilation method	Data assimilated	Period*
ECMWF-ORAS4 (Balmaseda et al., 2013)	NEMOv3 (Madec, 2008), 1° × 1°, 42 levels	–	Heat, momentum, and freshwater fluxes from ERA-40 (Uppala et al., 2005), ERA-Interim (Dee et al., 2011), and the ECMWF operational archive (Balmaseda et al., 2013)	3D-Var FGAT (Mogensen et al., 2012)	*T* and *S* profiles from EN3 (XBT, CTD, TAO, TRITON, PIRATA, RAMA, ARGO and APB), altimetry, SST from ERA-40, NCEP OI v2 (Reynolds et al., 2002) and OSTIA SST (Stark et al., 2007), SIC from ERA-40	1957–2011
CFSR (Saha et al., 2010)	MOM4 (Griffies et al., 2008), 0.5° × 0.5°, 40 levels	GDFL sea-ice simulator (Griffies et al., 2008)	NCEP operational global forecast system model (Saha et al., 2010)	3D-Var (Saha et al., 2010)	*T* profiles (XBT, TAO, TRITON, PIRATA, RAMA and ARGO), synthetic *S* profiles, OI SST (Reynolds et al., 2007) and HadISST (Rayner et al., 2003), SIC	1979–2009
MyOcean-UR025.4 (Ferry et al., 2012)	NEMOv3.2, 0.25° × 0.25°, 75 levels	LIM2 (Fichefet and Morales-Maqueda, 1997)	Heat, momentum, and freshwater fluxes from ERA-Interim	OI FOAM-NEMO system (Storkey et al., 2010)	*T* and *S* profiles from EN3 (ARGO, XBT, CTD, TAO and PIRATA), altimetry, SST from ICOADS (Worley et al., 2005; Woodruff et al., 2011), SIC	1993–2010
ECCO2 (Menemenlis et al., 2008)	MITgcm (Marshall et al., 1997), 0.25° × 0.25°, 50 levels	MITgcm sea-ice model (Marshall et al., 1997)	Surface forcing from JRA-25 (Onogi et al., 2007)	Green's function (Menemenlis et al., 2005)	*T* and *S* profiles (CTD, TAO, ARGO, XBT), altimetry, SST, SIC	1992–2010
SODA 2.2.4 (Carton and Giese, 2008; Giese and Ray, 2011)	POP2 (Smith et al., 1992), 0.5° × 0.5°, 40 levels	–	Heat, momentum, and freshwater fluxes from 20CRv2 (Compo et al., 2011).	OI (Carton and Giese, 2008)	*T* and *S* profile from WOD09 (XBT, MBT, CTD, TAO, TRITON and ARGO), SST from ICOADS 2.5 and AVHRR	1871–2010

* The reanalysis products were restricted from 1980 to the end of the simulation.

for European Modelling of the Ocean (NEMO) version 3 (Madec, 2008). The method of data assimilation used is 3D-Var (three-dimensional variation; Mogensen et al., 2012). ECMWF ORAS4 assimilates the temperature and salinity profiles from EN3, sea-level anomalies and sea surface temperature (SST). The sea-ice concentration (SIC) data are from ERA-40 (ECMWF 40 Years Reanalysis), and they are only used to correct the SST values (Balmaseda et al., 2013). The ocean model is driven by daily fluxes of heat, momentum and freshwater from the ERA-40 (prior to 1989), the ERA-Interim (from 1989 to 2010) and ECMWF's operational archive (after 2010; Balmaseda et al., 2013). This ocean reanalysis product is hereafter referred to as ECMWF.

The Climate Forecast System Reanalysis (CFSR) is a coupled atmosphere–ocean–land surface–sea-ice data assimilation. The ocean system is based on the Modular Ocean Model version 4p0d (MOM4), which is coupled with an interactive ice model (Griffies et al., 2008). CFSR uses 3D-Var as the data assimilation method (Saha et al., 2010). The reanalysis system assimilates temperature profiles from XBT (expendable bathythermograph), moorings, Argo floats and SST only in the top 750 m (Xue et al., 2011). CFSR also assimilates synthetic salinity profiles (Xue et al., 2011) and SIC (Saha et al., 2010). The atmospheric model is based on the previous National Center for Environmental Prediction (NCEP) operational global forecast system (Saha et al., 2010). It is important to note that the CFSR ocean reanalysis uses a combination of six data streams, each from a different initial condition (Saha et al., 2010). This segmentation leads to serious discontinuity in the deep ocean, which has consequences for decadal prediction (Xue et al., 2011). For this reason, CFSR assessment was only performed for climatological analysis and was not included in the evaluation of time series properties.

The MyOcean University of Reading (UR025.4) reanalysis product is performed with the ocean model NEMO version 3.2 coupled with Louvain-la-Neuve ice model version 2 (LIM2; Fichefet and Morales-Maqueda, 1997). It includes an annual estimation of Antarctica ice sheet melt in the oceanic model (Ferry et al., 2012). The assimilation system used in UR025.4 is an Optimal Interpolation (OI) scheme based on the UK Met Office operational FOAM–NEMO system (Storkey et al., 2010). It assimilates in situ and satellite SST data, satellite sea level data, satellite SIC data, and in situ temperature and salinity profile data from the EN3 data set. Surface atmospheric forcing is obtained from the ERA-Interim, and bulk fluxes are calculated as suggested by Large and Yeager (2009). Hereafter, the UR025.4 reanalysis product is referred to as MyOcean.

The Estimating the Circulation and Climate of the Ocean, Phase II (ECCO2) reanalysis system is based on the global ocean model of the Massachusetts Institute of Technology general circulation model (MITgcm; Marshall et al., 1997) in a cube–sphere grid. MITgcm is coupled to a sea-ice model that computes ice thickness, ice concentration, and snow cover. ECCO2 uses Green's function as an assimilation system (Menemenlis et al., 2005). It assimilates sea surface height anomalies, SST, temperature and salinity profiles, and sea-ice concentration, motion and thickness. We used the solution "cube 92" with a 0.25° regular latitude–longitude grid here (hereafter referred to as ECCO2). The surface forcing of this solution is provided by the Japanese 25-year Reanalysis (JRA-25; Onogi et al., 2007). We restricted the analysis to the period 1992–2005 because ECCO2 has poorly represented the water properties in most of the Southern Ocean during the last 6 years (2005–2010) of the simulation (Azaneu et al., 2014), showing abrupt changes in the properties of deep-water masses. These authors showed that after the opening of the oceanic polynya in 2004, the thermohaline values of Weddell Sea water masses became unrealistic, most likely because of strong, open-ocean deep convection simulated in the Weddell Sea.

Simple Ocean Data Assimilation version 2.2.4 (hereafter SODA) is a global reanalysis system based on the Parallel Ocean Program version 2.0.1 (Smith et al., 1992). The assimilation system used in SODA is an OI multivariate sequential-type scheme (Carton and Giese, 2008). It assimilates in situ temperature and salinity profiles and in situ and satellite SSTs. The ocean model is forced by fluxes of heat, momentum and freshwater from the Twentieth Century Reanalysis Project version 2 (20CRv2; Compo et al., 2011). SODA does not use a sea-ice model, although the surface heat flux is modified when the surface temperature reaches the freezing point of seawater.

3 Observational data sets, reanalysis outputs and methods used for reanalysis evaluation

The in situ θ and S were selected from two WOCE (World Ocean Circulation Experiment) hydrographic repeat sections in the Weddell Sea (Table 2; Fig. 1) as follows: (i) section WOCE A12 (also referred to as WOCE SR2 in the literature) along the Greenwich Meridian, with a sampling period spanning from 1984 to 2010 (e.g., Fahrbach et al., 2011); and (ii) section WOCE SR4 between Joinville Island and Kapp Norvegia, with a sampling period spanning between 1989 and 2010 (e.g., Fahrbach et al., 2004). Section WOCE A12 was restricted to latitudes higher than 60° S. All observed θ and S data were collected by high-accuracy CTDs. Those sections were chosen to be evaluated in the Weddell Sea because of the availability of historical data nearby, because of their importance in regional circulation and the export of deep waters (e.g., Naveira Garabato et al., 2002; Klatt et al., 2005; Kerr et al., 2012a), and because they are representative of the entire Weddell Basin. Reanalysis grid points closer to the geographical location of the in situ hydrographic stations (observations) were selected from the monthly mean fields corresponding in time to the period of the in situ measurements.

Table 2. Overview of the observed hydrographic sections used for the validation of the reanalyses. Details of the observed data can be found in Whitworth and Nowlin (1987), Fahrbach et al. (2001, 2004, 2007, 2011), Fahrbach and De Baar (2010) and Rohardt et al. (2011).

Expedition	Cruise period (dd/mm/yyyy)	WOCE section
AJAX (leg 2)	16/01/1984–29/01/1984	A12
ANT-VIII/2	06/09/1989–31/10/1989	SR4
ANT-IX/2	16/11/1990–30/12/1990	SR4
ANT-X/4	21/05/1992–30/07/1992	A12
ANT-X7	03/12/1992–23/01/1993	SR4
ANT-XIII/4	17/03/1996–20/05/1996	A12/SR4
ANT-XV/4[a]	28/03/1998–23/05/1998	A12/SR4
ANT-XVI/2	09/01/1999–16/03/1999	A12
ANT-XVIII/3	05/12/2000–12/01/2001	A12
ANT-XX/2	24/11/2002–23/01/2003	A12
ANT-XXII/3	21/01/2005–06/04/2005	A12/SR4
ANT-XXIV/3	06/02/2008–16/04/2008	A12/SR4
ANT-XXVII/2[b]	28/11/2010–05/02/2011	A12/SR4

[a] Does not extend all of the way to the shelf in the eastern Weddell Sea. [b] Only 2010 data are used.

Because the ocean reanalysis data sets have different vertical resolutions and because the position of observed stations varies between occupations, we linearly interpolated the observational data sets to the vertical grid for each reanalysis to allow direct comparisons among the ocean reanalysis products and observations. Horizontally, the reanalysis and the observational data sets were spatially interpolated and gridded with 0.5° latitude and 1° longitude for the WOCE A12 and WOCE SR4 sections, respectively.

The structure of the water column was evaluated using classical θ–S diagram comparisons, and simple differences in the hydrographic properties of the sections between the reanalysis results and field observations were calculated. We used the root-mean-square error (RMSE) criteria following Heuzé et al. (2013) to evaluate which ocean reanalysis product better represented the entire water column. In addition, the statistical patterns of the hydrographic fields were evaluated using a normalized Taylor diagram (Taylor, 2001) for a more robust comparison of the reanalysis products being evaluated. Briefly, the normalized Taylor diagram combines statistical parameters (correlation coefficient – r, normalized standard deviation – σ_n, and normalized centered root-mean-square error – CRMSE) to compare the spatial patterns from the ocean reanalyses and the observed hydrographic fields. The CRMSE is used as a measure of the difference between values predicted by a model and values that are observed, minimizing the effect of the model mean bias. Further details and equations are presented in Taylor (2001). We performed the statistical analysis considering the entire water column and used the field observations as the reference data set. The reanalysis fields that showed better concordance with the observations lie closer to the reference point in the Taylor

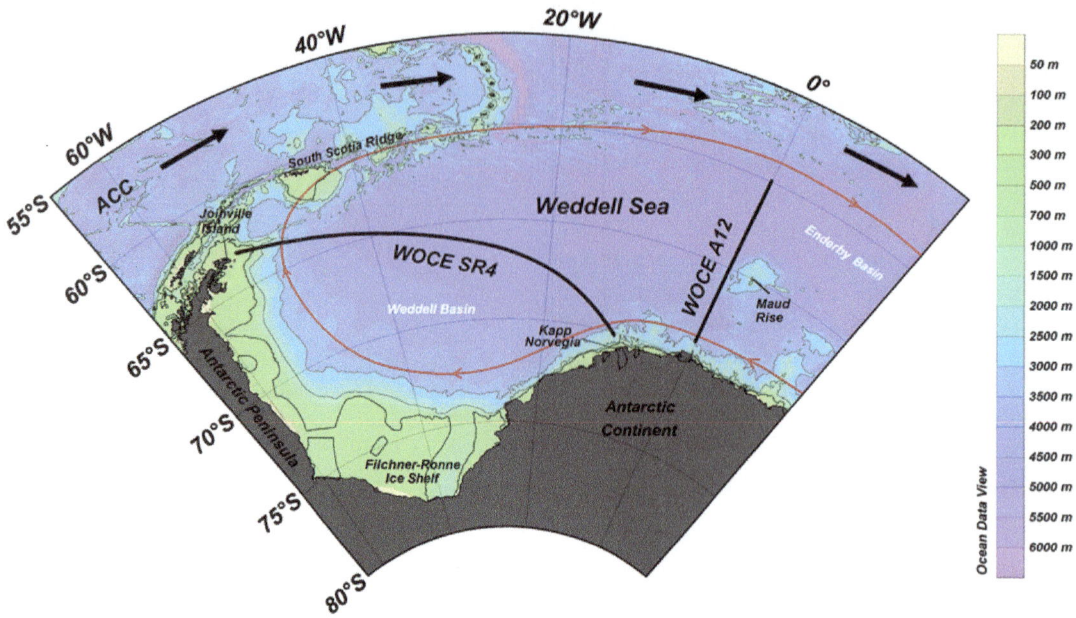

Figure 1. Schematic locations of the hydrographic sections. WOCE A12 is found along the Greenwich Meridian, and WOCE SR4 lies between Joinville Island and Kapp Norvegia. The arrows indicate the direction of the Antarctic Circumpolar Current (ACC; black) and the Weddell Gyre (red) flows. The thin black and gray lines represent the 500, 1500, and 3000 m isobaths.

diagram (i.e., had low CRMSE and high r and σ_n values close to 1).

For the time series of water mass properties, we show the range of values for the station data sets, excluding the station set of 1998 at WOCE SR4. We used the original resolution of each data set (i.e., a monthly time series) to evaluate patterns of variability in the hydrographic properties, and the annual linear trend was fit for each time series of hydrographic properties.

3.1 Definition of the water masses

The Weddell Sea water masses were defined based on the γ^n isopycnal surfaces for all data sets. The interface between the surface and intermediate layers was defined as proposed by Franco et al. (2007). We used the definition of Orsi et al. (1999) to distinguish between the deep and bottom layers in the inner Weddell Sea. Thus, we separated the water mass layers from the surface to the bottom using the γ^n isopycnals of 28.1, 28.27, and 28.4 kg m^{-3}, corresponding to the AASW–WDW, WDW–WSDW, and WSDW–WSBW interfaces, respectively.

4 Results

4.1 Water column structure and simple differences in hydrographic properties

In general, all of the ocean reanalysis products that were evaluated captured the main structure of the water masses in the Weddell Sea (Figs. 2, 3). The lighter AASW could be observed lying above the warm and salty intermediate water (WDW), with θ and S decreasing with depth and marking the dense deep (WSDW) and bottom (WSBW) waters of the Weddell Sea. The ECMWF, MyOcean and SODA products had the θ–S structures that most closely approximated the observations of both the WOCE A12 and WOCE SR4 sections, especially when considering the intermediate and deep layers (WDW, WSDW, and WSBW). ECCO2 showed a similar θ–S structure when compared with observation data. However, its dense WSBW layer was ~0.2 °C warmer than the in situ data. The CFSR product captured the stratification of the water masses along the water column, but its θ–S structure was displaced by warmer (except for the WDW layer) and fresher hydrographic properties, consequently making most layers lighter than the in situ data.

All the ocean reanalysis products evaluated had difficulty representing the AASW hydrographic values (Figs. 2–9). The mismatch between the data and the surface water representations was most likely a consequence of difficulties the products faced in reproducing several complex processes and fluxes acting on the ocean surface, which are seasonally influenced by physical processes at the air–sea and sea ice–ocean interfaces (Whitworth et al., 1998). The majority of the reanalyses showed the mean differences colder by 0.002–0.16 °C than those actually observed in the data recorded in situ along the WOCE A12 line (Fig. 4). Conversely, for the WOCE SR4 section, all the reanalyses revealed mean differences that were 0.06–0.4 °C warmer than the observations (Fig. 5). For S, most of the reanalyses overestimated this field

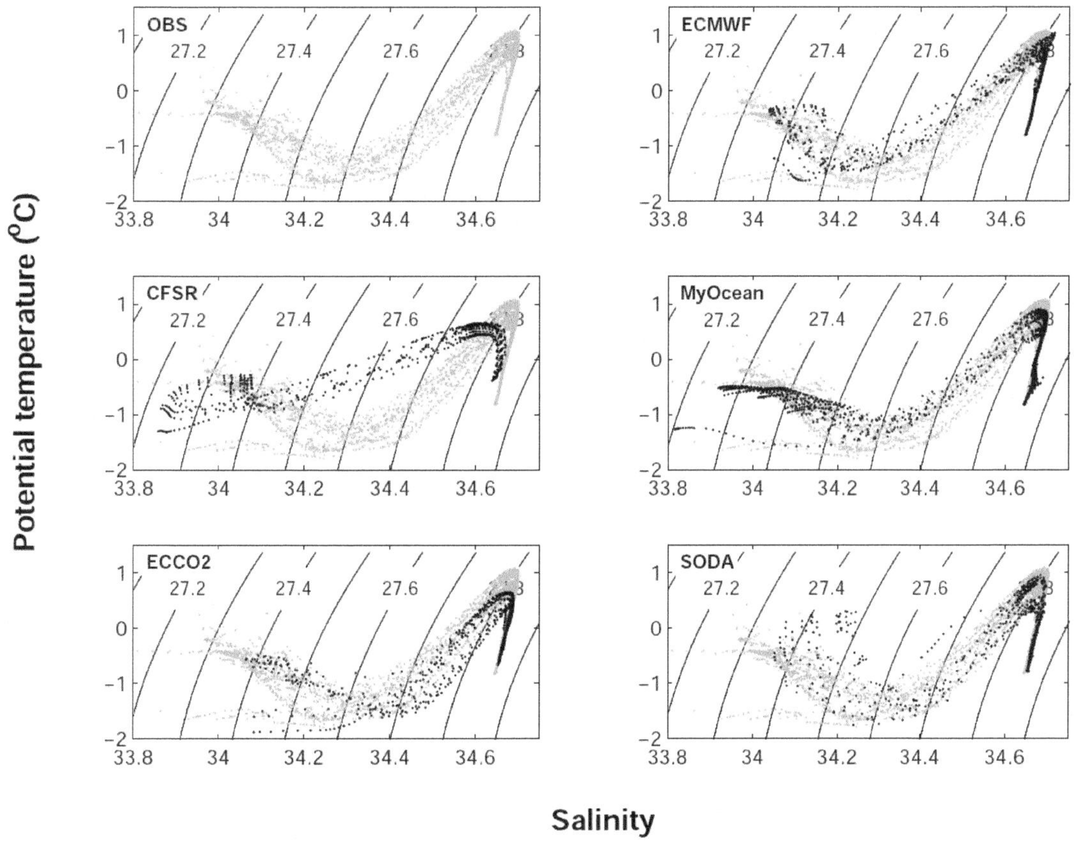

Figure 2. θ–S diagrams from section WOCE A12. Gray (black) dots are observation (reanalysis) data. Solid lines are potential density isopycnals.

by a mean difference of 0.01–0.11 in both sections (Figs. 6, 7), whereas for the γ^n field, the majority of the reanalyses showed a mean underestimation of 0.004–0.3 kg m^{-3} (Figs. 8, 9). For the S and γ^n fields, CFSR showed the greater mean differences (-0.3 ± 0.17 for salinity and -0.3 ± 0.16 kg m^{-3} for neutral density), whereas MyOcean detected minor differences (an overestimation of 0.003 ± 0.09 for S and an underestimation of 0.004 ± 0.08 kg m^{-3} for γ^n). The greatest mean differences were found primarily at the AASW–WDW interface and near the continental boundaries (Figs. 4–9). Note that along the WOCE sections, depending on the product, the simulated values both underestimated and overestimated the observations (Figs. 4–9).

In general, the majority of the ocean reanalysis products showed cold and fresh waters relative to the observations at intermediate depths (Figs. 4–7). However, the ECMWF, CFSR and ECCO2 products did not appear to follow this general pattern for θ below a depth of ~ 1000 m. The θ difference varied from ~ 0.05 °C for the MyOcean product to more than ~ 0.2 °C for the CFSR and ECCO2 products (Figs. 4, 5). The differences in the S field were greater than 0.05 for CFSR and ECMWF (near the Antarctic Margin) in both the WOCE A12 (Fig. 6) and WOCE SR4 (Fig. 7) sections. The MyOcean reanalysis had the smallest differences in the S field compared to the observed data (< 0.003) at the intermediate layer among all of the reanalysis products evaluated here. In this layer, the γ^n field was clearly dependent on the S differences found in each ocean reanalysis product. The differences in the reanalyses' intermediate layers were most likely advected with WDW into the Weddell Gyre because the colder and fresher WDW variety could be observed in section WOCE A12 (Figs. 4, 6).

At the deep layer, the reanalysis showed two distinct patterns of θ differences. The MyOcean and SODA reanalysis products were generally colder than observations by less than 0.05° C, whereas CFSR and ECCO2 were warmer for both the WOCE A12 (Fig. 4) and WOCE SR4 sections (Fig. 5). These latter reanalysis products overestimated θ values (~ 0.2–0.4 °C warmer), leading to the representation of bottom waters that did not reach the WSBW temperature threshold (i.e., -0.7 °C; Carmack and Foster, 1975) or its corresponding neutral density (28.4 kg m^{-3}). In contrast, the ECMWF product was warmer than observations at section WOCE A12 (Fig. 4) and colder at WOCE SR4 (Fig. 5). This reanalysis also showed a temperature overestimation of ~ 0.3 °C near the Antarctic margins at 1000–2000 m depth (Figs. 4, 5) due to a representation of the WDW inflow core deeper and closer to the Antarctic continent than indicated

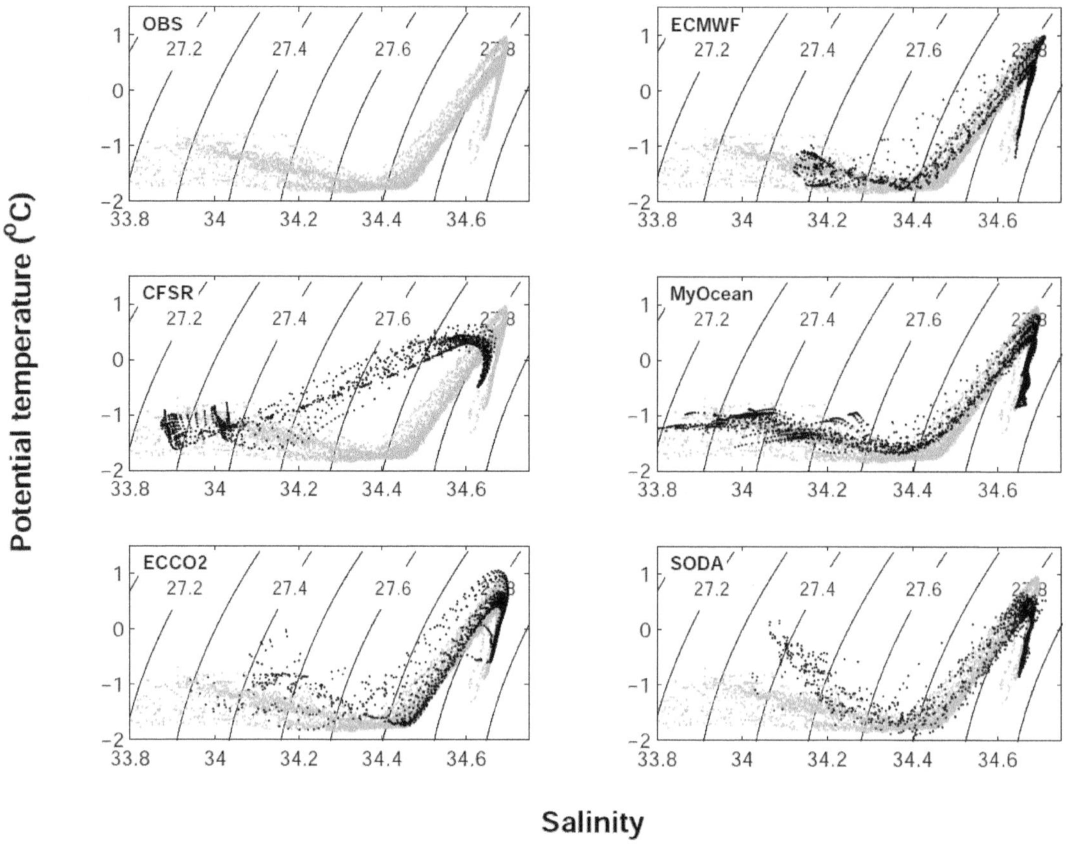

Figure 3. Same as in Fig. 2, but for section WOCE SR4.

by the observations. At this layer, the S field showed smaller differences than those of the whole upper structure of the water column, as demonstrated by the underestimation of S in almost all the sections by all the products, except for the WOCE SR4 section by the CFSR product (Figs. 6, 7). The θ value had greater influence on the γ^n field at this layer because of the smaller differences in S. The salinity differences in the CFSR results (0.025–0.05; Figs. 6, 7), associated with its warm ocean representation ($> 0.2\,°C$; Figs. 4, 5), helped to increase the difference in density with respect to observations, resulting in differences of $\sim -0.1\,kg\,m^{-3}$ (Figs. 8, 9). None of the reanalysis systems evaluated represented the downslope flow of the WSBW plume in the western continental slope of the WOCE SR4 section during the period analyzed (Fig. 9) as expected from observations (e.g., Fig. 02 of Fahrbach et al., 2001).

4.2 Statistical representation of the hydrographic spatial fields

We used the RMSE to quantify the accuracy of the ocean reanalysis products in representing the hydrographic fields. A reanalysis was considered accurate if, for each hydrographic parameter, the RMSE was smaller than the mean RMSE of the five reanalysis products. Table 3 summarizes the accuracy threshold for each parameter. The use of RMSE criteria revealed that ECMWF and MyOcean could be considered accurate in their representation of almost all parameters except for WOCE SR4 θ and WOCE A12 S, respectively (Table 3). For SODA, only S and γ^n were accurate in both sections. ECCO2 was accurate in its representation of all hydrographic fields at WOCE SR4, but only γ^n was acceptable at WOCE A12 (Table 3). Conversely, the RMSE criteria showed that CFSR was not accurate in representing any of the variables analyzed in any section (Table 3), with all of its RMSEs above the mean RMSE of the five reanalyses. However, the mean RMSE could be influenced by the CFSR results, given that CFSR does not assimilate the WOCE deep-ocean data. It is fairer to judge the accuracy of the reanalysis outputs without CFSR RMSE, because using this approach the four reanalyses compared assimilate WOCE data. When CFSR was not considered, the mean RMSE decreased, and ECMWF was no longer accurate for most of the hydrographic properties (Table 3). In this sense, only MyOcean could be considered to accurately represent all the hydrographic properties of the Weddell Sea. Table 3 also shows that the RMSE for salinity was higher in WOCE A12 than in WOCE SR4. This difference could be associated with more intense hydrodynamic processes occurring closest to the WOCE A12 section (e.g.,

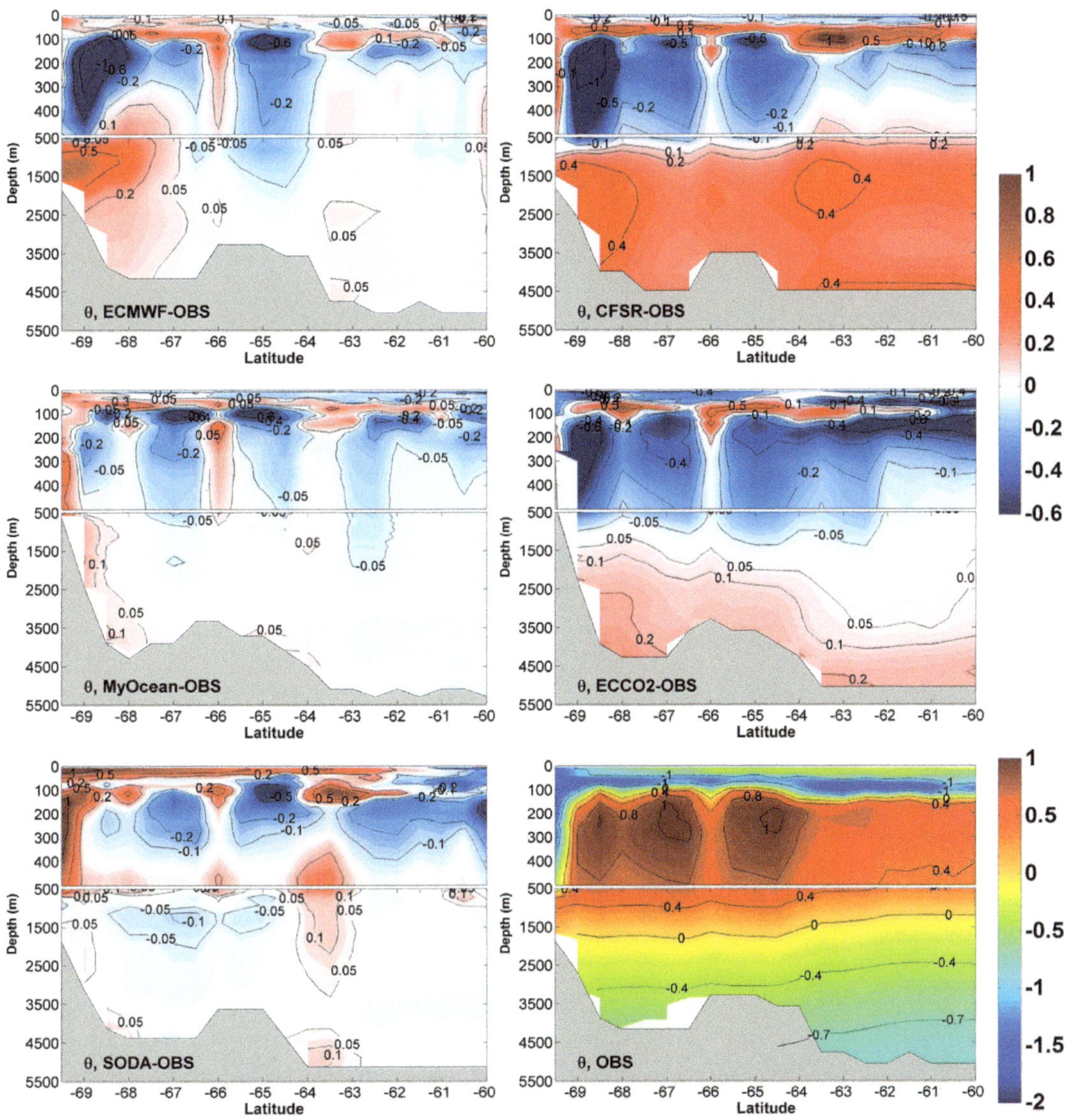

Figure 4. Section WOCE A12. Representation of the time-averaged in situ data and the differences between the reanalyzed and observed data for θ. ECMWF, CFSR, MyOcean, ECCO2, and SODA results are indicated in each figure. The upper 500 m is expanded to show more detail.

Table 3. RMSE results used to evaluate the accuracy of the reanalyses, as established by Heuzé et al. (2013). The mean RMSEs of the five reanalyses assessed are in parentheses, and the mean RMSEs, considering all products except CFSR, are in italics. Reanalyses that have an RMSE lower than the mean RMSE are considered accurate in the intercomparison.

	A12			SR4		
	θ °C	S	γ^n kg m^{-3}	θ °C	S	γ^n kg m^{-3}
	(0.2838)	(0.2366)	(0.0779)	(0.2919)	(0.0995)	(0.1078)
	(0.2534)	*(0.2288)*	*(0.0512)*	*(0.2794)*	*(0.0662)*	*(0.0631)*
ECMWF	0.2631	0.2101	0.0576	0.2977	0.0826	0.0768
CFSR	0.4018	0.2676	0.1846	0.3420	0.2326	0.2421
MyOcean	0.1704	0.2424	0.0325	0.1710	0.0533	0.0516
ECCO2	0.2889	0.2610	0.0701	0.2883	0.0647	0.0598
SODA	0.2948	0.2018	0.0447	0.3606	0.0642	0.0608

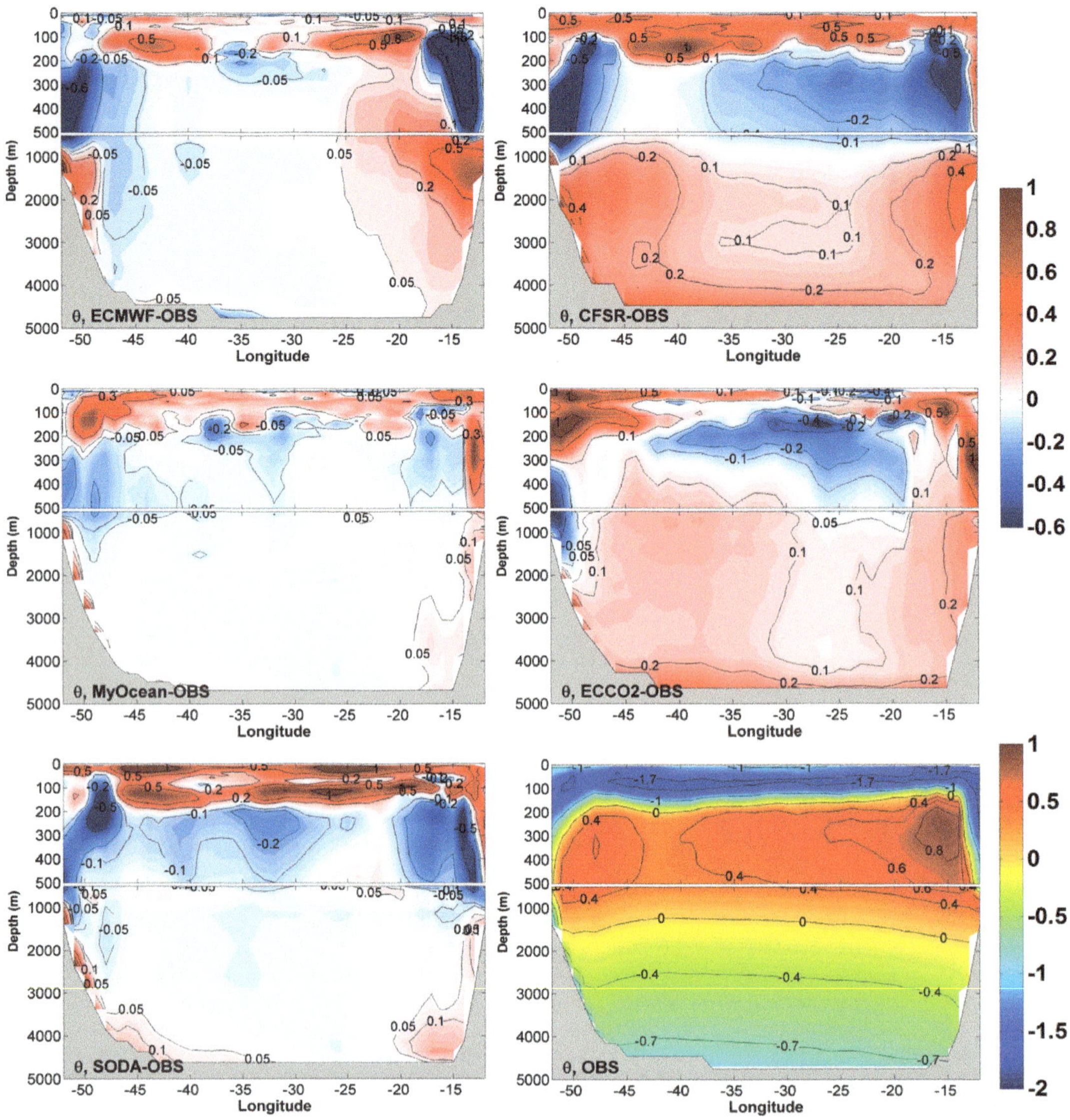

Figure 5. Same as in Fig. 4, but for section WOCE SR4.

Klatt et al., 2005) than in WOCE SR4, implying that that region is more sensitive to changes in the water column structure. The dynamic impacts on the three dimensional oceanic flow field all the way up through the water column is accentuated by the presence of the Maud Rise seamount (Holland, 2001). In less dynamic regions – e.g., WOCE SR4 (which is dampened by the Weddell Gyre circulation) – most of the reanalysis methods were able to represent the *S* field, including MyOcean and ECCO2, despite not meeting the accuracy criteria for WOCE A12. We also evaluated the representation of the hydrographic properties of each reanalysis using a more robust statistical analysis through the standardized Taylor diagram (Fig. 10).

Generally, all of the ocean reanalysis products evaluated in this study represented the γ^n and θ fields better than the *S* field throughout the water column in both sections (Fig. 10). In WOCE A12, a good representation of the γ^n field was correlated with a good representation of the θ field (Fig. 10a), whereas in WOCE SR4, both the *S* and θ reanalysis fields were responsible for good γ^n statistical measurements (Fig. 10b).

In the WOCE A12 section (Fig. 10a), the reanalysis systems that gave results closest to the reference point were MyOcean and SODA, both with CRMSEs of less than 0.25 and correlation coefficients (rs) of 0.99 for γ^n. ECMWF and ECCO2 had CRMSEs of $\sim$0.25 and rs of $\sim$0.97; however, the former system had a slightly better r and a normalized standard deviation of ~ 1, which implied that ECMWF provides a better representation than ECCO2. For θ, MyOcean was also close to the reference (CRMSE $\sim$0.25 and $r > 0.95$), but followed by ECMWF and then ECCO2 and SODA. The MyOcean product was the closest to the

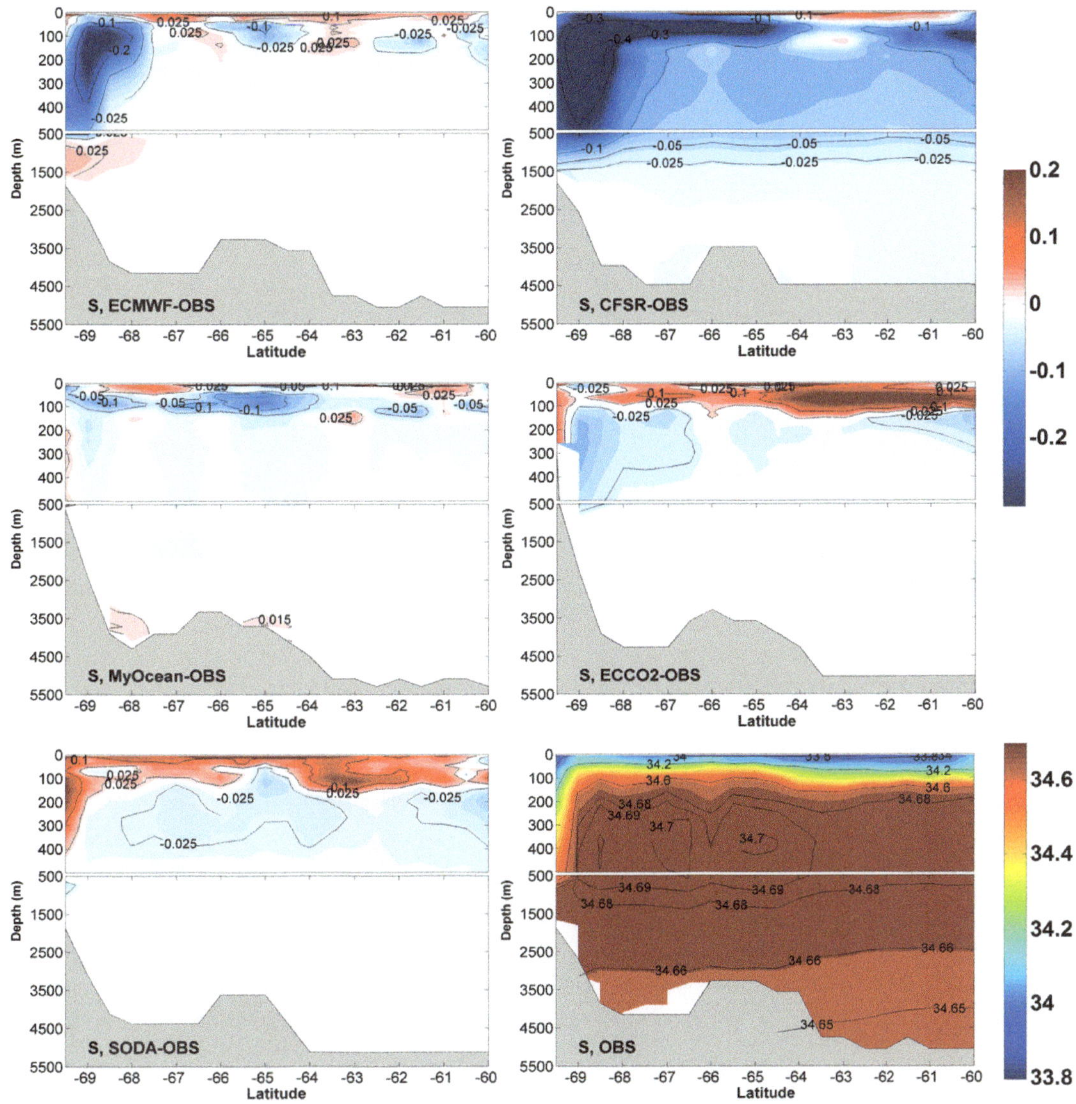

Figure 6. Section WOCE A12. Representation of the time-averaged in situ data and the differences between the reanalyzed and observed data for S. ECMWF, CFSR, MyOcean, ECCO2, and SODA results are indicated in each figure. The upper 500 m is expanded to show more detail.

reference for salinity (CRMSE ~ 0.60 and $r \sim 0.80$), followed by ECMWF and SODA. For all fields, the CFSR product was the furthest from the reference point in this section.

In WOCE SR4 (Fig. 10b), MyOcean produced the closest reanalysis results to the reference, with a CRMSE < 0.25 and $r \sim 0.98$ for all fields. In this section, ECCO2 had CRMSE < 0.32 and $r > 0.94$ for all hydrographical properties, which made it the second most accurate reanalysis system. SODA showed γ^n and S fields closer to the reference point than ECMWF; however, θ in the latter reanalysis had a better CRMSE and r than in the first system. As for WOCE A12, CFSR was the furthest from the reference point (except for θ). CFSR showed a good θ distribution pattern (Fig. 10), although it had significant differences in absolute values (Figs. 5, 7, 9).

4.3 Variability and trends in deep water masses

Ocean reanalysis products are powerful tools that can be used in climate studies because of their generally high temporal resolution. To make use of this property, we also assessed the temporal variability and trends of the deep water masses represented by each reanalysis product.

4.3.1 Warm Deep Water

Most of the reanalysis systems evaluated represented the WDW layer ($28.1 \leq \gamma^n < 28.27\,\mathrm{kg\,m^{-3}}$) as warmer and saltier in WOCE A12 (Fig. 11) than in WOCE SR4 (Fig. 12). This difference occurs because WDW advection towards the inner Weddell Sea is associated with cooling and freshening

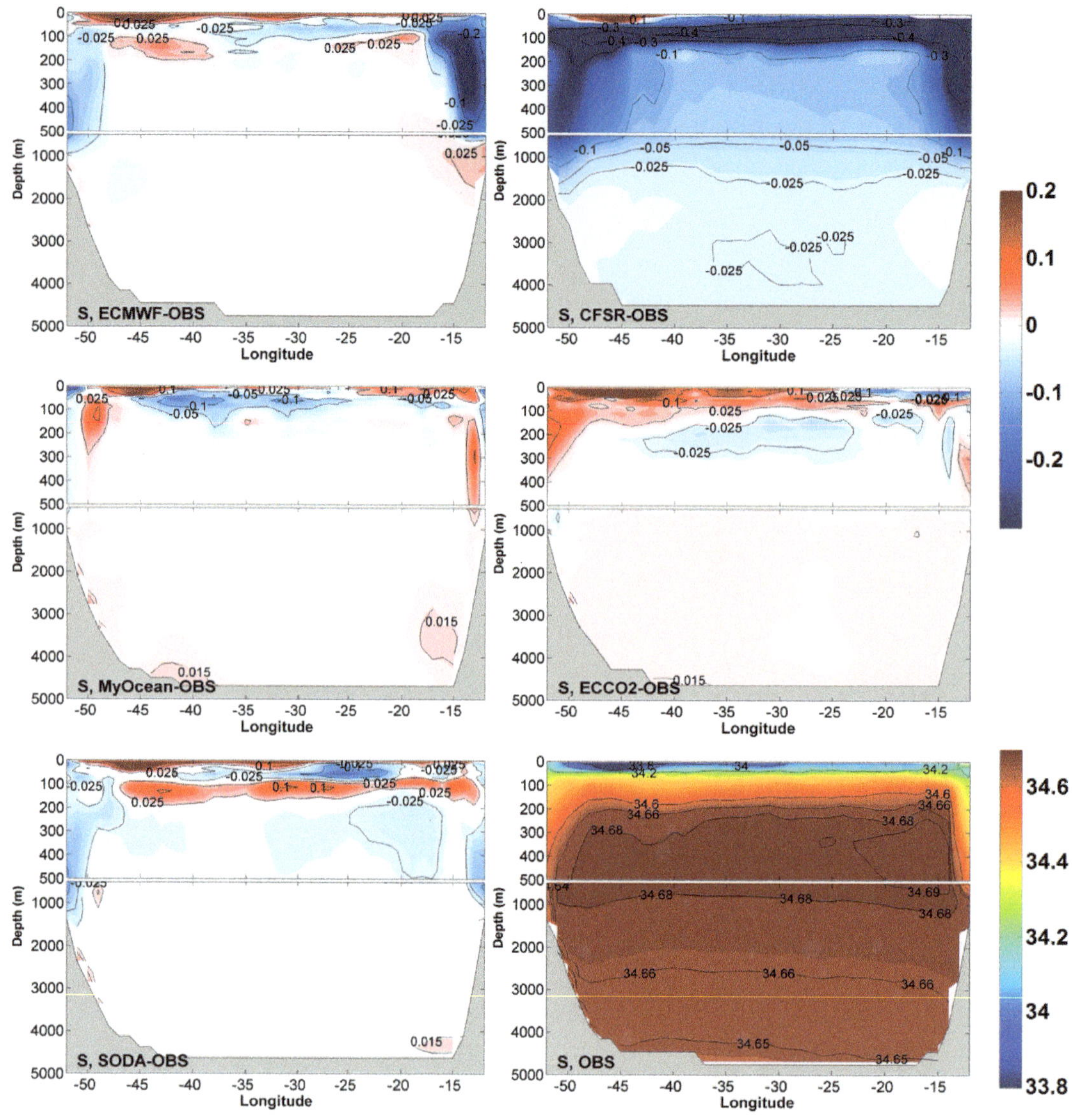

Figure 7. Same as in Fig. 6, but for section WOCE SR4.

of this water mass through mixing processes with upper waters (e.g., Schröder and Fahrbach, 1999). Moreover, the differences between A12 and SR4 support that the mixing processes along the WDW pathway are being captured by most of the reanalysis products.

The ECMWF, MyOcean and ECCO2 θ and S fields showed a decreasing trend for both of the areas analyzed during the following periods: 1980–2011, 1993–2004, and 1992–2004 (Table 4). These trends were observed clearly beyond the 1990s. Although MyOcean and ECCO2 had hydrographic fields represented until 2010, here they were only evaluated until 2004 because of the anomalous variability observed in both reanalysis systems beyond 2004 (Figs. 11, 12). ECCO2 showed a clearly negative trend in θ and S in both sections, which began in the second half of the 1990s and intensified beyond 2004 (Figs. 11, 12) due the opening of an oceanic polynya near the prime meridian that led to injection of dense water directly at great depths (Azaneu et al., 2014). In contrast, SODA showed increasing trends of θ and S (1980–2010) for this water mass in both sections, although these trends were not statistically significant at WOCE A12 (Table 4). In addition, this was the only reanalysis that captured a significant decreasing trend in the γ^n field throughout the period analyzed, which was associated with its increased potential temperature in section WOCE SR4 (Table 4). SODA also showed a marked temporal variability in γ^n throughout the entire series (Figs. 11, 12).

4.3.2 Weddell Sea Deep Water

The ECMWF reanalysis product showed a stable period in the WOCE A12 section between 1987 and 2007 (Fig. 13),

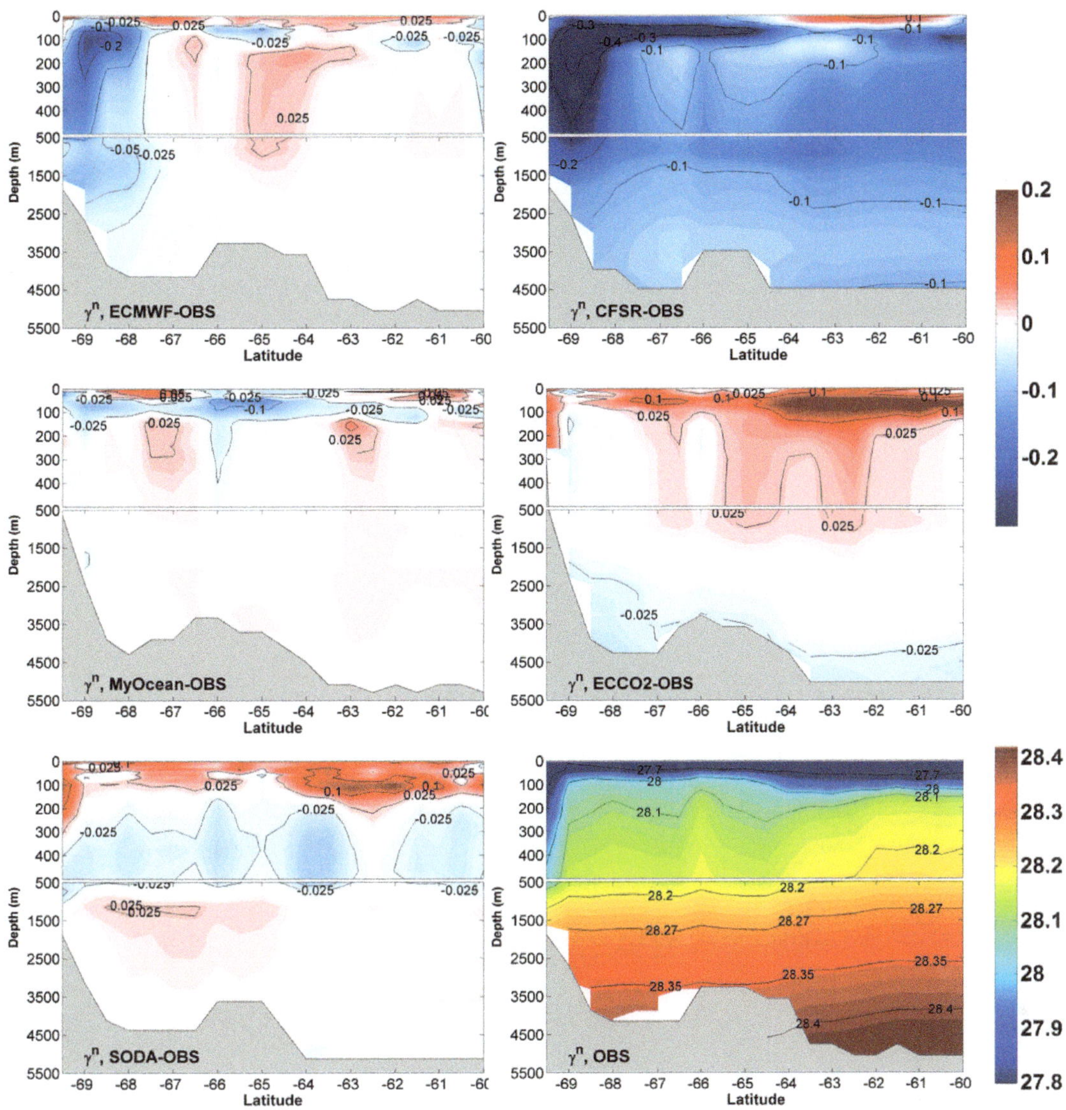

Figure 8. Section WOCE A12. Representation of the time-averaged in situ data and the differences between the reanalyzed and observed data for γ^n. ECMWF, CFSR, MyOcean, ECCO2, and SODA results are indicated in each figure. The upper 500 m is expanded to show more detail.

whereas before 1987 and after 2007, anomalous variability patterns were observed; in WOCE SR4, ECMWF showed low levels of variability throughout the entire period (Fig. 14). The MyOcean and ECCO2 products showed an anomalous period after 2004 in both sections (Figs. 13, 14). These anomalous periods in ECMWF (in section WOCE A12), MyOcean and ECCO2 were not considered when we calculated the trends for the hydrographic properties of the WSDW layer ($28.27 \leq \gamma^n < 28.4\,\mathrm{kg\,m^{-3}}$).

ECMWF, MyOcean and SODA showed cooling, freshening and increasing density trends in section WOCE A12 (Fig. 13) for the periods 1987–2007, 1993–2004 and 1980–2010, respectively (Table 4). For 2000–2007, ECMWF showed a decline in γ^n that was associated with its increasing θ. In the same period, S also exhibited a slight increase that could also be observed in the in situ data (Fig. 13). In contrast, ECCO2 showed warming, freshening and lightening trends for the same section from 1992 to 2004 (Fig. 13) although only the S trend was statistically significant (Table 4).

In WOCE SR4 (Fig. 14), ECMWF and SODA showed cooling and freshening trends (Table 4), with the former unveiling a lightening trend and the latter showing the opposite pattern. In this section, there was a colder and fresher pattern in 2008 that was only captured in ECMWF, but it was intensified in comparison to the observations (Fig. 14). The MyOcean and ECCO2 products showed warming, increasing salinity and lightening trends until 2004 (Table 4), but only the latter reanalysis had significant results. After 2005,

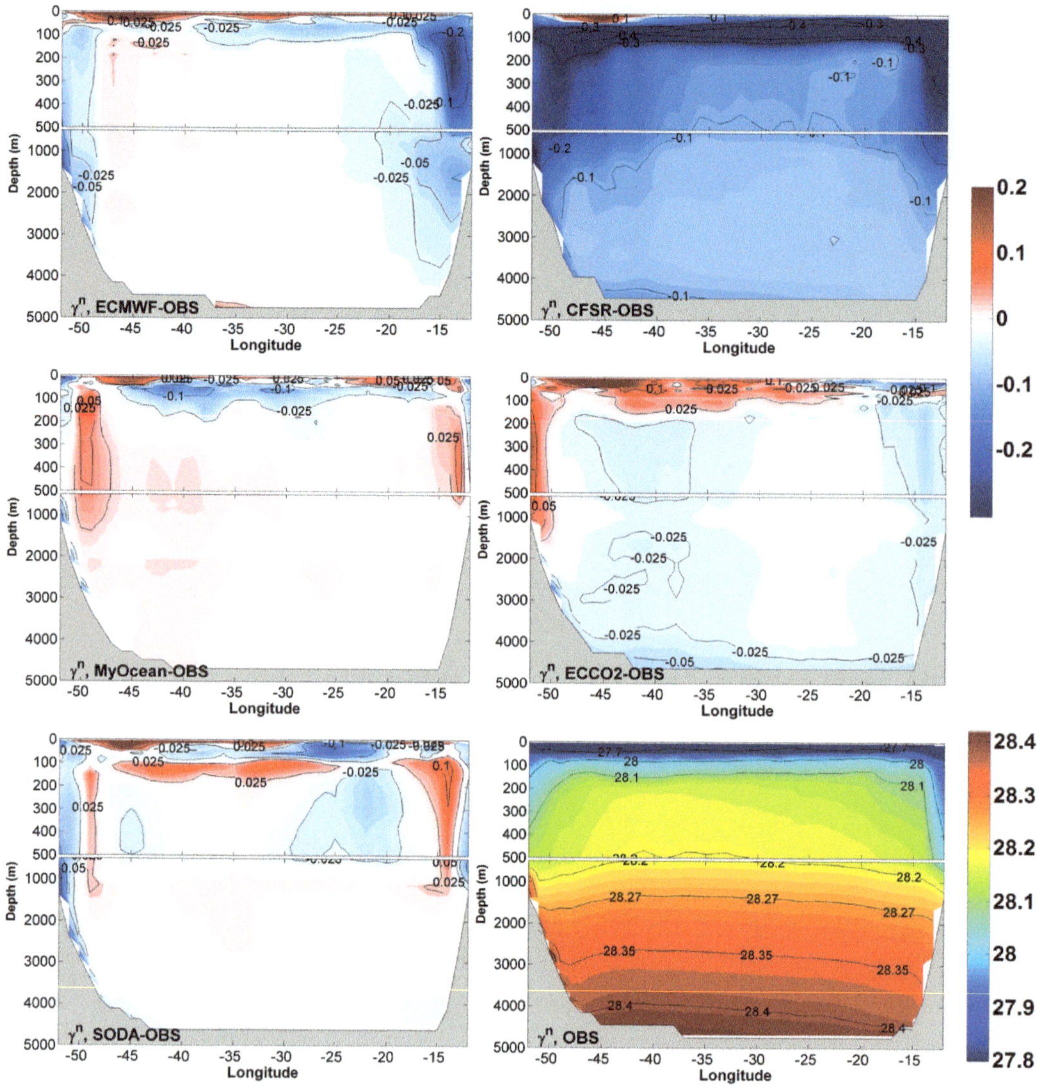

Figure 9. Same as in Fig. 8, but for section WOCE SR4.

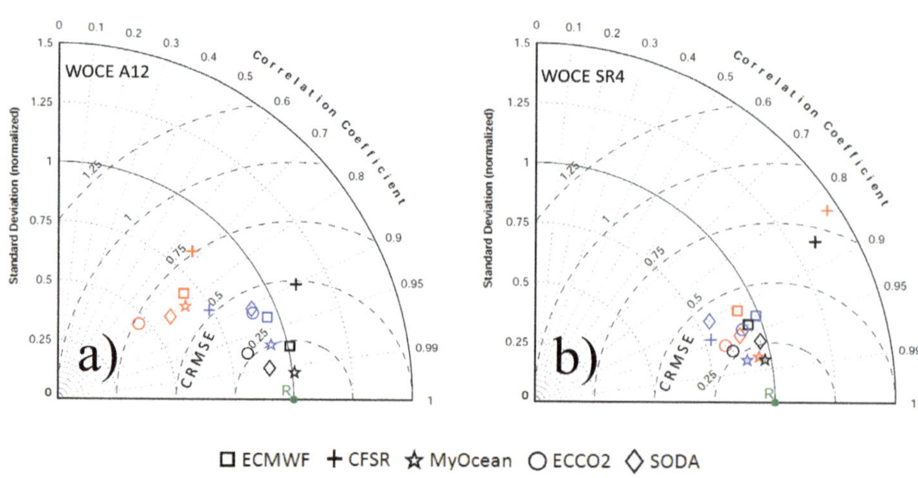

Figure 10. Taylor diagrams of sections WOCE A12 **(a)** and WOCE SR4 **(b)**. θ (blue), S (red), and γ^n (black). ECMWF (square), CFSR (plus sign), MyOcean (pentagram), ECCO2 (circle), and SODA (diamond). Observed data serve as a reference (green R).

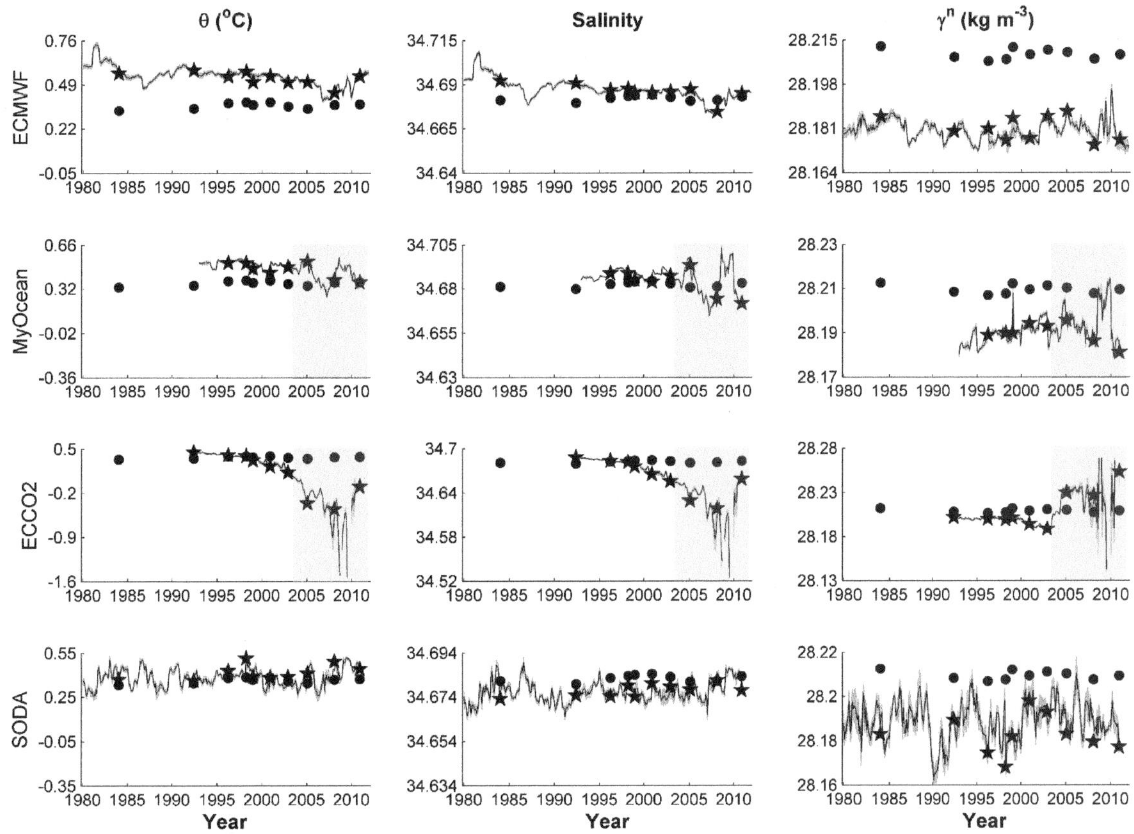

Figure 11. Time series of monthly mean hydrographic properties of WDW in ECMWF, MyOcean, ECCO2, and SODA (top to bottom) averaged along the WOCE A12 section line. From left to right: θ, S, and γ^n. The gray shading indicates the standard deviation due to variation caused by the different station locations in the different years of the hydrographic cruises. The dots indicate the values derived from the observed data, and the pentagrams are the values from the corresponding reanalysis data. Note that the scales are different to show the variability in the time series. The gray rectangles denote the period in which the hydrographic properties showed anomalous variability.

an anomalous cooling and freshening occurred in ECCO2 in response to a polynya that opened in the Weddell Sea (Azaneu et al., 2014), and WSDW increased in density (Figs. 13, 14).

4.3.3 Weddell Sea Bottom Water

For WSBW ($\gamma^n \geq 28.4\,\mathrm{kg\,m^{-3}}$), ECMWF revealed a warming trend beyond the second half of the 1990s (Figs. 15, 16). However, when the entire period (1980–2011) was considered, a cooling trend was observed in both sections (Table 4). In addition, ECMWF exhibited a freshening trend throughout the entire period (Table 4; Figs. 15, 16). In section WOCE SR4 (Fig. 16), ECMWF also modeled 2008 as a year subject to cooling and freshening, which was also observed in the in situ data. The γ^n decreased in both sections throughout the entire period (Table 4) and was clearly observed beyond the 1990s (Figs. 15, 16).

The MyOcean product showed an increasing trend in θ and S and an opposite trend in γ^n in the 1993–2010 period in WOCE A12 (Table 4). Its WSBW annual mean results were similar to the in situ data (Figs. 15, 16). In WOCE SR4 (Fig. 16), there was an anomalous increase in all hydrographic properties after 2010. In the period 1993–2009, the MyOcean reanalysis showed warming and lightning trends (Table 4).

SODA's monthly mean values were clearly warmer and saltier than those observed by in situ data (Figs. 15, 16). In section WOCE A12 (Fig. 15), θ and S exhibited an increasing trend throughout the 1980–2010 period, and a decrease in γ^n was observed (Table 4). However, in section WOCE SR4 (Fig. 16), only S had a statistically significant trend, indicating long-term freshening.

5 Discussion and conclusions

The ocean reanalysis products evaluated here (ECMWF, CFSR, MyOcean, ECCO2 and SODA) have few common characteristics. Instead, their ocean models, spatial resolutions (both horizontal and vertical), assimilation methods, observed data sets being assimilated, couplings with sea-ice models, and physics applied to ocean and sea-ice models all

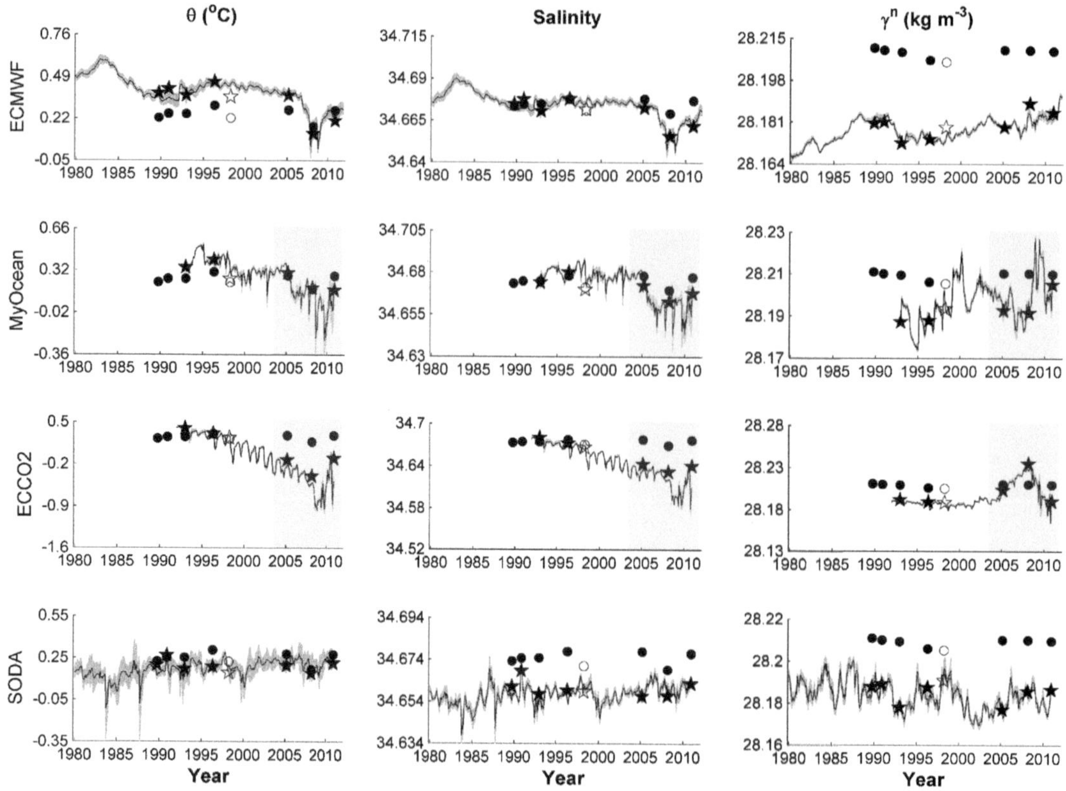

Figure 12. Time series of monthly mean hydrographic properties of WDW in ECMWF, MyOcean, ECCO2, and SODA (top to bottom) averaged along the WOCE SR4 section line. From left to right: θ, S, and γ^n. The gray shading indicates the standard deviation due to variation caused by the different station locations in the different years of the hydrographic cruises (excluding the set of stations of 1998). The dots indicate the values derived from the observed data, and the pentagrams are the values from the corresponding reanalysis data. The filled markers denote full sections, and the open markers denote section 1998, which does not extend over the entire eastern margin of the Weddell Sea. Note that the scales are different to show the variability in the time series. The gray rectangles denote the period in which the hydrographic properties showed anomalous variability.

differ (Table 1). Thus, the evaluation of some features can be represented in distinct ways because the ocean model dynamics can respond to different assimilations procedures, parameterizations, and initial forcing fields (Figs. 2–9). For example, all of the reanalyses represented the inflow/outflow cores of WDW in the WOCE SR4 repeat section, but the core average depths, shapes and spatial extensions differed among all of the products evaluated (not shown). It is important to evaluate the ocean reanalysis systems because if some common biases exist, dynamically complex regions will be difficult to capture in the ocean models.

Moreover, it is important to highlight that the WOCE data set cannot be considered a fully independent test of the reanalysis products because this data set is assimilated during the simulation phase by most of the models. In this sense, it is reasonable that CFSR deviates from the WOCE observations much more than any of the other products (Figs. 2–9). However, in the absence of any independent observations, the WOCE data set remains the best choice for such evaluation because of its comprehensive nature, high resolution and the representativeness of the Weddell Sea water mass structure and spatial distribution.

The ocean surface layer was the location of the major differences among the ocean reanalysis products (Figs. 4–9). A good representation of the surface ocean is vital for climate studies, but even the products coupled with a sea-ice model (i.e., CFSR, MyOcean, and ECCO2) did not correctly represent the surface properties. The errors in the representation of surface water can also be observed in the θ–S diagrams (Figs. 2, 3) and may be consequences of the difficulties faced when reproducing the complex processes acting on the surface ocean, such as the processes and fluxes at the air–sea and ice–ocean interfaces. One exception to this trend in misfit was the MyOcean reanalysis, which provided similar levels of variability in hydrographic properties compared to the observations in both sections, despite the persistence of differences in the absolute values.

In contrast to the representations of the surface layer, the deep-ocean representations deviated less from the observed data in terms of absolute values (Figs. 2–9). In this layer, the ECMWF, MyOcean and SODA products provided the most

Table 4. Linear fits and confidence bounds (95 % confidence) of the annual time series of hydrographic properties. Bold font indicates a statistically significant trend ($P \leq 0.05$). The period considered to determine the trends depends on the water mass in each reanalysis.

			WDW	WSDW	WSBW
WOCE A12	ECMWF	Period of the trend	1980–2011	1987–2007	1980–2011
		θ ($^\circ$C yr^{-1})	**−0.0040** (±0.0016)	**−0.0052** (±0.0021)	**−0.0009** (±0.0003)
		S (yr^{-1})	**−0.0004** (±0.0001)	**−0.0006** (±0.0002)	**−0.0003**(±0.00004)
		γ^n (kg m^{-3} yr^{-1})	−0.0001 (±0.0001)	**+0.0003** (±0.0002)	**−0.0001**(±0.00003)
	MyOcean	Period of the trend	1993–2004	1993–2004	1993–2010
		θ ($^\circ$C yr^{-1})	−0.0033 (±0.0041)	**−0.0022** (±0.0017)	**+0.0022**(±0.0003)
		S (yr^{-1})	−0.0001 (±0.0004)	**−0.0002** (±0.0002)	**+0.0002**(±0.0001)
		γ^n (kg m^{-3} yr^{-1})	**+0.0009** (±0.0005)	**+0.0002** (±0.0002)	**−0.0003**(±0.0001)
	ECCO2	Period of the trend	1992–2004	1992–2004	–
		θ ($^\circ$C yr^{-1})	**−0.0374** (±0.0120)	+0.0003 (±0.0019)	–
		S (yr^{-1})	**−0.0032** (±0.0007)	**−0.0002** (±0.00007)	–
		γ^n (kg m^{-3} yr^{-1})	+0.0002 (±0.0012)	−0.0004 (±0.0004)	–
	SODA	Period of the trend	1980–2010	1980–2010	1980–2010
		θ ($^\circ$C yr^{-1})	+0.0010 (±0.0020)	**−0.0005** (±0.0004)	**+0.0001**(±0.0001)
		S (yr^{-1})	+0.0001 (±0.0002)	−0.00001 (±0.00003)	**+0.00001** (±0.000004)
		γ^n (kg m^{-3} yr^{-1})	+0.0001 (±0.0003)	**+0.0001** (±0.00004)	**−0.00002**(±0.00002)
WOCE SR4	ECMWF	Period of the trend	1980–2011	1980–2011	1980–2011
		θ ($^\circ$C yr^{-1})	**−0.0080** (±0.0026)	−0.0004 (±0.0006)	**−0.0005** (±0.0003)
		S (yr^{-1})	**−0.0006** (±0.0002)	**−0.0001** (±0.00005)	**−0.0002** (±0.00004)
		γ^n (kg m^{-3} yr^{-1})	**+0.0003** (±0.0001)	−0.0001 (±0.0001)	**−0.0002** (±0.00005)
	MyOcean	Period of the trend	1993–2004	1993–2004	1993–2009
		θ ($^\circ$C yr^{-1})	**−0.0127** (±0.0100)	+0.0014 (±0.0051)	**+0.0014** (±0.0010)
		S (yr^{-1})	−0.0002 (±0.0004)	+0.0001 (±0.0004)	−0.00002 (±0.0001)
		γ^n (kg m^{-3} yr^{-1})	**+0.0018** (±0.0012)	−0.00005 (±0.0005)	**−0.0004** (±0.0001)
	ECCO2	Period of the trend	1992–2004	1992–2004	–
		θ ($^\circ$C yr^{-1})	**−0.0516** (±0.0062)	**+0.0049** (±0.0016)	–
		S (yr^{-1})	**−0.0038** (±0.0003)	**+0.0003** (±0.0001)	–
		γ^n (kg m^{-3} yr^{-1})	+0.0001 (±0.0004)	**−0.0008** (±0.0002)	–
	SODA	Period of the trend	1980–2010	1980–2010	1980–2010
		θ ($^\circ$C yr^{-1})	**+0.0041** (±0.0014)	**−0.0003** (±0.0003)	−0.0001 (±0.0001)
		S (yr^{-1})	**+0.0002** (±0.0001)	−0.00001 (±0.00003)	**−0.00001** (±0.00001)
		γ^n (kg m^{-3} yr^{-1})	**−0.0003** (±0.0002)	**+0.0001** (±0.00004)	+0.00001 (±0.00001)

accurate absolute values among all of the reanalyses evaluated. Considering the hydrographic properties analyzed, γ^n was best represented for all of the reanalyses, which reflects the fact that θ and S can compensate for each other to better represent the deep layers in ocean reanalysis products (Fig. 10). Kerr et al. (2012b) reported a good representation of the deep ocean structure and water mass contribution in the Weddell Sea and Weddell–Scotia Confluence through an investigation of the earlier version of the SODA product (version 1.4.2). The same authors reported that SODA version 1.4.2 represented the S field for the deep ocean poorly. The SODA product version 2.2.4 analyzed here showed an improved S field due to some modifications from the previous version (e.g., an increase in assimilated salinity data).

The horizontal resolution among the ocean reanalyses evaluated here varied from 1° (e.g., ECMWF) to 1/4° (e.g., ECCO2 and MyOcean; Table 1). Although increased horizontal resolution is important to better simulate hydrographic representations, we highlight that improvements in parameterization, such as advection schemes and subgrid-scale mixing processes, may have as much impact on the reanalyses as modest increases in horizontal resolution (e.g., Renner et al., 2009). For example, for the deep layers, the ECCO2 product (1/4°) had greater differences in absolute values (in relation to the in situ data) than the ECMWF product (1°; Figs. 2–9). One must examine the model biases (because the deep ocean layers are normally poorly sampled) and assimilation methods used. Furthermore, as reported by Dee (2005), all data

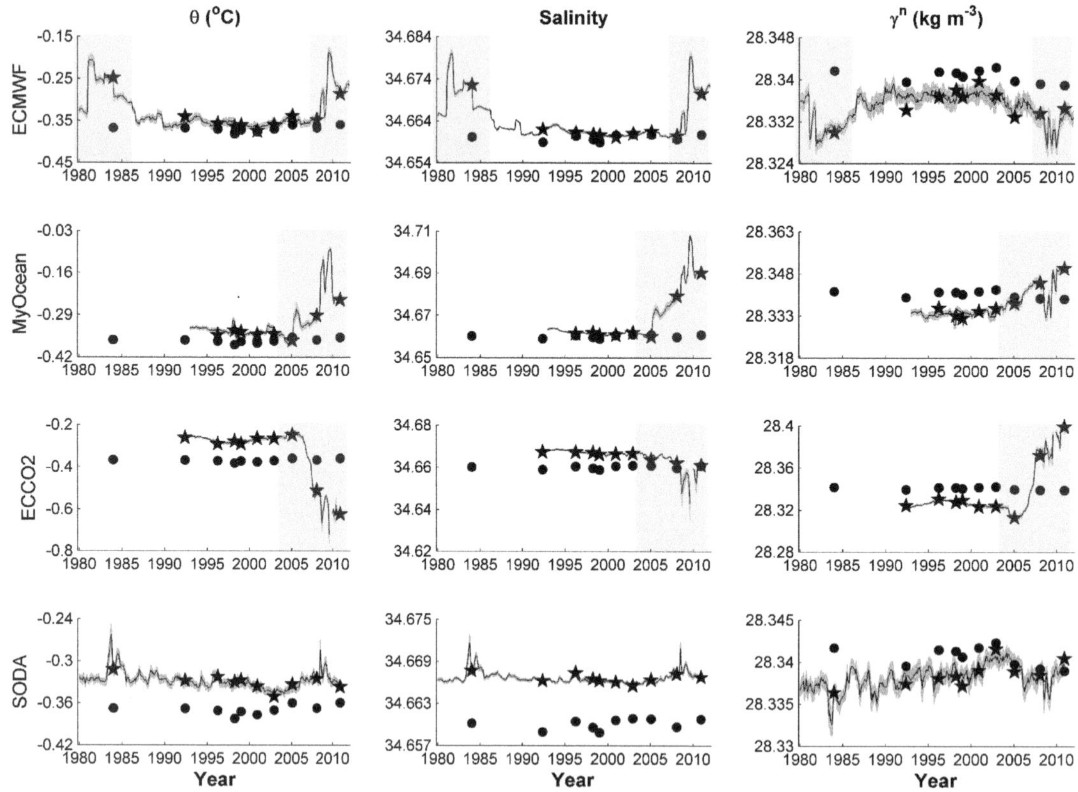

Figure 13. Same as in Fig. 11, but for WSDW.

assimilation systems are affected by systematic errors associated with the following: (i) problems with input data, (ii) approximations relative to the in situ observations, (iii) limitations of the assimilating models, and (iv) the assimilation methodology itself. Because these errors are intrinsic to each reanalysis, our validation could have been biased by at least one of these points. Moreover, comprehensive data quality control before data assimilation is an essential step in assessing reanalysis quality. In fact, increasing the deep-ocean observations available for assimilation by ocean reanalysis products is required to better represent this region of the oceans.

Three of the five reanalysis products evaluated here were coupled with sea-ice models (i.e., CFSR, MyOcean, and ECCO2). Coupling with a sea-ice model is essential for reproducing the deep-water properties in ocean circulation models (e.g., Kerr et al., 2009b) because both dynamic and thermodynamic sea-ice processes play significant roles in the Southern Ocean's climate variability and bottom water formation (e.g., Jacobs and Comiso, 1989; Venegas and Drinkwater, 2001). Although CFSR and ECCO2 both contained sea-ice models, they portrayed deep water masses as being warmer than what was provided in the observations, and CFSR also showed fresher waters at the deep layer (Figs. 2–9). Neither of these two reanalysis systems was able to properly reproduce the WSBW layer ($\gamma^n \geq 28.40\,\mathrm{kg\,m^{-3}}$). However, ECCO2 represented the spatial variability and water mass distribution well with respect to the in situ data (Fig. 10). Conversely, the SODA and ECMWF reanalyses represented absolute values of θ, S and γ^n of the deep waters that were close to observations despite the absence of a suitably coupled sea-ice model. This result highlights the fact that surface data assimilation in those products are responding satisfactorily to represent the processes and exchanges at the air–sea interface.

None of the reanalyses represented the downslope flow of dense water in the western slope of section WOCE SR4 (Figs. 5, 7, 9). Those limitations are expected for z level models (Winton et al., 1998), which could lead to excessive diapycnal mixing and poor representation of downslope flows (Willebrand et al., 2001). A simple way to improve the representation of denser varieties of AABW is to use certain procedures to directly inject dense water from the continental shelf to the deep ocean (e.g., Briegleb et al., 2010). Recently, Heuzé et al. (2013) showed that the process of AABW formation was not represented accurately in climate models, leading to extensive areas of deep ocean convection. Kerr et al. (2012a) investigated a high-resolution (1/12°) simulation of the OCCAM model and noted that despite the absence of a dense shelf break and slope plume in the model, the presence of deep ocean convection could explain the relatively good AABW export rates to the global ocean from

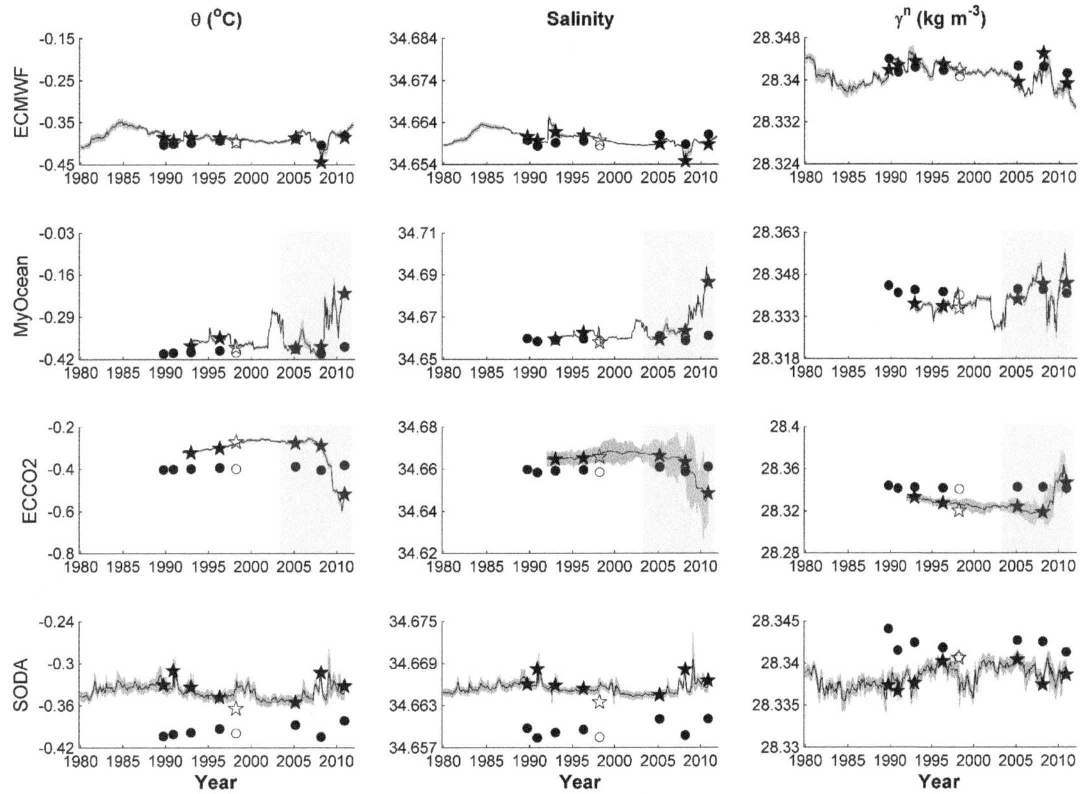

Figure 14. Same as in Fig. 12, but for WSDW.

the Weddell Sea. Heuzé et al. (2013) suggested that a super-parameterization scheme, perhaps based on a high-resolution isopycnal model, might improve the downslope-flow representation. In addition, the use of a low-vertical-resolution model for deep ocean layers is not effective in representing water masses such as WSBW, which is less than 1 km thick (e.g., Fahrbach et al., 2001; Kerr et al., 2009a). Thus, increasing the vertical resolution of the deep and bottom layers in ocean models, which is frequently performed for the surface ocean, could be further considered to more consistently represent the deep ocean structure and make future ocean models more accurate.

Adding ice shelves in a coupled sea ice–ocean model improves the simulation of the sea-ice cover and alters the hydrography in the Weddell Sea with global effects, as shown by Hellmer (2004) and Wang and Beckmann (2007). Kerr et al. (2009b) and Renner et al. (2009) also noted the need for adequate sea-ice models and the inclusion of ice-shelf processes to improve simulations of global ocean circulation models. More recently, Meccia et al. (2013), using a regional ocean model, showed that the representation of Ice Shelf Water was improved in their experiment that included ice-shelf thermodynamic parameterization in the Weddell Sea. None of the reanalyses discussed here included the ice shelves in their simulations, even though such shelves are key elements in the formation of Ice Shelf Water – water masses that are directly involved in the formation of WSBW (Foldvik et al., 1985). MyOcean simulations included the effects of the Antarctica ice sheet melt in its oceanic model, and this inclusion may have improved the representation of its surface salinity absolute values (Figs. 6, 7). Ice-shelf effects on the ocean structure could be inserted in ocean reanalysis products if observations near those areas became available. The inclusion of ice shelves in models is a factor that must be strongly considered for the optimization of deep-ocean representation in future reanalysis results, but more observations under permanent ice shelves would also most likely result in better ocean reanalysis outputs in the Weddell Sea.

Ocean reanalysis products are powerful tools for climate studies because of their generally high-temporal resolution. Thus, the horizontal and vertical spatial-average representations and their temporal variability should be assessed. Over the last decade, several studies have highlighted the variability and trends in the hydrographic properties of the Weddell Sea. The most prominent trend is the WDW warming from the 1970s to the 2000s (Robertson et al., 2002; Smedsrud, 2005; Fahrbach et al., 2004, 2011). Considering the products investigated here, only the SODA reanalysis showed a statistically significant WDW warming trend ($+0.0041\,^{\circ}\mathrm{C\,yr^{-1}}$; Table 4) in the WOCE SR4 section during the 1980–2010 period. This trend found by the SODA product was less than that observed at the WDW inflow in the

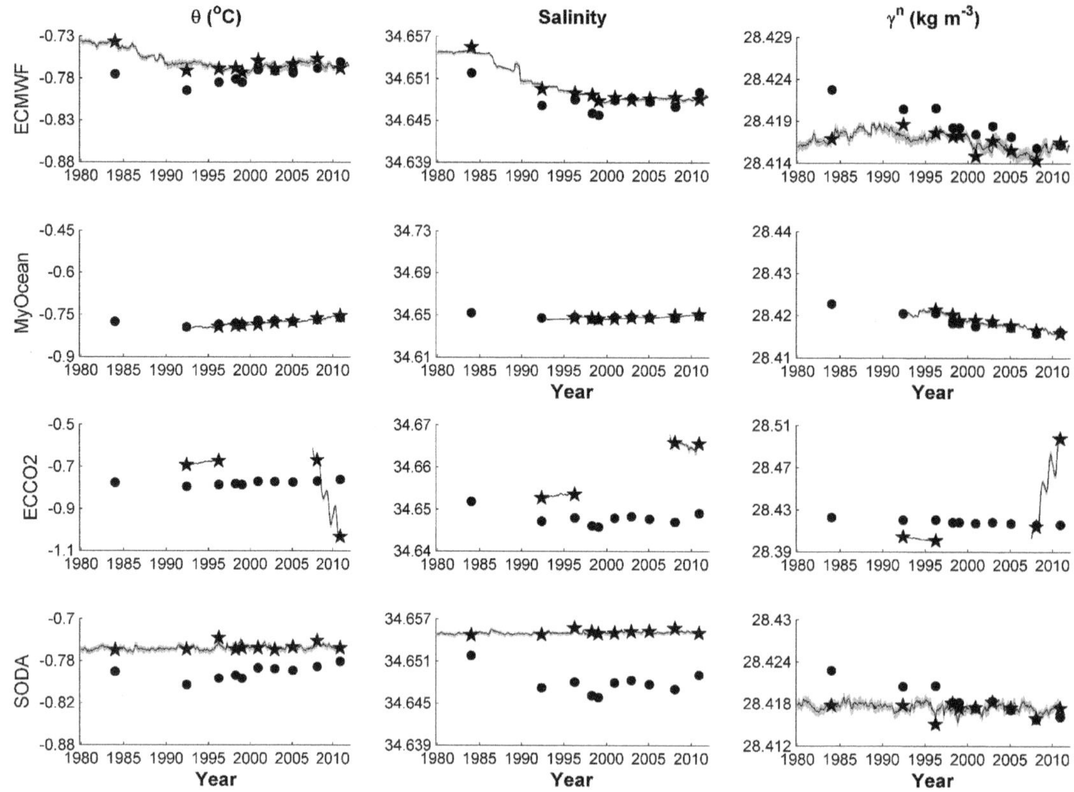

Figure 15. Same as in Fig. 11, but for WSBW.

Weddell Sea from 1971 to 2000 (+0.012 °C yr^{-1}; Robertson et al., 2002), which could be associated with the evaluation of the entire section here. In contrast, the ECMWF, MyOcean and ECCO2 products showed statistically significant cooling trends (Table 4) in both sections for the periods 1980–2011, 1993–2004 and 1992–2004, respectively. Following the cooling shown in ECMWF, MyOcean, and ECCO2, the products showed a freshening trend, whereas in SODA, S increased with time (Table 4). According to observed data, WDW freshened during the 1971–2000 period (see, for instance, Fig. 9 from Robertson et al., 2002). However, sparse temporal data showed an increase in salinity between the 1980s and the 2000s (Fahrbach et al., 2004, 2011). In association with the θ trends, γ^n showed an increasing trend in ECMWF and MyOcean (and a decreasing trend in SODA). Robertson et al. (2002) clearly showed that WDW density decreased during the period 1971–2000.

Fahrbach et al. (2011) reported warming and increasing salinity trends for the WSDW layer between the 1980s and 2000s in section WOCE A12. In this context, all reanalysis products that had statistically significant trends showed cooling and freshening trends for this section, which were associated with an increase in density (Table 4). This discrepancy may have been caused by the different criteria used to define the AABW varieties. Robertson et al. (2002) also found a warming trend from the 1970s to 1990s in the WOCE SR4 section, although their results were not statistically significant given the interannual variability of their data. In this section, a warming trend (+0.0049 °C yr^{-1}) was reproduced by ECCO2 (1992–2004) followed by an increase in salinity (+0.0003 yr^{-1}). The MyOcean reanalysis (1993–2004) also showed an increase in θ and S, but none of the properties considered were statistically significant (Table 4). All reanalyses examined showed a γ^n decreasing trend for WOCE SR4 (except for the SODA product; Table 4). AABW observations showed a warming trend for the global AABW exported from the Southern Ocean and a reduction in its volume (e.g., Purkey and Johnson, 2010, 2012; Azaneu et al., 2013). The latter may be a consequence of the downward trend in AABW density, although no sign of freshening has been found in the inner Weddell Sea (Azaneu et al., 2013). Recently, Jullion et al. (2013) reported a significant freshening of AABW of −0.004 decade^{-1} in the Drake Passage, with no significant decrease in its thickness. In section WOCE A12, the reanalyses showed a freshening trend in WSDW of the same order as that found by Jullion et al. (2013) and Azaneu et al. (2013) for AABW. In section WOCE SR4, only ECMWF showed a freshening trend that corresponded with the one found by Azaneu et al. (2013, Table 4).

Throughout the Weddell Sea, bottom waters have been warming with no clear change in salinity (e.g., Robertson et al., 2002; Fahrbach et al., 2004, 2011; Purkey and Johnson,

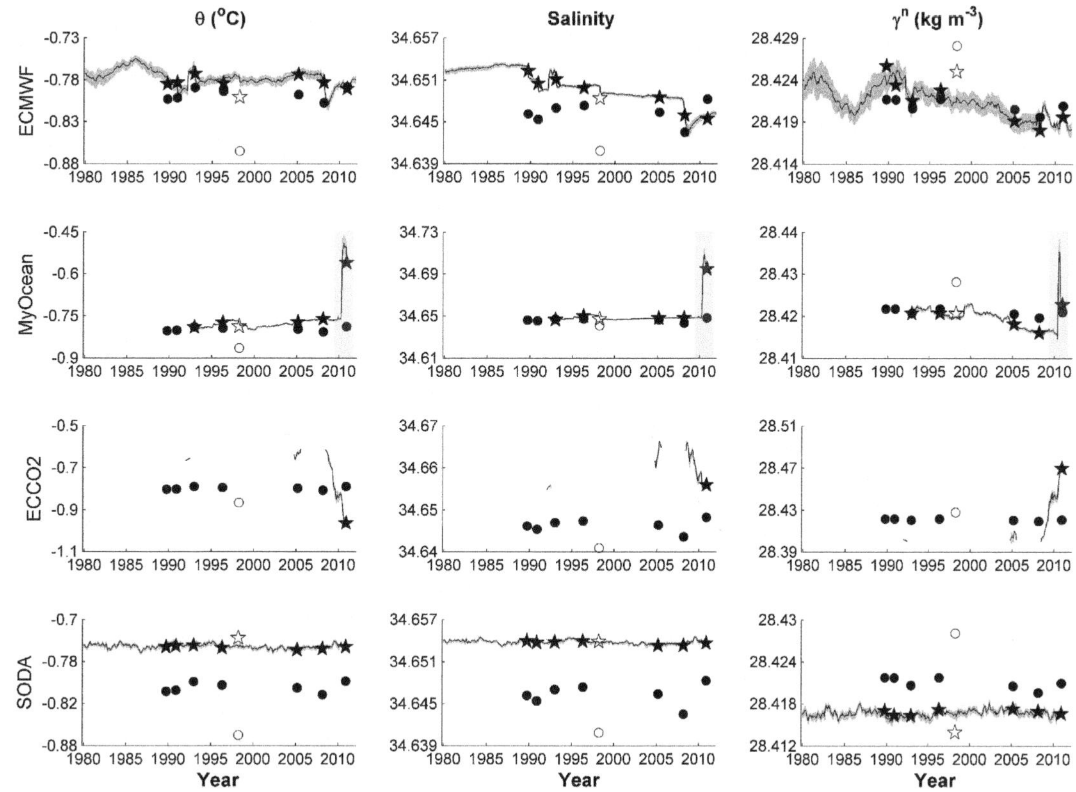

Figure 16. Same as in Fig. 12, but for WSBW.

2010; Azaneu et al., 2013). Considering all statistically significant results of the ocean reanalysis time series, the density of WSBW is declining, but the causes are not clear because both cooling/freshening and warming/increasing salinity were reproduced (Table 4). The ocean reanalyses that showed a warming trend (MyOcean and SODA) also presented an increase in salinity, while ECMWF showed cooling and freshening trends (Table 4). A freshening trend of the WSBW would be expected as a result of shelf water freshening (e.g., Hellmer et al., 2011; Azaneu et al., 2013) because this bottom water results from shelf water mixing. However, clear salinity changes have not yet been observed in the deep layers of the Weddell Sea (Azaneu et al., 2013); the causes are not yet clearly identified but could be the result of the opposing effects of source waters and dynamic processes masking signal identification.

In general, it appears that all of the reanalysis products require considerable development to improve their representation of the Weddell Sea water masses, particularly in their representation of dense-water production and overflows. The choice of the reanalysis product that best reproduces the main regional oceanographic patterns depends upon one's particular interest. For example, if the primary interest is in describing prominent oceanographic features, e.g., water mass distribution and thermohaline values, MyOcean produces results closest to the observational data; thus, MyOcean is considered the best product for this purpose. If trends and temporal variability in the water masses are the major motivations, the choice of reanalysis product is dependent on the water mass evaluated. For WDW, SODA captured the warming, the salinity increase and the density decrease patterns that were found in the observations. For WSDW, the ECCO2 results were closer to the observational estimates, reproducing the freshening trends in WOCE SR4 and the warming in WOCE SR4 and WOCE A12. Finally, for WSBW, MyOcean and SODA were able to capture the warming pattern recorded in the data set, although the simulated salinity increase in WOCE A12 and decrease in WOCE SR4 are not clear in the real ocean. Caution is advised regarding the anomalous periods that were seen in most of the reanalyses that were assessed.

Overall, our results suggest that the MyOcean and SODA products are valuable tools for studying the states of deep layers in the Weddell Sea because these tools closely reproduce the hydrographic absolute values of the observational data and capture the main regional features. More effort is needed to address surface layers because several complex processes acting on the ocean surface – e.g., atmosphere–ocean–cryosphere interactions – may not be correctly reproduced, causing large differences in absolute values. A good representation of the surface layer is also vital for the representation of the deep layers because deep water masses are dependent on the thermohaline characteristics of surface water masses (Foster and Carmack, 1976) that result from these atmosphere–ocean–cryosphere interactions (Whitworth et al., 1998). The variability and trends represented by ocean reanalyses may still have some biases because the Southern Ocean suffers from a lack of in situ data and is biased by summer observations. A better representation of ocean features and hydrographic properties by ocean reanalysis will be useful for long-term studies in polar regions to better understand the connections between ocean variability and possible implications for the global climate.

Acknowledgements. We thank the research groups involved in each ocean reanalysis system for making their simulations freely available. ECMWF ORAS4 was provided by ECMWF (http://www.ecmwf.int/); CFSR was provided by the Environmental Modeling Center, National Centers for Environmental Prediction, National Weather Service, NOAA, US Department of Commerce from its website http://rda.ucar.edu/datasets/ds093.2; MyOcean UR025.4 was provided by MyOcean Products from its website http://www.myocean.eu/; ECCO2 was provided by the ECCO2/MAP/NASA (http://ecco2.jpl.nasa.gov/); and SODA was provided from the SODA program (http://www.atmos.umd.edu/~ocean/). T. S. Dotto and M. Azaneu acknowledge financial support from the CAPES Foundation. This study is a contribution to the activities of the Brazilian High Latitudes Oceanography Group (GOAL; www.goal.furg.br) and the Brazilian National Institute of Science and Technology of Cryosphere (INCT-CRIOSFERA; 573720/2008-8). GOAL has been funded by the Brazilian Antarctic Program (PROANTAR), the Brazilian Ministry of the Environment (MMA), the Brazilian Ministry of Science, Technology and Innovation (MCTI), the Council for Research and Scientific Development of Brazil (CNPq grants no. 520189/2006-0; 556848/2009-8), and grant 2013/02111-4 from the Sao Paulo Research Foundation (FAPESP). We acknowledge two anonymous reviewers for their constructive comments, which greatly improved the manuscript. This work is dedicated to the memory of Eberhard Fahrbach, who died on 21 April 2013. Without his vision and efforts, this most valuable data set of the Weddell Sea covering more than two decades would not exist.

Edited by: A. Schiller

References

Azaneu, M., Kerr, R., Mata, M. M., and Garcia, C. A. E.: Trends in the deep Southern Ocean (1958–2010): Implications for Antarctic bottom water properties and volume export, J. Geophys. Res., 118, 4213–4227, doi:10.1002/jgrc.20303, 2013.

Azaneu, M., Kerr, R., and Mata, M. M.: Assessment of the ECCO2 reanalysis on the representation of Antarctic Bottom Water properties, Ocean Sci. Discuss., 11, 1023–1091, doi:10.5194/osd-11-1023-2014, 2014.

Balmaseda, M. A., Mogensen, K., and Weaver, A. T.: Evaluation of the ECMWF ocean reanalysis system ORAS4, Q. J. Roy. Meteorol. Soc., 139, 1132–1161, doi:10.1002/qj.2063, 2013.

Briegleb, B., Danabasoglu, G., and Large, W.: An overflow parameterization for the ocean component of the Community Climate System Model, Tech. Note NCAR/TN-481+ STR, 84 pp., 2010.

Bromwich, D. H., Nicolas, J. P., and Monaghan, A. J.: An assessment of precipitation changes over Antarctica and the Southern Ocean since 1989 in contemporary global reanalyses, J. Climate, 24, 4189–4209, doi:10.1175/2011JCLI4074.1, 2011.

Carmack, E. C. and Foster, T. D.: On the flow of water out of the Weddell Sea, Deep Sea Res., 22, 711–724, doi:10.1016/0011-7471(75)90077-7, 1975.

Carton, J. A. and Giese, B. S.: A Reanalysis of Ocean Climate Using Simple Ocean Data Assimilation (SODA), Mon. Weather Rev., 136, 2999–3017, doi:10.1175/2007MWR1978.1, 2008.

Carton, J. A. and Santorelli, A.: Global decadal upper-ocean heat content as viewed in nine analyses, J. Climate, 21, 6015–6035, doi:10.1175/2008JCLI2489.1, 2008.

Compo, G. P., Whitaker, J. S., Sardeshmukh, P. D., Matsui, N., Allan, R. J., Yin, X., Gleason, B. E., Vose, R. S., Rutledge, G., Bessemoulin, P., Brönnimann, S., Brunet, M., Crouthamel, R. I., Grant, A. N., Groisman, P. Y., Jones, P. D., Kruk, M. C., Kruger, A. C., Marshall, G. J., Maugeri, M., Mok, H. Y., Nordli, Ø., Ross, T. F., Trigo, R. M., Wang, X. L., Woodruff, S. D., and Worley, S. J.: The twentieth century reanalysis project, Q. J. Roy. Meteorol. Soc., 137, 1–28, doi:10.1002/qj.776, 2011.

Dee, D. P.: Bias and data assimilation, Q. J. Roy. Meteorol. Soc., 131, 3323–3343, doi:10.1256/qj.05.137, 2005.

Dee, D. P., Uppala, S. M., Simmons, A. J., Berrisford, P., Poli, P., Kobayashi, S., Andrae, U., Balmaseda, M. A., Balsamo, G., Bauer, P., Bechtold, P., Beljaars, A. C. M., van de Berg, L., Bidlot, J., Bormann, N., Delsol, C., Dragani, R., Fuentes, M., Geer, A. J., Haimberger, L., Healy, S. B., Hersbach, H., Hólm, E. V., Isaksen, L., Kållberg, P., Köhler, M., Matricardi, M., McNally, A. P., Monge-Sanz, B. M., Morcrette, J.-J., Park, B.-K., Peubey, C., de Rosnay, P., Tavolato, C., Thépaut, J.-N., and Vitart, F.: The ERA Interim reanalysis: Configuration and performance of the data assimilation system, Q. J. Roy. Meteorol. Soc., 137, 553–597, doi:10.1002/qj.828, 2011.

Fahrbach, E. and De Baar, H.: The Expedition of the Research Vessel Polarstern to the Antarctic in 2008 (ANT-XXIV/3). Berichtezur Polar- und Meeresforschung, Vol. 606, Alfred-Wegener-Institut, Bremerhaven, 228 pp., 2010.

Fahrbach, E., Rohardt, G., Scheele, N., Schröder, M., Strass, V., and Wisotzki, A.: Formation and discharge of deep and bottom water in the northwestern Weddell Sea, J. Marine Res., 53, 515–538, doi:10.1357/0022240953213089, 1995.

Fahrbach, E., Harms, S., Rohardt, G., Schröder, M., and Woodgate, R. A.: Flow of bottom water in the northwestern Weddell Sea, J. Geophys. Res., 106, 2761–2778, doi:10.1029/2000JC900142, 2001.

Fahrbach, E., Hoppema, M., Rohardt, G., Schröder, M., and Wisotzki, A.: Decadal-scale variations of water mass properties in the deep Weddell Sea, Ocean Dynam., 54, 77–91, doi:10.1007/s10236-003-0082-3, 2004.

Fahrbach, E., Rohardt, G., and Sieger, R.: 25 Years of Polarstern hydrography (1982–2007), WDC-MARE Reports 5, Alfred-Wegener-Institut, Bremerhaven, 94 pp., 2007.

Fahrbach, E., Hoppema, M., Rohardt, G., Boebel, O., Klatt, O., and Wisotzki, A.: Warming of deep and abyssal water masses along the Greenwich meridian on decadal time scales: The Weddell gyre as a heat buffer, Deep Sea Res. Pt. II, 58, 2509–2523, doi:10.1016/j.dsr2.2011.06.007, 2011.

Ferry, N., Barnier, B., Garric, G., Haines, K., Masina, S., Parent, L., Storto, A., Valdivieso, M., Guinehut, S., and Mulet, S.: Nemo: The modelling engine of global ocean reanalysis, Mercator Ocean Quarterly Newsletter, 46, 46–59, 2012.

Fichefet, T. and Morales-Maqueda, M. A.: Sensitivity of a global sea ice model to the treatment of ice thermodynamics and dynamics, J. Geophys. Res., 102, 12609–12646, doi:10.1029/97JC00480, 1997.

Foldvik, A., Gammelsrød, T., and Tørrensen, T.: Circulations and Water Masses on the Southern Weddell Sea Shelf, Antarctic Research Series, 43, 5–20, 1985.

Foster, T. and Carmack, E.: Temperature and salinity structure in the Weddell Sea, J. Phys. Oceanogr., 6, 36–44, doi:10.1175/1520-0485(1976)006<0036:TASSIT>2.0.CO;2, 1976.

Franco, B., Mata, M. M., Piola, A., and Garcia, C. A. E.: Northwestern Weddell Sea deep outflow into the Scotia Sea during the austral summers of 2000 and 2001 estimated by inverse methods, Deep Sea Res. Pt. I, 54, 1815–1840, doi:10.1016/j.dsr.2007.06.003, 2007.

Giese, B. S. and Ray, S.: El Niño variability in simple ocean data assimilation (SODA), 1871–2008, J. Geophys. Res., 116, C02024, doi:10.1029/2010JC006695, 2011.

Gouretski, V. and Danilov, A.: Weddell Gyre: structure of the eastern boundary, Deep Sea Res. Pt. I, 40, 561–582, doi:10.1016/0967-0637(93)90146-T, 1993.

Griffies, S. M., Harrison, M. J., Pacanowski, R. C., and Rosati, A.: A Technical Guide to MOM4, GFDL Ocean Group Technical Report No. 5, NOAA/Geophysical Fluid Dynamics Laboratory, available at: www.gfdl.noaa.gov (last access: 12 February 2014), 2008.

Hellmer, H. H.: Impact of Antarctic ice shelf basal melting on sea ice and deep ocean properties, Geophys. Res. Lett., 31, L10307, doi:10.1029/2004GL019506, 2004.

Hellmer, H. H., Schodlok, M. P., Wenzel, M., and Schröter, J. G.: On the influence of adequate Weddell Sea characteristics in a large-scale global ocean circulation model, Ocean Dynam., 55, 88–99, doi:10.1007/s10236-005-0112-4, 2005.

Hellmer, H. H., Huhn, O., Gomis, D., and Timmermann, R.: On the freshening of the northwestern Weddell Sea continental shelf, Ocean Sci., 7, 305–316, doi:10.5194/os-7-305-2011, 2011.

Heuzé, C., Heywood, K. J., Stevens, D. P., and Ridley, J. K.: Southern Ocean bottom water characteristics in CMIP5 models, Geophys. Res. Lett., 40, 1–6, doi:10.1002/grl.50287, 2013.

Holland, D. M.: Transient sea-ice polynya forced by oceanic flow variability, Progr. Oceanogr., 48, 403–460, doi:10.1016/S0079-6611(01)00010-6, 2001.

Huhn, O., Hellmer, H. H., Rhein, M., Rodehacke, C. B., Roether, W., Schodlok, M. P., and Schroder, M.: Evidence of deep and bottom water formation in the western Weddell Sea, Deep Sea Res. Pt. II, 55, 1098–1116, doi:10.1016/j.dsr2.2007.12.015, 2008.

Huhn, O., Hellmer, H. H., Rhein, M., Rodehacke, C. B., Roether, W., Schodlok, M. P., and Schröder, M.: Evidence of deep and bottom water formation in the western Weddell Sea, Deep Sea Res. Pt. II, 55, 1098–1116, doi:10.1016/j.dsr2.2007.12.015, 2008.

Jackett, D. R. and McDougall, T. J.: A neutral density variable for the world's oceans, J. Phys. Oceanogr., 27, 237–263, doi:10.1175/1520-0485(1997)027<0237:ANDVFT>2.0.CO;2, 1997.

Jacobs, S. S. and Comiso, J. C.: Sea ice and oceanic processes on the Ross Sea continental shelf, J. Geophys. Res., 94, 18195, doi:10.1029/JC094iC12p18195, 1989.

Jullion, L., Naveira Garabato, A. C., Meredith, M. P., Holland, P. R., Courtois, P., and King, B. A.: Decadal freshening of the Antarctic Bottom Water exported from the Weddell Sea, J. Climate, 26, 8111–8125, doi:10.1175/JCLI-D-12-00765.1, 2013.

Kerr, R., Mata, M. M., and Garcia, C. A. E.: On the temporal variability of the Weddell Sea Deep Water masses, Antarctic Science, 21, 383–400, doi:10.1017/S0954102009001990, 2009a.

Kerr, R., Wainer, I., and Mata, M. M.: Representation of the Weddell Sea deep water masses in the ocean component of the NCAR-CCSM model, Antarctic Sci., 21, 301–312, 2009b.

Kerr, R., Heywood, K. J., Mata, M. M., and Garcia, C. A. E.: On the outflow of dense water from the Weddell and Ross Seas in OCCAM model, Ocean Sci., 8, 369–388, doi:10.5194/os-8-369-2012, 2012a.

Kerr, R., Wainer, I., Mata, M. M., and Garcia, C. A. E.: Quantifying Antarctic deep waters in SODA reanalysis product, Pesquisa Antártica Brasileira, 5, 47–59, 2012b.

Klatt, O., Fahrbach, E., Hoppema, M., and Rohardt, G.: The transport of the Weddell Gyre across the Prime Meridian, Deep Sea Res. Pt. II, 52, 513–528, doi:10.1016/j.dsr2.2004.12.015, 2005.

Large, W. G. and Yeager, S. G.: The global climatology of an interannually varying air–sea flux data set, Clim. Dynam., 33, 341–364, doi:10.1007/s00382-008-0441-3, 2009.

Lumpkin, R. and Speer, K.: Global Ocean Meridional Overturning, J. Phys. Oceanogr., 37, 2550–2562, doi:10.1175/JPO3130.1, 2007.

Madec, G.: NEMO ocean engine, Note du Pôle de modélisation de l'Institut Pierre-Simon Laplace, 27, 217 pp., 2008.

Marshall, J., Adcroft, A., Hill, C., Perelman, L., and Heisey, C.: A finite-volume, ncompressible Navier Stokes model for studies of the ocean on parallel computers, J. Geophys. Res., 102, 5753–5766, doi:10.1029/96JC02775, 1997.

Meccia, V., Wainer, I., Tonelli, M., and Curchitser, E.: Coupling a thermodynamically active ice shelf to a regional simulation of the Weddell Sea, Geosci. Model Dev., 6, 1209–1219, doi:10.5194/gmd-6-1209-2013, 2013.

Menemenlis, D., Fukumori, I., and Lee, T.: Using Green's functions to calibrate an ocean general circulation model, Mon. Weather Rev., 133, 1224–1240, doi:10.1175/MWR2912.1, 2005.

Menemenlis, D., Campin, J.M., Heimbach, P., Hill, C., Lee, T., Nguyen, A., Schodlok, M., and Zhang, H.: ECCO2: High Resolution Global Ocean and Sea Ice Data Synthesis, Mercator Ocean Quarterly Newsletter, 31, 13–21, 2008.

Mogensen, K., Alonso, M., and Weaver, A.: The NEMOVAR ocean data assimilation system as implemented in the ECMWF ocean analysis for System 4, Technical Memorandum, 668, 59 pp., 2012.

Naveira Garabato, A. C., McDonagh, E. L., Stevens, D. P., Heywood, K. J., and Sanders, R. J.: On the export of Antarctic Bottom Water from the Weddell Sea, Deep Sea Res. Pt. II, 49, 4715–4742, doi:10.1016/S0967-0645(02)00156-X, 2002.

Nicholls, K. W., Østerhus, S., Makinson, K., Gammelsrød, T., and Fahrbach, E.: Ice-Ocean processes ocer the continental shelf of the southern Weddell Sea, Antarctica: A review, Rev. Geophys., 47, RG3003, doi:10.1029/2007RG000250, 2009.

Ohshima, K. I., Fukamachi, Y., Williams, G. D., Nihashi, S., Roquet, F., Kitade, Y., Tamura, T., Hirano, D., Herraiz-Borreguero, L., Field, I., Hindell, M., Aoki, S., and Wakatsuchi, M.: Antarctic Bottom Water production by intense sea-ice formation in the Cape Darnley polynya, Nat. Geosci., 6, 235–240, doi:10.1038/ngeo1738, 2013.

Onogi, K., Tsutsui, J., Koide, H., Sakamoto, M., Kobayashi, S., Hatsushika, H., Matsumoto, T., Yamazaki, N., Kamahori, H., Takahashi, K., Kodokura, S., Wada, K., Kato, K., Oyama, R., Ose, T., Mannoji, N., and Taira, R.: The JRA-25 reanalysis, J. Meteorol. Soc. Jpn., 85, 369–432, 2007.

Orsi, A. H., Nowlin Jr., W. D., and Whitworth III, T.: On the circulation and stratification of the Weddell Gyre, Deep Sea Res. Pt. I, 40, 169–203, doi:10.1016/0967-0637(93)90060-G, 1993.

Orsi, A. H., Whitworth III, T., and Nowlin Jr., W. D.: On the meridional extent and fronts of the Antarctic Circumpolar Current, Deep Sea Res. I, 42, 641–673, doi:10.1016/0967-0637(95)00021-W, 1995.

Orsi, A. H., Johnson, G. C., and Bullister, J. L.: Circulation, mixing, and production of Antarctic Bottom Water, Progr. Oceanogr., 43, 55–109, doi:10.1016/S0079-6611(99)00004-X, 1999.

Purkey, S. G. and Johnson, G. C.: Warming of Global Abyssal and Deep Southern Ocean Waters between the 1990s and 2000s: Contributions to Global Heat and Sea Level Rise Budgets, J. Climate, 23, 6336–6351, doi:10.1175/2010JCLI3682.1, 2010.

Purkey, S. G. and Johnson, G. C.: Global Contraction of Antarctic Bottom Water between the 1980s and 2000s, J. Climate, 25, 5830–5844, doi:10.1175/JCLI-D-11-00612.1, 2012.

Rayner, N. A., Parker, D. E., Horton, E. B., Folland, C. K., Alexander, L. V., Rowell, D. P., Kent, E. C., and Kaplan, A.: Global analyses of sea surface temperature, sea ice, and night marine air temperature since the late nineteenth century, J. Geophys. Res., 108, 4407, doi:10.1029/2002JD002670, 2003.

Renner, A. H. H., Heywood, K. J., and Thorpe, S. E.: Validation of three global ocean models in the Weddell Sea, Ocean Modell., 30, 1–15, doi:10.1016/j.ocemod.2009.05.007, 2009.

Reynolds, R. W., Rayner, N. A., Smith, T. M., Stokes, D. C., and Wang, W.: An improved in situ and satellite SST analysis for climate, J. Climate, 15, 1609–1625, doi:10.1175/1520-0442(2002)015<1609:AIISAS>2.0.CO;2, 2002.

Reynolds, R. W., Smith, T. M., Liu, C., Chelton, D. B., Casey, K. S., and Schlax, M. G.: Daily high-resolution-blended analyses for sea surface temperature, J. Climate, 20, 5473–5496, 2007.

Robertson, R., Visbeck, M., Gordon, A. L., and Fahrbach, E.: Long-term temperature trends in the deep waters of the Weddell Sea, Deep Sea Res. Pt. II, 49, 4791–4806, doi:10.1016/S0967-0645(02)00159-5, 2002.

Rohardt, G., Fahrbach, E., and Wisotzki, A.: Physical oceanography during POLARSTERN cruise ANT-XXVII/2, Alfred Wegener Institute, Helmholtz Center for Polar and Marine Research, Bremerhaven, doi:10.1594/PANGAEA.772244, 2011.

Saha, S., Moorthi, S., Pan, H.-L., Wu, Z., Wang, J., Nadiga, S., Tripp, P., Kistler, R., Woollen, J., Behringer, D., Liu, H., Stokes, D., Grumbine, R., Gayno, G., Wang, J., Hou, Y.-T., Chuang, H.-Y., Juang, H.-M. H., Sela, J., Iredell, M., Treadon, R., Kleist, D., Van Delst, P., Keyser, D., Derber, J., Ek, M., Meng, J., Wei, H., Yang, R., Lord, S., Van Den Dool, H., Kumar, A., Wang, W., Long, C., Chelliah, M., Xue, Y., Huang, B., Schemm, J.-K., Ebisuzaki, W., Lin, R., Xie, P., Chen, M., Zhou, S., Higgins, W., Zou, C.-Z., Liu, Q., Chen, Y., Han, Y., Cucurull, L., Reynolds, R. W., Rutledge, G., and Goldberg, M.: The NCEP climate forecast system reanalysis, Bull. Am. Meteorol. Soc., 91, 1015–1057, doi:10.1175/2010BAMS3001.1, 2010.

Schröder, M. and Fahrbach, E.: On the structure and the transport of the eastern Weddell Gyre, Deep-Sea Res. Pt. II, 46, 501–527, doi:10.1016/S0967-0645(98)00112-X, 1999.

Smedsrud, L. H.: Warming of the deep water in the Weddell Sea along the Greenwich meridian: 1977–2001, Deep Sea Res. Pt. I, 52, 241–258, doi:10.1016/j.dsr.2004.10.004, 2005.

Smith, R., Dukowicz, J., and Malone, R.: Parallel ocean general circulation modeling, Physica D: Nonlinear Phenomena, 60, 38–61, doi:10.1016/0167-2789(92)90225-C, 1992.

Stark, J. D., Donlon, C. J., Martin, M. J., and McCulloch, M. E.: OSTIA: An operational, high resolution, real time, global sea surface temperature analysis system, In: Oceans'07 IEEE Aberdeen, conference proceedings. Marine challenges: coastline to deep sea. Aberdeen, Scotland, IEEE., 1–4, 2007.

Storkey, D., Blockley, E. W., Furner, R., Guiavarc'h, C., Lea, D., Martin, M. J., Barciela, R. M., Hines, A., Hyder, P., and Siddorn, J. R.: Forecasting the ocean state using NEMO: The new FOAM system, J. Operational Oceanogr., 3, 3–15, 2010.

Talley, L. D.: Closure of the global overturning circulation through the Indian, Pacific, and Southern Oceans: Schematics and transports, Oceanography, 26, 80–97, doi:10.5670/oceanog.2013.07, 2013.

Taylor, K. E.: Summarizing multiple aspects of model performance in a single diagram, J. Geophys. Res., 106, 7183–7192, doi:10.1029/2000JD900719, 2001.

Uppala, S. M., KÅllberg, P. W., Simmons, A. J., Andrae, U., Bechtold, V. D. C., Fiorino, M., Gibson, J. K., Haseler, J., Hernandez, A., Kelly, G. A., Li, X., Onogi, K., Saarinen, S., Sokka, N., Allan, R. P., Andersson, E., Arpe, K., Balmaseda, M. A., Beljaars, A. C. M., Berg, L. V. D., Bidlot, J., Bormann, N., Caires, S., Chevallier, F., Dethof, A., Dragosavac, M., Fisher, M., Fuentes, M., Hagemann, S., Hólm, E., Hoskins, B. J., Isaksen, L., Janssen, P. A. E. M., Jenne, R., Mcnally, A. P., Mahfouf, J.-F., Morcrette, J.-J., Rayner, N. A., Saunders, R. W., Simon, P., Sterl, A., Trenberth, K. E., Untch, A., Vasiljevic, D., Viterbo, P., and Woollen, J.: The ERA-40 re-analysis, Q. J. Roy. Meteorol. Soc., 131, 2961–3012, doi:10.1256/qj.04.176, 2005.

van Sebille, E., Spence, P., Mazloff, M. R., England, M. H., Rintoul, S. R., and Saenko, O. A.: Abyssal connections of Antarctic Bottom Water in a Southern Ocean State Estimate, Geophys. Res. Lett., 40, 2177–2182, doi:10.1002/grl.50483, 2013.

Venegas, S. A. and Drinkwater, M. R.: Sea ice, atmosphere and upper ocean variability in the Weddell Sea, Antarctica, J. Geophys. Res., 106, 16747–16766, doi:10.1029/2000JC000594, 2001.

Wang, C. and Beckmann, A.: Investigation of the impact of Antarctic ice-shelf melting in a global ice–ocean model (ORCA2-LIM), Ann. Glaciol., 46, 78–82, doi:10.3189/172756407782871602, 2007.

Whitworth III, T. and Nowlin Jr., W. D.: Water masses and currents of the Southern Ocean at the Greenwich meridian, J. Geophys. Res., 92, 6462–6476, doi:10.1029/JC092iC06p06462, 1987.

Whitworth III, T., Orsi, A. H., Kim, S. J., Nowlin Jr., W. D., and Locarnini R. A.: Water masses and mixing near the Antarctic slope front, in: Ocean, Ice, and Atmosphere: Interactions at the Antarctic Continental Margin, Antarct. Res. Ser., 75, edited by: Jacobs, S. S. and Weiss, R. F., 26 pp., AGU, Washington, DC, 1998.

Willebrand, J., Barnier, B., Böning, C., Dieterich, C., Killworth, P. D., Le Provost, C., Jia, Y., Molines, J.-M., and New, A. L.: Circulation characteristics in three eddy-permitting models of the North Atlantic, Progr. Oceanogr., 48, 123–161, doi:10.1016/S0079-6611(01)00003-9, 2001.

Winton, M., Hallberg, R., and Gnanadesikan, A.: Simulation of Density-Driven Frictional Downslope Flow in Z-Coordinate Ocean Models, J. Phys. Oceanogr., 28, 2163–2174, doi:10.1175/1520-0485(1998)028<2163:SODDFD>2.0.CO;2, 1998.

Woodruff, S. D., Worley, S. J., Lubker, S. J., Ji, Z., Freeman, J. E., Berry, D. I., Brohan, P., Kent, E. C., Reynolds, R. W., Smith, S. R., and Wilkinson, C.: ICOADS Release 2.5: Extensions and enhancements to the surface marine meteorological archive, Int. J. Climatol., 31, 951–967, doi:10.1002/joc.2103, 2011.

Worley, S. J., Woodruff, S. D., Reynolds, R. W., Lubker, S. J., and Lott, N.: ICOADS release 2.1 data and products, International J. Climatol., 25, 823–842, doi:10.1002/joc.1166, 2005.

Xue, Y., Huang, B., Hu, Z.-Z., Kumar, A. A., Wen, C., Behringer, D., and Nadiga, S.: An assessment of oceanic variability in the NCEP climate forecast system reanalysis, Clim. Dynam., 37, 2511–2539, doi:10.1007/s00382-010-0954-4, 2011.

12

Equilibrator-based measurements of dissolved nitrous oxide in the surface ocean using an integrated cavity output laser absorption spectrometer

I. Grefe[1,*] and J. Kaiser[1]

[1]School of Environmental Sciences, University of East Anglia, Norwich Research Park, Norwich, UK
[*]now at: Dalhousie University, Department of Oceanography, 1355 Oxford Street, Halifax NS B3H 4R2, Canada

Correspondence to: J. Kaiser (j.kaiser@uea.ac.uk)

Abstract. Dissolved nitrous oxide (N_2O) concentrations are usually determined by gas chromatography (GC). Here we present laboratory tests and initial field measurements using a novel setup comprising a commercially available laser-based analyser for N_2O, carbon monoxide and water vapour coupled to a glass-bed equilibrator. This approach is less labour-intensive and provides higher temporal and spatial resolution than the conventional GC technique. The standard deviation of continuous equilibrator or atmospheric air measurements was 0.2 nmol mol^{-1} (averaged over 5 min). The short-term repeatability for reference gas measurements within 1 h of each other was 0.2 nmol mol^{-1} or better. Another indicator of the long-term stability of the analyser is the standard deviation of the calibrated N_2O mole fraction in marine air, which was between 0.5 and 0.7 nmol mol^{-1}. The equilibrator measurements were compared with purge-and-trap gas chromatography–mass spectrometry (GC-MS) analyses of N_2O concentrations in discrete samples from the Southern Ocean and showed agreement to within the 2 % measurement uncertainty of the GC-MS method. The equilibrator response time to concentration changes in water was from 142 to 203 s, depending on the headspace flow rate. The system was tested at sea during a north-to-south transect of the Atlantic Ocean. While the subtropical gyres were slightly undersaturated, the equatorial region was a source of nitrous oxide to the atmosphere, confirming previous findings (Forster et al., 2009). The ability to measure at high temporal and spatial resolution revealed submesoscale variability in dissolved N_2O concentrations. Mean sea-to-air fluxes in the tropical and subtropical Atlantic ranged between -1.6 and 0.11 μmol m^{-2} d^{-1} and confirm that the subtropical Atlantic is not an important source region for N_2O to the atmosphere, compared to global average fluxes of 0.6–2.4 μmol m^{-2} d^{-1}. The system can be easily modified for autonomous operation on voluntary observing ships (VOS). Future work should include an interlaboratory comparison exercise with other methods of dissolved N_2O analyses.

1 Introduction

Nitrous oxide (N_2O) is an important trace gas in the atmosphere, influencing Earth's climate as well as stratospheric chemistry. It is currently the third most important greenhouse gas in terms of 100 year global warming potential after CO_2 and CH_4 (Ravishankara et al., 2009). Furthermore, it is the main precursor of stratospheric NO_x, which catalytically destroys ozone (Crutzen, 1970). As chlorofluorocarbons (CFCs) are phased out under the Montreal Protocol, N_2O is the most important, currently emitted substance involved in stratospheric ozone depletion (Ravishankara et al., 2009). Atmospheric concentrations are rising at a rate of 0.26 % per year with the ocean contributing about 30 % to total emissions (Forster et al., 2007). Bacterial nitrification and denitrification are assumed to be the main production pathways for N_2O in the ocean, while denitrification can also act as a sink under suboxic conditions (Elkins et al., 1978; Cohen and Gordon, 1978; Knowles, 1982). Even though nitrification is an aerobic process, N_2O production is enhanced as oxygen concentrations decrease (Goreau et al.,

1980; Yoshida et al., 1989; Yoshinari, 1976). Nitrifier denitrification is an alternative pathway for N_2O production by ammonia-oxidising bacteria, which appears to be important for the near-surface N_2O source (Poth and Focht, 1985; Popp et al., 2002; Sutka et al., 2006; Sutka et al., 2004). Recently, the importance of N_2O production by archaeal ammonia oxidation was discovered, potentially accounting for a significant part of the oceanic N_2O source (Löscher et al., 2012; Wuchter et al., 2006). The estimated source of rivers and coastal regions range currently from 0.5 to 2.7 $Tg\,a^{-1}$ (in N equivalents) and from 1.8 to 5.8 $Tg\,a^{-1}$ for the open ocean (Denman et al., 2007). In the light of the uncertainties in the marine N_2O source and potential future emission increases due to ocean deoxygenation (Codispoti, 2010), accurate observations in space and time are important to give a better estimate of regional sources and global budgets.

The most common technique for N_2O concentration measurements is injection of a gas sample onto a gas chromatographic column coupled to an electron-capture detector (GC-ECD) (Weiss, 1981; Weiss et al., 1992; Butler et al., 1989; Walter et al., 2004). Here we present an alternative method using a laser-based optical absorption analyser (Baer et al., 2002) that in combination with an equilibrator enables continuous N_2O analyses at ambient levels in seawater. The system is low maintenance, can be easily calibrated and allows for higher measurement frequency than GC-ECD methods. It has the potential to facilitate observations over long time series, revealing variability and trends, as is already happening for CO_2 measurement systems, e.g. on ships of opportunity. Laboratory tests and results from field deployments of the analyser in combination with an equilibrator are presented.

2 Materials and methods

2.1 Laboratory tests

The N_2O/CO analyser (Los Gatos Research, LGR, model N_2O/CO-23d) used in this study, measures gas-phase mole fractions of N_2O, carbon monoxide (CO) and water vapour (H_2O) using off-axis integrated cavity output spectroscopy (ICOS). Test results and environmental data for N_2O are reported in this study. The analyser was connected to a 1.7 L percolating packed glass bed equilibrator as described by Cooper et al. (1998). The analyser's internal membrane pump was used to circulate the gas phase through the equilibrator. Water was pumped through the equilibrator at a flow rate of 1.8 $L\,min^{-1}$. Two 4-port 2-position valves (Vici) allowed for fast switching between the equilibrator headspace and a 6-port multiposition valve (Vici), connecting to other gas lines e.g. atmospheric air and references (Fig. 1). A similar setup has been described by Gülzow et al. (2011) for dissolved CO_2 and CH_4 measurements using an ICOS analyser and by Becker et al. (2012) for measurements of $\delta^{13}C(CO_2)$ and fCO_2 using continuous wave cavity ring-down spectroscopy.

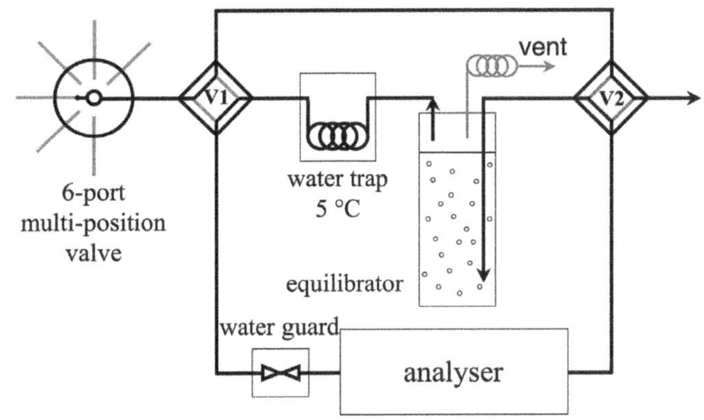

Figure 1. Setup for laboratory tests. V1 and V2: 4-port 2-position valves. Arrows indicate gas flow through the equilibrator. The vent is a 3 m long coiled 1/8 in. plastic tube to allow for volume and pressure changes of the equilibrator headspace due to variations in gas tension.

A water trap was installed downstream of the equilibrator to reduce the amount of water vapour in the headspace gas. The trap consisted of a thermoelectric cool box (T08 DC, Mobicool), held at 5 °C, and a miniature filter with manual drain (Norgren) to collect the condensing water. A custom-built safety valve ("water guard") was installed upstream of the analyser as an additional protection against water entering the measurement cell. The water guard consists of a stainless steel tee (Swagelok) with electrodes and a solenoid valve downstream of the sensor. Water in the gas line closes the electric circuit in the water guard, triggering the closure of the valve and cutting off the gas supply to the analyser. This is really just a safety precaution as the "water-guard" was neither triggered during the laboratory tests, nor at sea. The instrument's water vapour measurements are used by the software to calculate N_2O dry mole fractions (Eq. 1). $x(N_2O)$ is the N_2O dry mole fraction, $x_{meas}(N_2O)$ and $x_{meas}(H_2O)$ are the measured N_2O and H_2O mole fractions:

$$x\,(N_2O) = \frac{x_{meas}\,(N_2O)}{1 - x_{meas}\,(H_2O)}. \quad (1)$$

Furthermore, line broadening due to changing water vapour concentrations is accounted for by the instrument's software. The instrument is factory calibrated for H_2O using a dewpoint generator. In order to validate this water vapour correction, measurements of dry air were compared to calculated values for the dry mole fraction of humidified air. A cylinder with dry air was connected to the analyser via the multiposition valve. The gas line from the cylinder was split with one line going directly to the valve and the other one passing first through the water-filled cold trap to humidify the air. H_2O mole fractions were between 1 and 1.4 % for the humidified gas and below the analyser's detection limit (around 0.2 % for H_2O) for dry gas. The calculated dry mole fraction of the humidified gas was compared to that of the dry gas.

In addition, to test for analyser variability and drift, dry cylinder gas was measured for 24 h. To test for leaks within the analyser, laboratory air (325 nmol mol^{-1} dry mole fraction, uncalibrated) and zero grade air (O_2 and N_2, ~ 84 nmol mol^{-1} N_2O dry mole fraction uncalibrated, BOC) were mixed in a sample loop to obtain lower mole fractions than in ambient air. The mixtures with 215.1 nmol mol^{-1} N_2O were recirculated through the analyser for 12 and 21 min. Any leaks are expected to be noticed as an increase in N_2O mole fractions caused by ambient air. The valve board was leak-checked separately by pressurising the gas lines of the equilibrator loop with compressed air to just below 120 kPa. The equilibrator itself was bypassed during this test because it communicates with the atmosphere at three points: the water inlet, the water outlet and the pressure vent. As the LGR analyser keeps the pressure in the measurement cell constant at 11 kPa and the measurement range of the built-in pressure sensor ends at 22 kPa, a Li-Cor CO_2 instrument with built-in pressure gauge was used instead for monitoring pressure changes over time.

The response time of the coupled system of the ICOS (integrated cavity output spectroscopy) analyser and equilibrator was characterised in further laboratory tests. The equilibrator time constant τ, i.e. the time during which a concentration difference between the gas- and the water phase declines to $1/e$ (36.8 %) with regard to the start value, was determined as described in Gülzow et al. (2011). τ was only evaluated for N_2O as CO background concentrations were too variable in the laboratory where tests took place. Two 100 L reservoirs, open to the atmosphere, were filled with fresh water from the tap, which is supersaturated in N_2O. Mole fractions between 694 and 1065 nmol mol^{-1} were measured in the equilibrator headspace during five experiments with supersaturated water. One of the reservoirs was then equilibrated with ambient air by recirculation; the other was kept at elevated N_2O concentrations. For the experiments, the water was pumped through the equilibrator from the bottom of the reservoirs at a flow rate of 1.8 L min^{-1}, starting with the equilibrated reservoir, and then changing to water with high N_2O concentrations. After the measured dry mole fraction reached a plateau (x_{max}), water in equilibrium with ambient air was pumped through the equilibrator. τ was then calculated as described in Gülzow et al. (2011), recording the decay of N_2O dry mole fractions (x_t) back to ambient values (x_{min}). The observed x_t values (Fig. 2) were fitted to an exponential equation:

$$x_t = x_{min} + (x_{max} - x_{min})\, e^{-\frac{t}{\tau}}. \quad (2)$$

By rearranging Eq. (2), τ can be inferred from the slope of a linear regression of $-\ln[(x_t - x_{min})/(x_{max} - x_{min})]$ against time.

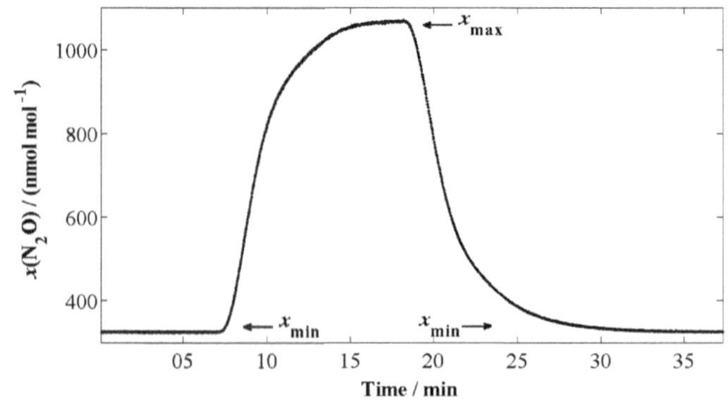

Figure 2. N_2O concentrations during the step experiment for determination of τ. Equilibrated water flowing through the equilibrator is replaced with water containing higher N_2O concentrations.

Initially, it was attempted to use the analyser in combination with a semi-permeable membrane (Membrana, MiniModule). For this purpose, the gas flow through the cavity was reduced to 100 mL min^{-1} (293 K, 100 kPa) by inserting a needle valve between the internal diaphragm pump and a check valve downstream of the measurement cell. The yield of dissolved gases extracted over the membrane was too low to sustain the analyser's operating pressure in the measurement cell. Therefore, an equilibrator was used instead of the semi-permeable membrane. As the reduced gas flow did not lead to problems with the equilibrator setup, the valve was retained during the first field test (see Sect. 2.2). Without the throttle valve, the flow rate increased to approximately 400 mL min^{-1} (293 K, 100 kPa). The time constant τ was determined for both headspace flow rates.

2.2 Field deployment of equilibrator and ICOS analyser

The N_2O analyser was tested at sea during cruise AMT20 of the Atlantic Meridional Transect project from Southampton, UK, to Punta Arenas, Chile (12 October to 25 November 2010), on board RRS *James Cook*. Figure 3 shows the setup for underway measurements during the cruise, Fig. 4 the cruise track.

The equilibrator was connected to the ship's pumped underway seawater supply, drawing water from a depth of approximately 5 m. Filters (Vacu-guard, part number 6722-5000, Wheaton) were inserted at the gas inlet and outlet of the equilibrator to protect pump and measurement cells of the analyser from seawater. Temperatures in the equilibrator were measured with two Pt-100 temperature probes (Omega Engineering Limited), calibrated against a mercury reference thermometer with an accuracy of 0.06 °C. This measurement uncertainty leads to a 0.17 % (at 30 °C) to 0.25 % (at 0 °C) uncertainty in the derived surface seawater concentrations. For future studies, a temperature calibration to better than 0.01 °C is recommended to further reduce measurement

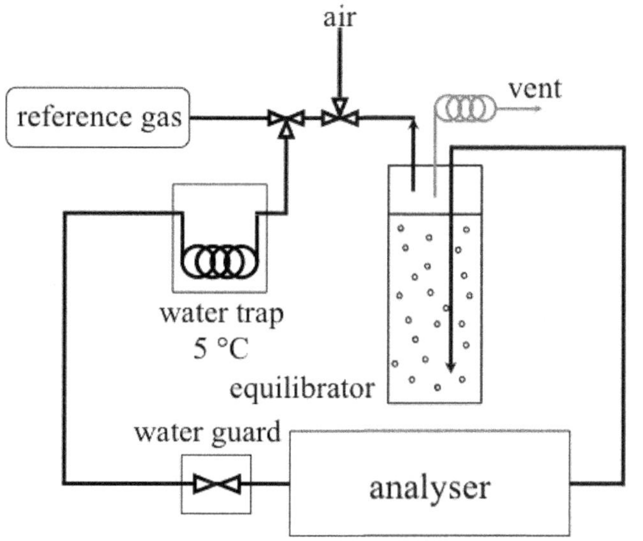

Figure 3. Underway setup for field deployment during AMT 20. Two manual 3-port valves allow switching between measurements of the equilibrator headspace, marine air, and dry air.

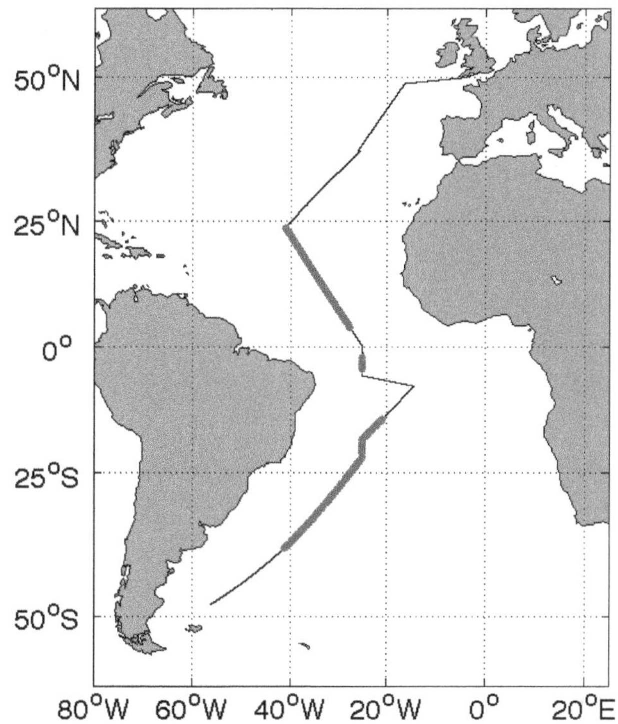

Figure 4. Cruise track of AMT20. Thick grey lines indicate positions of measurements with the N_2O analyser.

uncertainties. The water flow through the equilibrator was set to approximately 1.8 L min^{-1} via a valve. Flow was verified by regular (at least hourly) monitoring of the constancy of the water level in the equilibrator and manual measurement of the rate of water flow from the bottom of the equilibrator. Flow stagnation was observed on four occasions during the cruise, and attributed to pressure fluctuations in the underway system caused by other users. Affected data were removed from the results presented here, as well as the first 15 min after readjustment of the water flow. For subsequent field deployments, a flow restrictor will be used to stabilise the pumped seawater supply to the equilibrator. Two three-way valves (part number SS-41GXS2, Swagelok) allowed changing between sample gas stream from the equilibrator and marine air, drawn from the ship's bow (Fig. 3). Dried air with 323.7 nmol mol^{-1} N_2O was used as a working reference, calibrated against IMECC/NOAA (Infrastructure for the Measurement of the European Carbon Cycle/National Oceanic and Atmospheric Administration) primary standards. Every 8 h, the analyser was calibrated by switching from equilibrator headspace to the reference gas and then marine air for 20 min each. This was followed by another reference measurement after 40 min to assess short-time drift. Only the last 5 min of each measurement were analysed to allow for complete flushing of the measurement cell. Correspondingly, the first 15 min after switching back to the equilibrator headspace or to air measurements were not used for evaluation to allow for flushing of air and reference gas residuals from the measurement cavity and equilibrator.

N_2O concentrations (c) were calculated from dry mole fractions (x) using the solubility function F at equilibrator temperature T_{eq} (Weiss and Price, 1980):

$$c = xF\left(T_{eq}, S\right) p_{eq}, \tag{3}$$

where T_{eq} and p_{eq} are equilibrator temperature and pressure (assumed to be equal to ambient atmospheric pressure, p_{air} at sea level and assuming 100 % relative humidity) and S is salinity. Note that the unit of F in Eq. (8) of Weiss and Price (1980) (mol kg^{-1} or mol L^{-1}) is inconsistent with the unit of F in their tables (mol kg^{-1} atm^{-1} or mol L^{-1} atm^{-1}). We use F as parameterised by Eq. (13) in Weiss and Price (1980), multiplied by the equilibrator pressure and the dry mole fraction, to derive c.

Seawater saturations (s) were based on equilibrium values for measured atmospheric mole fractions x_{air} and mole fractions in seawater, corrected for temperature differences between equilibrator and seawater intake (T_{in}):

$$s = \frac{xF\left(T_{eq}, S\right)}{x_{air}F\left(T_{in}, S\right)}. \tag{4}$$

The air–sea flux (Φ) was calculated from the gas transfer coefficient (k_w) and the difference between N_2O concentrations in seawater c and air equilibrium concentrations (c_{air}):

$$\Phi = k_w\left(c - c_{air}\right) = k_w\left[c - x_{air}F\left(T_{in}, S\right) p_{air}\right]. \tag{5}$$

A water vapour correction is not required because it is already included in the parametrisation of F (Weiss and Price, 1980). F requires the equilibrator pressure to be close to $p^\circ = 1\,\text{atm} = 1013.25\,\text{hPa}$. The correction term for deviations of the equilibrator pressure p_{eq} from p° is given in Eq. 6 and is between 0.9998 at $p_{eq} = 970\,\text{hPa}$ and 1.0002 at $p_{eq} = 1050\,\text{hPa}$ for temperatures between −2 and 30 °C. This correction of less than ±0.02 % is negligible.

$$c(N_2O) = x(N_2O)\,F\left(N_2O, T_{eq}, S\right)\,p_{eq}\frac{1-\frac{p_{vap}(H_2O,T_{eq},S)}{p_{eq}}}{1-\frac{p_{vap}(H_2O,T_{eq},S)}{p^\circ}} \tag{6}$$

k_w was calculated using the parameterisation of Nightingale (2000) and converted to units of $\text{m}\,\text{d}^{-1}$, where u is wind speed at 10 m above sea level (Eq. 7).

$$\frac{k_w}{\text{m}\,\text{d}^{-1}} = 0.24\left[0.222\left(\frac{u}{\text{m}\,\text{s}^{-1}}\right)^2 + 0.333\frac{u}{\text{m}\,\text{s}^{-1}}\right]\left(\frac{Sc}{600}\right)^{-0.5} \tag{7}$$

This relationship shows an intermediate dependence on wind speed compared to the other frequently used parameterisations of Liss and Merlivat (1986) and Wanninkhof (1992), and gives fluxes in very good agreement (on average $0.004\,\mu\text{mol}\,\text{m}^{-2}\,\text{d}^{-1}$ difference) with the parameterisation of Sweeney et al. (2007). k_w was adjusted for N_2O with the Schmidt number Sc calculated following Wanninkhof (1992). The wind speed at 10 m height was taken from the 6-hourly operational analysis data set of the European Centre for Medium-Range Weather Forecasts (ECMWF) and interpolated to the time and position of the respective measurement (download from BADC at http://badc.nerc.ac.uk/view/badc.nerc.ac.uk__ATOM__dataent_ECMWF-OP).

Instantaneous values for k_w and Φ were compared to those using 30-day wind-speed-weighted averages to give a more robust estimate of gas exchange in light of variable wind speed over time (Reuer et al., 2007). The calculation takes into account that wind speed varies over time at any given location and weighs it appropriately according to mixed layer depth, assuming a constant N_2O production/influx rate in the mixed layer. Differences between both estimates of sea–air exchange were small. In the following, we discuss only instantaneous fluxes for consistency with previous studies of N_2O air-sea exchange (Fig. 5).

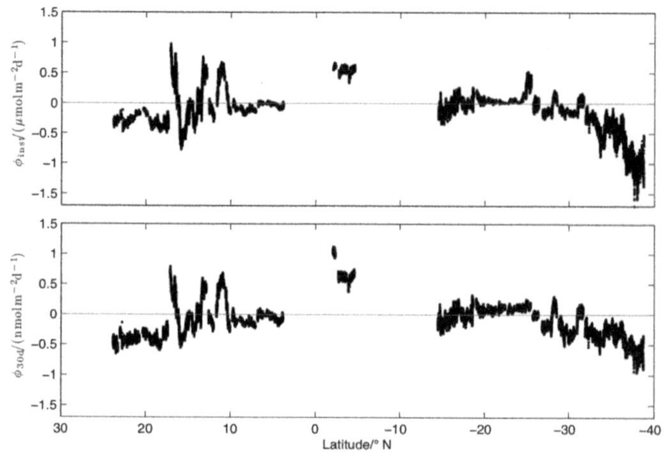

Figure 5. Comparison of N_2O flux calculated from instantaneous (top panel) and 30-day-averaged wind speeds (bottom panel).

2.3 GC-MS measurements of discrete samples

During a subsequent cruise to the Weddell Sea on board RRS *James Clark Ross* from 20 January to 2 February 2012 (JR255A), rosette water samples were collected. These were analysed for N_2O concentrations using purge-and-trap gas chromatography–mass spectrometry (GC-MS), with the data and comparisons to simultaneous equilibrator ICOS measurements presented in Sect. 3.2. The majority of the sea surface measurements on this cruise will be presented elsewhere.

Rosette seawater samples for isotope analysis were collected in 500 mL (nominal value) serum bottles (Wheaton). Triplicate samples were taken immediately after recovery of the CTD (conductivity, temperature and depth sensor) and were allowed to overflow at least three times the bottle's volume. Sample bottles were closed with butyl stoppers and aluminium crimp seals and poisoned with 1 mL saturated mercuric chloride solution. Of the sample, 1 mL was replaced with CP grade helium (BOC) to reduce the risk of leaks due to temperature driven volume changes of the water during transport and storage. The setup of the GC-MS follows McIlvin and Casciotti (2010). Samples are loaded manually; dissolved gases are quantitatively extracted with a helium purge stream and trapped with liquid nitrogen. Water vapour in the gas stream is removed with a Nafion dryer (Perma Pure), CO_2 is trapped on Carbosorb (Merck). The sample is then injected into a continuous-flow GC-MS system. Late eluting substances are removed with a precolumn as described in Röckmann et al. (2003) before N_2O is separated from residual CO_2 on the PoraPlot Q analytical column. The sample enters the mass spectrometer (Thermo Scientific, MAT 253) via an open split and is analysed for m/z 44, 45 and 46. The N_2O amount in the sample can be calculated from the peak area with a precision of 2 % (based on analysis of replicates) and the sample volume, which is determined by sample weight, water temperature and salinity, with a precision of 0.02 %. The overall precision of 2 % is comparable to the 1.8 % concentration uncertainty achieved for GC-ECD measurements by Walter et al. (2006) and 2.6 % for GC-MS measurements by McIlvin and Casciotti (2010).

3 Results and discussion

3.1 Laboratory tests

To improve the precision of individual data points retained for further analysis, 10 s averages were calculated from measurements at 1 Hz. Since the headspace gas was in contact with the water phase in the equilibrator, water vapour concentrations were high. The cold trap only removed water to a dew point of 5 °C, corresponding to about 8.7 mmol mol^{-1}. Correction from measured values to dry mole fractions is therefore required for the evaluation of dissolved N_2O concentrations in seawater.

Compressed air directly from the cylinder had a measured N_2O mole fraction of (332.7 ± 0.2) nmol mol^{-1}, while H_2O mole fractions were below the detection limit of 2 mmol mol^{-1}. The measured N_2O mole fraction in humidified air with 10.6 mmol mol^{-1} H_2O was (329.7 ± 0.2) nmol mol^{-1}. Since the H_2O mole fraction in compressed air was below the detection limit of 2 mmol mol^{-1}, we assume it to be in the range from 0 to 2 mmol mol^{-1}. The corresponding dry mole fraction of compressed air is therefore $(332.7^{+0.8}_{-0.2})$ nmol mol^{-1}, where the error estimate in the positive direction corresponds to a H_2O mole fraction of 2 mmol mol^{-1}, that in the negative direction corresponds to the statistical uncertainty and a H_2O mole fraction of 0. The calculated dry mole fraction of humidified air was (333.1 ± 0.2) nmol mol^{-1}. This value is within measurement uncertainties of the corrected mole fraction of compressed air directly from the cylinder. The H_2O vapour dilution correction is considered to be sufficient; no further corrections for line broadening were applied.

The stability of the analyser at low N_2O mole fractions was assessed by measuring a gas cylinder over 24 h. The standard deviation was 0.2 nmol mol^{-1} for a mean N_2O mole fraction on 48.7 nmol mol^{-1}; slightly better than the 5 h-stability of ± 0.6 nmol mol^{-1} reported by Arévalo-Martínez et al. (2013). Minimum and maximum values measured during this period were 48.2 and 49.4 nmol mol^{-1}, respectively. An Allan variance test for this low N_2O mole fraction showed a minimum Allan deviation of (25 ± 1) pmol mol^{-1} for an integration interval of 90–180 s (1.5–3 min).

An Allan variance test for a reference gas with an N_2O mole fraction closer to ambient values (299 nmol mol^{-1}) gave a minimum Allan deviation of (33 ± 6) pmol mol^{-1} for an integration interval of 80–105 s. These results are comparable to the value of 37 pmol mol^{-1} over 2 min reported by Arévalo-Martínez et al. (2013).

The highest observed increase of N_2O mole fractions during the two leak tests for the LGR analyser was 0.024 nmol mol^{-1} min^{-1}. The gas volume of the 400 mL measurements cell at a pressure of 11 kPa corresponds to 40 mL while the tubing of the circular gas path for this test is assumed to be at atmospheric pressure, resulting in a volume of approximately 40 mL. The total gas volume during the test is therefore 80 mL. The leak rate was calculated as the increase in N_2O concentrations, divided by the difference between background N_2O and circulating gas mixture and multiplied by the total gas volume. The resulting leak rate for the N_2O analyser used for recirculating air, e.g. through an equilibrator, is 0.017 mL min^{-1} or 0.29×10^{-3} mL s^{-1}. This leak rate is likely to be due to the pump head (on the order of 10^{-3} mL s^{-1}, KNF Neuberger (UK) Ltd., personal communication, 2013). Pressure in the valve board was stable over 10 min at $119\,731 \pm 0.006$ Pa as recorded by the Li-Cor's pressure gauge, indicating the absence of leaks.

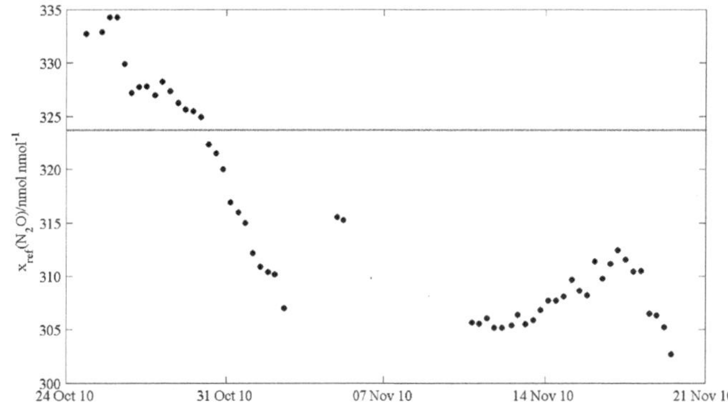

Figure 6. Measured values for the reference gas during the transect. Nominal value of 323.7 nmol mol^{-1} indicated by grey line.

The response time of the coupled analyser-equilibrator system to concentration changes in the water phase is described by the equilibration time constant τ, which was determined for conditions experienced during field deployment. For a gas flow of 400 mL min^{-1} through the measurement cell, τ equalled (142 ± 1) s for N_2O ($n = 5$). The 95 % relaxation time (i.e. 3τ) is therefore about 7 min. Reducing the gas flow to 100 mL min^{-1} increased τ to (203 ± 1) s ($n = 3$), without changing the measured mole fraction. In the limit where the water flow rate is much higher than the gas exchange rate, the value for τ depends on the transfer coefficient k (Rafelski et al., 2012). Presumably, at the higher gas flow rate enhanced turbulence increased the efficiency of gas transfer between water and gas. Therefore, the needle valve will be removed during future deployments to reduce delays in the system's response to changing N_2O concentrations in the environment. Values for τ are similar to those reported for N_2O of 180 s by Arévalo-Martínez et al. (2013), using a similar measurement setup.

3.2 Precision and comparison with GC-MS

Under field conditions during AMT20, the difference between two subsequent calibration measurements, spaced 40 min apart, was on average 0.2 nmol mol^{-1} or better. This is comparable to the short-term stability of 0.2 nmol mol^{-1} over 3 min reported by Arévalo-Martínez et al. (2013). During AMT20, substantial long-term drift was encountered

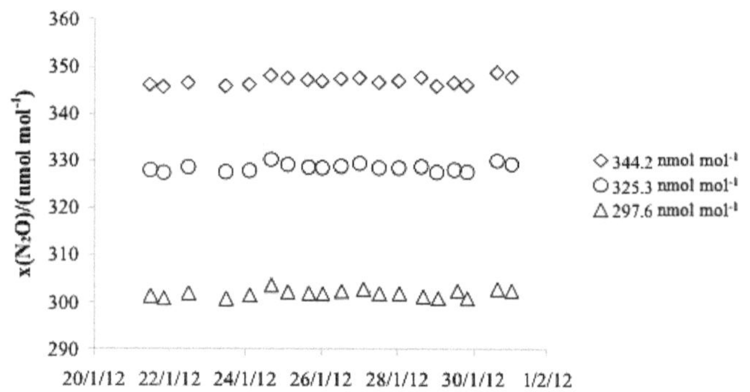

Figure 7. N_2O mole fractions for the three reference gases used during JR255A over time.

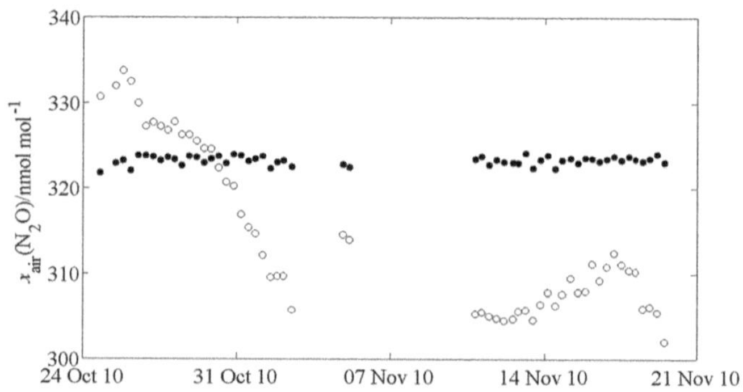

Figure 8. N_2O mole fractions measured during AMT20 in tropospheric air. Open circles are raw measurement values before drift and offset corrections, filled circles are fully corrected values.

(Fig. 6). It was assumed that instrument drift was linear between reference gas measurements, and the linear dynamic range of the analyser passed through the origin. As only one reference gas was used during AMT20, this could not be investigated any further. During cruise JR255A three reference gases were used (Fig. 7). The stability of the dry mole fractions of these three gases over the length of the field campaign was 0.9 nmol mol^{-1} (0.3 %) or better ($n = 19$). The slope for the dynamic range was 1.04 and the y axis offset -15 nmol mol^{-1}. This would have a negligible influence for the range of N_2O mole fractions encountered during AMT20 due to their small variations around the nominal value of the reference gas (323.7 nmol mol^{-1}): at the lower end (near 97 % saturation), it would lead to positive bias of 0.5 nmol mol^{-1} (or less than 0.2 %); at the upper end (near 103 % saturation), it would lead to a negative bias of -0.4 nmol mol^{-1}. Therefore, the reference gas measurements could be used to fully correct air and equilibrator measurements for analyser offset and drift, using linear interpolation between calibrations (see also Fig. 8). The instrument drift was caused by a gradual change of the laser tuning (R. Provencal, personal communication, 2010). The laser was replaced after AMT20 and the analyser has been stable since. The use of three reference gases, as done for JR255A, is recommended to improve the calibration of analyser data during future field deployments.

Measurements of atmospheric N_2O mole fractions during AMT20 and JR255A were used for comparison with data of selected Advanced Global Atmospheric Gases Experiment (AGAGE) stations (Prinn et al., 2000). Dry mole fractions were corrected for instrument drift and offset with the calibration measurements. During AMT20, mole fractions of N_2O measured in marine background air were (323.2 ± 0.5) nmol mol^{-1} throughout the cruise. An interhemispheric difference of slightly less than 1 nmol mol^{-1} was expected (Butler et al., 1989; Rhee et al., 2009), but did not show in the data. This small difference might have not been captured due to the analyser drift described above, and the small number of atmospheric measurements ($n =$ 26 in the Northern Hemisphere and $n = 29$ in the Southern Hemisphere). This undersampling in combination with natural variability (e.g. seasonality, source of air masses) could account for the absence of the expected interhemispheric gradient of (1.2 ± 0.8) nmol mol^{-1} in this data set. During future measurement campaigns, more frequent atmospheric air measurements or a dedicated analyser may be required to reliably detect the interhemispheric N_2O mole fraction gradient. Nevertheless, the measured atmospheric mole fractions of (323.2 ± 0.5) nmol mol^{-1} agree within measurement uncertainties with mean values for October and November of the AGAGE stations Mace Head for the Northern Hemisphere ((324.1 ± 0.7) nmol mol^{-1}) and Cape Grim for the Southern Hemisphere ((322.9 ± 0.3) nmol mol^{-1}, data from http://agage.eas.gatech.edu/data_archive) (Fig. 6). During JR255A, measured N_2O mole fractions in air were (324.0 ± 0.7) nmol mol^{-1} ($n = 11$), which is consistent with the value of (323.9 ± 0.5) nmol mol^{-1} measured at Cape Grim in January 2012. Neither during AMT20 nor during JR255A, were measurements of atmospheric N_2O mole fractions influenced by wind direction.

The equilibrator type used for seawater measurements is described in Cooper et al. (1998). The authors found no systematic differences for CO_2 measurements made with this equilibrator compared to shower-head equilibrators. As the solubility characteristics of N_2O are similar to CO_2 (Weiss and Price, 1980), no bias is expected to be introduced by using this equilibrator type and the equilibrator efficiency should be similar. Correcting equilibrator measurements for instrument drift and offset using the reference gas measurements resulted in a precision of 0.2 nmol mol^{-1} over a 5 min average.

Table 1 shows a comparison between equilibrator measurements and GC-MS analyses of discrete water samples during cruise JR255A. Concentrations measured with GC-MS were (1.3 ± 0.9) % higher than those obtained with the LGR N_2O analyser. Although the values match within the 2 % measurement uncertainty associated with the GC-MS

Table 1. Comparison between ICOS and GC-MS measurements during JR255A. Sampling time (GMT, Greenwich Mean Time) and position for the three depth casts overlapping with analyser measurements. c_{eq} is the equilibrium concentration for atmospheric N_2O based on sea surface temperature (θ_0), salinity (S_0) and atmospheric pressure (p_{atm}).

Date & time	Latitude/ °N	Longitude/ °E	θ_0/ °C	S_0	p_{atm}/ kPa	$c(N_2O, LGR)$/ $(nmol\,L^{-1})$	$c(N_2O, GC\text{-}MS)$/ $(nmol\,L^{-1})$	$c_{eq}(\theta_0, S_0\, p_{atm})$/ $(nmol\,L^{-1})$
23/01/2012, 19:53	−63.4	−53.0	0.56	34.31	101	15.44 ± 0.02	15.77 ± 0.3	14.90 ± 0.2
24/01/2012, 03:48	−63.5	−52.1	0.24	34.02	101	15.36 ± 0.02	15.58 ± 0.3	15.12 ± 0.2
25/01/2012, 06:35	−63.3	−53.3	0.46	34.28	100	15.52 ± 0.02	15.58 ± 0.3	14.96 ± 0.2

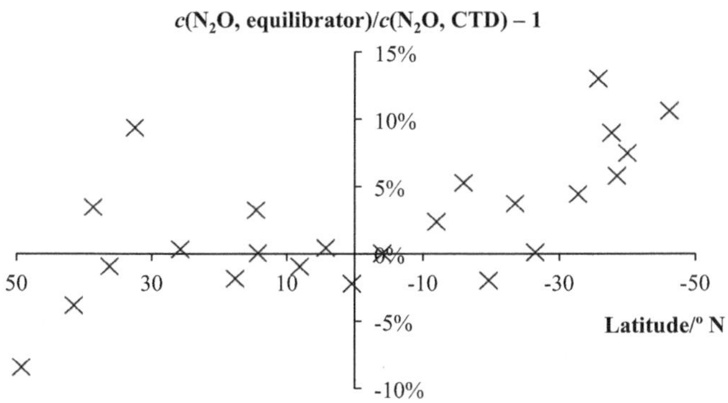

Figure 9. Difference between equilibrator and discrete N_2O measurements from the CTD rosette (Rhee et al., 2009).

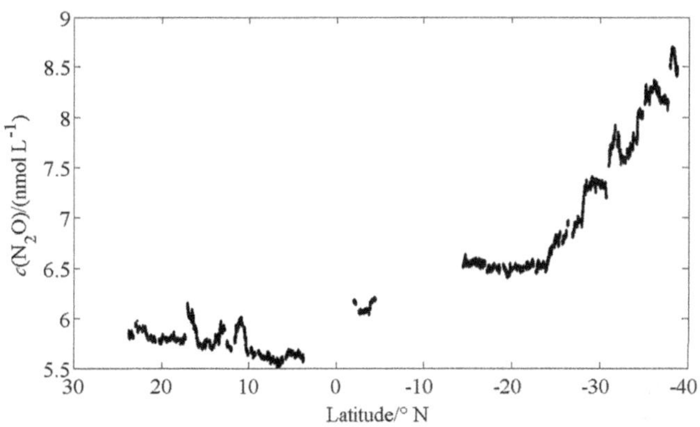

Figure 10. N_2O concentrations in surface waters during the cruise.

measurements, this could point towards a potential offset, either due to the analytical method or source of the water. In subsequent tests, the flow at the equilibrator vent was measured as $\pm 0.3\,mL\,min^{-1}$. As the gas flow through the equilibrator is much higher (100 and 400 $mL\,min^{-1}$ respectively during AMT20 and JR255A), this cannot explain the lower N_2O concentrations measured by the equilibrator. Also, the N_2O mole fraction in laboratory air is close to or slightly above that of dissolved air, so the minor vent flow cannot explain the small, statistically insignificant differences between equilibrator and discrete measurements. Another potential explanation could be the consumption of N_2O in anoxic biofilms within the ship's seawater pipes (Juranek et al., 2010). Foster et al. (2009) also cite a personal communication by Hermann Bange, who found loss of methane in underway sampling systems, but apparently no problems for N_2O.

There is actually further evidence for offsets between discrete and underway N_2O measurements during a previous AMT cruise (AMT7, 1998) in the paper by Rhee et al. (2009). On average, the difference of equilibrator to discrete CTD measurements is $(2.2 \pm 5.7)\,\%$, but the range of deviations is from -8 to $+13\,\%$, with lower values nearer the beginning of the cruise at 50° N (Fig. 9).

In future, GC-MS or GC-ECD samples from the pumped seawater supply should be compared with rosette samples to check that there are no biases due to the condition of the ship's seawater intake.

3.3 Surface ocean measurements

The coupled system of the N_2O analyser and equilibrator was tested in the field during AMT20 in boreal autumn 2010. The system worked well initially, but two problems occurred: (1) stagnant water flow through the equilibrator on several occasions, and (2) the measured values for the N_2O reference gas drifted (Fig. 6).

Measurements of dissolved N_2O in the surface ocean were collected between 24° N and 39° S. Due to the problems with the laser no data was collected between 4° N and 2° S and between 5 and 14° S (Fig. 4). N_2O concentrations in surface waters ranged from 5.5 to 8.6 $nmol\,L^{-1}$, with lowest average concentrations measured in the North Atlantic Gyre between 24 and 11° N (Fig. 10, Table 2). Surface waters were slightly undersaturated. However, towards the southern limb of the gyre, an increase in N_2O saturations above mean values of 99.0 % were observed on three occasions between 20 and 11° N. These periods lasted 14, 4 and 12 h, respectively, while saturations increased up to 104 % (Fig. 11b, arrows).

Potential sources for N_2O could be entrainment of deep waters into the mixed layer or advection from the oxygen minimum zone of the Mauritanian upwelling. However, no correlation between N_2O saturation and sea surface temperature or O_2 saturation was observed. Therefore, deep-water entrainment is unlikely to be the cause of the relatively high N_2O saturations. Another potential source could be in situ

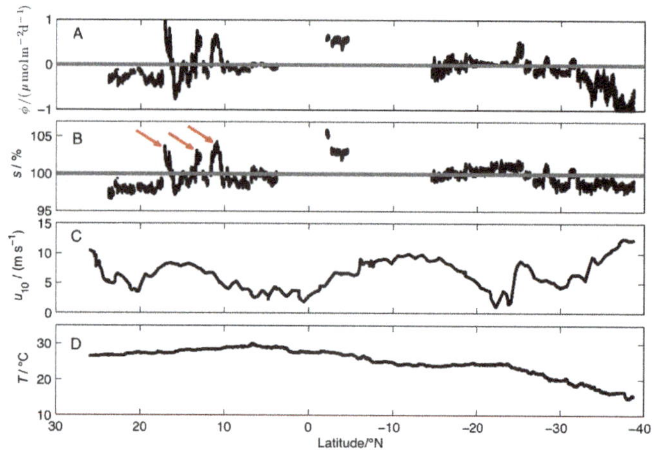

Figure 11. **(A)**: Sea-to-air N_2O flux, grey line denotes zero flux. Positive values indicate fluxes from sea to air. **(B)**: Surface water saturations, equilibrium saturation 100 % indicated by grey line, arrows mark saturation-peaks between 20 and 10° N (see text for details). **(C)**: Wind speed at 10 m height from ECMWF climatology for AMT20. **(D)**: Sea surface temperature at inlet.

Table 2. Mean N_2O concentration, saturation and air–sea flux for the northern gyre (24–11° N), equatorial region (11° N–5° S, with gaps between 4° N and 2° S) and southern gyre (14–39° S).

Latitude	c (N_2O)/ (nmol L^{-1})	s (N_2O)/ %	Φ/ (μmol m^{-2} d^{-1})
24–11° N	5.8 ± 0.1	99.0 ± 1.6	−0.14 ± 0.31
11° N–5° S	5.8 ± 0.2	100.4 ± 1.8	0.11 ± 0.26
14–39° S	7.1 ± 0.7	99.7 ± 1.0	−0.16 ± 0.33

production by nitrification or nitrifier denitrification. High rates of nitrogen fixation were previously observed in this region (Moore et al., 2009) and could provide a substrate for N_2O-producing bacteria. Forster et al. (2009) found average saturations of 104 % during spring but 97 % saturation during autumn in the latitude band between 26 and 11° N, comparable to the 99 % mean saturation measured in autumn for this study. More data are needed to confirm whether there is a real seasonal trend towards lower N_2O saturations in the North Atlantic Gyre during boreal autumn.

The highest saturations of up to 107 % were found close to the Equator. However, average saturations for the equatorial region between 11° N and 5° S were only 100.4 %. Surface saturations of on average 104–109 % were previously reported for this region (Walter et al., 2004; Forster et al., 2009; Oudot et al., 2002). Rhee et al. (2009) found maximum saturations of 110 % at the Equator. Unfortunately, no data could be collected directly at the Equator, due to analyser maintenance between 5° N and 2° S. High surface saturations can be expected due to equatorial upwelling of N_2O-rich waters. N_2O saturations of 99.3 % and the lowest N_2O concentrations were measured between 10.6–5.8° N and 27.5–31.5° W,

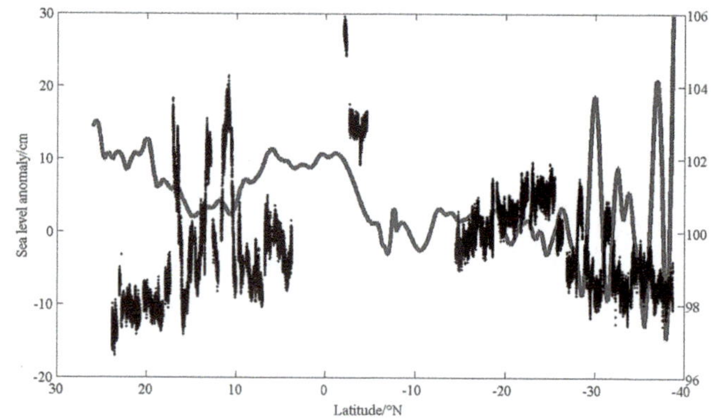

Figure 12. Meridional variations in sea level height anomaly (grey line, left axis) and N_2O saturation (black dots, right axis).

associated with low salinities. Walter et al. (2004) observed similarly low saturations of about 100 % in this region and related it to a retroflection of the North Brazil Current, advecting low-salinity Amazon plume waters into the North Equatorial Countercurrent (NECC).

N_2O concentrations increased south of 14° S and reached mean values of 7.7 nmol L^{-1} between 25 and 39° S. While surface waters in the latitudinal band of 14–25° S were on average in equilibrium with the atmosphere, saturations decreased south of 25° S (Fig. 11b) as water temperature decreased. Mean saturations between 14 and 39° S were 99.7 %, similar to 101 % saturation observed in austral spring (Forster et al., 2009; Rhee et al., 2009), while average saturations in austral autumn were higher (104 %, Forster et al., 2009). This was attributed to accumulated N_2O production during spring and summer.

Eddies can bring thermocline waters with higher nutrient and N_2O concentrations into the mixed layer (McGillicuddy et al., 2007). This might stimulate in situ N_2O production from remineralisation of additional biomass as well as increase the mixed layer inventory simply due to mixing with deeper waters with higher N_2O concentrations. Satellite altimeter products from Ssalto/Duacs (gridded sea level anomalies, 1/3° × 1/3° grid; http://www.aviso.oceanobs.com/en/data/products/sea-surface-height-products/global.html) were used to trace eddies. No clear relationship between sea level anomalies (SLA) and saturations was observed in the northern gyre (Fig. 12). Between 28 and 36° S, however, higher N_2O saturations seem to be associated with negative SLAs, while lower saturations occur with positive SLAs. This could point towards upwelling eddies, introducing waters with higher N_2O concentrations, possibly originating from the Benguela upwelling, to the South Atlantic Gyre region.

Generally, oligotrophic gyres are expected to be weak N_2O sinks, especially in winter, due to thermal effects, with a potential for weak sources in summer while coastal and equatorial upwelling zones are sources of N_2O (Nevison et al.,

1995; Suntharalingam and Sarmiento, 2000). The gyres in both hemispheres were acting as sinks for atmospheric N_2O at the time of the survey, due to slightly undersaturated surface waters (Fig. 11a). Average fluxes for the region between 24 and 11° N were $(-0.14 \pm 0.31)\,\mu mol\,m^{-2}\,d^{-1}$ and $(-0.16 \pm 0.33)\,\mu mol\,m^{-2}\,d^{-1}$ for 14–39° S (Table 2). For comparison, Forster et al. (2009) observed negative sea-to-air N_2O fluxes between −0.02 and $-0.04\,\mu mol\,m^{-2}\,d^{-1}$ between 26 and 11° N during autumn, corresponding to weak N_2O uptake. Fluxes in spring were positive. The southern gyre was found to be a source of N_2O at all times. They pointed out, that emissions in spring were four times higher between 6–40° S than in autumn, rather due to varying N_2O inventories in the mixed layer than to changing wind speeds. This points towards remarkable interannual differences in the oceanic N_2O source and, as shown in this study, also intra-annual variation.

The equatorial region was a source of N_2O to the atmosphere where high surface saturations coincided with relatively high wind speeds. The average flux was $0.53\,\mu mol\,m^{-2}\,d^{-1}$ between 10° and 3° N, which is most likely an underestimation, due to analyser downtime. Comparable values of $0.52\,\mu mol\,m^{-2}\,d^{-1}$ were previously observed for the latitudinal band between 12 and 1.5° N (Walter et al., 2004). For latitudes between 11° N and 5° S, the Western Tropical Atlantic Longhurst province, emissions of $(0.11 \pm 0.26)\,\mu mol\,m^{-2}\,d^{-1}$ were measured. For comparison, 0.16–0.33 $\mu mol\,m^{-2}\,d^{-1}$ were previously reported from the Atlantic Ocean during austral spring (Forster et al., 2009). There might be seasonal variability, as higher emissions were observed during austral autumn (1.17–2.13 $\mu mol\,m^{-2}\,d^{-1}$; Forster et al., 2009).

Generally, saturations reported here are within the lower range of previously published values for the tropical and subtropical Atlantic which might be due to N_2O consumption in anoxic biofilms within the pumped seawater system as discussed above.

4 Conclusions and outlook

Laboratory and field tests showed that the Los Gatos N_2O/CO analyser can be coupled with an equilibrator to reliably measure both, atmospheric and marine N_2O concentrations. Small-scale changes of concentrations could be observed, giving a very detailed picture of the marine N_2O budget. This is important for monitoring environments that are highly variable in space and time, such as fronts, upwelling and coastal regions.

The system is virtually ready for deployment on platforms of opportunity as shown for a similar setup for methane and carbon dioxide measurements (Gülzow et al., 2011). It records high-resolution data (N_2O mole fractions recorded at a frequency of $1\,s^{-1}$, $\tau = 203\,s$) while operation is low maintenance and can be easily automated. This is an advantage to labour-intensive discrete sampling techniques. Calibration can be automated and atmospheric and marine dissolved gas measurements can be analysed in alternation using the same instrument. The resolution of the described system is good (relaxation time of 140 s for a headspace flow rate of 400 mL min^{-1}), but response times could be reduced further by increasing the headspace flow, decreasing the headspace volume or, potentially, by decreasing the instrument operating pressure and measurement cell volume. The instrument is protected from water by a cold trap and a solenoid valve connected to a humidity detector. The cold trap currently requires manually draining every 2–3 days, but this could be automated as for CO_2 analysers. Another remaining operator-dependent task is cleaning of the seawater flow regulator, which could be avoided by using screens and prefilters. However, this is a problem common to all equilibrator techniques. Measurements of depth profiles would require large sample sizes or a pumped CTD due to the relatively long relaxation time of the equilibrator. For depth profiles, headspace sampling and GC analysis of discrete samples would therefore be the preferred option.

In addition, an effective underway system should comprise the following characteristics:

- A calibration for temperature probes measuring seawater temperature in the equilibrator to a precision better than 0.01 °C.
- Use of a flow restrictor or flow regulator to control water flow through the equilibrator.
- Use of three reference gases for improved calibration and drift correction in case the difference between seawater concentrations and atmospheric values are is higher than experienced during AMT20.

The laser drift experienced during AMT20 does not relate to the experimental setup and the instrument has been deployed successfully after laser replacement without further issues. Saturation values from measurements with the analyser-equilibrator setup are comparable to previous studies using gas chromatography and the agreement with atmospheric AGAGE measurements is excellent. First comparisons with GC-MS measurements show agreement between the two methods within measurement uncertainties. Precision for the ICOS analyser under field conditions after correction for analyser offset and drift are

- 0.2 nmol mol^{-1} over 40 min for reference gas injections
- 0.5 to 0.7 nmol mol^{-1} for atmospheric measurements
- 0.2 nmol mol^{-1} over 5 min for equilibrator measurements.

Further data comparison with GC-ECD measurements in the laboratory and during field campaigns should be implemented in the future, as well as testing for differences between rosette and underway samples. Another interesting test would be comparing the performance of this coupled equilibrator-analyser setup with other equilibrator types and laser-based N_2O analysers of different manufacturers, e.g. as part of the European FP7 InGOS project (http://www.ingos-infrastructure.eu), the SCOR Working Group on dissolved N_2O and CH_4 measurements and the SOLAS and COST-sponsored CH_4-N_2O database MEMENTO (http://memento.geomar.de).

Acknowledgements. We would like to thank captain and crew of RRS *James Cook* and principal scientist Andy Rees for their support during AMT20, as well as BODC for underway sea surface data. Helpful discussions with Grant Forster and Sunke Schmidtko (UEA) are gratefully acknowledged. We would also like to thank Doug Baer and Robert Provencal (Los Gatos Research) for support with the N_2O/CO analyser and Dorothee Bakker for providing the equilibrator used in this study. Special thanks go to Gareth A. Lee for invaluable help with the laboratory experiments and fieldwork preparation. This study was supported by the European Community's Seventh Framework Programme (FP7/2007-2013) under grant agreement number 237890 (Marie Curie Initial Training Network "INTRAMIF") and UK Natural Environment Research Council National Capability funding to Plymouth Marine Laboratory and the National Oceanography Centre, Southampton. This is contribution number 222 of the AMT programme.

Edited by: M. Hoppema

References

Arévalo-Martínez, D. L., Beyer, M., Krumbholz, M., Piller, I., Kock, A., Steinhoff, T., Körtzinger, A., and Bange, H. W.: A new method for continuous measurements of oceanic and atmospheric N_2O, CO and CO_2: performance of off-axis integrated cavity output spectroscopy (OA-ICOS) coupled to non-dispersive infrared detection (NDIR), Ocean Sci., 9, 1071–1087, doi:10.5194/os-9-1071-2013, 2013.

Baer, D. S., Paul, J. B., Gupta, M., and O'Keefe, A.: Sensitive absorption measurements in the near-infrared region using off-axis integrated-cavity-output spectroscopy, Appl. Phys. B, 75, 261–265, 2002.

Becker, M., Andersen, N., Fiedler, B., Fietzek, P., Körtzinger, A., Steinhoff, T., and Friedrichs, G.: Using cavity ringdown spectroscopy for continuous monitoring of $\delta^{13}C(CO_2)$ and fCO_2 in the surface ocean, Limnol. Oceanogr., 10, 752–766, 2012.

Butler, J. H., Elkins, J. W., Thompson, T. M., and Egan, K. B.: Tropospheric and dissolved N_2O of the west Pacific and east Indian Oceans during the El Nino Southern Oscillation event of 1987, J. Geophys. Res., 94, 14865–14814, 1989.

Codispoti, L. A.: Interesting times for marine N_2O, Science, 327, 1339, 2010.

Cohen, Y. and Gordon, L. I.: Nitrous oxide in the oxygen minimum of the eastern tropical North Pacific: Evidence for its consumption during denitrification and possible mechanisms for its production, Deep Sea Res., 25, 509–524, 1978.

Cooper, D. J., Watson, A. J., and Ling, R. D.: Variation of pCO_2 along a North Atlantic shipping route (UK to the Caribbean): A year of automated observations, Marine Chem., 60, 147–164, 1998.

Crutzen, P. J.: The influence of nitrogen oxides on the atmospheric ozone content, Q. J. R. Meteorol. Soc., 96, 320–325, 1970.

Denman, K. L., Brasseur, G., Chidthaisong, A., Ciais, P., Cox, P. M., Dickinson, R. E., Hauglustaine, D., Heinze, C., Holland, E., Jacob, D., Lohmann, U., Ramachandran, S., da Silva Dias, P. L., Wofsy, S. C., and Zhang, X.: Couplings Between Changes in the Climate System and Biogeochemistry, in: Climate Change 2007: The Physical Science Basis, Contribution of Working Group I to the Fourth Assessment Report of the Intergovernmental Panel on Climate Change, edited by: Solomon, S., Qin, D., Manning, M., Chen, Z., Marquis, M., Averyt, K. B., Tignor, M., and Miller, H. L., Cambridge University Press, Cambridge, United Kingdom and New York, NY, USA, 2007.

Elkins, J. W., Wofsy, S. C., McElroy, M. B., Kolb, C. E., and Kaplan, W. A.: Aquatic sources and sinks for nitrous oxide, Nature, 275, 602–606, 1978.

Forster, G., Upstill-Goddard, R. C., Gist, N., Robinson, C., Uher, G., and Woodward, E. M. S.: Nitrous oxide and methane in the Atlantic Ocean between 50 degrees N and 52 degrees S: Latitudinal distribution and sea-to-air flux, Deep-Sea Res. Pt. II, 56, 964–976, doi:10.1016/j.dsr2.2008.12.002, 2009.

Forster, P., Ramaswamy, V., Artaxo, P., Berntsen, T., Betts, R., Fahey, D. W., Haywood, J., Lean, J., Lowe, D. C., Myhre, G., Nganga, J., Prinn, R., Raga, G., Schulz, M., and Van Dorland, R.: Changes in Atmospheric Constituents and Radiative Forcing, in: Climate Change 2007: The Physical Science Basis. Contribution of Working Group I to the Fourth Assessment Report of the Intergovernmental Panel on Climate Change, edited by: Solomon, S., Qin, D., Manning, M., Chen, Z., Marquis, M., Averyt, K. B., Tignor, M., and Miller, H. L., Cambridge University Press, Cambridge, United Kingdom and New York, NY, USA, 2007.

Goreau, T. J., Kaplan, W. A., Wofsy, S. C., McElroy, M. B., Valois, F. W., and Watson, S. W.: Production of NO_2^- and N_2O by nitrifying bacteria at reduced concentrations of oxygen, Appl. Environ. Microbiol., 40, 526–532, 1980.

Gülzow, W., Rehder, G., Schneider, B., Deimling, J. S., and Sadkowiak, B.: A new method for continuous measurement of methane and carbon dioxide in surface waters using off-axis integrated cavity output spectroscopy (ICOS): An example from the Baltic Sea, Limnol. Oceanogr. Methods, 9, 176–184, 2011.

Juranek, L. W., Hamme, R. C., Kaiser, J., Wanninkhof, R., and Quay, P. D.: Evidence of O_2 consumption in underway seawater lines: Implications for air-sea O_2 and CO_2 fluxes, Geophys. Res. Lett., 37, L01601, doi:10.1029/2009GL040423, 2010.

Knowles, R.: Denitrification, Microbiol. Rev., 46, 43–70, 1982.

Liss, P. S. and Merlivat, L.: Air-sea gas exchange rates: Introduction and synthesis, The role of air-sea exchange in geochemical cycling, 185, 113–127, 1986.

Löscher, C. R., Kock, A., Könneke, M., LaRoche, J., Bange, H. W., and Schmitz, R. A.: Production of oceanic nitrous oxide by ammonia-oxidizing archaea, Biogeosciences, 9, 2419–2429, doi:10.5194/bg-9-2419-2012, 2012.

McGillicuddy, D. J., Anderson, L. A., Bates, N. R., Bibby, T., Buesseler, K. O., Carlson, C. A., Davis, C. S., Ewart, C., Falkowski, P. G., and Goldthwait, S. A.: Eddy/wind interactions stimulate extraordinary mid-ocean plankton blooms, Science, 316, 1021–1026, 2007.

McIlvin, M. R. and Casciotti, K. L.: Fully automated system for stable isotopic analyses of dissolved nitrous oxide at natural abundance levels, Limnol. Oceanogr. Methods, 8, 54–66, 2010.

Moore, C. M., Mills, M. M., Achterberg, E. P., Geider, R. J., LaRoche, J., Lucas, M. I., McDonagh, E. L., Pan, X., Poulton, A. J., and Rijkenberg, M. J. A.: Large-scale distribution of Atlantic nitrogen fixation controlled by iron availability, Nat. Geosci., 2, 867–871, 2009.

Nevison, C. D., Weiss, R. F., and Erickson, D. J.: Global oceanic emissions of nitrous oxide, J. Geophys. Res., 100, 15809–15820, 1995.

Nightingale, P. D., Malin, G., Law, C. S., Watson, A. J., Liss, P. S., Liddicoat, M. I., Boutin, J., and Upstill-Goddard, R. C.: In situ evaluation of air-sea gas exchange parameterizations using novel conservative and volatile tracers, Global Biogeochem. Cy., 14, 373–387, 2000.

Oudot, C., Jean-Baptiste, P., FourrÈ, E., Mormiche, C., Guevel, M., Ternon, J. F., and Le Corre, P.: Transatlantic equatorial distribution of nitrous oxide and methane, Deep Sea Res. Pt. I, 49, 1175–1193, 2002.

Popp, B. N., Westley, M. B., Toyoda, S., Miwa, T., Dore, J. E., Yoshida, N., Rust, T. M., Sansone, F. J., Russ, M. E., Ostrom, N. E., and Ostrom, P. H.: Nitrogen and oxygen isotopomeric constraints on the origins and sea-to-air flux of N_2O in the oligotrophic subtropical North Pacific gyre, Global Biogeochem. Cy., 16, 12-1–12-10, doi:10.1029/2001gb001806, 2002.

Poth, M. and Focht, D. D.: ^{15}N kinetic analysis of N_2O production by Nitrosomonas europaea: an examination of nitrifier denitrification, Appl. Environ. Microbiol., 49, 1134–1141, 1985.

Prinn, R., Weiss, R., Fraser, P., Simmonds, P., Cunnold, D., Alyea, F., O'Doherty, S., Salameh, P., Miller, B., and Huang, J.: A history of chemically and radiatively important gases in air deduced from ALE/GAGE/AGAGE, J. Geophys. Res. Atmos. (1984–2012), 105, 17751–17792, 2000.

Rafelski, L. E., Paplawsky, B., and Keeling, R. F.: An equilibrator system to measure dissolved oxygen and its isotopes, J. Atmos. Ocean. Technol., 30, 361–377, doi:10.1175/JTECH-D-12-00074.1, 2012.

Ravishankara, A. R., Daniel, J. S., and Portmann, R. W.: Nitrous oxide (N_2O): the dominant ozone-depleting substance emitted in the 21st century, Science, 326, 123–125, doi:10.1126/science.1176985, 2009.

Reuer, M. K., Barnett, B. A., Bender, M. L., Falkowski, P. G., and Hendricks, M. B.: New estimates of Southern Ocean biological production rates from O_2/Ar ratios and the triple isotope composition of O_2, Deep Sea Res. Pt. I, 54, 951–974, 2007.

Rhee, T. S., Kettle, A. J., and Andreae, M. O.: Methane and nitrous oxide emissions from the ocean: A reassessment using basin-wide observations in the Atlantic, J. Geophys. Res., 114, D12304, doi:10.1029/2008JD011662, 2009.

Röckmann, T., Kaiser, J., Brenninkmeijer, C. A. M., and Brand, W. A.: Gas chromatography/isotope-ratio mass spectrometry method for high-precision position-dependent ^{15}N and ^{18}O measurements of atmospheric nitrous oxide, Rapid Communications Mass Spectrometry, 17, 1897–1908, doi:10.1002/rcm.1132, 2003.

Suntharalingam, P. and Sarmiento, J. L.: Factors governing the oceanic nitrous oxide distribution: Simulations with an ocean general circulation model, Global Biogeochem. Cy., 14, 429–454, 2000.

Sutka, R. L., Ostrom, N. E., Ostrom, P. H., and Phanikumar, M. S.: Stable nitrogen isotope dynamics of dissolved nitrate in a transect from the North Pacific Subtropical Gyre to the Eastern Tropical North Pacific, Geochim. Cosmochim. Acta, 68, 517–527, doi:10.1016/S0016-7037(03)00483-6, 2004.

Sutka, R. L., Ostrom, N. E., Ostrom, P. H., Breznak, J. A., Gandhi, H., Pitt, A. J., and Li, F.: Distinguishing nitrous oxide production from nitrification and denitrification on the basis of isotopomer abundances, Appl. Environ. Microbiol., 72, 638–644, doi:10.1128/aem.72.1.638-644.2006, 2006.

Sweeney, C., Gloor, E., Jacobsen, A. R., Key, R. M., McKinley, G., Sarmiento, J. L., and Wanninkhof, R.: Constraining global air–sea gas exchange for CO_2 with recent bomb ^{14}C measurements, Global Biogeochem. Cy., 21, GB2015, doi:10.1029/2006GB002784, 2007.

Walter, S., Bange, H. W., and Wallace, D. W. R.: Nitrous oxide in the surface layer of the tropical North Atlantic Ocean along a west to east transect, Geophys. Res. Lett., 31, L23S07, doi:10.1029/2004GL019937, 2004.

Walter, S., Bange, H. W., Breitenbach, U., and Wallace, D. W. R.: Nitrous oxide in the North Atlantic Ocean, Biogeosciences, 3, 607–619, doi:10.5194/bg-3-607-2006, 2006.

Wanninkhof, R.: Relationship between wind speed and gas exchange, J. Geophys. Res, 97, 7373–7382, 1992.

Weiss, R. F.: Determinations of carbon dioxide and methane by dual catalyst flame ionization chromatography and nitrous oxide by electron capture chromatography, J. Chromatogr. Sci., 19, 611–616, 1981.

Weiss, R. F. and Price, B. A.: Nitrous oxide solubility in water and seawater, Marine Chem., 8, 347–359, 1980.

Weiss, R. F., Van Woy, F. A., and Salameh, P. K.: Surface water and atmospheric carbon dioxide and nitrous oxide observations by shipboard automated gas chromatography: Results from expeditions between 1977 and 1990, Oak Ridge National Lab., TN (United States), Carbon Dioxide Information Analysis Center, 1992.

Wuchter, C., Abbas, B., Coolen, M. J. L., Herfort, L., Van Bleijswijk, J., Timmers, P., Strous, M., Teira, E., Herndl, G. J., and Middelburg, J. J.: Archaeal nitrification in the ocean, Proc. Natl. Aca. Sci., 103, 12317–12322, doi:10.1073/pnas.0600756103, 2006.

Yoshida, N., Morimoto, H., Hirano, M., Koike, I., Matsuo, S., Wada, E., Saino, T., and Hattori, A.: Nitrification rates and ^{15}N abundances of N_2O and NO_3^- in the western North Pacific, Nature, 341, 895–897, 1989.

Yoshinari, T.: Nitrous oxide in the sea, Marine Chem., 4, 189–202, 1976.

13

Modelling seasonal circulation and thermohaline structure of the Caspian Sea

M. Gunduz[1] **and E. Özsoy**[2]

[1]Institute of Marine Sciences and Technology, Dokuz Eylul University, Izmir, Turkey
[2]Institute of Marine Sciences, Middle East Technical University, Erdemli, Turkey

Correspondence to: M. Gunduz (murat.gunduz@deu.edu.tr)

Abstract. The wind- and buoyancy-driven seasonal circulation of the Caspian Sea is investigated for a better understanding of its basin-wide and mesoscale dynamics, mixing and transport. The model successfully reproduces the following basic elements of the circulation: the southward-flowing current systems along the eastern and western coasts, the upwelling along the eastern coast, the cyclonic circulation in the Middle Caspian Sea (MCS), especially in winter, and the cyclonic and anticyclonic cells of circulation in the South Caspian Sea (SCS). The observed seasonal variability of sea level and sea surface temperature (SST) is well reproduced. Mesoscale structures, not always evident from hydrographic observations, are identified.

1 Introduction

The Caspian Sea (36° to 47° N and 46° to 54° E) is a totally enclosed, landlocked body of water on the Euro-Asian continent. It is a fertile inland sea located between vast deserts of Central Asia, extensive northern plains and the Caucasian high mountain range. Large uncertainties exist with respect to climate change in this zone of sharp transition between climates of the neighbouring Euro-Mediterranean, Siberian, Indian Ocean and Central Asian domains (Bengtsson, 1998; Arpe and Roeckner, 1999; Arpe et al., 2000; Özsoy, 1999; Özsoy et al., 2001). Updated information on the environmental status of the sea can be found at http://www.caspianenvironment.org.

Large variations in sea level, often of abrupt nature, are well documented in hydrological and sedimentary records. Consequent flooding of the adjacent flat land in the past has had detrimental effects on neighbouring societies, for instance on the ancient Khazar state that once occupied its northern shores (Rodionov, 1994). Despite its overwhelming importance, sea level has consistently defied prediction, because of a poor understanding of the water cycle in the extensive catchment basin and, more generally, the specific response of the region to climate. Yet sea level change is often a cumulative indicator of a series of other changes leading up to possible regime shifts between alternative states of the ecosystem. Large variations have been observed in winter mixing, varying from shallow to deep convection, with strong cases of deep ventilation (Kosarev and Yablonskaya, 1994; Dumont, 1998). In one such case, following the abrupt drop of sea level in 1977, dramatic changes in stratification and nutrient regimes are known to have occurred (Kosarev and Yablonskaya, 1994).

Like the neighbouring Black Sea, the Caspian Sea is an enclosed basin with large fresh-water inputs from rivers, the most significant such contribution being from the Volga, with others including the Ural, Kura and Terek. Input from rivers are known to be capable of creating buoyancy-driven shelf currents, as shown by a number of detailed studies (e.g. Kourafalou et al., 1996; Lazure and Jegou, 1998; Chen et al., 1999).

It is challenging to capture and to explain the observed complex features of the Caspian Sea circulation arising from wind and atmospheric/riverine buoyancy forcing acting on this elongated, enclosed sea and worthwhile to study the rather large contrasts between its basins, the variability of flow and mixing processes along its coasts.

The general circulation inferred from hydrographic survey data from the end of the 19th century till the 1950s has been described as being cyclonic in the Middle Caspian Sea (MCS). Instrumental surveys along the western coast of the MCS, carried out in 1935–1937 (Stockman, 1938; Baidin and Kosarev, 1986), have shown predominantly southward currents along the western coast of the basin. Data from oceanographic observations since the 1950s in coastal areas have confirmed southward currents along the western coast of the MCS (Klevtsova, 1967) and a semi-permanent anticyclonic vortex south of the Abşeron peninsula (Tsytsarev, 1967). Southward surface currents along the eastern coast of the MCS (Klevtsova, 1967; Bondarenko, 1993) evident in these observations at first seem to contradict the above-mentioned general cyclonic circulation of the MCS. However, below the surface circulation, northward currents with seasonal variation are known to exist along the eastern coast of the basin, supporting the proposed cyclonic circulation (Kosarev and Yablonskaya, 1994). A southward current east of the Abşeron Peninsula passing to the South Caspian Sea (SCS) has been inferred from hydrographic observations, with a current speed of 50–60 cm s^{-1} at the surface, decreasing to 40–50 cm s^{-1} near the bottom (Kosarev and Yablonskaya, 1994). By making use of satellite-derived SST and chlorophyll data, Sur et al. (2000) have revealed the southward flow of cold water from the north along the west coast, and the northward intrusion of warm water from the SCS to the MCS near the eastern coast. In the shallow North Caspian Sea (NCS), the circulation appears to be almost totally controlled by local winds (Bondarenko, 1993; Terziev et al., 1992). Based on the above-mentioned features, Lednev (1943) has created the schematic representation of the Caspian Sea circulation shown in Fig. 1.

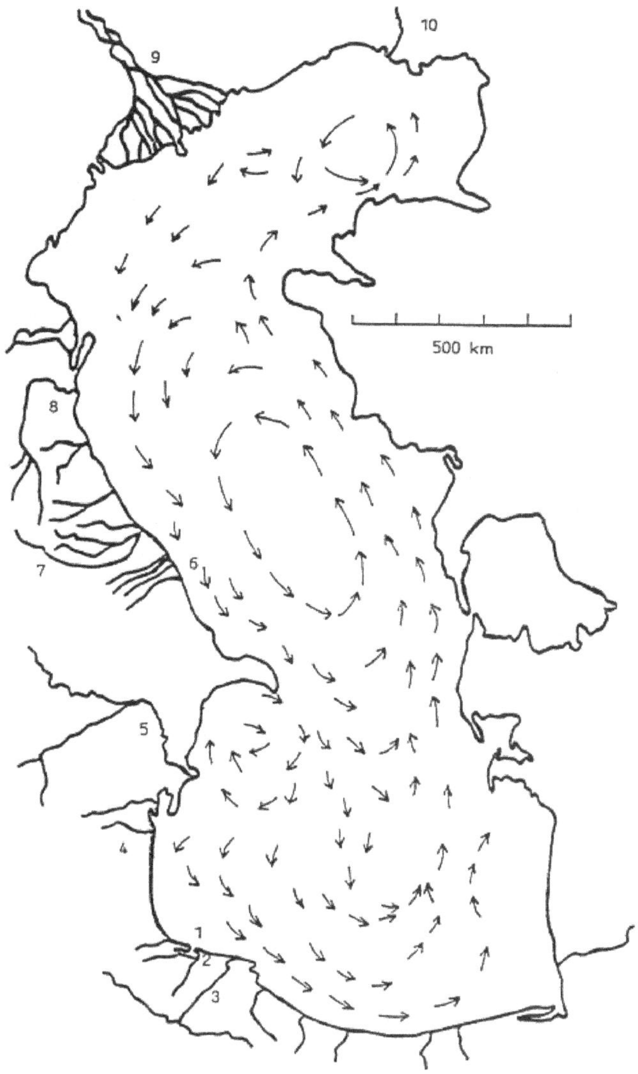

Figure 1. Schematic representation of the Caspian Sea currents (Lednev, 1943).

The construction of the missing elements of the above description of seasonal circulation could hopefully be achieved through details supplied by numerical model studies performed from the mid-1970s until the present. The diagnostic model developed by Sarkisyan et al. (1976) indicated a mainly wind-driven circulation and the effects of thermal stratification on the establishment of this circulation. Diagnostic models of Trukhchev et al. (1995) and Tuzhilkin et al. (1997) have indicated dipole structures, consisting of a cyclonic and anticyclonic pair in the northwest and southeast, respectively, of the MCS, and an anticyclonic and cyclonic pair in the northwest and southeast, respectively, of the SCS basins.

Later, a review and a series of computations of the seasonal circulation based on a coupled sea hydrodynamics, air–sea interaction and sea ice thermodynamics model of the Caspian Sea was made by Ibrayev et al. (2010) establishing the importance of seasonal changes from cyclonic to anticyclonic circulation patterns, topographic generation and steering of currents along the eastern and western coasts. A persistent northward transport by subsurface currents was shown along the eastern shelf slope, despite the surface currents often being directed southward in the same region, confirming the features reviewed above, as well as displaying horizontally and vertically layered structures of the boundary currents over a mild slope. Recently, by assimilating climatological temperature and salinity into a primitive-equation ocean circulation model, Knysh et al. (2008) obtained a cyclonic circulation in the MCS in winter, with the above-mentioned features of a southward surface current and a northward subsurface current along the eastern shelf. Knysh et al. (2008) found a highly variable circulation corresponding to an intense, mostly barotropic circulation in February and a relatively less intense baroclinic circulation in the March–June period.

Faced with an insufficient understanding of the circulation with its various details hidden from current knowledge, our primary focus is to recreate the 3-D seasonal circulation under climatological wind, surface and river buoyancy forcing to discover basin-wide features and mesoscale details and check their consistency with available observations.

2 Bathymetry

Each of the three basins of the Caspian Sea has distinct physical and biological properties (Fig. 2). The NCS is extremely shallow, with an average depth of about 5 m, the Ural Furrow of 8–10 m depth cutting through the otherwise flat region. The NCS is separated from the MCS by an abrupt depth transition at the shelf edge. The depth in the MCS is about 190 m on average, reaching a maximum value of 788 m at the Derbent depression on its western side. The western slope of the MCS is quite steep, whereas the eastern slope is milder. The Garabogazköl, a small enclosure with less than 10 m depth, originally connected to the Caspian Sea and adjacent to the eastern deserts, was recently been dammed to reduce water loss from the Caspian Sea by evaporation. The deep basins of the MCS and the SCS are separated by the Abşeron sill of 150 m maximum depth. The SCS is the deepest basin, with a maximum depth of 1025 m and shallower depressions of 900 to 1000 m depth. The wide continental shelf extending along the eastern side of the SCS is connected to the narrow shelf continuing along the eastern side of the MCS.

3 Model features and setup

Features of the Caspian Sea, such as the weak stratification, deep mixing and strong topographic gradients at continental slopes and sills make the HYbrid Coordinate Ocean Model (HYCOM) (Bleck, 2006) ideally suited for our studies.

The HYCOM, a generalized (hybrid, isopycnal/terrain-following σ/z level) coordinate primitive-equation model, has been widely used for global and regional studies and tested for performance. To cite a few recent studies, the choices of vertical coordinates have been evaluated with regard to the Atlantic Ocean by Chassignet et al. (2003); performance in the Kuroshio Extension region has been tested by Kelly et al. (2007); and the sensitivity to atmospheric forcing in the Black Sea has been investigated by Kara et al. (2005).

The hybrid coordinates extend the range of applicability of the model so as to allow better representation of shallow coastal or weakly stratified waters. The transition from isopycnal coordinates in stratified water to fixed-depth z levels in unstratified surface mixed layer and to terrain-following σ levels in shallow water occurs smoothly by making use of a layer continuity equation (Bleck, 2002).

A Mercator grid (204 × 354) with meridional resolution of 4.4 km and zonal resolution of 0.04 × cos (lat) × 111.2 km corresponding to 3.0–3.3 km is used. HYCOM is capable of migrating between isopycnal, z level and σ level. In this study, the combination of only z level and σ level are used. Thirty hybrid layers consisting of 14 sigma levels to a depth of 50 m and 16 z levels below that depth are used in the vertical.

Table 1. Climatological mean flow values obtained from RivDIS (see online at http://daac.ornl.gov/RIVDIS/rivdis.shtml) for the river discharge into the Caspian Sea.

	Volga ($m^3 s^{-1}$)	Kura ($m^3 s^{-1}$)	Ural ($m^3 s^{-1}$)
Jan	3917.5	467.9	60.9
Feb	3960.2	493.9	56.0
Mar	4032.8	553.0	70.1
Apr	7488.5	813.3	899.5
May	24 021.5	1067.2	1377.3
Jun	20 075.3	850.8	409.7
Jul	8163.6	451.7	198.5
Aug	5590.6	313.7	131.9
Sep	5036.9	326.7	105.3
Oct	5358.7	387.9	98.4
Nov	5450.0	423.8	90.8
Dec	3948.0	461.0	63.4

River runoff is first added to a single ocean grid point and smoothed over surrounding ocean grid points, treated as a contribution to precipitation. Monthly discharges of the Volga River, which account for about 80 % of the climatological mean river discharge of 250 $km^3 yr^{-1}$ (Kosarev and Yablonskaya, 1994), and the two other major rivers Ural and Kura, are applied, based on the Global River Discharge (RivDIS) climatology (Vorosmarty et al., 1997). Table 1 shows the monthly discharge of the rivers from RivDIS data set. The peak discharges of these rivers typically occur in April–July, decreasing during November–December. Exchange with Garabogazköl, currently almost entirely disconnected from the Caspian Sea, has not been taken into consideration.

The gridded temperature and salinity climatological data (0.25° × 0.2° resolution and 0, 5, 10, 15, 20, 25, 30, 50, 75, 100, 125, 150, 200, 250, 300, 400, 500, 600 and 800 m depth levels) of Ibrayev et al. (2001) are used to initialize the model, after interpolation to the model grid.

The atmospheric forcing is based on the seasonal average ECMWF Re-Analysis (ERA-40) (resolution 1.125° × 1.125°) of the European Centre for Medium-Range Weather Forecasts (ECMWF), encompassing 40 years of data (1958–1998). The surface forcing climatology is constructed for the following variables; zonal and meridional components of wind stress, wind speed at 10 m above the sea surface, air temperature, air mixing ratio at 2 m above the sea surface, precipitation, net shortwave radiation and net longwave radiation. The high frequency (6-hourly) component has then been added to the wind stress in the same way as described by Wallcraft et al. (2003), choosing data from the September 1994–September 1995 period as a typical year that would most closely match annual cyclicity in winds.

The model is initialized with monthly mean temperature and salinity climatology and forced with climatological mean seasonal wind stress and surface heat fluxes, with surface

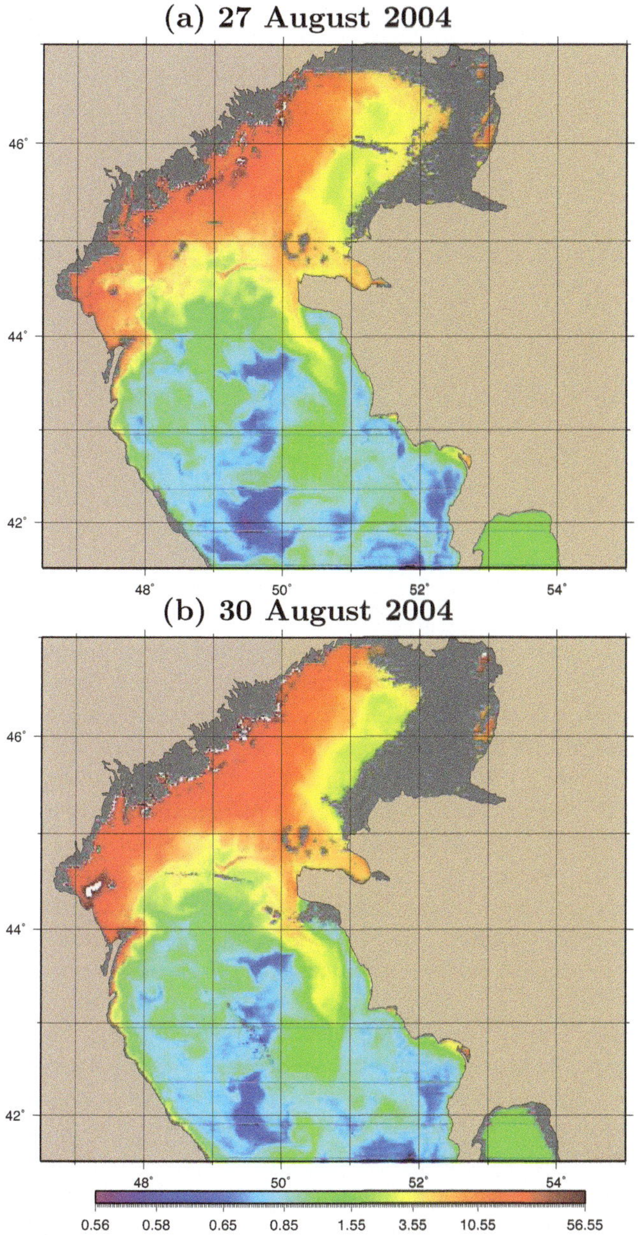

Figure 2. The Caspian Sea model bathymetry.

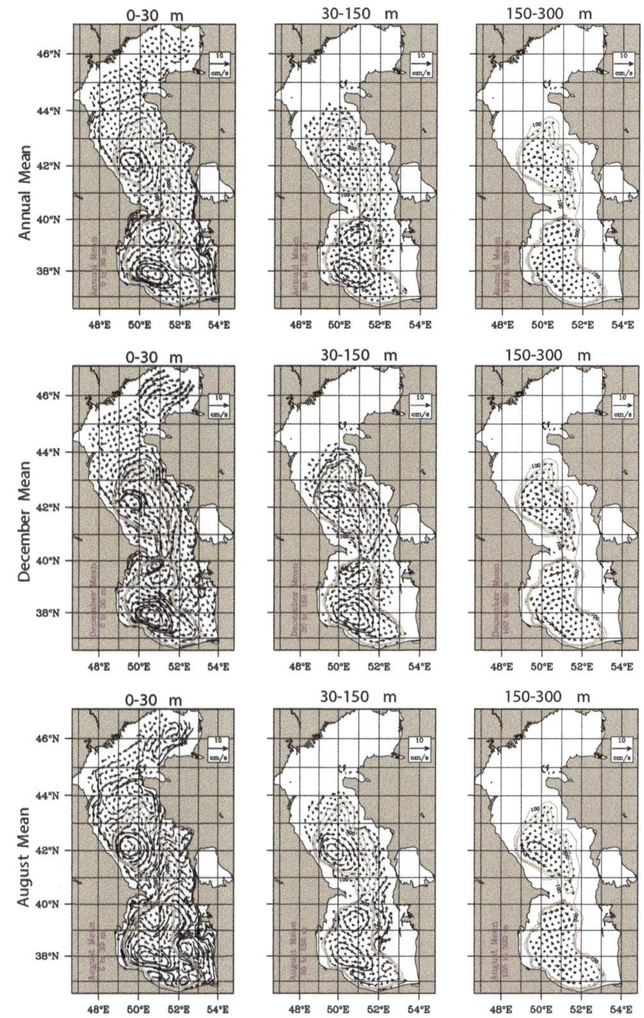

Figure 3. Model-derived mean currents (cm s^{-1}) from a control run averaged over depth intervals of 0–30 m, 30–150 m and 150–300 m. Contour lines shows the 100, 200 and 300 m bathymetry contours. Upper row: annual; middle row: December; lower row: August.

salinity relaxed to the observed seasonal mean values. A run period of about eight model years was sufficient to reach statistical equilibrium (i.e. a perpetual annual cycle, with negligible temporal drift). The model results presented in the following sections are based on either daily or monthly means that were constructed from the last four years of the simulations.

4 Seasonal circulation

Figure 3 (upper) shows the annual mean circulation of the Caspian Sea averaged over 0–30, 30–150 and 150–300 m depth intervals. Annual mean currents are generally found to be weak, with maximum values of 7–8 cm s^{-1} along coastal areas. Although surface currents driven by daily winds display significant variability, this is not reflected in the annual average circulation.

The circulation in the shallow Northern Caspian Sea is especially sensitive to wind direction and speed. Because there are no stable current patterns observed here, the magnitude of the annual mean current is low. The annual mean indicates northward flow in general, with considerable variation from one season to another.

Although the current model has an ice model incorporated into it, the amount of ice predicted by the model was usually less than the observed levels (not shown). The inefficiency of ice formation in the model could possibly be attributed to insufficient heat loss from the shallow northern Caspian Sea

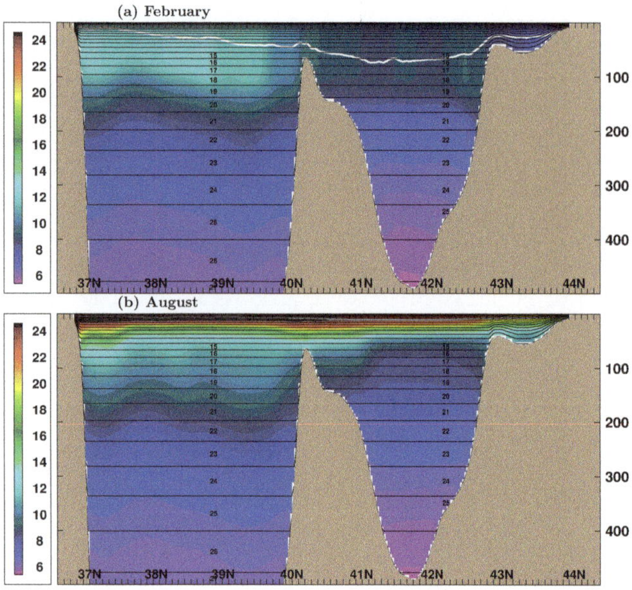

Figure 4. Model-derived monthly mean temperature at 50° E ((**a**) February (**b**) August).

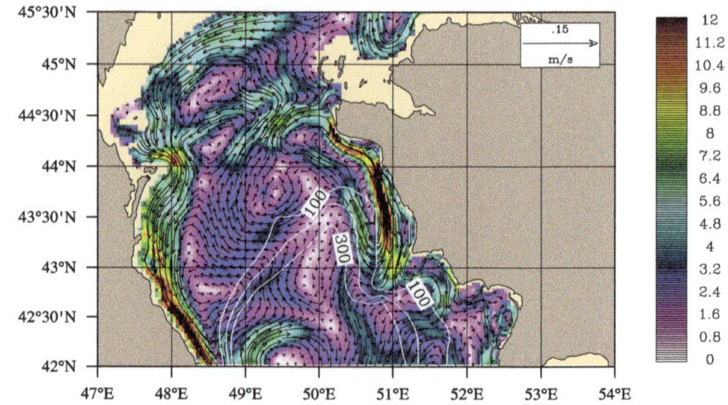

Figure 5. Model-derived current over the MCS in in July. The southward flow along the northeastern coast of the MCS is evident in this figure.

estimated by the model bulk formulae. Firstly, the heat fluxes were obtained from a coarse-resolution atmospheric model used by the ECMWF reanalysis. Secondly, careful evaluations of heat flux estimates compared with the observations (Pettenuzzo et al., 2010) have shown that the ERA-40 reanalysis data set used in our study is unable to close the surface budgets of heat in the case of the Mediterranean Sea unless some corrections are applied. Because we did not apply such corrections, which must be carefully computed for the Caspian Sea, we had to accept the fluxes given by the existing module of HYCOM, although we note that the ice effects are underrepresented in winter months in the NCS as is the case in December (Fig. 3 middle)

The most striking feature of the Caspian Sea circulation is the presence of an anticyclonic gyre to the west of the MCS at about 42° N and the cyclonic and anticyclonic pair of gyres in the SCS, respectively at 38° N and 39.5° N. The intensive anticyclonic gyre in the Derbent depression of the MCS and the two circulation cells located in topographic depressions of the SCS are mostly barotropic, with well-preserved features from the surface to the bottom. All three cells are persistent throughout the year with increasing intensity in summer, consistent with the seasonal winds.

The southward surface currents reproduced by the model along either side of the Caspian Sea are in good agreement with the available observations reviewed above. The southward current seems to be confined to shallow water (depth $<$ 20 m) along both coasts. The intensity of the current along the western coast decreases south of 42° N, as part of the current becomes incorporated into the anticyclonic gyre there and, after recirculating around it, the current enters the SCS on the western side of the Abşeron sill. The southward surface current along the eastern coast of the MCS changes direction below the surface layer to become a northward current at depth, a well-known feature from observations and earlier model results. This current immediately below the surface layer supports the existence of a basin-wide subsurface cyclonic circulation following the coasts of the MCS. The persistent feature identified as the Derbent anticyclone in the western part of the basin is integrated within and interacts with the described cyclonic cell.

Throughout the year, southward flow is evident throughout the water column at the western side of the Abşeron sill offshore of the Abşeron Peninsula, in agreement with observations (Kosarev and Yablonskaya, 1994). In contrast to the southward current on its western side, a northward current occurs on the eastern side of the sill, transporting warm water of the southern basin to the north. This northward current is evident from satellite SST data, often as a vein of warm water originating from the SCS and flowing towards the MCS along the eastern coast. The sill topography appears to be an important factor controlling the exchange between the two basins; the exchange in turn responds to the structure of the seasonal circulation in the adjacent basins.

In the South Caspian Sea, a persistent dipole pattern and a basin-wide cyclonic circulation confined to the 100–300 m bathymetric contours are observed. The cyclonic shallow circulation encircles the inner basin dipole pattern. With regard to the dipole structure, the cyclonic cell is located in the southwestern part of the basin, while the anticyclonic cell is located to its north. The anticyclonic cell, identified as the "Kura anticyclone" by Trukhchev et al. (1995) and Tuzhilkin et al. (1997), was first described by Lednev (1943). These authors also found the cyclonic cell in the southeastern part of the SCS. The dipole pattern is a permanent feature of the SCS, evident during the whole year, with a mainly barotropic structure extending from the surface to the bottom. The southward current originating from the MCS passes the sill offshore of the Abşeron Peninsula, progressing in the

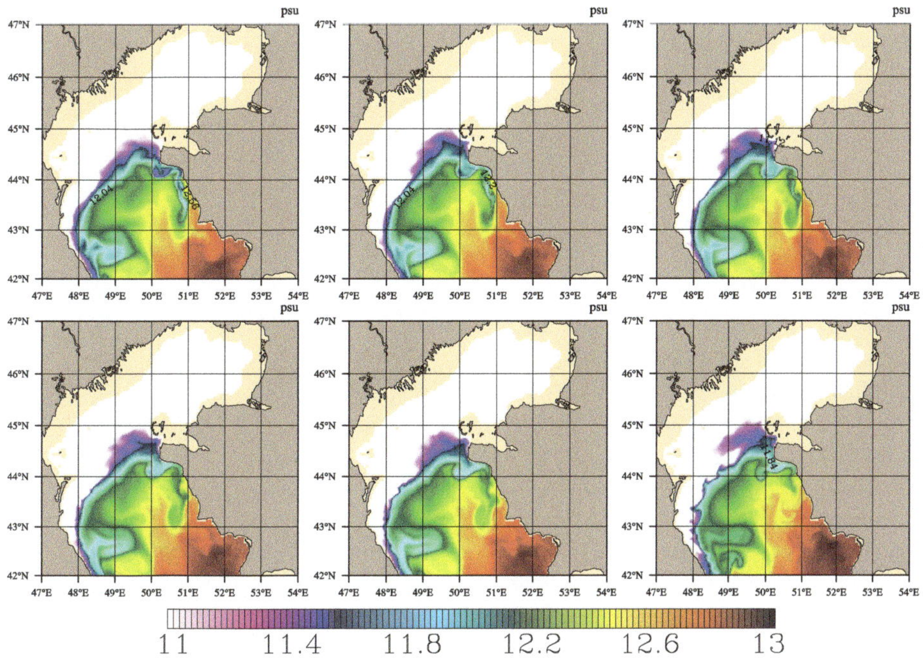

Figure 6. Sea surface salinity (SSS) in practical salinity unit (psu) for 28/30/31 July (upper panel) and 1/2/8 August (lower panel).

form of a coastal jet along the western coast of the SCS. This coastal current reaches its maximum intensity after becoming confined to the steep inner shelf along the west coast of the SCS. The coastal jet follows the shallow bathymetry to turn clockwise around the SCS along the 100–300 m bathymetric contours of the continental slope, finally arriving at the eastern side of the Abşeron sill. Outside this cyclonic circulation cell, southerly flow occurs on the wide eastern continental shelf of the SCS. The double gyre SCS seasonal circulation appears more stable compared to the circulation patterns in the other two basins of the Caspian Sea, in confirmation of Peeters et al. (2000), who showed enhanced vertical stability of this basin, based on the Brunt–Väisälä frequency.

December and August monthly mean currents averaged over depth intervals of 0–30 m, 30–150 m and 150–300 m, are plotted in the middle and lower panels of Fig. 3, respectively, to represent winter and summer regimes of circulation. A cyclonic surface circulation dominates the Caspian Sea from December to March, when the wind blows mainly from the east and northeast directions. Weak currents are found in most of the MCS, except within the coastal jets and the mid-basin anticyclonic cell. The three-cell circulation preserves its structure in winter.

In summer, westward Ekman surface currents across the MCS and upwelling along the eastern coast are driven by strong north to northwesterly prevailing winds. Concurrent with the summer surface circulation, a northward-flowing intermediate depth countercurrent exists along the eastern coast of the MCS. The Derbent anticyclone on the western side of the basin strengthens in summer. This anticyclone has similarly been predicted by Schrum et al. (unpublished data) using the HAMburg Shelf-Ocean-Model(HAMSOM) forced by surface buoyancy and momentum fluxes computed from six hourly ECMWF ERA-15 atmospheric reanalysis data set. The southward advection of fresh water from the Volga river by the southward-flowing coastal jet and the downwelling created at the west coast by the Ekman currents create favourable conditions for the intensification of the Derbent anticyclone in summer. There are two dominant wind regimes in the Caspian Sea. During the winter, wind blows to the west, and in summer the wind direction changes dramatically and the wind starts to blow to the south. Gunduz (2014) has shown that variation of the wind speed and direction (Fig. 5 in the paper) may have an influence on the changing of the cyclonic circulation in winter to the anticyclonic one in summer in the MCS.

From October to January, the northward current flowing along the SCS and the MCS turns west upon reaching the northern continental slope. Then, following the western coast with increased velocity, it completes a basin-wide cyclonic circuit. The northward flow along the eastern coast transports warm water originating from the SCS into the MCS in autumn and winter months. In summer, this flow appears diminished because of the strong Ekman circulation that creates upwelling along the eastern coast. The Derbent anticyclone is weaker in winter and almost diminished in April but rejuvenates in June.

It is evident from the above description that the Caspian Sea general circulation is dominated by a basin-scale cyclonic "rim current" following its narrow coastal topography and continental slopes wherever these slopes are sharply defined as well as cyclonic/anticyclonic gyres present within this circulation system in the deep basins of the MCS and SCS. The topography of the basins and the Abşeron sill play important roles in steering this circulation. Coastal jet flows arise especially along the slopes of the western coast, only to be interrupted by the sharp protrusion of the Abşeron Peninsula. The circulation along the wider continental shelves and continental slopes of the corrugated eastern coast (see Fig. 2) is more complex, showing a reversal of direction in the vertical and a complicated horizontal structure. The general circulation appears to be mainly wind driven, while sources of buoyancy also play significant roles.

The most challenging task for the modelling of enclosed seas like the Caspian Sea is to properly represent the annual thermal cycle, since the thermal regime changes from an entirely thermally mixed one in winter to a strongly stratified one in summer (see Beletsky and Schwab (2001), who detected the same sequence in the case of Lake Michigan). Our simulations with the HYCOM Caspian Sea model were able to reproduce the following main features of the Caspian Sea thermal structure.

Figure 4 shows the temperature cross sections along 50° E in February and August. In this figure, the white line shows the base of the mixed layer. In February, there are considerable differences in mixed layer depth (MLD) between the two deep basins of the Caspian Sea. While the MLD in MCS is about 100 m in winter, it only reaches to 20 m in the SCS. The most striking feature of the temperature stratification is the sharp contrast between the vertical gradients and the relative levels of temperature in the MCS and SCS basins, which results from the difference in mixing and the effect of the Abşeron sill separating the two basins.

Since the density in the Caspian Sea is largely determined by temperature (in the absence of a strong salinity stratification), the temperature contrasts between basins reflect pressure gradients developing as a result of density differences. The largest temperature differences at sill level develop in late winter and early spring when differential mixing and cooling occurs in the two basins. As a result, the dense water in the MCS is expected to flow south across the Abşeron sill to the SCS. The weakness of thermal stratification in the SCS during the following periods appears partly to be a result of the sill overflow and ensuing thermocline-level injections of dense water from the MCS into the SCS. Temperature isolines of colder water on the Abşeron sill area connected to those at the thermocline level of the SCS provide limited evidence in this direction.

A well-developed thermocline generally persists throughout the summer (Fig. 4b), when the mixed layer depth typically decreases to about 15 m. As the stratification develops in either the MCS and the SCS basins the level of the thermocline becomes distinctly different between the basins. While the 11° C temperature occurs at about 50 m depth in MCS in August, it occurs at a depth of about 130 m in the SCS.

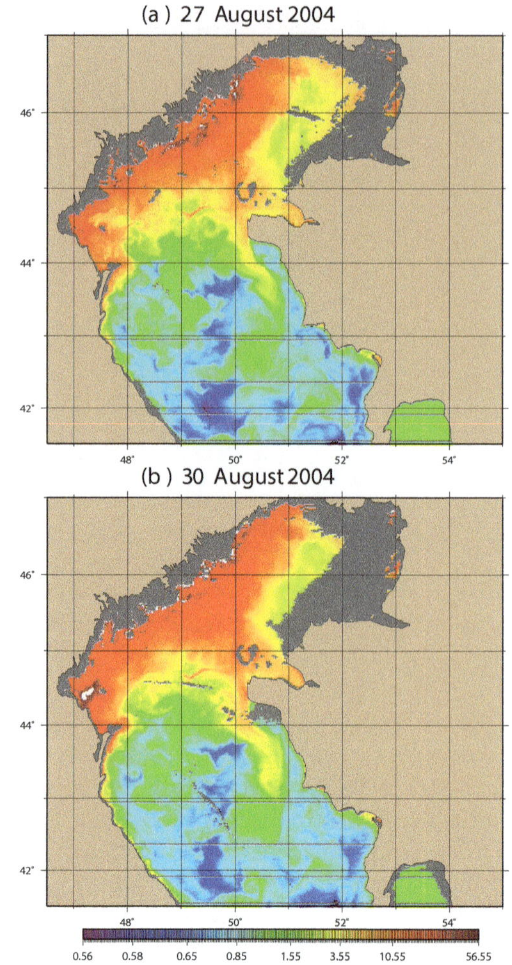

Figure 7. Chlorophyll concentration (in mg m^{-3}) on **(a)** 27 August 2004 and **(b)** 30 August 2004. The chlorophyll concentration contours are plotted with different colours and intervals for a better visualization of the structure.

5 Mesoscale circulation

The coastal jets described above and intense circulation cells often become unstable and create eddies and other mesoscale activity responsible for local mixing and transport. A similar role of the rim current is well known in the case of the neighbouring Black Sea (Rachev and Stanev, 1997). The circulation elements also influence preconditioning for wintertime convection contributing to deep/intermediate-water formation.

Model-derived and satellite-detected sea surface temperature (SST), height (SSH) and salinity (SSS) and other fields are often dominated by mesoscale features with patterns that are quite different from the mean circulation patterns discussed in the last section. While the mean fields are dominated by the coastal jets, rim currents and basin-scale gyres, the snapshots are dominated by eddies with horizontal scales

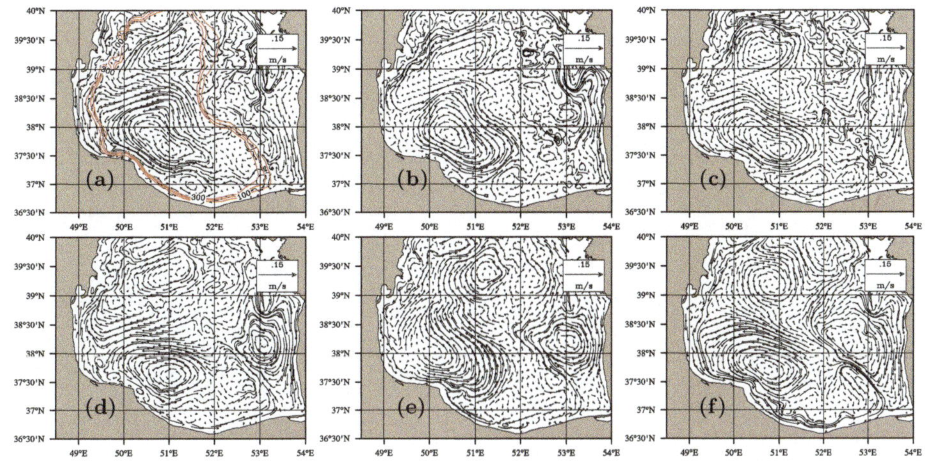

Figure 8. The evolution of the surface flow field in the SCS on 13 July, 12 August and 11 September (upper panels) and 11 October, 10 November and 10 December (lower panels); the red line in the first figure shows the 100, 200 and 300 m bathymetry contour.

of 10–100 km, especially in the upwelling region, along the coastal boundaries and near dense-water formation sites.

Figure 5 shows 0–30 m average currents in the MCS for the month of July, where a complex pattern of basin circulation is displayed, together with intense southerly coastal jet flows along the western and northeastern coasts. While the flow along the western coast is a relatively steady feature of the circulation, the jet flow near the eastern coast of the MCS which later separates from the coast towards the south is a seasonal feature that starts forming in June and continues to intensify until August. At the headland just north of 43° N on the eastern coast, the flow separates from the coast to divide into two branches: while one branch turns to the north and follows the topographic slope between the 100 and 300 m depth contours, the other branch flows west towards the interior of the sea. This coastal jet advects relatively fresh water of high chlorophyll content from the NCS towards the MCS along the eastern coast. Later, in September, the branch that flows into the interior of the sea increases in intensity. The jet completely disappears in winter, replaced by a northward current joining the basin cyclonic circulation in the same area (see Fig. 3 lower).

Sea surface salinity (SSS) on 28/30/31 July and 1/2/8 August in Fig. 6 shows the coastal jet along the eastern coast of the MCS advecting cold, low-salinity water towards the south. A mushroom like pattern associated with coastal convergence north of the anticyclonic cell on the west coast pumps low-salinity coastal waters towards the interior of the Sea.

Analysis of satellite-derived chlorophyll distribution in Fig. 7 shows southward transport along the western coast of the Caspian Sea in selected cases from August 2004. The development of the southward coastal jet in the northeastern part of the MCS is also evident in the sequence of images in the same figure, in agreement with the seasonal model simulations.

To display mesoscale features in the SCS, we present snapshots of the flow field on 13 July, 12 August, 11 September, 11 October, 10 November and 10 December in Fig. 8. In these figures one can immediately distinguish the basic cyclonic and anticyclonic gyres permanently occupying the south and north of the deep basin, separated by a westward-flowing jet flow, and an anticyclonic circulation in the shallow eastern shelf, all of which are strongly modified with time.

Following the 100–300 m bathymetric contours (drawn in red in Fig. 8a) one can identify basin-wide cyclonic flow in the form of a coastal jet closely following the western and southern coasts. This flow later turns north along the outer edge of the wide eastern shelf, completing a rim-current pattern following the continental slope. The rim current follows topography, which acts like a barrier limiting exchange between coastal/shelf and deep waters and at the same time gives rise to a series of instabilities in the form of mesoscale eddies, wave-like oscillations and filaments. These features in Fig. 8 correspond to a rich field of mesoscale activity.

There are a few modifications over the course of time to the rim current and the embedded simple three-cell circulation of the SCS outlined above. The cyclonic and anticyclonic subbasin gyres display important changes in size and strength, and, as shown in Fig. 8e, the southern cyclonic gyre at times extends to take over part of the anticyclonic circulation to its north and later recedes south. The circulation interacts with the coast and reversals of currents occur in shallow areas landward of the continental slope. Such reversals occur in the northwest coastal sector of the SCS, particularly near the headland around 50.2° E, 37.5° N, leading to the development of an anticyclonic vortex near 52° E, 37° N, sheltered from the cyclonic gyre. The flow along the eastern shelf is also highly variable. The intense southerly flow observed in the shallow inner shelf of the southeastern corner in Fig. 8a–c later becomes organized in the form of an anticyclonic gyre

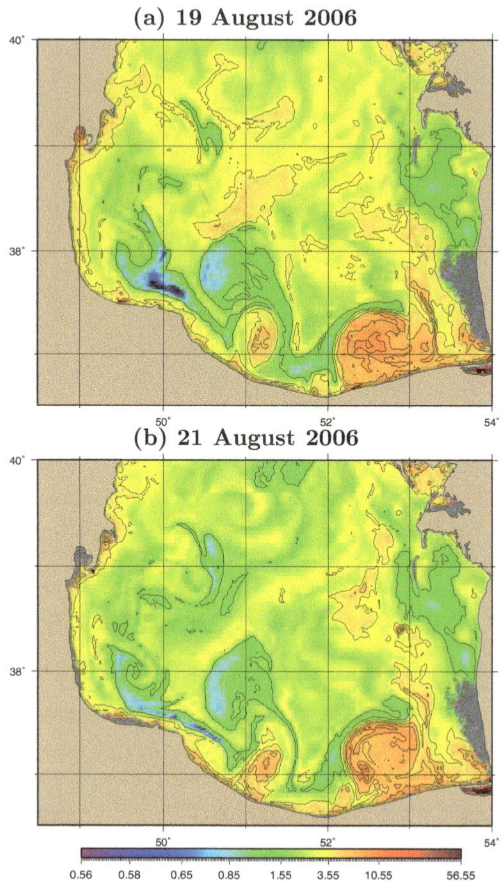

Figure 9. Chlorophyll concentration (in mg m^{-3}) on **(a)** 19 August 2006 and **(b)** 21 August 2006. The chlorophyll concentration contours are plotted with different colours and intervals for a better visualization of the structure.

east of the continental slope marked by the 100–300 m depth contours in Fig. 8d–f. The region outside the inner shelf but east of the continental slope displays numerous small eddies and oscillating currents most of the time and especially when the anticyclonic eddy in the southeast is not present.

Some of the features described above are easily detected from satellite data. Chlorophyll concentration obtained from the Moderate-resolution Imaging Spectroradiometer (MODIS)-Aqua satellite sensors in Fig. 9 reveals the rim-current structure described above, with a maximum of chlorophyll within the coastal jet outlining the southwestern coast of the SCS and then moving along the continental slope bounding the eastern continental shelf, almost exactly along the 100–300 m contours of bathymetry shown in Fig. 8. The other place where a maximum of chlorophyll concentration is found is along a line extending west from the eastern shelf slope region, which is reminiscent of the jet between the cyclonic and anticyclonic gyres of the western SCS found in our seasonal simulations. High chlorophyll values are found confined to two anticyclonic eddies along the southern coast, similar to the small vortices found there in our simulations.

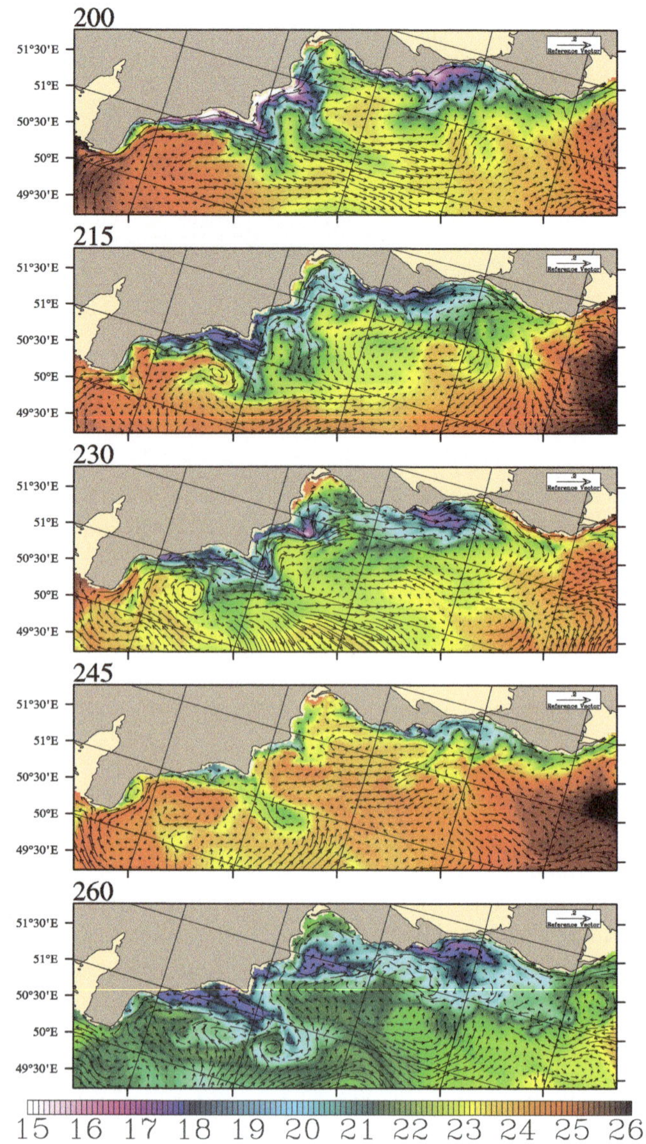

Figure 10. Snapshot of the model SST of the eighth model year along the eastern coast of the Sea. The model days are shown at the upper left side of the each figure.

6 Upwelling

In summer through to early fall there is continuous upwelling along the eastern coast of the Caspian Sea, studied by numerous authors (Baidin and Kosarev, 1986; Kosarev, 1975, 1990; Terziev et al., 1992; Sur et al., 2000). Satellite SST data often show relatively colder waters along the coast and filaments penetrating offshore from the upwelling region. The upwelling region extends from 41° to 44° N along the eastern coast with a temperature anomaly of 2–3 °C in a region extending 5–20 km from the coast. The upwelling at the central part of the eastern coast appears more active in this image, but a review of similar data shows that the upwelling

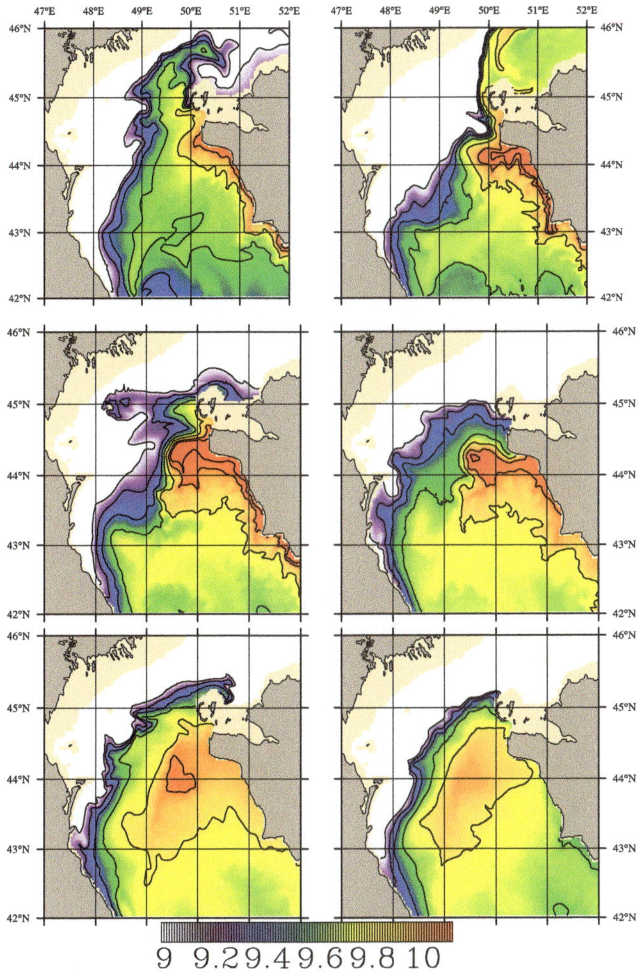

Figure 11. Potential density in sigma-theta units of deepest depth on 2 January and 20 January (upper row), 9 February and 28 February lower (middle row), and 20 March and 1 April (lower row). White areas show a density of less than 9.

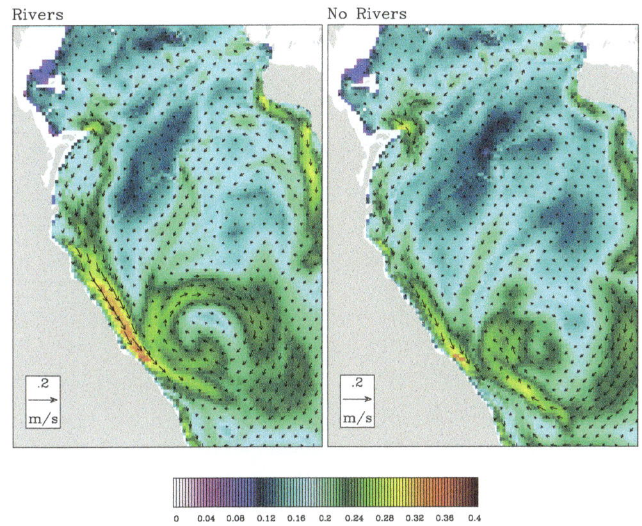

Figure 12. Model surface velocity vector mean and rms vector amplitude over a year's simulation, with rivers and without rivers. Means are denoted with arrows while rms amplitude is colour contoured. Every fourth vector is shown in the along-shore and across-shore direction.

region shifts to the north or south during other times. The eastern coast upwelling is closely related with the predominant northerly winds in summer. The upwelling distribution evidently changes with the wind intensity and direction.

Figure 10 shows snapshots of model SST at days 200, 215, 230, 245 and 260 of the model annual cycle at the upwelling region along the eastern coast of the Caspian Sea. Mesoscale dynamic features such as cold filaments and single and mushroom-shaped dipole eddies are often evident in the SST patterns throughout the whole model year. The upwelling at times covers the entire eastern coast, while at other times (e.g. day 245), it almost vanishes. Irregularities along the coast, such as headlands and changes in orientation or slope of the complex topography affect the structure of the upwelled waters. Filaments of upwelled cold water reach the interior of the basin following the generally westward Ekman drift currents at the surface. Upwelling features in the model-simulated SST are consistent with satellite observations based on the Pathfinder SST data set (not shown).

Compared to the other biologically active upwelling regions of the world ocean, plankton blooms with high chlorophyll concentration are, strangely, not evident in satellite data of the upwelling region along the eastern coast of the Caspian Sea.

An important consequence of Caspian Sea upwelling is the advection of high-salinity water onto the shelf by compensation currents along the coast, creating favourable conditions (preconditioning) for dense-water formation during the later winter period, which then leads to intense slope convection that can ventilate the deeper waters of the Middle Caspian Sea.

7 Dense-water formation

Ventilation of the deep waters of the Caspian Sea occurs through a series of complex processes. The rate of formation of dense water is controlled by the rate of cooling in the shallow continental shelf areas in the northern part of the Caspian Sea, ice formation in the North Caspian Sea, saline water ejection by ice, the background stratification and incidence of fronts, fresh-water input, surface fluxes, and mesoscale circulation. The fate of the sinking dense water depends on topography through the latter's influence on the circulation. As dense bottom currents flow down the continental slope and into the sea interior, they entrain lower-density overlying water, thus becoming relatively less negatively buoyant, influencing interior ocean properties (Papadakis et al., 2003).

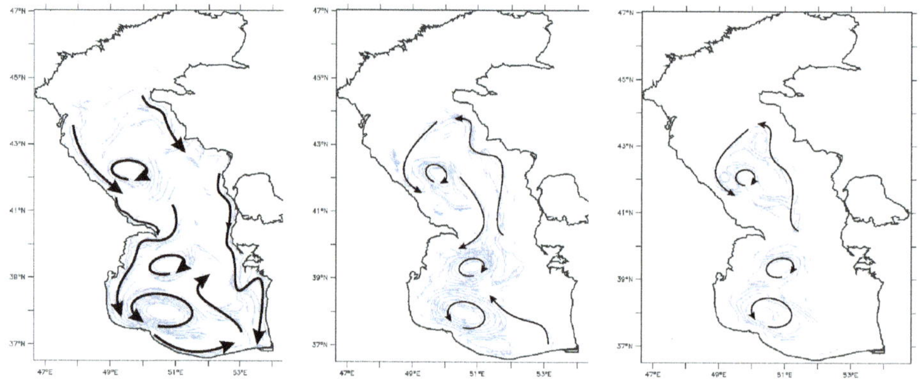

Figure 13. Schematic representation of the annual mean circulation of the Caspian Sea for the depth ranges (left panel) surface (0–10 m), (middle panel) intermediate (50–100 m) and (right panel) deep (200–1000 m).

Dense-water formation contributes to renewal of the deep water and thereby affects the circulation of the sea. The mechanism for dense-water formation is not clear in the Caspian Sea. Rodionov (1994) has suggested two possible mechanisms for the sinking of dense water: (1) in summer, evaporation increases the density of sea water at the surface, which then sinks deeper, especially along the eastern coast; (2) in winter, salt ejection by freezing results in saline surface water in the NCS which could be incorporated into the dense water formed. In addition, surface waters flowing northward along the eastern coast of the MCS come into contact with the ice on the northern slopes of the MCS, become denser by cooling and sink to greater depths of the MCS (Ferronsky et al., 2003). Kosarev and Yablonskaya (1994) have also noted that the conditions for the development of wintertime vertical motions are more favourable along the eastern coast compared to other regions.

Based on two cruises undertaken in September 1995 and September 1996 (Peeters et al., 2000) have suggested a similar mechanism for dense-water formation, whereby water transported north along the eastern part of the Caspian Sea becomes colder by encountering the ice and less saline cold water in the north and sinks to the deeper parts of the Middle Caspian Sea. Climatological data show a cold water mass with high density near the eastern coast at 44° N.

In our model experiments we found dense-water formation in the northeastern part of the MCS as suggested by previous authors. Figure 11 shows the evolution of density of the deepest depth over the dense-water formation site. Beginning with January, day 20, dense water is evident along the eastern coast of the sea, mostly confined along the eastern coast of the sea. The density of waters along the eastern coast gradually increases in February and begins to move to the interior of the sea. In March dense water flows down at the continental slope and fills the interior of the sea. Beginning in April the density of deepest depth become similar to the density of the surrounding waters, and dense-water generation is completed along the eastern coast.

8 Role of the rivers

With fresh-water inputs from a large river such as the Volga, the river buoyancy contribution is commonly expected to be an important driver of circulation. In the case of the Caspian Sea, the observed southward flow along the western coast is anticipated to be a major component of this buoyancy-driven circulation. To test this statement, an experiment was designed in which the river runoff is turned off, leaving climatological surface fluxes and wind stress as the only forcing.

Turning rivers off, it is found that main features of the circulation are preserved, though weaker. It is also noteworthy that not only the surface currents, but also the deeper currents decrease in intensity when the rivers are turned off. Figure 12 compares the annual mean surface currents in the MCS with and without rivers in better detail. There is significant weakening of the west coast southerly jet and the Derbent anticyclone fed by this current when the river inflow is turned off. The other feature noted in Fig. 12 is the decrease in intensity of the current near the northeastern coast of the basin. The similarity of the other features with and without the river effect illustrates the roles of the wind and surface buoyancy forcing in maintaining these circulation features. The fresh-water discharge appears primarily responsible for maintaining the pronounced southward-flowing coastal jet and the anticyclone along the west coast, as well as the surface coastal jet on the eastern coast of the MCS. The effect of major rivers in the north is of consequence in the SCS: the cyclonic gyre in the northwest of the SCS becomes much weaker, while the cyclonic gyre in the southeast becomes more intense and covers almost the entire basin in the case without river runoff.

9 Conclusions

A high resolution eddy-resolving hydrodynamic model was used to investigate the dynamics of the Caspian Sea circulation. The model, forced by climatological mean seasonal atmospheric forcing and river inputs of fresh water, successfully reproduces the rather well-known features of the Caspian Sea circulation and reveals the mesoscale variability of the circulation.

Based on the model results, the circulation at the surface, at intermediate depth below the mixed layer and at deeper levels is schematized in Fig. 13 based on the annual circulation patterns presented in Fig. 3. These schematic patterns were constructed by eliminating currents of magnitude greater than $2.5\,\mathrm{cm\,s^{-1}}$ at the surface, $1\,\mathrm{cm\,s^{-1}}$ at intermediate depth and $0.3\,\mathrm{cm\,s^{-1}}$ at the deep level and by using the remaining current vectors.

The surface circulation (Fig. 13, left panel) consists of narrow coastal jets originating in the north and flowing south along western and eastern coasts. Part of the flow escaping from these jets near headlands is incorporated into other elements of the circulation. In this way, the coastal jets and the anticylonic/cyclonic gyres in the individual basins are interconnected by thin streams of currents which follow the coast and the continental slope. In a number of cases, a strong current following a stream is diverted by a nearby feature, running in the opposite direction to its first orientation: for example, at the headland on the northeastern coast, along the western coast in the proximity of the anticyclonic gyre of the MCS, at the southern coast and on the eastern side of the dipole in the SCS.

The intermediate depth circulation, Fig. 13, middle panel, is much simpler, and consists of the anticyclone in the MCS, the cyclone/anticyclone pair in the SCS and a south to north current along the main eastern basin topographic slope (which extends to the mid-basin in the SCS and appears closer to the eastern coast in the MCS). This northward current then interconnects to the cyclonic/anticyclonic cells on the western side of the basin.

The deep circulation, Fig. 13, right panel, consists only of the anticyclonic cell in the MCS and the cyclonic/anticyclonic pair in the SCS, without appreciable slope currents joining them. Overall, the three major cells of circulation and shelf currents displayed in Fig. 13 appear to have significant barotropic components.

Acknowledgements. We would like to thank R. Ibrayev for numerous discussions and data support. The numerical HYCOM simulations were performed under the National Center for High Performance Computing (NCHPC) at the Istanbul Technical University, Istanbul Turkey. This research is funded by Multidisciplinary Analysis of the Caspian Sea Ecosystem (MACE) project.

Edited by: E. J. M. Delhez

References

Arpe, K. and Roeckner, E.: Simulation of the hydrological cycle over Europe: Model validation and impacts of increasing greenhouse gases, Adv. Water Resour., 23, 105–119, 1999.

Arpe, K., Bengtsson, L., Golitsyn, G. S., Mokhov, I. I., Semenov, V. A., and Sporyshev, P. V.: Connection between Caspian sea level variability and ENSO, Geophys. Res. Lett., 27, 2693–2696, 2000.

Baidin, S. S. and Kosarev, A. N. (Eds.): The Caspian Sea, Hydrology and Hydrochemistry, Nauka, 1986.

Beletsky, D. and Schwab, D. J.: Modeling circulation and thermal structure in Lake Michigan: Annual cycle and interannual variability, J. Geophys. Res.-Oceans, 106, 19745–19771, 2001.

Bengtsson, L.: Climate modelling and prediction, Achievements and challenges, in: WMO/IOC/ICSU Conference On The WCRP Climate Variability And Predictability Study (CLIVAR), UNESCO, Paris, 1998.

Bleck, R.: An oceanic general circulation model framed in hybrid isopycnic-Cartesian coordinates, Ocean Model., 4, 55–88, 2002.

Bleck, R.: On the use of hybrid vertical coordinates in ocean circulation modeling, in: Ocean Weather Forecasting: An Integrated View of Oceanography, edited by: Chassignet, E. and Verron, J., pp. 109–126, international Summer School of Oceanography, Agelonde in Lalonde-Les Maures, Agelonde, FRANCE, 20 September–1 October 2004, 2006.

Bondarenko, A. L. (Ed.): Currents of the Caspian Sea and Formation of Salinity of the Waters of the North Part of the Caspian Sea, Nauka, Moscow, 1993.

Chassignet, E., Smith, L., Halliwell, G., and Bleck, R.: North Atlantic Simulations with the Hybrid Coordinate Ocean Model (HYCOM): Impact of the vertical coordinate choice, reference pressure, and thermobaricity, J. Phys. Oceanogr., 33, 2504–2526, 2003.

Chen, C., Zheng, L., and Blanton, J.: Physical processes controlling the formation, evolution, and perturbation of the low-salinity front in the inner shelf off the southeastern United States: A modeling study, J. Geophys. Res.-Oceans, 104, 1259–1288, 1999.

Dumont, H.: The Caspian Lake: History, biota, structure, and function, Limnol. Oceanogr., 43, 44–52, 1998.

Ferronsky, V., Brezgunov, V., Vlasova, L., Polyakov, V., Froehlich, K., and Rozansky, K.: Investigation of Water-Exchange Processes in the Caspian Sea on the Basis of Isotopic and Oceanographic Data, Water Resour., 30, 10–22, 2003.

Gunduz, M.: Caspian Sea surface circulation variability inferred from satellite altimeter and sea surface temperature, J. Geophys. Res.-Oceans, 119, 1420–1430, 2014.

Ibrayev, R., Sarkisyan, A., and Trukhchev, D.: Seasonal variability of the circulation of the Caspian Sea reconstructed from mean multi-year hydrological data, Izvestiya, Atmos. Ocean. Phys., 37, 103–111, 2001.

Ibrayev, R. A., Özsoy, E., Schrum, C., and Sur, H. İ.: Seasonal variability of the Caspian Sea three-dimensional circulation, sea level and air-sea interaction, Ocean Sci., 6, 311–329, doi:10.5194/os-6-311-2010, 2010.

Kara, A., Hurlburt, H., Wallcraft, A., and Bourassa, M.: Black sea mixed layer sensitivity to various wind and thermal forcing products on climatological time scales, J. Climate, 18, 5266–5293, 2005.

Kelly, K. A., Thompson, L., Cheng, W., and Metzger, E. J.: Evaluation of HYCOM in the Kuroshio Extension region using new metrics, J. Geophys. Res.-Oceans, 112, doi:10.1029/2006JC003614, 2007.

Klevtsova, N. (Ed.): Sea current regime near east coast of the middle and south Caspian basin., Vol. 3, Proceeding of Hydro-meteo observatory of Baku, 1967.

Knysh, V., Ibrayev, R., Korotaev, G., and Inyushina, N.: Seasonal variability of climatic currents in the Caspian Sea reconstructed by assimilation of climatic temperature and salinity into the model of water circulation, Izvestiya, Atmos. Ocean. Phys., 44, 236–249, 2008.

Kosarev, A. (Ed.): Hydrology of the Caspian and Aral Seas, Mosk Gos Univ, Moscow, 1975.

Kosarev, A. (Ed.): The Caspian Sea, Water structure and dynamics, Nauka, Moscow, 1990.

Kosarev, A. and Yablonskaya, E.: The Caspian Sea, SPB Academic Publishing, The Hague, 1994.

Kourafalou, V., Oey, L., Wang, J., and Lee, T.: The fate of river discharge on the continental shelf .1. Modeling the river plume and the inner shelf coastal current, J. Geophys. Res.-Oceans, 101, 3415–3434, 1996.

Lazure, P. and Jegou, A.-M.: 3D modelling of seasonal evolution of Loire and Gironde plumes on Biscay Bay continental shelf, Oceanol. Acta, 21, 165–177, 1998.

Lednev, V. A. (Ed.): Currents of the Northern and Central Parts of the Caspian Sea, Moscow, 1943.

Özsoy, E.: Sensitivity to Global Change in Temperate Euro-Asian Seas (The Mediterranean, Black Sea and Caspian Sea): A Review, in: The Eastern Mediterranean as a Laboratory Basin for the Assessment of Contrasting Ecosystems, edited by: Malanotte-Rizzoli, P. and Eremeev, V., Vol. 51 of NATO Science Series, 281–300, Springer Netherlands, 1999.

Özsoy, E., Kubilay, N., Nickovic, S., and Moulin, C.: A hemispheric dust storm affecting the Atlantic and Mediterranean in April 1994: Analyses, modeling, ground-based measurements and satellite observations, J. Geophys. Res.-Atmos., 106, 18439–18460, 2001.

Papadakis, M., Chassignet, E., and Hallberg, R.: Numerical simulations of the Mediterranean sea outflow: impact of the entrainment parameterization in an isopycnic coordinate ocean model, Ocean Model., 5, 325–356, 2003.

Peeters, F., Kipfer, R., Achermann, D., Hofer, M., Aeschbach-Hertig, W., Beyerle, U., Imboden, D., Rozanski, K., and Frohlich, K.: Analysis of deep-water exchange in the Caspian Sea based on environmental tracers, Deep-Sea Res. Pt. I, 47, 621–654, 2000.

Pettenuzzo, D., Large, W., and Pinardi, N.: On the corrections of ERA-40 surface flux products consistent with the Mediterranean heat and water budgets and the connection between basin surface total heat flux and NAO, J. Geophys. Res.-Oceans, 115, C06022, doi:10.1029/2009JC005631, 2010.

Rachev, N. H. and Stanev, E. V.: Eddy Processes in Semienclosed Seas: A Case Study for the Black Sea, J. Phys. Oceanogr., 27, 1581–1601, 1997.

Rodionov, S.: Global and regional climate interaction: The Caspian Sea experience, Kluer Academic Publ., Dordrecht, The Netherlands, 1994.

Sarkisyan, A. S., Zaripov, B. R., Kosarev, A. N., and Rzheplinski, D. G.: Diagnostic calculations of currents in the Caspian Sea., Izvestiya, Atmos. Ocean. Phys., 12, 1106–1110, 1976.

Stockman, W.: Investigation of current kinematics at the western shore of the central part of the Caspian Sea, Vol. 1, Transactions of the Azerbaidjan Scientific-Investigative Fishery station, 1938.

Sur, H. I., Özsoy, E., and Ibrayev, R.: Chapter 16 Satellite-derived flow characteristics of the Caspian Sea, in: Satellites, oceanograghy and society, edited by: Halpern, D., Vol. 63 of Elsevier Oceanography Series, 289–297, Elsevier, 2000.

Terziev, F., Kosarev, A., and Keromov, A. (Eds.): Hydrometeorology and hydrochemistry of Seas, Vol. VI, The Caspian Sea, Gidrometeoizdat, Leningrad, 1992.

Trukhchev, D., Kosarev, A., Ivanova, D., and Tuzhilkin, V.: Numerical Analysis of the general circulation in the Caspian Sea, Comptes Rendus de l'Academie Bulgare des Sciences, 35–38, 1995.

Tsytsarev, A. N. (Ed.): Peculiarities of drift currents near Kura river mouth, 5, Proceeding of Baku Hydrometeobservatory, 1967.

Tuzhilkin, V., Kosarev, A. N., Trukhchev, D. I., and Ivanov, V. P.: Seasonal features of General Circulation in the Caspian Deepwater, Meterorol. Hydrol., 1, 91–99, 1997.

Vorosmarty, C., Sharma, K., Fekete, B., Copeland, A., Holden, J., Marble, J., and Lough, J.: The Storage and Aging of Continental Runoff in Large Reservoir Systems of the World, Ambio, 26, 210–219, 1997.

Wallcraft, A., Kara, A., Hurlburt, H., and Rochford, P.: The NRL Layered Global Ocean Model (NLOM) with an embedded mixed layer submodel: Formulation and tuning, J. Atmos. Ocean. Tech., 20, 1601–1615, 2003.

14

Residual circulation and freshwater transport in the Dutch Wadden Sea: a numerical modelling study

M. Duran-Matute[1,*], T. Gerkema[1], G. J. de Boer[2], J. J. Nauw[1], and U. Gräwe[3]

[1]Dept. of Physical Oceanography, Royal NIOZ, P.O. Box 59, 1790 AB Den Burg (Texel), the Netherlands
[2]Deltares, P.O. Box 177, 2600 MH Delft, the Netherlands
[3]Leibniz Institute for Baltic Sea Research Warnemünde (IOW), Seestraße 15, 18119 Rostock, Germany
[*]currently at: Department of Applied Physics, Eindhoven University of Technology, P.O. Box 513, 5600 MB Eindhoven, the Netherlands

Correspondence to: M. Duran-Matute (matias.duran-matute@nioz.nl)

Abstract. The Dutch Wadden Sea is a region of intertidal flats located between the chain of Wadden Islands and the Dutch mainland. We present numerical model results on the tidal prisms and residual flows through the tidal inlets and across one of the main watersheds. The model also provides insight into the pathways of fresh water originating from the two sluices at the Afsluitdijk (enclosure dike) through the use of passive tracers. All these results are obtained from three-dimensional numerical simulations carried out with the General Estuarine Transport Model (GETM), at a horizontal resolution of 200 m and with terrain-following vertical coordinates with 30 layers. We concentrate on the years 2009–2010, for which we impose meteorological forcing, freshwater discharge, and boundary conditions for tidal forcing and storm surges. Results from the model show an excellent agreement with various observational data sets for sea surface height, temperature, salinity and transport through the Texel Inlet. The simulations show that although tides are responsible for a characteristic pattern of residual transport through the inlets, the wind imposes a large variability on its magnitude and can even invert its direction during strong southwesterly winds. Even though these events are sporadic, they play an important role in the flushing of the Dutch Wadden Sea, as they strongly diminish the flushing time of fresh water. In addition, wind can force a residual transport across the Terschelling watershed of the same order, if not larger, than through any of the main tidal inlets, despite the fact that its tidal prism is much smaller than any of those of the inlets. For the pathways of fresh water, the Terschelling watershed turns out to be more important than was previously thought, while the opposite holds for the Vlie Inlet.

1 Introduction

The Wadden Sea is located along the coast of the Netherlands, Germany, and Denmark between the European mainland and a chain of barrier islands, and it is connected to the North Sea by a series of tidal inlets found in between the islands. It is the largest system of interconnected intertidal flats in the world and was declared a World Heritage site by UNESCO in 2009. It possesses a large biodiversity, serves as feeding ground for migratory birds (Reise et al., 2010, and references therein), and also hosts economic activities such gas extraction and fishing. In terms of water movement, it presents a complex estuarine dynamic due to the combined effects of the bathymetry, tides, wind surges, and freshwater flows.

The Dutch Wadden Sea extends from the Texel Inlet (the westernmost inlet of the Wadden Sea) at the west up to the Ems-Dollard estuary, which marks the beginning of the German Wadden Sea to the east. Figure 1 shows the configuration of the Dutch Wadden Sea without including the Ems-Dollard Estuary since it does not form part of the present study. There are five barrier islands in the Dutch Wadden Sea with a watershed behind each of them. These watersheds divide the Wadden Sea into a series of tidal basins (Oost et al., 2012). The current paper addresses two basic questions, which have been of interest for several decades, about the hydrodynamics of the Dutch Wadden Sea. What is the residual circulation pattern in the Wadden Sea? What is the fate of the fresh water discharged into it? (In other words, what are the main gateways for exchange with the North Sea?)

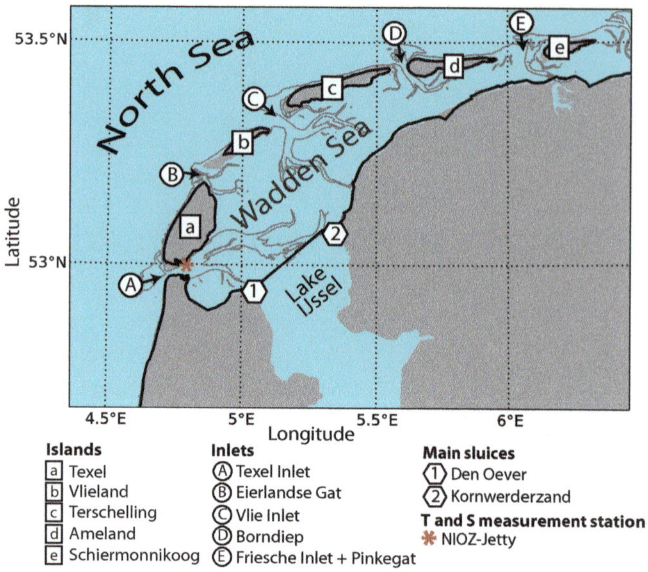

Figure 1. Map of the Dutch Wadden Sea up to the beginning of the Ems-Dollard estuary. The names of the islands and inlets are indicated. The position and name of the two major sluices are also indicated; more details on these and other sluices are given in Sect. 2.4. The grey line indicates the −5 m isobath.

Two inlets have long been regarded as the principal gateways: the Texel Inlet (label A in Fig. 1) and Vlie Inlet (label C in Fig. 1). They have the largest tidal prisms of the Dutch Wadden Sea, estimated to be 1054 and 1078 × 10^6 m^3, respectively. The values for the other inlets are much smaller: Eierlandse Gat, 207 (label B in Fig. 1); Borndiep, 478 (label D in Fig. 1); Pinkegat, 100 (label E in Fig. 1); Friesche Inlet, 200 (label E in Fig. 1). These empirical estimates were listed by Louters and Gerritsen (1994) and are all in 10^6 m^3.

The residual circulation in the western Dutch Wadden Sea (WDWS) has been previously studied, both through numerical simulations and measurements. This region spans from the Texel Inlet to the Terschelling watershed (the watershed between the mainland and the island of Terschelling). Ridderinkhof (1988a) used a vertically averaged numerical shallow-water model with a 500 m horizontal resolution from the Texel Inlet to the Ameland watershed. The forcing consisted of tides alone, i.e. wind forcing and freshwater discharge were not included. He found a residual circulation consisting of an inflow through the Vlie Inlet (1145 $m^3\ s^{-1}$), and an outflow through the Texel Inlet (820 $m^3\ s^{-1}$), the Eierlandse Gat (250 $m^3\ s^{-1}$) and the Terschelling watershed (55 $m^3\ s^{-1}$). Using an analytical model, this circulation was explained mainly by differences in water-level amplitude between the Texel and Vlie inlets (Ridderinkhof, 1988b). However, it was later shown, by expanding the analytical model to include wind effects, that the residual circulation between the Texel and Vlie inlets is highly variable due mainly to local wind stresses (Buijsman and Ridderinkhof, 2007b).

Between 2002 and 2004, Rijkswaterstaat (part of the Dutch Ministry of Infrastructure and Environment) carried out three 13 h ADCP (Acoustic Doppler Current Profiler) campaigns around the Texel Inlet (Elias et al., 2006). They found an average outflow rate of about 2300 $m^3\ s^{-1}$ within a tidal period. In 2012, two 13 h transects were carried out at the Vlie Inlet (Gerkema et al., 2014). They measured net imports through the main channel with an average flow rate of 1300 $m^3\ s^{-1}$ and 3900 $m^3\ s^{-1}$ within the two tidal periods sampled; the difference seems to be associated with wind speed and direction. However, these values are unlikely to be representative of the long-term mean since they are taken during very specific conditions; such measurements are rarely taken during extreme weather events that might be of high importance to the residual flow. The only long-term semi-continuous transport measurements are obtained by the ADCP on board the ferry crossing the Texel Inlet (Buijsman and Ridderinkhof, 2007a, b; Nauw et al., 2014). Buijsman and Ridderinkhof (2007a) estimated with these measurements an average volume outflow rate through the Texel Inlet of 2910 $m^3\ s^{-1}$ for the period 1998–2002.

The fate of the fresh water discharged into the WDWS has been a subject of study for even longer than the residual circulation (Postma, 1950; Zimmerman, 1976). This is probably due to the fact that the measurements of the freshwater distribution are easier to perform. On the other hand, no previous numerical results on this topic exist. The fresh water traversing the WDWS originates mainly from two sources (Fig. 1): the sluices at Den Oever and Kornwerderzand in the Afsluitdijk (the dike closing off the inner part of the basin, which was completed in 1932). On a decadal timescale, salinity in the WDWS shows only a slight decreasing trend (van Aken, 2008a), so for the period we are concerned with here (2009–2010), the mass of fresh water entering from the sluices must be nearly identical to the mass leaving the basin through all inlets combined.

Postma (1950) analysed several measurements of salinity at slack tides, including ones during which one sluice was closed. He concluded that all the fresh water from Den Oever leaves via the Texel Inlet, whereas the fresh water from Kornwerderzand is evenly split over the two main inlets, Texel and Vlie. Postma (1950) also emphasized that salinity cannot be used as a tracer indicative of water movements because of the strong mixing in the Wadden Sea. The latter aspect was explored further by Zimmerman (1976). On the basis of new data and a box model, he modified the earlier estimate by Postma (1950) for the fate of the discharge from Kornwerderzand, to 1/3 going towards the Texel Inlet and 2/3 into the Vlie basin (Zimmerman, 1976). Flushing times were estimated to be 15 tidal periods for Den Oever and 31 tidal periods for Kornwerderzand. As for the residual circulation, the residual transport of fresh water is highly variable. This was already pointed out by Zimmerman (1976) as he noticed an influence of the wind in the freshwater distribution and observed a clear decrease in the volume of fresh water inside the WDWS during a large storm surge.

With respect to the fate of the fresh water, two important aspects have still to be addressed. The first aspect is the quantification of the actual export of fresh water through the different tidal inlets. Although it was previously shown that the fresh water goes towards the Texel and Vlie inlets (Zimmerman, 1976), no estimates (from measurements or numerical models) of the net export exist. The model results to be presented here will shed light on the residual transport through the inlets and across the Terschelling watershed. In general, this exchange can be expected to show a lot of variability, depending on the freshwater discharge rate, wind direction and speed, water level surges, and on the phase in the spring–neap cycle. This variability also makes the application of the concept of flushing time less obvious, as will be discussed below. A second aspect concerns the unsuitability of salinity as a tracer, which is compounded by the fact that the fresh water from the two sluices is fundamentally indistinguishable.

In the current paper, we present answers to the questions at hand based on results from three-dimensional simulations of the Dutch Wadden Sea for the years 2009 and 2010 performed using the General Estuarine Transport Model (GETM). The model solves the primitive hydrostatic equations in three dimensions using a realistic bathymetry, freshwater discharge, tides and meteorological forcing. In the simulations, the fresh water is distinguished by tagging it with a different Eulerian passive tracer for each of the main sluices. These realistic simulations allow us to compute the residual water circulation and the residual flow of fresh water in a consistent and inherently mass-conservative way while obtaining a full three-dimensional synoptic view of the Dutch Wadden Sea. In other words, they allow us to revisit the results from the previous work described above, and to quantify the variability of the system by virtue of sufficiently long time series and a good temporal and spatial resolution.

The paper is organized as follows. Section 2 describes the setup: the domain, the bathymetry, and the forcing. For validation, a comparison with different observational data sets is carried out in Sect. 3. The tidal prisms at every inlet are presented in Sect. 4. Results on the residual circulation and the fate of the fresh water from the sluices at Den Oever and Kornwerderzand are presented in Sects. 5 and 6, respectively. Finally, the conclusions are outlined in Sect. 7.

2 Methods

Fully three-dimensional numerical simulations are carried out using GETM for the years 2009–2010. The numerical model GETM is an open-source code that has been developed specifically for coastal areas. It is now commonly used, and hence, detailed descriptions are abundant in the literature (see e.g. Stips et al., 2004 and Stanev et al., 2003). In addition, the documentation and the code are freely available at http://www.getm.eu. For these reasons, a complete description of the model will not be made here, but only the components and specific details relevant for the setup at hand are given in Appendix A.

The simulations include the most realistic possible (in terms of available data and computational resources) values for the bathymetry, atmospheric forcing, boundary conditions, and freshwater discharge. The simulations are initialized from rest on 1 November 2008. For salinity and temperature, climatological values for the month of November are used. Hence, two months are used to spin the model up in both the barotropic and baroclinic sense. This is sufficiently long given the typical flushing time of the basins of about one month (see Sect. 6.2).

2.1 Numerical domain and bathymetry

A rectangular domain rotated 17° in the anticlockwise direction with respect to the east–west axis was defined with its corners located at (4.4693° E, 52.4975° N); (4.0260° E, 53.3269° N); (6.3871° E, 53.7607° N); (6.7896° E, 52.9232° N) (see Fig. 2a). Within this domain, we defined an equidistant grid with a 200 m resolution using the *Rijksdriehoek* projection. This projection is the standard projection used by the Dutch Government, and hence it is convenient to use it here.

The Dutch coast is continuously monitored by the Dutch Government – in particular, by Rijkswaterstaat. As part of this monitoring programme, the bathymetry along the whole Dutch coast, from dry land until the 20 m isobath in the North Sea, is measured regularly. Every section of the Dutch Coast, including the Wadden Sea, is sampled at least once every seven years, with certain zones sampled more frequently due to their importance for navigation and other activities. These measurements are processed and made available in a 20 m grid (Wiegman et al., 2005). This data set is known as the *vaklodingen*, and the whole historical record is freely available (at http://opendap.deltares.nl). To construct the bathymetric map, we considered only the measurement closest in time (either in the past or in the future) to the years to be simulated: 2009–2010. Figure 2b shows the year during which the bathymetry at the different regions was measured. As can be seen, most of the data were measured around the simulated years with some small exceptions.

The process of constructing the bathymetry is as follows. First a 20 m grid with the same orientation as the model grid was constructed. An interpolation of the bathymetry from the *vaklodingen* was made onto this high-resolution grid. Then, the average depth of the area covered by 10 × 10 points (i.e. 200 m × 200 m) of the high-resolution grid was assigned to the equivalent point in the 200 m grid. For areas of the grid not covered by the *vaklodingen* (i.e. outside the 20 m isobath in the North Sea), a simple linear interpolation from a 200 m

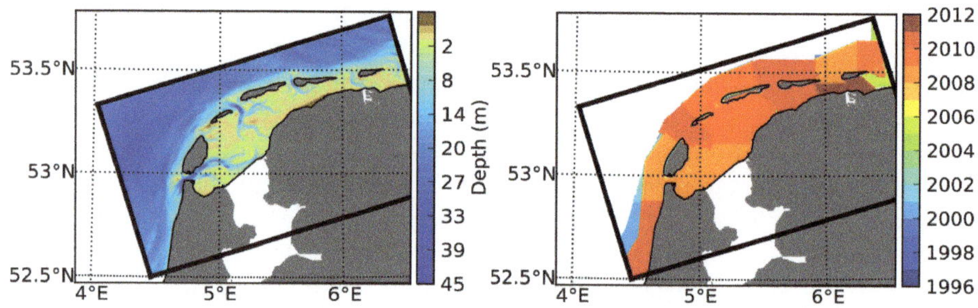

Figure 2. **(a)** Numerical domain and bathymetry. **(b)** Dates of the measurements used to compose the bathymetry up to the 20 m isobath in the North Sea.

resolution bathymetric map compiled by TNO and the Dutch Navy was used.

The bathymetric map was further smoothed for the stability of the model and to avoid numerical noise. This was done in two steps: (1) application of a weighted average filter and (2) using a heuristic linear approach (Dutour Sikirić et al., 2009) (see Appendix B for more details).

2.2 Lateral boundary conditions

At the open boundaries of the numerical domain, boundary conditions for the sea surface height (SSH), vertically integrated velocities and vertical profiles of salinity and temperature were applied. On the watershed at the eastern boundary (south of Rottumerplaat), a wall was placed since it is difficult to match the position of the channels in this region with the boundary conditions.

The Flather type boundary conditions are used for the barotropic mode. The vertically integrated velocities and sea surface elevation at the boundaries are taken from results of a two-dimensional model with data assimilation run by Rijkswaterstaat to predict the SSH along the Dutch coast (Plieger, 1999; Verlaan and Heemink, 1998). The extracted data imposed at the boundary have a temporal resolution of 10 min.

Salinity and temperature at the boundaries were obtained from simulations with a set of nested setups. The simulations for these setups were also performed with GETM. First, a two-dimensional North Atlantic setup with 4 nm resolution with only wind forcing was used to model surges. The results of this setup together with the Oregon State University Tidal Prediction Software (OSU-TPS), and climatology for salt and temperature were used for the boundary conditions of a three-dimensional North Sea setup with a 1 nm resolution and 30 vertical layers. In this setup, salinity, temperature, freshwater discharge and complete meteorological forcing are all taken into account. Finally, another three-dimensional southern North Sea setup with a 600 m resolution and 42 vertical layers was used. The salinity and temperature data at a temporal resolution of two hours was extracted for the boundary conditions. In addition, temporal relaxation and a four-point sponge layer are used for the salinity and temperature boundary conditions. (More details are given in Appendix D.)

We are aware of the fact that the usage of boundary conditions from two different models might cause slight inconsistencies. However, due to data assimilation, the Rijkswaterstaat tide/surge model shows a superior accuracy in the tidal and intratidal variability.

A complete validation and description of the nested setups, and in particular the 600 m southern North Sea, will not be presented in the current paper. The quality of the boundary conditions obtained is evaluated only through the quality of the results within the numerical domain considered in the current paper as shown in Fig. 2a.

2.3 Meteorological forcing and surface boundary conditions

Meteorological data were obtained from the operational forecast model of the German Weather Service (DWD). The available variables are wind speed and direction, air temperature, precipitation, cloudiness and dew point, which are discretized in a grid with a resolution of $1/16°$ and have a temporal resolution of three hours.

The surface stresses τ_s^x and τ_s^y are calculated from the wind velocity and are used to set the dynamic boundary condition at the surface:

$$(\nu_t + \nu)\partial_z u = \alpha \tau_s^x, \quad (1)$$

$$(\nu_t + \nu)\partial_z v = \alpha \tau_s^y. \quad (2)$$

(See also Appendix A for definitions.)

The solar radiation as a function of depth $I(z)$ – needed in the transport equation for the potential temperature (Eq. A9) – is calculated according to the expression

$$I(z) = I_0[a\,e^{-\eta_1 z} + (1-a)e^{-\eta_2 z}], \quad (3)$$

where I_0 the albedo-corrected radiation normal to the sea surface, a is a weighting parameter and η_1^{-1} and η_2^{-1} are the attenuation lengths (Paulson and Simpson, 1977). The parameters a, η_1 and η_2 depend on the amount of suspended material in the water. For the simulations presented in the current paper, $a = 0.57$, $\eta_1 = 0.37\,\mathrm{m}^{-1}$ and $\eta_2 = 3.54\,\mathrm{m}^{-1}$. This is

Table 1. Average and maximum freshwater discharge in $m^3\ s^{-1}$ at the different sluices for the years 2009–2010. The sluice numbers correspond to those in Fig. 3.

Number	Name	Yearly averaged discharge ($m^3\ s^{-1}$)	Maximum instantaneous discharge ($m^3\ s^{-1}$)
1	Den Oever	241.18	2517.81
2	Kornwerderzand	200.00	1954.32
3	Lake Lauwers	37.60	1594.00
4	Helsdeur	8.99	99.91
5	Harlingen	4.39	124.35
6	Oostoever	2.78	40.50
7	Dijkmanshuizen	0.56	3.03
8	Prins Hendrik	0.49	1.66
9	Krassekeet	0.48	3.24
10	Eierland	0.31	3.04
11	De Schans	0.08	1.85
12	De Zandkes	0.01	0.06

equivalent to a Jerlov coastal water type 3 (Jerlov, 1976) according to a non-linear least-squares fit (Stips, 2010).

The boundary conditions at the water surface for potential temperature and salinity are computed using the heat fluxes, precipitation, and evaporation following the work by Kondo (1975).

2.4 Freshwater discharge

Fresh water is discharged into the Dutch Wadden Sea from several sluices. Compared to this discharge, the contribution of precipitation minus evaporation is negligible (Zimmerman, 1976). The main sluices in terms of discharge are Den Oever and Kornwerderzand, located in the Afsluitdijk (enclosing dike). This dike separates the Wadden Sea from Lake IJssel (a freshwater lake). To include the freshwater discharge from the sluices in the simulations, the discharge rate for each one of them has to be specified. The location of the sluices implemented in the model are presented in Fig. 3, and the yearly average and maximum freshwater discharge for each of the sluices are presented in Table 1.

For the two main sluices, a record of the total discharge per tidal cycle, the times at which the gates are open and closed, and the water level inside and outside the sluices are available. To implement the discharge in the model, it is important to take into account that discharge only takes place when the gates are open and the water level at the Wadden Sea side is lower than the water level on the inside of the sluice. In other words, considering the average discharge per tidal period would probably not give the correct salinity distribution and variability. Using the available information and assuming that the Bernoulli equation holds for the discharge velocity, we reconstructed the discharge rate with a 10 min temporal resolution (see Appendix C).

For the sluices at Harlingen and Lake Lauwers, a record of discharge rate at a resolution of 15 min is kept by the authorities. In these cases, a simple linear interpolation is done to obtain the discharge rate at a resolution of 10 min. For the sluices at Helsdeur and Oostoever, values for daily average discharge are available. In these two cases, the discharge is spread over the periods when the SSH at the tidal gauge in the port of Den Helder (next to the Helsdeur sluice) is below 0.5 m. This is not the exact condition for discharge to occur, but it reproduces part of the variability. For all other sluices, the daily average discharge is interpolated to a time series with a resolution of 10 min. The modelled discharge rates for these small sluices are not as realistic as for the large sluices, but this is immaterial as they account (all together) for less than 0.5 % of all the fresh water discharged into the Dutch Wadden Sea.

2.5 Numerical schemes and computational details

The equations are solved on an Arakawa C-grid. The advection schemes for momentum are TVD-P2-PDM (third order) in the horizontal and TVD-SUPERBEE (second order) in the vertical. For salinity and temperature, the scheme is TVD-SUPERBEE for both horizontal and vertical advection. The same holds for the advection of turbulent kinetic energy and dissipation rate. GETM uses a micro time step for the external mode and a macro time step for the internal mode. The chosen micro time step is 4 s, and the macro time step is 40 s (i.e. a splitting factor of 10). Internal pressure is calculated using the approach by Shchepetkin and McWilliams (2003). In addition, a background horizontal momentum diffusion of $5\ m^2\ s^{-1}$ is used. In general, conservative model parameters (e.g. time step) were chosen to keep the model from crashing during extreme events (storms, large freshwater discharges). However, less conservative values can be used for most of the time.

GETM is an efficient code for parallel runs. The domain is subdivided into 137 subdomains. The simulations are then carried out in a computing cluster with eight nodes of which each has two processors (AMD Opteron 6272 2.6 GHz) with 16 cores each. This gives a total of 256 cores of which 138 are used in the simulations. With this configuration, 11 to 12 h are required to simulate one month.

3 Comparison with measurements

In order to estimate the accuracy of the simulations, their results are compared with several observational data sets. Although the Dutch Wadden Sea is monitored extensively, this is mainly driven by ecological reasons, storm surge safety and navigation. Hence, few of the data sets available are suitable for a comparison with results from the simulations. In addition, most of the measurements of physical variables are concentrated around the Texel Inlet, while little information

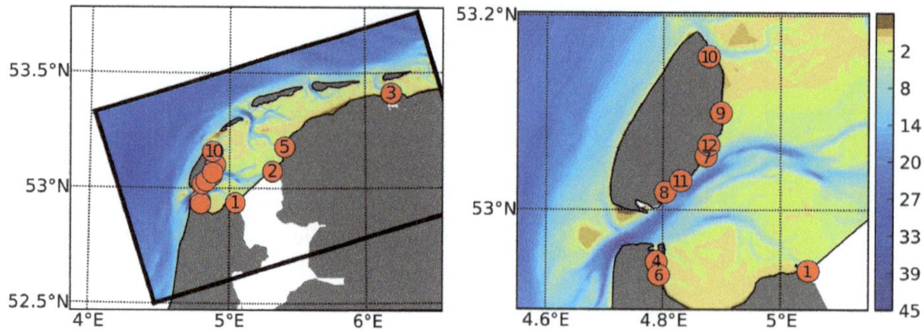

Figure 3. Location of sluices discharging fresh water into the Dutch Wadden Sea. The sluice numbers correspond to those in Table 1.

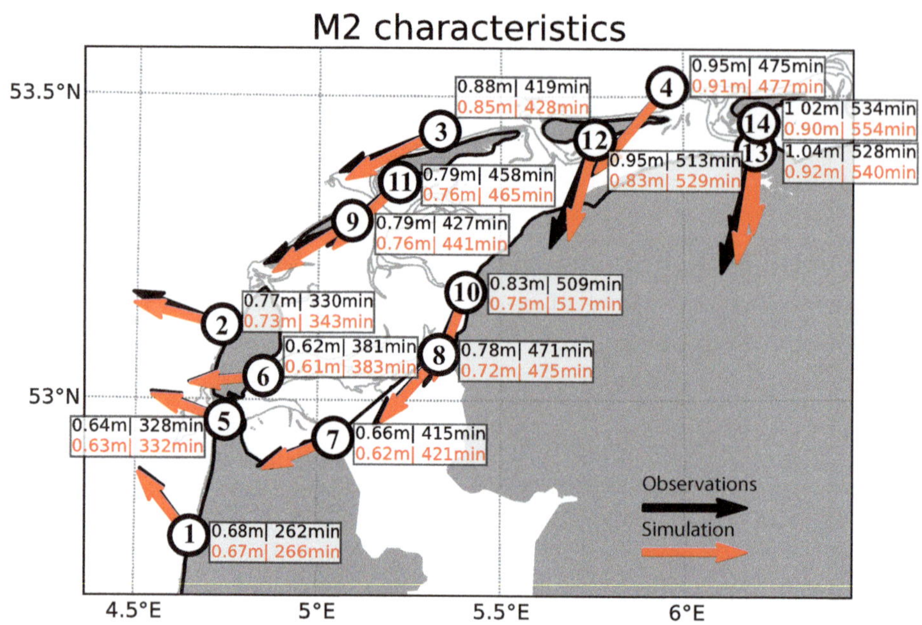

Figure 4. Comparison between the simulated and observed characteristics of the M2 tidal constituent at the 14 tidal stations in the domain. The position of each station is represented by a black circle, and the number inside the circle serves to identify the station. The length of the arrow represents the amplitude of the M2 tidal constituent, and the angle of the arrow represents its phase lag in degrees. The numerical values for the amplitude and the phase lag (in minutes with respect to Greenwich) are presented next to every station both for the observations and the simulation.

on salinity and temperature is found elsewhere. In the current paper, the comparison is therefore restricted to: (1) SSH measurements at 14 tidal stations, (2) time series of salinity and temperature at one station with 30 min temporal resolution, and (3) gross transport through the Texel Inlet as measured by the ferry across the Texel Inlet. In all cases, the accuracy of the simulations are quantified using the coefficient of determination R^2 and the root-mean-square (rms) error ϵ_{rms} obtained by comparing the measured data and the simulated data.

3.1 Sea surface elevation

Within the domain studied, there are 14 tidal gauges recording the instantaneous sea surface elevation every 10 min for the whole simulated period (2009–2010). To make a comparison, we extract a time series of sea surface elevation at the grid point closest to the position of every tidal gauge. In addition to calculating R^2 and ϵ_{rms}, we performed a harmonic analysis of both the simulated and the observed SSH time series using *t_tide* (Pawlowicz et al., 2002), and we compared the amplitude and phase of the different tidal constituents at all the tidal stations.

Figure 4 shows the position of the tidal stations, and the amplitude and phase lag (in minutes with respect to Greenwich) for the M2 tidal constituent, which is the dominant tidal constituent in the region. Station numbers from 1 to 4 correspond to stations in the North Sea; station numbers from 5 to 11 correspond to stations inside the WDWS; and numbers from 12 to 14 correspond to stations in the eastern Dutch Wadden Sea (EDWS). The tidal wave travels along the Dutch

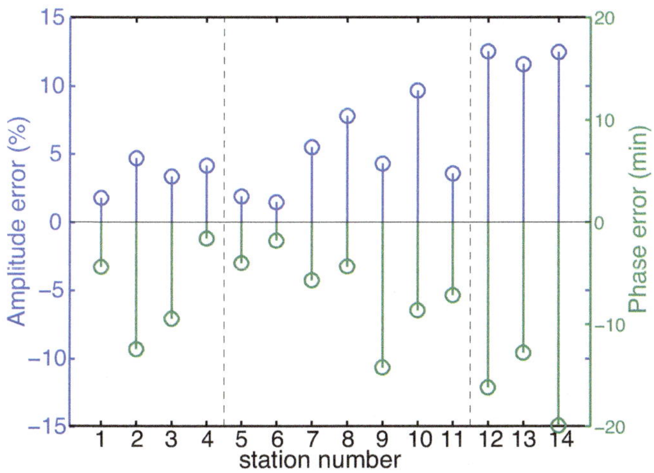

Figure 5. Amplitude error (in percentage of the measured amplitude) and phase error (in minutes) of the simulated M2 tidal constituent when compared with the measurements. The dashed lines divide North Sea, WDWS and EDWS stations.

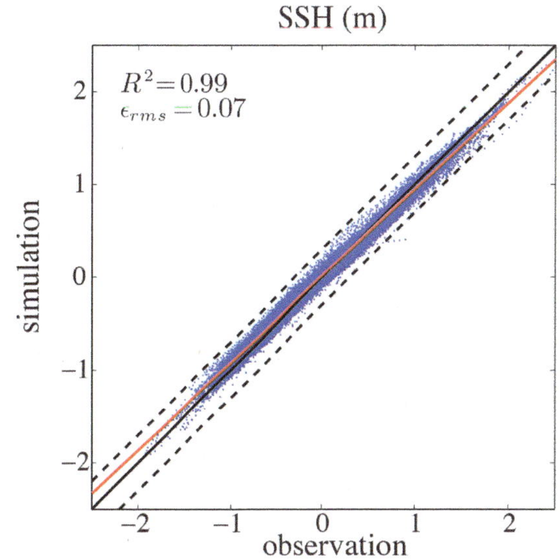

Figure 6. Scatter plot of observed vs. simulated SSH (in metres) at station 8 (Kornwerderzand). The red line represents the linear regression given by $a_{\text{coef}} + b_{\text{coef}} \times \text{SSH}_{\text{observation}}$. The centre black line represents the ideal linear regression (simulation = observation), while the side dashed black lines are the ideal regression ± 0.3 m. The coefficient of determination R^2 and the rms error ϵ_{rms} are also indicated.

coast from Southwest to Northeast with a (measured) phase lag of 262 min in station 1 to 475 min at station 4. In other words, it takes about 3.5 h to travel from station 1 to station 4. In general, a good agreement is found between the measured and the simulated characteristics of the M2 component.

Figure 5 presents in more detail the error of the simulated characteristics of the M2 constituent. All stations have an error in the amplitude between 2 % to 12 % and an error in phase between -1 min and -20 min (0.2–3 % of the tidal period). A positive error in the amplitude means that the amplitude is underestimated by the simulations, while a negative phase error means that the tidal wave in the simulations is ahead of the measured wave. The largest errors are found for the EDWS stations. This is probably due to such stations being in a location (port/channel) that is not well resolved in the simulation with the current resolution, and to the *wall* that was placed as a boundary condition across the watershed at the eastern boundary of the domain. However, the error is still within an acceptable range. A similar analysis was performed for the five most dominant tidal constituents (M2, S2, N2, O1, M4) with similar results. For all five components the largest errors are found for the stations in the EDWS. However, in the WDWS, the error in phase remains below 15 min. In terms of error in phase, this is about the best possible result considering that the temporal resolution of the observed SSH time series is 10 min with the time-centring not being specified. The error in the amplitude of the S2, N2, and O1 components in the WDWS remains below 10 %. For the M4 component, the error in amplitude increases in certain stations (stations: 3, 7, 8, 9, 10, 11) to about 20 %.

The comparison discussed so far is based on a harmonic analysis, in which the tidal constituents are extracted from the full signal, which also includes both set-ups and set-downs due to wind. In shallow areas like the Wadden Sea, the propagation of each constituent is affected by these changes in the sea level. This renders the notion of tidal harmonic *constants* somewhat dubious. In this respect, it makes sense to also compare the full signals.

A correlation analysis between the measured and simulated SSH, and a linear regression was done for all 14 stations. As an example, we present the scatter plot of the observed vs. simulated SSH at station 8 (outside the Kornwerderzand sluice) in Fig. 6. The linear regression (red line) is very close to the ideal result (black line). The slight angle between the two lines means that the simulated SSH range is slightly smaller than the real range. This is consistent with the fact that the amplitude of the M2 component is slightly underestimated by the simulations. The spread around the line is due to the phase difference between the observed and simulated variables. In this case, $R^2 = 0.99$ and $\epsilon_{\text{rms}} = 0.07$ m, indicating an excellent agreement between simulated and measured SSH. The rms error remains below 0.12 m for all North Sea and WDWS stations, while it remains below 0.18 m in the EDWS. For all stations, $R^2 > 0.96$.

3.2 Salinity and temperature time series

Time series of temperature and salinity are compared against measurements at the NIOZ jetty, which is located in the Texel Inlet just off the southern coast of the island of Texel (4.7892° E, 53.0018° N) as shown in Fig. 1. These time series were described and analysed by van Aken (2008a, b). The

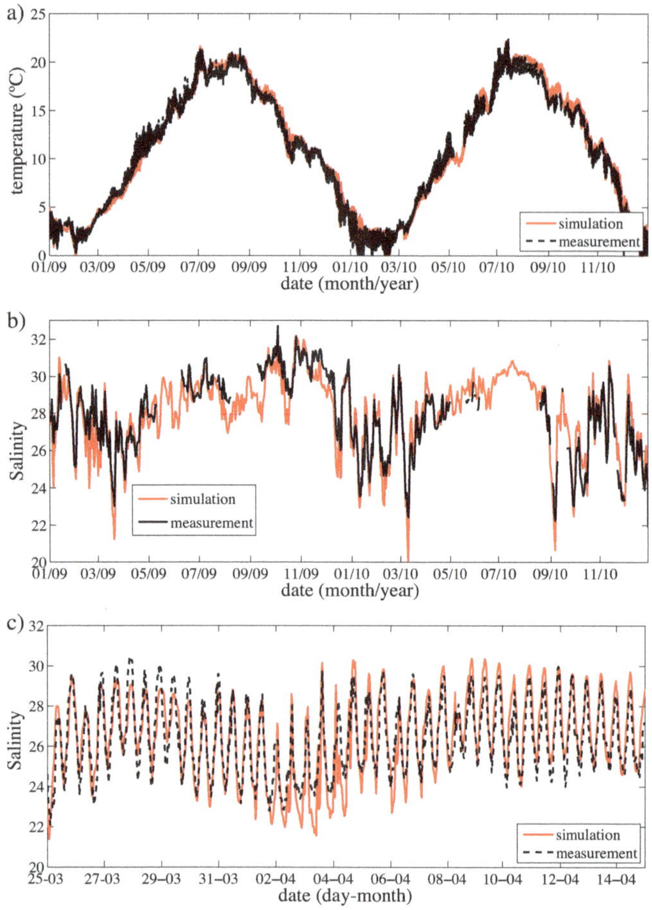

Figure 7. Comparison of temperature and salinity measurements at the NIOZ jetty station against results from the numerical simulations. **(a)** Time series of temperature with a 30 min resolution for the years 2009–2010. **(b)** Time series daily averaged salinity for the years 2009–2010. **(c)** Time series of salinity with a 30 min resolution for the days from 25 March 2009 to 15 April 2009. Gaps in the salinity time series are due to malfunctioning of the sensor or biofouling. The numerical results were interpolated to a fixed depth of 1.75 m below mean sea level.

measurements are carried out continuously using an Aanderaa conductivity/temperature sensor 3211 about 10 m off the Texel sea dike and at a depth of between 1.5 and 2 m below mean sea level; the actual depth has varied in time within this range. The raw data, with a fixed temporal resolution of 12 s, are smoothed by considering 1 min medians. For the final available product, which is used for the comparison, the 1 min medians are stored only every 30 min. Salinity data are sometimes missing due to instrument fouling even though the instrument was serviced about once a month. For their comparison with the measured data, the numerical results of the closest grid cell were interpolated in the vertical to a fixed depth of 1.75 m below mean sea level.

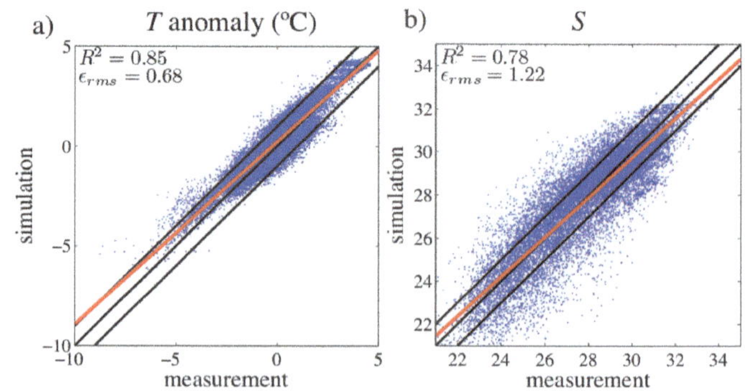

Figure 8. Scatter plots of measured vs. simulated **(a)** temperature (every 30 min), **(b)** salinity (every 30 min). The red lines correspond to a linear regression. The centre black line represents the ideal linear regression (simulation = measurement), while the side black lines are the ideal regression ±1. For every plot the coefficient of determination (R^2) and the rms error ϵ_{min} are indicated.

Figure 7a shows the time series of temperature at the NIOZ jetty station in 2009 and 2010 for both measurements and simulations. The seasonal cycle is well reproduced. Figure 7b presents the time series of the daily averaged salinity on the practical salinity scale (PSS). The daily average was computed to observe more clearly the trends over the year. In general, the time series of the simulated data follows well the trends of the measured data with fresher water during late winter/early spring and saltier water towards the end of summer. It can also be seen that the water in the autumn of 2009 is much saltier than the autumn in 2010. Figure 7c shows the time series of salinity at a 30 min resolution for a period of 20 days. It can be seen that the semi-diurnal variability in salinity is well reproduced by the model with some differences in the range of about unity, but without bias.

To quantify the differences between the measured and simulated values for both salinity and temperature, we performed a correlation of the time series. However, for the temperature time series, we compare the temperature anomaly which is obtained by removing the yearly cycle. This is done by fitting a sinusoidal function to the measured temperature time series, and then subtracting it to both the measured and simulated temperature time series. In this way we focus on the variability in weekly, daily and tidal time scales, which are much harder to reproduce than the seasonal variability.

Figure 8 shows the scatter plots of measured vs. simulated values of the temperature anomaly and salinity at the NIOZ jetty for the full years of 2009 and 2010. In the case of temperature, there is a very good agreement with $R^2 = 0.85$ and $\epsilon_{\mathrm{rms}} = 0.68$ °C. The difference between the measured and simulated values rarely exceeds 1 °C. This indicates that the tidal and intratidal variability are also well reproduced. The largest deviations were found during the winter at the end of 2010. This might be due to the presence of very low temperature from the end of November, and consequently, sea

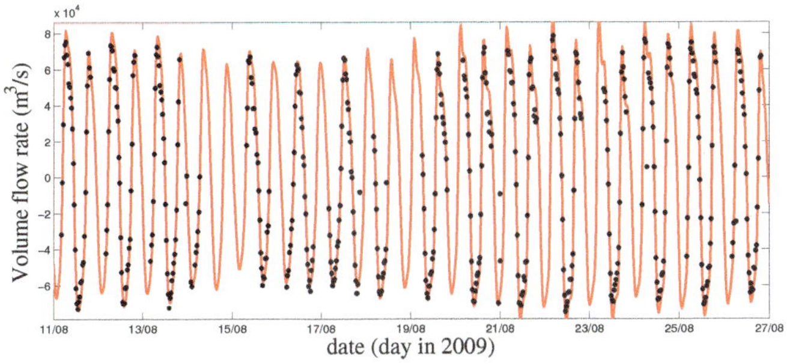

Figure 9. Time series of volume flow rate across the Texel Inlet as obtained from the TESO ferry measurements (black dots) and from the numerical simulations (red line) for 16 days in August 2009.

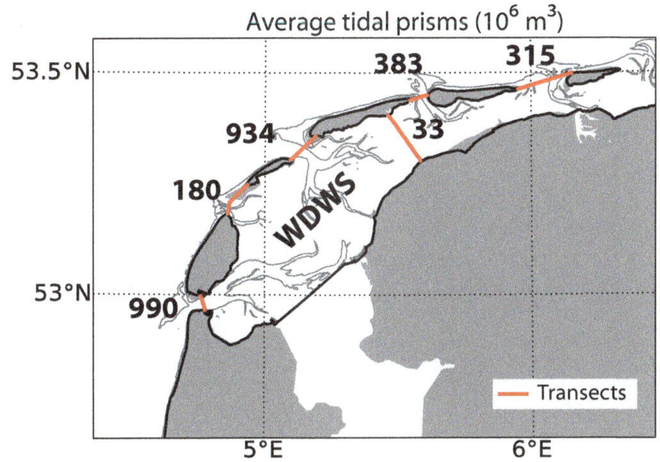

Figure 10. Average tidal prisms (10^6 m^3) of each inlet and the Terschelling watershed. The red lines mark the position of the transects used to compute the tidal prisms and transport. The numbers represent the tidal prism in 10^6 m^3.

ice, which is not taken into account in the model. For salinity, the spread is larger ($R^2 = 0.78$ and $\epsilon_{\mathrm{rms}} = 1.22$), which is usually the case in numerical simulations of estuaries since salinity depends on the complex pathways of the fresh water, while temperature is forced more locally. However, the linear regression is close to the ideal case.

3.3 Transport through the Texel Inlet

Between the port of Den Helder and the island of Texel, across the Marsdiep channel, a ferry passes every 30 min from 06:00 to 22:00 LT. This ferry carries an ADCP and a CTD (conductivity, temperature, depth), which measure current velocities, salinity and temperature. This data set has previously been used to study the dynamics of the Marsdiep channel and was described by Buijsman and Ridderinkhof (2007a) and Nauw et al. (2014). We compare the water volume flow rate through the Texel Inlet as obtained from the simulations and the calculations based on the velocity measurements for the year 2009. A vertical velocity profile with a vertical resolution of 1 m is measured every 1.3 s. The velocity is decomposed into its components perpendicular and parallel to the ferry's trajectory. Then the perpendicular component is integrated over the vertical and along the ferry's trajectory. This approach is more direct than the one previously used by Buijsman and Ridderinkhof (2007a) and Nauw et al. (2014) since it avoids the projection into a regular grid, and hence involves fewer assumptions. However, it still assumes that the data obtained during one crossing are taken at the same time, while in reality, the journey takes about 15 to 20 min.

The comparison between the simulation results and the ferry measurements is presented in Fig. 9. The coefficient of determination is $R^2 = 0.98$. For the rms error, we obtained $\epsilon_{\mathrm{rms}} = 8.00 \times 10^3$ which is less than 10 % of the maximum volume flow rate.

Due to the irregularity (both temporal and spatial) of the ferry's trajectory, it is difficult to make a detailed one-to-one comparison for individual locations along the transect. This detailed comparison is therefore left for a later paper.

The agreement between observations and simulations demonstrates that the model reproduces well the hydrodynamics of the modelled area in all its aspects (sea surface height, current velocities, temperature and salinity). Based on this, we are confident about describing, in the following sections, the exchange through the inlets between Dutch Wadden Sea and the North Sea using the numerical results.

4 Tidal prisms

As the tides rise and fall in the North Sea, large volumes of water are exchanged through the various inlets. To quantify this exchange, we compute the instantaneous volume flow rate at transects across the inlets and the Terschelling watershed (indicated by red lines in Fig. 10) by integrating the instantaneous velocity over the cross-sectional area. The tidal prism (the water volume going in or going out during a tidal cycle) at every inlet is calculated by integrating over time the absolute value of the flow rate from an instant with zero flow rate to the next. In addition, we define the average tidal prism as the average of the tidal prisms (both during flood and ebb) for every tidal period during the period 2009–2010.

The values for the average tidal prisms at all the transects – as obtained by integrating the velocity across the cross section of the transect – are shown in Fig. 10. The two major inlets in terms of their tidal prism are the Texel Inlet and the Vlie Inlet with about 10^9 m^3 each. In fact, each of them has a larger tidal prism than all the minor inlets combined. As expected, the tidal prism at the Terschelling watershed is much smaller than that at any of the inlets. The values for the inlets confirm the previous estimates listed by Louters and Gerritsen (1994) within a maximum difference of 20 % for the Borndiep. However, empirical estimates vary considerably,

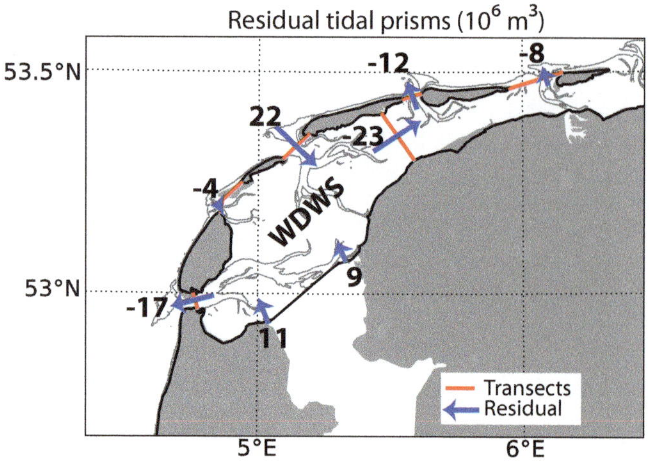

Figure 11. Average residual tidal prisms (flood − ebb) (10^6 m^3) of each inlet and the Terschelling watershed. The average input of fresh water per tidal period through the Den Oever and Kornwerderzand sluices is also presented. The numbers (in 10^6 m^3) and the length of blue arrows represent the magnitude of the residual tidal prism. The direction of the arrow represents the direction of the average residual tidal prism. The red lines mark the position of the transects used to compute the tidal prisms and transport.

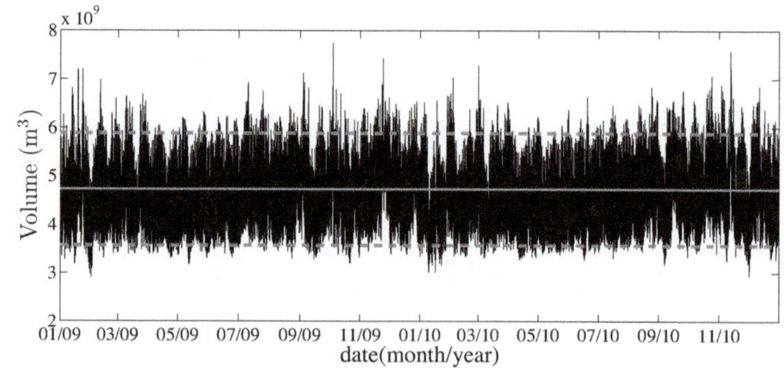

Figure 12. Time series of the volume inside the WDWS (black line). The grey line represents the mean volume over the period 2009–2010. The dashed grey lines represent the average maximum and minimum volumes over the period 2009–2010.

first and foremost because they depend on the conditions during which the measurements are made, such as wind direction and spring–neap cycles. For example, Postma (1982) mentions 430×10^6 m^3 as the tidal prism for the Borndiep, which (coincidentally or not) is closer to our model value. Given the accuracy of the model results in reproducing the records of the tidal gauges and the transport measurements by the TESO ferry (as demonstrated above), it is reasonable to regard the model results on tidal prisms as reliable, too. In fact, the model results allow us to calculate the long-term average tidal prism including a large variety of conditions, something which obviously cannot be achieved by the usual short-term (13 h) measurements.

Table 2 presents the average tidal prisms for the inlets and the Terschelling watershed. Here we distinguish between the flood and ebb tidal prisms, which are, in general, different. The difference is indicative of the average residual flow (also indicated in Table 2 and Fig. 11). For convenience, *flood* at the watershed is defined as having a flow from east to west; i.e. into the WDWS. It can then be seen that there is a net inflow at the Vlie Inlet and outflow at every other inlet. In the WDWS, the inflow at the Vlie is supplemented with an average inflow of fresh water through the sluices of about 20×10^6 m^3 per tidal cycle. Remarkably, the residual tidal prism for the watershed (-23×10^6 m^3) is of the same order of magnitude as those for the Texel and Vlie inlets (18×10^6 m^3 and 21×10^6 m^3, respectively) even though the average tidal prism itself is two orders of magnitude smaller.

Table 2 also presents the standard deviation of the tidal prism at every transect which gives a measure of their variability. This variability is partly due to the spring–neap cycle and partly due to wind speed and direction. For all the inlets, the standard deviation is of the order of 20 % of the average tidal prism. Note that the estimates obtained through the measurements mentioned by Louters and Gerritsen (1994) and Postma (1982) fall well within the range of possible values obtained in the numerical simulations. It should also be noted that these empirical values were obtained several decades ago; in the meantime, the prism may have changed somewhat. For the Terschelling watershed, the standard deviation reaches about 200 % of the average tidal prism during flood. The variability is in all cases much larger than the *residual* tidal prisms. As we will discuss below, the wind (rather than exclusively tides) is a determining factor in the transport across the watershed. In fact, although the *tidal prism* is defined as the difference between the volumes at high and low water, its actual value at every tidal period is strongly influenced by non-tidal effects, notably wind forcing, which causes it to be highly variable. In conclusion, the average tidal prisms for ebb and flood give a good indication of the long-term net exchange of water volumes through the inlets, but in the short run, the variability overshadows these average residuals. For this reason, we study more in detail the temporal variability of the residual water transport in the next section.

As a complementary view to tidal prisms, we consider the volume within the basin. In a basin with several inlets (as the one studied here), flood and ebb at different inlets occur at different times (see phases in Fig. 4), so the tidal prisms at the inlets cannot be simply added up to get the change in volume. Here, we consider the volume of the WDWS basin, which is delimited by the transects across the Texel Inlet, the Eierlandse Gat, the Vlie Inlet, and the Terschelling watershed. Figure 12 shows the time series of that volume for the years 2009–2010 at a 30 min resolution. This volume can be computed from the model output since it contains the water depth at every grid point. The average water volume in this basin for the 2 years of simulation is of

Table 2. Average, standard deviation (SD), and residual of the flood and ebb tidal prisms for all inlets and the Terschelling watershed. For convenience, *flood* at the watershed is defined as having a flow from east to west.

	Average (10^6 m^3)		SD (10^6 m^3)		Residual (10^6 m^3)
	flood	ebb	flood	ebb	flood – ebb
Texel Inlet	982	999	180	174	−17
Eierlandse Gat	178	182	45	38	−4
Vlie Inlet	945	923	173	165	22
Borndiep	377	389	70	79	−12
Pinkegat + Friesche	310	318	63	69	−8
Terschelling watershed	21	44	44	123	−23

4730×10^6 m^3, while the average maximum volume over the same period is 5724×10^6 m^3, and the average minimum volume is 3700×10^6 m^3; i.e. 43 % of the mean water volume inside the basin leaves and enters every tidal cycle. The difference between the average maximum and minimum volumes is 2024×10^6 m^3, which is smaller (by ~ 6 %) than the sum of the tidal prisms as calculated from the volume flow rate through the transects limiting this region. This difference reflects the phase shift of the in- and outflow between the inlets.

In addition, Fig. 12 shows that the maxima of the volume are highly variable (more so than the minima). The spring–neap cycle can be clearly observed. Besides, there are occasionally very large peaks, with the highest being 7341×10^6 m^3 or 1.65 times the average volume, on 4 October 2009. These extreme events can be ascribed to storm surges.

5 Variability of the residual circulation

For the calculation of residual transport it is, first of all, necessary to have a clear idea over what interval of time the residual is to be taken. At any one location (a tidal gauge, for example), this already poses a problem because the tidal period is not well defined: the time between low waters is, in general, unequal to the time between high waters, and both vary rather erratically over longer periods. This is true even if one were to consider a purely tidal signal (as in tidal predictions), disregarding the effects of wind. This is illustrated for data from tidal prediction at Vlieland Harbour (Gerkema et al., 2014). The underlying cause is that the tidal signal consists of tidal constituents that have incommensurable periods. A further complication is that subsequent low (or high) waters are unequal; in other words, the background state itself changes over the period. In the case at hand, the problem seems to be even more complicated since we have to deal not just with one location but with an entire area, involving differences in phases of the tide of up to 3 h.

In the previous section, where we computed local tidal prisms, the definition of the tidal period was based on the time between alternate slacks. Here, we will consider the multiple-inlet system as an entity, and the previous definition no longer works because of non-simultaneity of slacks. To define a tidal period valid for the whole region, we consider the time series of the water volume inside the WDWS shown in Fig. 12. We define the reference volume as the long-term average volume during the years 2009–2010. During a tidal cycle, the actual volume goes up and down, matching the reference volume at a certain moment during ebb and at a certain moment during flood. The tidal period is defined as the interval between two consecutive moments when the instantaneous volume matches the reference volume during flood. An advantage of this definition is that, for every period, the sum of all the residual inflows and outflows equals zero. (Strictly speaking, the calculation should be in terms of mass rather than volume, since volume is not conserved if a certain mass of fresher water enters and an equal mass of saltier water leaves, but this effect can be neglected here.) As can be seen in Fig. 12, for certain tidal periods the minimum is larger or the maximum is smaller than the average volume. This only means that the calculation of the residual transport will be done over a larger number of tidal periods (usually two or three at most) but still over an integer number of tidal periods according to our definition.

Figures 13a and 14a show the average residual volume flow rate for every tidal period as a function of time for the Texel Inlet, the Vlie Inlet, and the Terschelling watershed for the years 2009 and 2010, respectively. The Eierlandse Gat is omitted since, as already seen in Table 2 and Fig. 11, the residual flow there is much smaller. Most of the time, the residual circulation consists of an inflow through the Vlie Inlet and an outflow through the Texel Inlet, with a relatively small in- or outflow across the watershed. This is consistent with the differences between the average flood and ebb tidal prisms, and also with the results of Ridderinkhof (1988b). This occurs for 63 % of the tidal periods, but the situation can change completely due to wind effects. Figures 13b and 14b show the time series of wind speed and direction at the centre of the Texel Inlet as obtained from the forcing data. It can be seen that all periods with southwesterly to westerly winds in excess of 15 m s^{-1} result in an inverted residual circulation with an outflow rate through the Vlie Inlet and the watershed of at least 5×10^3 m^3 s^{-1}. These periods are

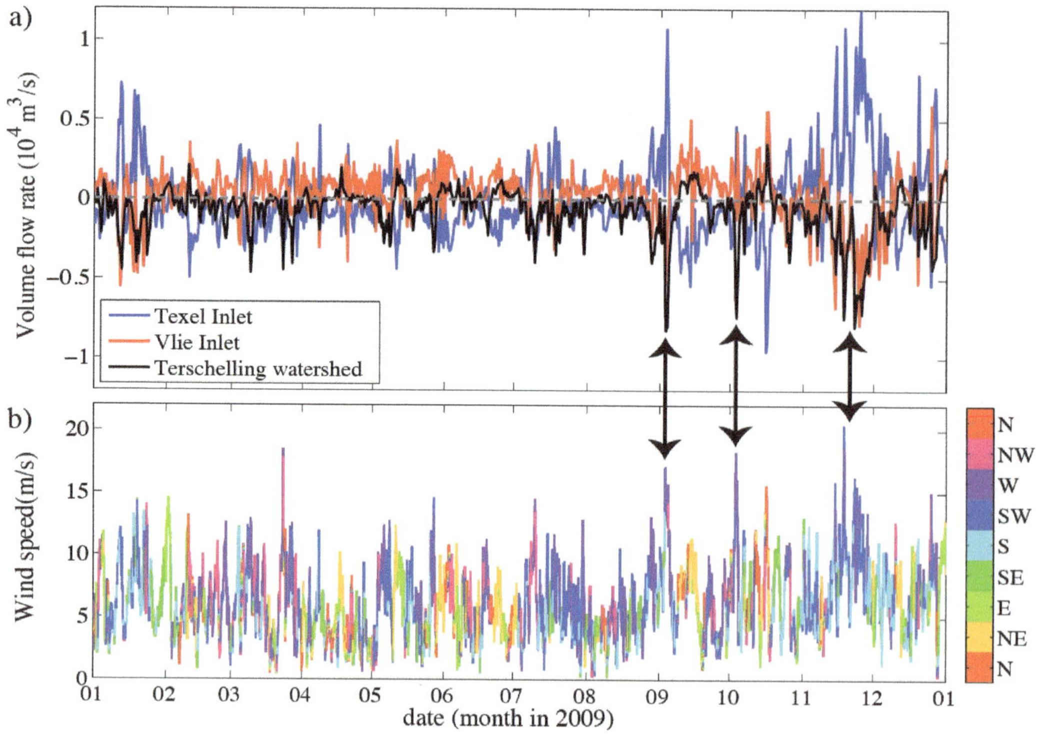

Figure 13. **(a)** Residual volume flow rate as a function of time for the Texel Inlet, the Vlie Inlet, and the watershed in 2009. Positive flow rate corresponds to an inflow, while negative flow rate corresponds to an outflow. **(b)** Wind speed and direction next to the Texel Inlet as a function of time. The wind velocity is obtained from the data used for the atmospheric forcing. The black arrows point to strong southwesterly to westerly wind events which correspond to periods when the residual circulation is inverted with respect to the typical case.

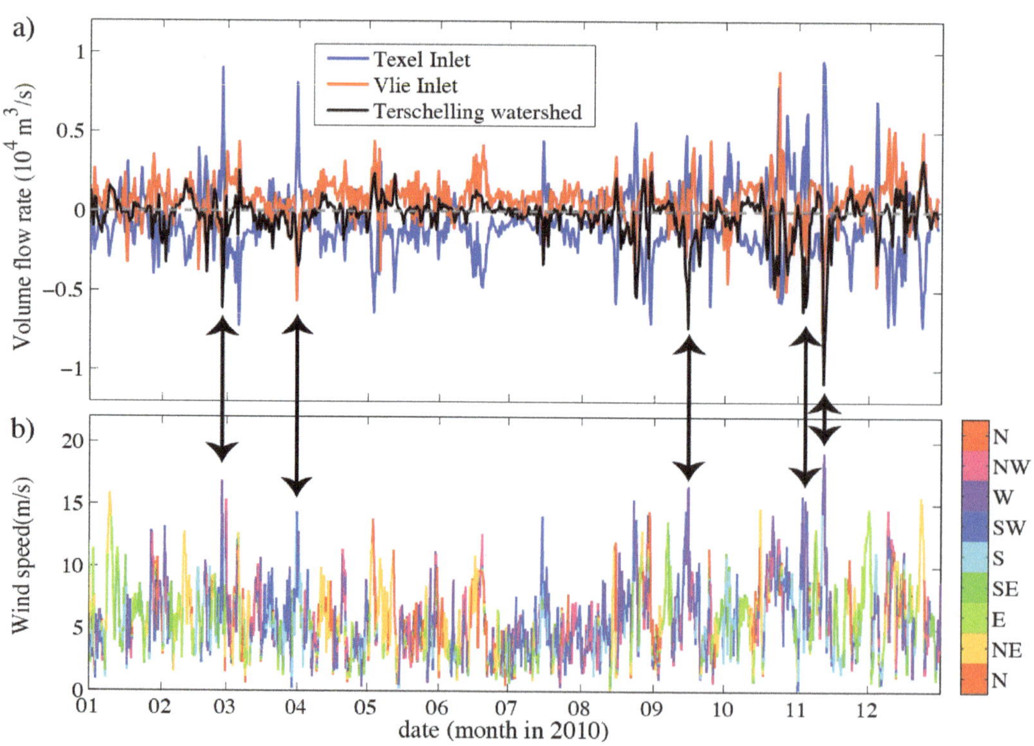

Figure 14. As Fig. 13, for the year 2010.

marked with arrows in Figs. 13 and 14. Of great interest during the simulated years is a period when strong southwesterly winds blew continuously during about two weeks in November 2009. During this period, a very strong inverted residual circulation took place with a large import of water through the Texel Inlet, and an outflow through the Vlie and across the watershed. Although such events seem to be infrequent (another similar event, although smaller, occurred in December 2011), they could be of importance for the general state and evolution of the Wadden Sea since up to 50 % of the average volume of the basin gets refreshed within a period of about one week. On the other hand, strong easterly winds enhance the residual flow from the Vlie Inlet to the Texel Inlet. This is in agreement with the work by Li (2013) who investigated the effects of wind in an idealized three-inlet system. He found that: (1) wind can significantly influence the residual circulation in a multiple inlet with an inflow usually in the upwind inlet and an outflow in the downwind inlet; (2) the weakest residual flows are found in the middle inlet; (3) wind can produce high-magnitude net flows that tides cannot produce. However, it is clearly important for the WDWS to include the Terschelling watershed, i.e. include at least one more inlet, into such an idealized model.

Since wind effects render the residual circulation highly variable, caution has to be exerted when defining the *normal* or *typical* conditions. For this, we present the histograms of the residual volume flow rate Q' through the three inlets of the WDWS and the Terschelling watershed as shown in Fig. 15. The histograms are constructed by binning the residual flow rate for every tidal period into 100 bins between -8×10^3 and $8 \times 10^3\,\mathrm{m^3\,s^{-1}}$. In addition, Table 3 presents the value for the median, mean, standard deviation and skewness for each of the inlets and the watershed. Particularly noticeable are the large dissimilarities between the mean and median values.

All four histograms present a clear asymmetry, meaning that the average residual flow over the simulated period is not indicative of *typical* conditions as it deviates from the peak in the histogram. Instead, the median values correspond well with the peak, giving a better indication of the typical conditions: an inflow through the Vlie of about 600–700 $\mathrm{m^3\,s^{-1}}$ and at the sluices of about 450 $\mathrm{m^3\,s^{-1}}$; and an outflow through the Texel Inlet of about 600–700 $\mathrm{m^3\,s^{-1}}$, and both through the Eierlandse Gat and the Terschelling watershed of about 100–200 $\mathrm{m^3\,s^{-1}}$. The order of magnitude and direction of these residual flow rates is in agreement with the results obtained by Ridderinkhof (1988b), particularly if we consider that the freshwater discharge was not taken into account in his study. Hence, we can assume that the median flow rates are representative of relatively calm winds and are mainly driven by tides. In addition, it can be noted that the typical values of the residual flow rates (including the freshwater discharge) add up to almost null.

As discussed above, much of the variability and the extreme events can be attributed to the varying winds that can enhance, weaken or even invert the tidally driven residual circulation depending on the wind speed and direction. The variability of the residual flow rate (which is represented by the standard deviation) is largest at the Texel Inlet, followed by the Vlie Inlet and the watershed, and finally, by the Eierlandse Gat. The fact that the variability is larger at the Texel and Vlie inlets can be attributed to the existence of a relatively good connection between the two inlets. On the other hand, the low variability at the Eierlandse Gat suggests that this inlet is rather isolated from the rest of the WDWS by the Texel and Vlieland watersheds. The large variability of the residual flow rate in the Terschelling watershed is also clearly associated with the wind.

The histograms of the residual flow for both the Texel Inlet and the Eierlandse Gat are positively skewed, while the histograms for the Vlie Inlet and the watershed are negatively skewed. This is a reflection of the predominant wind direction both in average and during most of the major storms, which results in more tidal periods where the typical tidally driven residual circulation with an inflow through the Vlie Inlet is weakened or reversed. Furthermore, note that the histogram for the residual flow at the watershed has a larger skewness and peakedness than those for the Texel Inlet and the Vlie Inlet. This shows that the transport through the watershed has a clear preferential direction, and explains the large net transport. However, even in the preferential direction, strong winds seem to be needed to drive a transport across this watershed.

6 Freshwater distribution and transport

6.1 Spatial distribution of fresh water

In order to study the freshwater distribution and transport, the fresh water from the two main sluices (Den Oever and Kornwerderzand) is tagged with an Eulerian passive tracer following a similar approach to Zhang et al. (2009) and Meier (2007). A different tracer is associated with each of the sluices, and the fresh water has a concentration equal to unity ($c_i = 1$, where i refers to the sluice). In this way, we can distinguish the fresh water from the main sluices from the fresh water from other sources (the Rhine, other sluices, and precipitation) in an unambiguous way.

For the advection of the passive tracers, the Framework for Aquatic Biological Modeling (FABM) passive tracer module (see http://fabm.sourceforge.net/) coupled to GETM is used. GETM solves for the advection of the passive tracers using the same methods as for salinity and temperature. The initial concentration in the domain is set to zero, and the model is spun up during two months. We consider this time to be sufficient since the typical transport timescales in the Wadden

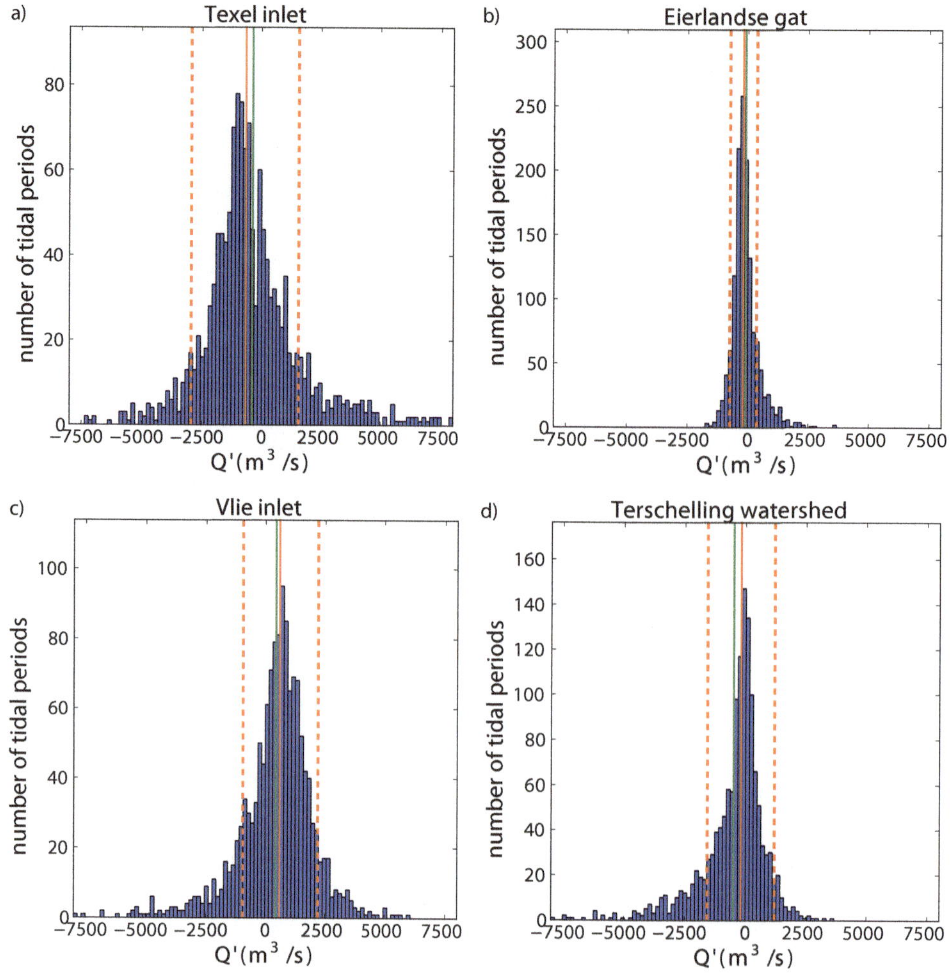

Figure 15. Histograms of the residual flow rate through the different inlets of the WDWS and the Terschelling watershed. The solid red line represents the median value, and the dashed lines represent the median value plus/minus one standard deviation. The green line represents mean value.

Table 3. Median, standard deviation and skewness of the residual flow rate at the inlets of the WDWS and the Terschelling watershed.

	Median ($m^3\ s^{-1}$)	Mean ($m^3\ s^{-1}$)	Standard deviation ($m^3\ s^{-1}$)	Skewness ($m^3\ s^{-1}$)
Texel Inlet	−683	−381	2256	1.07
Eierlandse Gat	−186	−98	561	1.52
Vlie Inlet	612	452	1592	−0.81
Watershed	−180	−494	1387	−1.88

Sea are shorter than a month. The flushing time of the fresh water will be discussed in more detail in Sect. 6.2.

In describing the fate of fresh water, we could mean two different things: how the water gets saltier along the way, or where the fresh water goes in the sense of Lagrangian tracers (e.g. drifters, buoys). The latter is purely advective, the former is diffusive as well. A tracer concentration, defined with a certain value at the sluice, but initially zero elsewhere, behaves identically to salinity if the diffusion coefficients are the same and if that sluice acts as the only source as in the work by Zhang et al. (2009). When dealing with multiple freshwater sources as in the Wadden Sea or Baltic Sea (Meier, 2007), a couple of points must be clarified.

1. Since there are multiple sources, the way the fresh water from one sluice gets saltier is influenced by the other sources. For example, if fresh water from Den Oever meets fresh water from Kornwerderzand there is no diffusion since the salinity is identical. If on the other hand,

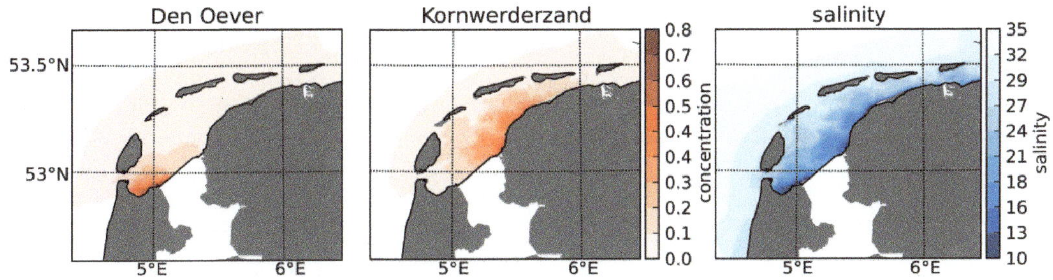

Figure 16. Average concentration in the uppermost layer of the passive tracer from **(a)** Den Oever and **(b)** Kornwerderzand for the month of April 2009. **(c)** Average salinity in the upper layer for the same month.

the Den Oever fresh water would instead meet North Sea water, there is a gradient in salinity.

2. Since the advection–diffusion equation is linear in salinity or tracer concentration, the working of the equation is the same for these quantities, even if they differ by a constant factor or by an added constant.

From the above two points, it follows that the behaviour of the tracer released in Den Oever would be identical to that of fresh water if the Den Oever sluices were the only source of fresh water. We can follow the tracer concentration, calculate the amount in a volume V:

$$V_i = \int\limits_V c_i \mathrm{d}V, \tag{4}$$

and calculate the transport through the transects delimiting the WDWS:

$$Q_{i,j} = \int\limits_{A_j} c_i u_\mathrm{n} \mathrm{d}A \tag{5}$$

with j denoting the transect of interest, u_n the velocity component normal to the transect, and A_j the cross-sectional area of that transect.

Figure 16a shows the average distribution of the concentration in the uppermost layer of the Den Oever tracer and Fig. 16b of the Kornwerderzand tracer for April 2009. In general, the Den Oever tracer remains in the southern part of the WDWS close to the Texel Inlet, while the Kornwerderzand tracer is distributed all the way from the Texel Inlet to the Friesche Inlet. It is clear that for this distribution to occur, the fresh water from Kornwerderzand has to be transported through the Terschelling watershed. Figure 16c shows the average salinity distribution in the uppermost layer. By comparing the salinity distribution to the tracer distribution, it can be observed that the regions of low salinity are a superposition of the regions of high tracer concentration.

High concentrations of the tracers on the tidal flats or other shallow areas are not equivalent to large volumes of fresh water. In other words, a map of concentration in such a complex bathymetry as the one of the Wadden Sea, gives little

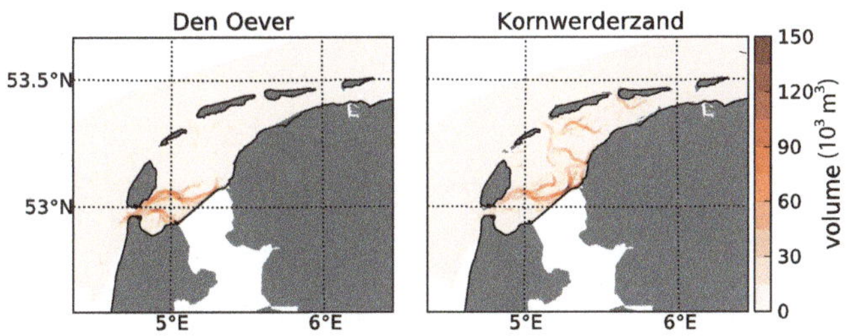

Figure 17. Average volume (10^3 m^3) per horizontal grid cell of the fresh water from **(a)** Den Oever and **(b)** Kornwerderzand for the month of April 2009.

information on the position and destination of the fresh water. Figure 17 shows the average distribution of the volumes of fresh water for the month of April 2009. While the concentrations are higher on certain tidal flats, the largest volumes of fresh water are located in the channels. The fresh water from Den Oever is located in the channels leading to the Texel Inlet, and the fresh water from Kornwerderzand is distributed along the channels in the whole of the WDWS and even around the Borndiep. This is in line with the findings of Postma (1950) and Zimmerman (1976), who found that the fresh water from Kornwerderzand gets distributed in almost equal parts between the Texel and Vlie basins, whereas the water from Den Oever travels mostly towards the Texel Inlet. We now examine where it actually *leaves* the WDWS.

Figure 18a shows the time series with a 30 min resolution of the instantaneous flow rate of the Kornwerderzand tracer through the Texel Inlet, the Vlie Inlet and across the Terschelling watershed. As can be seen, the magnitude of the instantaneous transport is higher at the inlets than at the watershed. The tracer changes its destination inlet as a function of time depending on the conditions (e.g. wind direction and freshwater discharge rate). Once the tracer reaches one of the inlets, it does not exit the Wadden Sea permanently. It goes in and out through the inlets with the tide, and a clear outgoing residual trend cannot be deduced from the instantaneous flow rate. To compute the residual flow of the tracers through the inlets, we use the same approach to the one used in the previous section to calculate the residual water transport.

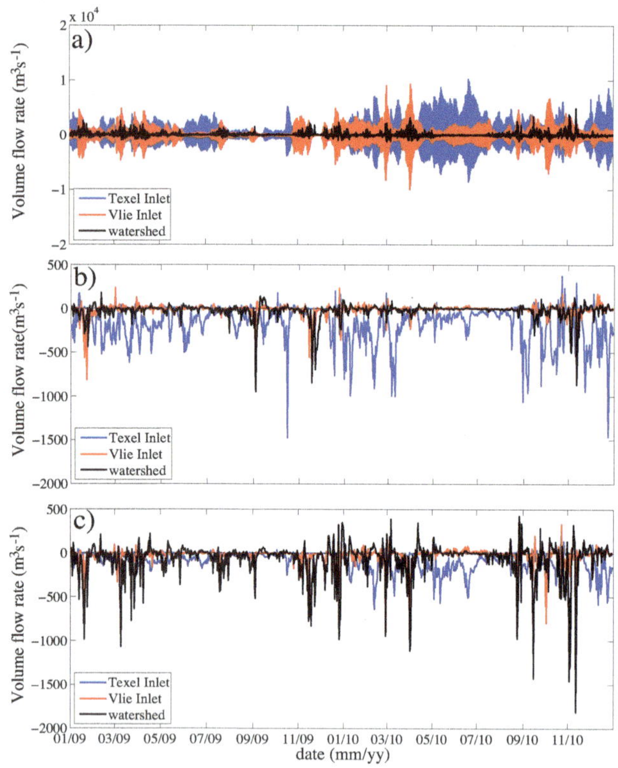

Figure 18. **(a)** Time series of the instantaneous volume flow rate of the tracer corresponding to water from Kornwerderzand at the Texel Inlet, the Vlie Inlet, and the Terschelling watershed. The temporal resolution is 30 min. **(b)** Tidally averaged volume flow rate of fresh water from Den Oever through the Texel Inlet, the Vlie Inlet and across the Terschelling watershed. **(c)** Tidally averaged volume flow rate of fresh water from Kornwerderzand through the Texel Inlet, the Vlie Inlet and across the Terschelling watershed.

Figure 18b and c show the residual transport through the Texel Inlet, the Vlie Inlet and across the Terschelling watershed of the tracers associated with the two major sluices. The tracer from Den Oever (Fig. 18b) exits mainly through the Texel Inlet. A few exceptions occur during strong southwesterly winds when there is an outflow through the Vlie Inlet and across the watershed. These periods correspond with the periods when the residual circulation in the WDWS is inverted such as was the case in November 2009 (see Fig. 13). During the period 2009–2010, the Den Oever tracer exited the WDWS 81 % through the Texel Inlet, 12 % through the watershed, 3 % through the Vlie Inlet and 4 % through the Eierlandse Gat.

For the water out of Kornwerderzand (Fig. 18c), there is a more balanced distribution of the residual outflow through the possible outlets. Remarkable, however, is the importance of the residual outflow across the Terschelling watershed, which is usually disregarded. In fact, 40 % of the Kornwerderzand tracer exits across the watershed freshening the water in the contiguous basin, which corresponds to the Borndiep Inlet. About an equal amount of the tracer from Kornwerderzand (43 %) exits through the Texel Inlet. Surprisingly, the Vlie is not an important outlet for this fresh water with only 10 % of the total outflow, which is comparable to the 7 % of the outflow through the Eierlandse Gat. The time series of the outflow of the tracer through the Terschelling watershed exhibits large peaks. A visual comparison between the residual outflow of the tracers and the wind (Figs. 13b–14b) shows that the large peaks are related to strong wind events.

For both tracers, the total averaged outflow across the transects was compared with the inflow at the sluices. The computed outflow is within 5 % of the inflow, giving confidence in the results. Note that in principle there is no reason for them to be identical since there is a lag between the inflow and the outflow.

6.2 Flushing frequency of fresh water

Transport timescales, such as the residence time and the flushing time, are used to estimate the renewal and ventilation of estuaries and other coastal areas, and are useful to explain several processes of importance in biology and ecology (see e.g. Abdelrhman, 2005; Monsen et al., 2002, and references therein). Note that actual names and definitions vary greatly within the literature (Abdelrhman, 2005). The flushing time T_f is commonly defined as the ratio of the mass of a constituent in a water body M_i to its rate of renewal (inflow) F_i, i.e.

$$T_f = \frac{M_i}{F_i}, \qquad (6)$$

where the subscript i is associated with origin of the specific constituent; see e.g. the work by Bolin and Rodhe (1973) – where the flushing time is actually referred to as *turn-over time* – and by Zimmerman (1976). However, Eq. (6) was derived for a steady-state condition, which is not the case for the Wadden Sea for the timescales relevant here. In addition, we are interested in studying the variability of the flushing time of fresh water and its dependence on the external forcing. Some of the limitations of considering Eq. (6) for highly variable systems have been previously discussed by Monsen et al. (2002).

For the steady-state case, the inflow and the outflow are identical, and so, the flushing time (Eq. 6) can be defined as the ratio of the mass of the constituent in the water body to the outflow. For the unsteady case, this definition means that the flushing time is the time it would take to empty the available mass at the current outward volume flow rate. For the case studied here, this still presents a difficulty since the outflow (and the inflow) are null during some periods, resulting in infinite flushing times. We find it useful to define the tidally averaged flushing frequency (the inverse of the flushing time) for every tidal period T as

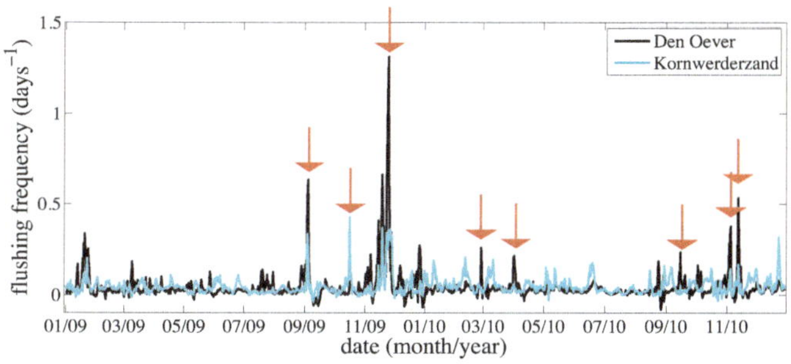

Figure 19. Flushing frequency (days^{-1}) of the fresh water from the sluices of Den Oever and Kornwerderzand.

$$\Omega_i = \frac{\int_T \sum_j Q_{i,j} dt}{\int_T V_i(t) dt}, \quad (7)$$

where a sum of the residual volume flow rate Q_i over all the inlets and the watershed (sum over j) is considered, and the volume of the specific constituent is used instead of the mass.

Figure 19 shows the tidally averaged flushing frequency Ω_i as a function of time for the tracers associated with Den Oever (DO) and Kornwerderzand (KDZ). Negative frequencies mean that there is a net import of fresh water from the North Sea or the adjacent basin to the East. This can occur, for example, after a large flushing event when a large portion of the fresh water is located outside the WDWS, especially if the wind rotates close to 180° as during the passage of cold fronts.

A typical value for the flushing frequency of the fresh water is given by the median frequencies: $\Omega_{DO} = 0.0363$ days^{-1} (equivalent to a flushing time $1/\Omega_{DO} = 27.5$ days) and $\Omega_{KDZ} = 0.0259$ days^{-1} (equivalent to a flushing time $1/\Omega_{KDZ} = 38.6$ days). These estimates for the flushing times are 2.5–3 times longer than the ones found by Zimmerman (1976) of about 8 days for the fresh water from Den Oever and 16 days for the fresh water from Kornwerderzand. Nonetheless, the values obtained by Zimmerman (1976) are still well within the possible range. The difference can be due to the conditions during the period that the measurements were taken since the flushing time is highly variable. In addition, in Zimmerman's box model, there is a sensitivity of the calculation of the freshwater volume to the choice of a reference salinity, which may explain discrepancies with our results.

There are important flushing episodes which are again associated with strong winds (Fig. 19). The arrows in Fig. 19 correspond to the same strong wind events signalled in Figs. 13–14. During the years 2009–2010, the most important flushing episode occurred in November 2009 when the flushing frequency reached a value larger than 1 day^{-1}. This event was characterized by strong persistent southwesterly winds (also discussed in Sect. 5) that forced a strong persistent outward flow that drastically reduced the volume of fresh water inside the WDWS. In fact, the volume of fresh water inside the basin decreased one order of magnitude in 10 days from 550×10^6 m^3 on 18 November 2009 to 65×10^6 m^3 on 27 November 2009. The end of this flushing period corresponds to some of the highest salinities registered at the NIOZ jetty, as shown in Fig. 7b.

7 Conclusions

In this paper, we have shown model results for the hydrodynamics of the Dutch Wadden Sea, with a focus on the western Dutch Wadden Sea, for the years 2009–2010. For this purpose, we used the GETM/GOTM model at a 200 m resolution in the horizontal and with 30 vertical layers. Boundary conditions impose tides, wind surges, freshwater discharges, wind stress, precipitation, evaporation and surface heat fluxes.

Comparisons with different kinds of observations show an excellent agreement. These include a comparison with the observed SSH at 14 tidal gauges, with transport flow rates through the Texel Inlet as derived from ADCP measurements beneath the TESO ferry, and with measurements of salinity and temperature at the NIOZ jetty. The agreement between these data sets and the simulations makes it reasonable to assume that the model results are close to reality also at places where no measurements are available to compare with. Consequently, we have created as a by-product a solid 2-year hindcast database of the hydrodynamic conditions of the Dutch Wadden Sea for further use by sediment transport, and especially, ecological studies that can benefit from an upgrade to an overall spatial resolution of 200 m and a temporal resolution of 30 min.

Qualitatively, conclusions from earlier studies are mostly confirmed by our results. For example, we also find that most of the fresh water discharged at the sluice at Den Oever leaves via Texel Inlet, in agreement with Postma (1950) and Zimmerman (1976). We find, on average, a residual flow from the Vlie Inlet to the Texel Inlet, confirming Ridderinkhof (1988b), and average values of the tidal prisms of the various inlets are close to earlier (empirical-based) estimates listed by Postma (1982) and Louters and Gerritsen (1994).

In addition, the present study sheds light on the key aspect of variability due to wind, which had previously been unapproachable for a quantitative analysis, either because measurements lack spatial coverage and are mostly biassed to mild weather, or because models lacked wind forcing. In fact, this is the first three-dimensional modelling study of the Dutch Wadden Sea that includes, besides tides, realistic wind forcing and freshwater discharge.

This has allowed us to calculate not only the tidal prisms for each inlet but also their variability (standard deviation), which turns out to be typically about 20 % of the average tidal prism. For the Terschelling watershed, the variability even exceeds the average tidal prism. In all cases, the variability in the tidal prism is much larger than the residual prism, the long-term average of the difference between ebb and flood prisms (Table 2).

We also examined the time series that show the episodes responsible for this variability. We find that, for each inlet, the probability distribution of the residual flow rate spreads around the typical value which is approximated better by the median than by the average. In fact, the median and mean are very dissimilar (Table 3) due to a clear skewness of the distributions dictated by the wind's direction. In fact, the three-inlet system of the WDWS is reminiscent of the idealized system studied by Li (2013) in which the outer inlets presented the largest response to wind and the middle inlet had the weakest. However, the large residual flow across the Terschelling watershed during strong southwesterly winds is a crucial component, whose role in the overall circulation in the Dutch Wadden Sea is much larger than was previously assumed.

To calculate the residual transport, we propose an alternative to the more common (tidal) harmonic analysis on the time series of transport flow rate or velocity. There, the residual of the harmonic analysis is assumed to be the residual flow. However, this method fails here because set-up by winds affects the tidal propagation, and hence, the total signal is not just the linear sum of independent components. This is particularly true at the Terschelling watershed, since due to its shallowness, it can be dry for several hours a day. This clearly renders the time series of the transport rate highly non-linear. For this reason, we adopted an alternative way by calculating the residuals directly from time integration of the volume flow rates. The definition of the time interval, i.e. the tidal period, is based on the volume inside the basin, and hence, it is valid for the whole basin.

With regard to the discharge of fresh water from the sluice at Kornwerderzand, it had never been established previously where it actually leaves the Wadden Sea, although it was presumed that the Vlie Inlet was a major exit route along with the Texel Inlet (Postma, 1950). It now becomes clear that the Vlie Inlet, in fact, plays only a minor role in the net exchange and that, instead, the Terschelling watershed together with the Texel Inlet are predominant.

Most of the time, the hydrodynamic state of the Dutch Wadden Sea, with regard to the residual circulation and freshwater flow patterns, is close to the typical, i.e. median state. However, at other times it is upset or reversed with a skewed tendency due to the wind direction. As a result, the yearly mean values of residual volume flow rates or flushing times are not necessarily representative. The variability here has an episodic character, and this is a key feature of the net transports and flushing times. This is the main outcome of this paper, along with the quantitative substantiation of it that the model provides.

Appendix A: Numerical model

GETM is a finite difference model solving the three-dimensional hydrostatic equations of motion with the Boussinesq approximation and the eddy viscosity assumption:

$$\partial_t u+\partial_z uw+\partial_z\left[(\nu_t+\nu)\partial_z u\right] +\alpha\left\{\partial_x u^2+\partial_y uv-\partial_x 2A_\mathrm{H}\partial_x u-\partial_y[A_\mathrm{H}(\partial_y u+\partial_x v)] -fv-\int_z^{\xi}\partial_x b\mathrm{d}z'\right\}=-g\partial_x\xi, \tag{A1}$$

$$\partial_t v+\partial_z vw+\partial_z\left[(\nu_t+\nu)\partial_z v\right] +\alpha\left\{\partial_x uv+\partial_y v^2-\partial_y 2A_\mathrm{H}\partial_y v-\partial_x[A_\mathrm{H}(\partial_y u+\partial_x v)] +fu-\int_z^{\xi}\partial_y b\mathrm{d}z'\right\}=-g\partial_y\xi, \tag{A2}$$

and the continuity equation for an incompressible fluid:

$$\partial_x u+\partial_y v+\partial_z w=0, \tag{A3}$$

where u, v and w are the velocity components in the x, y and z directions, respectively; ξ is the sea surface elevation; $f=2\Omega\cos(\phi)$ is the Coriolis parameters with Ω the rotation rate of the earth and ϕ the latitude; $g=9.81\,\mathrm{m^2\,s^{-1}}$ is the gravitational acceleration; b is the buoyancy; ν is the kinematic viscosity; ν_t is the vertical eddy viscosity; $A_\mathrm{H}(x,y)$ is the horizontal eddy viscosity; and α is parameter used for the drying–flooding as described below. The buoyancy b is defined as

$$b=-g\frac{\rho-\rho_0}{\rho_0}, \tag{A4}$$

with ρ the density and ρ_0 a reference density.

GETM has a *thin layer* flooding and drying algorithm. This means that at least a thin layer with thickness D_min (minimum depth) is present at every grid point. This is achieved by having the bottom drag coefficient that exponentially increases for decreasing water depths, and hence, the flow is stopped for very shallow depths. Drying and flooding of tidal flats is incorporated into Eqs. (A1) and (A2) through the non-dimensional parameter

$$\alpha=\min\left\{1,\frac{D(x,y)-D_\mathrm{min}}{D_\mathrm{crit}-D_\mathrm{min}}\right\}, \tag{A5}$$

where $D(t,x,y)=\xi(t,x,y)-H(x,y)$ is the total depth; D_crit is a parameter called *critical depth* which simply determines the depth at which some of the terms start to be switched off. In this way, $\alpha=1$ when $D\geq D_\mathrm{crit}$, and the usual full momentum equations are considered. When the layer becomes very shallow $D\leq D_\mathrm{crit}$, simplified physics are considered which tend to a balance between bottom friction and the horizontal pressure gradient as $D_\mathrm{crit}\rightarrow D_\mathrm{min}$. For the simulations presented in the current paper, $D_\mathrm{min}=10$ cm and $D_\mathrm{crit}=26$ cm were used.

In the vertical, terrain-following σ-coordinates are used such that the height of the layers h_k is given by

$$h_k=D(t,x,y)\sigma_k, \tag{A6}$$

where $\sigma_k=k/N-1$ with $k=1,2,\ldots,N$, and $N=30$ (i.e. 30 vertical layers). This entrains a modification of the physical condition at the bottom since the discrete velocity point nearest to the bottom is half a grid box from the real bottom. The estimates for the velocities at the centre of the bottom layer are then given by

$$u_\mathrm{b}=\frac{u_*^\mathrm{b}}{\kappa}\ln\left(\frac{0.5h_1+z_0}{z_0}\right), \tag{A7}$$

$$v_\mathrm{b}=\frac{v_*^\mathrm{b}}{\kappa}\ln\left(\frac{0.5h_1+z_0}{z_0}\right), \tag{A8}$$

where $\kappa=0.4$ the von Kármán constant; z_0 is the bottom roughness length, which in this case is given the constant value $z_0=1.7$ mm; and u_*^b, v_*^b are the friction velocities.

The transport equations for potential temperature T (in °C) and salinity S (in the practical salinity scale, PSS) are

$$\partial_t T+\partial_x uT+\partial_z wT-\partial_z(\nu_t'\partial_z T) -\partial_x(A_\mathrm{H}^T\partial_x T)-\partial_y(A_\mathrm{H}^T\partial_y T)=\frac{\partial_z I(z)}{c_p'\rho_0} \tag{A9}$$

$$\partial_t S+\partial_x uS+\partial_z wS-\partial_z(\nu_t'\partial_z S) -\partial_x(A_\mathrm{H}^T\partial_x S)-\partial_y(A_\mathrm{H}^T\partial_y S)=0 \tag{A10}$$

with ν_t' the vertical eddy diffusivity, A_H^T the horizontal eddy diffusivity, $I(z)$ the solar radiation at depth z, and c_p' the specific heat capacity of water. More details on how $I(z)$ is calculated are given in Sect. 2.3. The equation of state for seawater (TEOS-10 equation of state adopted for modelling according to Jackett et al., 2006) is used to calculate the density from the salinity, temperature and pressure.

The vertical eddy viscosity ν_t and the vertical eddy diffusivity ν_t' are parameterized using an open-source water column model: the General Ocean Turbulence Model (GOTM); for more details see www.gotm.net, Burchard (2002), Stips et al. (2004) and references therein. Within GOTM different turbulence closure models can be chosen. For the simulations presented in the current paper, a k–ϵ model is used. In addition, advection of the prognostic turbulence quantities k (kinetic turbulent energy) and ϵ (turbulent dissipation) is done within GETM.

Appendix B: Bathymetry smoothing

The weighted averaging filter is of the form

$$H^s_{i,j} = \frac{1}{KN+1}\left[(1-K)H_{i,j} + K\sum_{m=-1}^{1}\sum_{n=-1}^{1} H_{i+m,j+n}\right] \quad \text{(B1)}$$

with $K = 0.5$ the weight of the surrounding cells. To respect the mask, the land cells around $H_{i,j}$ are not taken into account, and N is the number of water cells around $H_{i,j}$. This first step is used to reduce gradients between contiguous measurements carried out in different years, and between those from the vaklodingen and those in the North Sea 200 m bathymetric data. For the heuristic linear approach (Dutour Sikirić et al., 2009), we define first a desired maximum slope between contiguous cells. In our case, the maximum slope is set to 1.5, i.e. all grid points have to meet the condition

$$\frac{H_{i,j} - H_{i+n,j+m}}{H_{i,j} + H_{i+n,j+m}} \leq 1.5$$

with $m = -1, 0, 1$ and $n = -1, 0, 1$. All the points that do not meet this condition are identified, and afterwards the smoothing algorithm only modifies the bathymetry at the necessary points to meet the imposed condition.

Using these two steps, the hypsometry and the volume of the basins are barely affected. For example, the volume in the Marsdiep–Vlie basin calculated while considering a water level at 0 m only varies 0.3 % between the smooth and the non-smooth bathymetries.

Appendix C: Reconstruction of the freshwater discharge rate to 10 min temporal resolution

Using the available information and assuming that the Bernoulli equation holds for the discharge velocity, we reconstruct the discharge rate with a 10 min temporal resolution. In order to do so, we first estimate the discharge velocity

$$v_{\mathrm{D}}(t) = \mathrm{sgn}(\xi_{\mathrm{in}}(t) - \xi_{\mathrm{out}}(t))\sqrt{2g|\xi_{\mathrm{in}}(t) - \xi_{\mathrm{out}}(t)|}. \quad \text{(C1)}$$

with ξ_{in} the water level in Lake IJssel and ξ_{out} the water level at sea. The analytical estimated discharge per tidal period (during the time that the gates are open) is then given by

$$V'_{\mathrm{D}} = \int_{t_{\mathrm{open}}}^{t_{\mathrm{close}}} W(t)v_{\mathrm{D}}(t)(H + \xi_{\mathrm{out}}(t))\mathrm{d}t, \quad \text{(C2)}$$

where W is the width of the open gates and H is the mean depth at that location. Finally, a correction for the interaction between the currents in the gates, density differences and other three-dimensional effects has to be made. The freshwater discharge is then given by

$$V_{\mathrm{D}} = \mu V'_{\mathrm{D}}, \quad \text{(C3)}$$

where μ is a correction factor known as the *discharge coefficient*. The value for μ is obtained by doing a linear regression of the analytical discharge estimate Q'_{D} vs. the measured discharged which is provided by Rijkswaterstaat. The discharge coefficient takes the values $\mu = 1.09$ for the Den Oever complex (composed of three groups with five gates each) and $\mu = 1.02$ for the complex at Kornwerderzand (composed of two groups with five gates each). Using these coefficients, the discharge rate is calculated using the water height measurements which are recorded every 10 min:

$$Q_{\mathrm{D}} = \mu W(t)v_{\mathrm{D}}(t)(H + \xi_{\mathrm{out}}(t)). \quad \text{(C4)}$$

Appendix D: Temporal relaxation of 3-D boundary conditions

The boundary conditions for salinity and temperature are relaxed for outgoing flow giving a more realistic situation. This is done by taking the value of the salinity and the temperature at the boundary (S_{b} and T_{b}, respectively) such that

$$S_{\mathrm{b}} = (1-C)S_{\mathrm{int}} + CS_{\mathrm{ext}} \quad \text{(D1)}$$

$$T_{\mathrm{b}} = (1-C)T_{\mathrm{int}} + CT_{\mathrm{ext}} \quad \text{(D2)}$$

with

$$C = \begin{cases} 0 & \text{if} \quad u_{\mathrm{n}} \geq u_{\mathrm{cut}} \\ 0.25\frac{u-u_{\mathrm{min}}}{u_{\mathrm{max}}-u_{\mathrm{min}}} & \text{if} \quad -u_{\mathrm{cut}}/4 < u_{\mathrm{n}} < u_{\mathrm{cut}} \\ 0.25 & \text{if} \quad u_{\mathrm{n}} \leq -u_{\mathrm{cut}}/4. \end{cases} \quad \text{(D3)}$$

S_{ext} and T_{ext} are the values at the boundary obtained from the nested setups; S_{int} and T_{int} are the values just inside the boundary; u_{n} is the velocity component normal to the boundary; and $u_{\mathrm{cut}} = 0.03\ \mathrm{m\ s^{-1}}$.

Acknowledgements. Gauge data, sea level boundary conditions, and bathymetric data were kindly supplied by Rijkswaterstaat via Deltares. Atmospheric data were provided by the German Weather Service. This work was supported through the project "PACE (The future of the Wadden Sea sediment fluxes: still keeping pace with sea level rise?)" by NWO, the Netherlands (ZKO-project #839.11.003), the German Federal Ministry of Research and Education (FKZ 03F0634A), and the Deltares Science Council. The authors are grateful to Karsten Bolding (Asperup, Denmark) for the code maintenance of GETM and technical support.

Edited by: M. Hoppema

References

Abdelrhman, M. A.: Simplified modeling of flushing and residence times in 42 embayments in New England, USA, with special attention to Greenwich Bay, Rhode Island, Estuar. Coast. Shelf S., 62, 339–351, 2005.

Bolin, B. and Rodhe, H.: A note on the concepts of age distribution and transit time in natural reservoirs, Tellus, 25, 58–62, 1973.

Buijsman, M. C. and Ridderinkhof, H.: Long-term ferry-ADCP observations of tidal currents in the Marsdiep inlet, Cont. Shelf Res., 57, 237–256, 2007a.

Buijsman, M. C. and Ridderinkhof, H.: Water transport at subtidal frequencies in the Marsdiep inlet, J. Sea Res., 58, 255–268, 2007b.

Burchard, H.: Applied Turbulence Modelling in Marine Waters, Lecture Notes in Earth Sciences, Vol. 100, Springer, Berlin, Heidelberg, New York, 2002.

Dutour Sikirić, M., Janeković, I., and Kuzmić, M.: A new approach to bathymetry smoothing in sigma-coordinate ocean models, Ocean Model., 29, 128–136, 2009.

Elias, E. P. L., Cleveringa, J., Buijsman, M. C., Roelvink, J. A., and Stive, M. J. F. : Field and model data analysis of sand transport patterns in Texel Tidal inlet (the Netherlands), Coast. Eng., 53, 505–529, 2006.

Gerkema, T., Nauw, J. J., and van der Hout, C. M.: Measurements on the transport of suspended particulate matter in the Vlie Inlet, Neth. J. Geosci., doi:10.1017/njg.2014.7, in press, 2014.

Jackett, D. R., McDougall, T. J., Feistel, R., Wright, D. G., and Griffies, S. M.: Algorithms for density, potential temperature, conservative temperature, and the freezing temperature of seawater, J. Atmos. Ocean. Tech., 23, 1709–1728, 2006.

Jerlov, N.: Marine Optics, Elsevier, Amsterdam, Oxford, New York, 1976.

Kondo, J.: Air–sea bulk transfer coefficients in diabatic conditions, Bound.-Lay. Meteorol., 9, 91–112, doi:10.1007/BF00232256, 1975.

Li, C.: Subtidal water flux through a multiple-inlet system: observations before and during a cold front event and numerical experiments, J. Geophys. Res.-Ocean., 118, 1877–1892, 2013.

Louters, T. and Gerritsen, F.: Het mysterie van de wadden: hoe een getijdesysteem inspeelt op de zeespiegelstijging (The mystery of the flats: how a tidal system responds to sea level rise), Tech. rep., Ministerie van Verkeer en Waterstaat, Directoraat-Generaal Rijkswaterstaat, Rijksinstituut voor Kust and Zee (RIKZ), The Hague, 1994 (in Dutch).

Meier, H. E. M.: Modeling the pathways and ages of inflowing salt- and freshwater in the Baltic Sea, Estuar. Coast. Shelf S., 74, 610–627, 2007.

Monsen, N. E., Cloern, J. E., Lucas, L. V., and Monismith, S. G.: A comment on the use of flushing time, residence time, and age as transport time scales, Limnol. Oceanogr., 47, 1545–1553, 2002.

Nauw, J. J., Merckelbach, L. M., Ridderinkhof, H., and van Aken, H. M.: Long-term ferry-based observations of the suspended sediment fluxes through the Marsdiep inlet using acoustic Doppler current profilers, J. Sea Res., 87, 17–29, 2014.

Oost, A. P., Hoekstra, P., Wiersma, A., Flemming, B., Lammerts, E. J., Pejrup, M., Hofstede, J., van der Valk, B., Kiden, P., Bartholdy, J., van der Berg, M. W., Vos, P. C., de Vries, S., and Wang, Z. B.: Barrier island management: lessons from the past and directions for the future, Ocean Coast. Manage., 68, 18–38, 2012.

Paulson, C. A. and Simpson, J. J.: Irradiance measurements in the upper ocean, J. Phys. Oceanogr., 7, 952–956, 1977.

Pawlowicz, R., Beardsley, B., and Lentz, S.: Classical tidal harmonic analysis including error estimates in MATLAB using T_TIDE, Comput. Geosci., 28, 929–937, 2002.

Plieger, R.: Een Kalman filter voor het Kuststrook model (A Kalman filter for the 'Kuststrook' model), Tech. rep., Ministerie van Verkeer en Waterstaat, Directoraat-Generaal Rijkswaterstaat, Rijksinstituut voor Kust and Zee (RIKZ), The Hague, 1999 (in Dutch).

Postma, H.: The distribution of temperature and salinity in the Wadden Sea, Tijdschrift van het Koninkijk Nederlandsch Aardrijkskundig Genootschap, 67, 34–42, 1950.

Postma, H.: Hydrography of the Wadden Sea: movements and properties of water and particulate matter, Balkema, 1982.

Reise, K., Baptist, M., Burbridge, P., Dankers, N., Ludwig, F., Flemming, B., Oost, A. P., and Smit, C.: The Wadden Sea – a universally outstanding tidal wetland, in: Wadden Sea Ecosystem, Common Wadden Sea Secretariat, Wilhelmshaven, Germany, 29, 7–24, 2010.

Ridderinkhof, H.: Tidal and residual flows in the western Dutch Wadden Sea I: numerical model results, Neth. J. Sea Res., 22, 1–21, 1988a.

Ridderinkhof, H.: Tidal and residual flows in the Western Dutch Wadden Sea II: an analytical model to study the constant flow between connected tidal basins, Neth. J. Sea Res., 22, 185–198, 1988b.

Shchepetkin, A. F. and McWilliams, J. C.: A method for computing horizontal pressure-gradient force in an oceanic model with a nonaligned vertical coordinate, J. Geophys. Res.-Ocean., 108, 3090, doi:10.1029/2001JC001047, 2003.

Stanev, E. V., Wolff, J.-O., Burchard, H., Bolding, K., and Flöser, G.: On the circulation in the East Frisian Wadden Sea: numerical modeling and data analysis, Ocean Dynam., 53, 27–51, 2003.

Stips, A. K.: Fitting measured irradiance of Jerlov water types to double exponential functions using R, Tech. rep., Joint Resarch Centre, European Commision, Ispra, 2010.

Stips, A. K., Bolding, K., Pohlmann, T., and Burchard, H.: Simulating the temporal and spatial dynamics of the North Sea using the new model GETM (general estuarine transport model), Ocean Dynam., 54, 266–283, 2004.

van Aken, H. M.: Variability of the water temperature in the western Wadden Sea on tidal to centennial time scales, J. Sea Res., 60, 227–234, 2008a.

van Aken, H. M.: Variability of the salinity in the western Wadden Sea on tidal to centennial time scales, J. Sea Res., 59, 121–132, 2008b.

Verlaan, M. and Heemink, A.: Kalman filtering in two and three-dimensional shallow water flow models, in: Proceedings of Hydroinformatics 98, Denmark, 1061–1068, 1998.

Wiegman, N., Perluka, R., Oude Elberink, S., and Vogelzang, J.: Vaklodingen: de inwintechnieken en hun combinaties. Vergelijking tussen verschillende inwintechnieken en de combinaties ervan ["Vaklodingen" (bathymetric surveys): data collection techniques and their combinations], Tech. rep., AGI-Rapport, AGI-2005-GSMH-012, Adviesdienst Geo-Informatica en ICT (AGI), Delft, 2005 (in Dutch).

Zhang, W. G., Wilkin, J. L., and Chant, R. J.: Modeling the pathways and mean dynamics of river plume dispersal in the New York Bight, J. Phys. Oceanogr., 39, 1167–1183, 2009.

Zimmerman, J. T. F.: Mixing and flushing of tidal embayments in the western Dutch Wadden Sea part I: Distribution of salinity and calculation of mixing time scales, Neth. J. Sea Res., 10, 149–191, 1976.

15

Quantification of octacalcium phosphate, authigenic apatite and detrital apatite in coastal sediments using differential dissolution and standard addition

J. F. Oxmann[1,*] **and L. Schwendenmann**[1]

[1]School of Environment, The University of Auckland, Auckland 1010, New Zealand
[*]currently at: GEOMAR Helmholtz Centre for Ocean Research Kiel, Marine Biogeochemistry, 24148 Kiel, Germany

Correspondence to: J. F. Oxmann (joxmann@geomar.de)

Abstract. Knowledge of calcium phosphate (Ca-P) solubility is crucial for understanding temporal and spatial variations of phosphorus (P) concentrations in water bodies and sedimentary reservoirs. In situ relationships between liquid- and solid-phase levels cannot be fully explained by dissolved analytes alone and need to be verified by determining particular sediment P species. Lack of quantification methods for these species limits the knowledge of the P cycle. To address this issue, we (i) optimized a specifically developed conversion–extraction (CONVEX) method for P species quantification using standard additions, and (ii) simultaneously determined solubilities of Ca-P standards by measuring their pH-dependent contents in the sediment matrix. Ca-P minerals including various carbonate fluorapatite (CFAP) specimens from different localities, fluorapatite (FAP), fish bone apatite, synthetic hydroxylapatite (HAP) and octacalcium phosphate (OCP) were characterized by XRD, Raman, FTIR and elemental analysis. Sediment samples were incubated with and without these reference minerals and then sequentially extracted to quantify Ca-P species by their differential dissolution at pH values between 3 and 8. The quantification of solid-phase phosphates at varying pH revealed solubilities in the following order: OCP > HAP > CFAP (4.5 % CO_3) > CFAP (3.4 % CO_3) > CFAP (2.2 % CO_3) > FAP. Thus, CFAP was less soluble in sediment than HAP, and CFAP solubility increased with carbonate content. Unspiked sediment analyses together with standard addition analyses indicated consistent differential dissolution of natural sediment species vs. added reference species and therefore verified the applicability of the CONVEX method in separately determining the most prevalent Ca-P minerals. We found surprisingly high OCP contents in the coastal sediments analyzed, which supports the hypothesis of apatite formation by an OCP precursor mechanism.

1 Introduction

Phosphorus cycles are influenced by P concentration and composition in both the liquid and solid phase. Although non-equilibrium behaviour is typical for natural systems (Carman and Rahm, 1997), precipitation and dissolution mechanisms may occasionally attain equilibrium between certain P minerals and dissolved inorganic P (DIP). Provided DIP levels are controlled by these mechanisms, P minerals may be qualitatively determined in natural (Avnimelech, 1983; Bray et al., 1973; Emerson and Widmer, 1978; Gaudette and Lyons, 1980) or experimental (Murrmann and Peech, 1969; Oxmann et al., 2008) environments using appropriate dissociation constants and apparent solubility products allowing for ionic strength effects and ion-pairing. Identification of particular mineral species using solubility criteria is, however, seldom reported for natural systems, mainly due to kinetic control of mineral reactions and/or adsorption-controlled DIP levels. For instance, P may be sorbed to Fe, Al and Mn oxides, iron oxyhydroxides, kaolinite and montmorillonite (Bar-Yosef et al., 1988; Chen et al., 1973; Hutchison and Hesterberg, 2004; Parfitt et al., 1975).

Whether adsorption effects or mineral formation and dissolution influence soluble P concentrations depends largely on sediment types and physical–chemical conditions. For example, the redox state affects DIP by promoting sorption to a FeOOH–phosphate complex in oxic sediments and release from this complex under reducing conditions (Einsele, 1936; Mortimer, 1941). A decrease of the redox potential (Eh) may therefore release P and could cause a shift from sorption to precipitation (Mortimer, 1971; Ruttenberg and Berner, 1993). However, soil incubation experiments by Murrmann and Peech (1969) indicated that pH also significantly affects adsorption–desorption and precipitation–dissolution mechanisms: whereas DIP was controlled by sorption below pH values of 8, DIP was in equilibrium with octacalcium phosphate (OCP) above pH 8. A similar pH-mediated change from adsorption to precipitation was confirmed for coastal sediments (Oxmann et al., 2008) and agrees with a reduced sorption capacity by (hydr)oxide minerals with increasing pH (Oh et al., 1999; Spiteri et al., 2008; Barrow et al., 1980; Naidu et al., 1990) and lower OCP solubility under progressively increasing alkaline conditions (Fig. 1; Lindsay et al., 1989). Hence, pH- and Eh-mediated effects are difficult to separate. Phosphorus release during reduction could also be caused by apatite solubilization (Golterman, 2001) because a change to anoxic conditions is often accompanied by pore water acidification (Jourabchi et al., 2005).

Due to decreasing apatite solubility with decreasing acidity (Fig. 1), calcium phosphate (Ca-P) concentrations of soils and sediments generally increase with pH (Lindsay et al., 1989; Oxmann et al., 2010; Tyler, 2002). Accordingly apatite frequently constitutes the main P fraction of marine sediments (e.g. Daesslé et al., 2004; Morse and Cook, 1978). Apatite is classified into biogenic and inorganic apatite. The formation of inorganic apatite by chemical precipitation and recrystallization processes has been subject of extensive research during past decades (e.g. Bentor, 1980; Sheldon, 1981; Tribble et al., 1995). Earlier research focused on the formation of carbonate fluorapatite (CFAP) bearing phosphorite deposits, which are referred to as such if they contain more than 5 wt% P_2O_5 (Ruttenberg and Berner, 1993). Phosphorites form mainly at continental shelves and have been therefore closely linked to upwelling phenomena (e.g. Kazakov, 1937; Parrish et al., 1986). Locally occurring saturation states that support a sustained CFAP precipitation were either explained by physical–chemical mechanisms, such as steady P supply by nutrient-rich deep ocean waters as well as temperature increase, pressure decrease, pCO_2 decrease and pH increase of ascending waters (Atlas, 1975) or by biochemical processes including the high productivity at upwelling zones and related P enrichment of interstitial waters. Concentrations of DIP may further increase due to additional diagenetic transformations such as P fixation in oxic and subsequent release in anoxic sediment (Ruttenberg and Berner, 1993). Even more factors, partly independent of upwelling processes, have been proposed for the concentration of pore water phosphate, such as river sources (e.g. Pevear, 1966) and P release by sulfur bacteria (Crosby and Bailey, 2012; Schulz and Schulz, 2005). Hence, different P enrichment

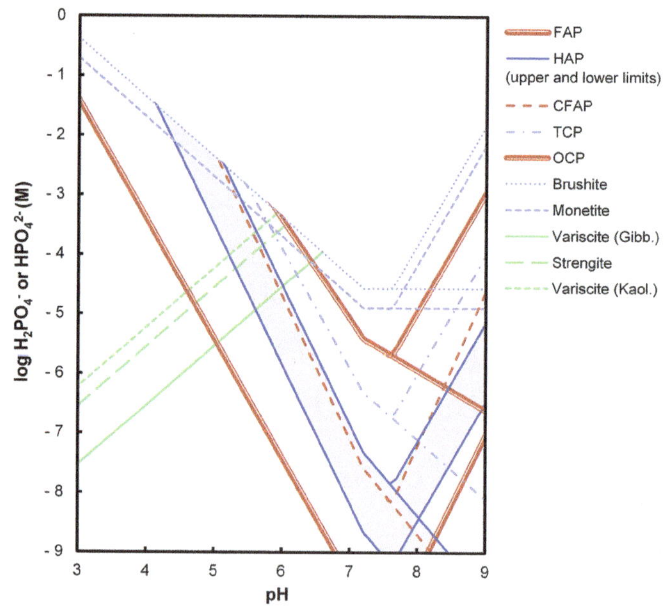

Figure 1. Apparent solubilities of calcium phosphates, strengite and variscite (in equilibrium with gibbsite or kaolinite) when Ca^{2+} is 10^{-2} M or is fixed by calcite. CO_2(g): 0.0003 atm; F^-: 10^{-4} M (equilibrium level of CaF_2); CO_3^{2-} at the given CO_2(g). Strengite and variscite are more insoluble under acidic conditions while calcium phosphate solubility decreases with increasing pH. At pH > 7.88 calcium phosphate solubility may increase due to Ca^{2+} depression caused by decreasing calcite solubility (see dividing curves at this pH for Ca^{2+} depression or maintained Ca^{2+} concentration). Calculations used apparent dissociation constants and thermodynamic solubility products reported in Lindsay et al. (1989). To exemplify the inherent uncertainties arising from variability among Ksp measurements, hydroxylapatite (HAP) solubility was calculated using the range of corresponding Ksp values reported in the literature (Baker et al., 1998). The solubility of carbonate fluorapatite (CFAP) was calculated using the apparent Ksp of 10^{-103} given by Perrone et al. (2002). See Lindsay et al. (1989) for details on calculations. Results for this system are similar to those computed by more-sophisticated algorithms (e.g. MEDUSA; Puigdomenech, 2000). See Atlas (1975) for the theoretical dependence of calcium phosphate solubility on salinity. FAP: fluorapatite; TCP: tricalcium phosphate; OCP: octacalcium phosphate.

mechanisms may increase the saturation state enough to enable spontaneous Ca-P precipitation for subsequent CFAP formation. However, the P precipitation in itself is chemically driven by the saturation state and kinetic factors without direct cellular control (e.g. Krajewski et al., 1994).

Solubility of CFAP is affected by the extent of carbonate inclusion in the crystal lattice and increases with carbonate content (Chen et al., 1973; Jahnke, 1984; LeGeros and LeGeros, 1984). Due to its extensive distribution in common sediments or deposits, CFAP has been suggested to be an important oceanic sink of reactive P (Ruttenberg, 1992; Ruttenberg and Berner, 1993). While numerous studies suggested that seawater is largely CFAP undersaturated despite the extensive CFAP distribution (Baturin, 1981; Bentor, 1980, and references therein), Atlas and Pytkowicz (1977) and Sheldon (1981) argued that seawater is close to CFAP saturation but noted the extremely slow kinetics of homo- or hetero-nucleation observed by incubation experiments. According to these kinetic considerations, Ca-P precipitation is apparently prevented in the water column (Atlas, 1975; Atlas and Pytkowicz, 1977). Moreover, direct nucleation of apatite was found to be inhibited by magnesium and, thus, spontaneous inorganic precipitation suggested to be kinetically impossible in marine environments (Golubev et al., 1999; Gunnars et al., 2004; Martens and Harriss, 1970). In contrast, magnesium has a lesser inhibitory effect on the precipitation of more soluble precursor phases and therefore selectively stabilizes precursor phases (Gunnars et al., 2004; Johnsson and Nancollas, 1992).

Given the kinetic constraints of apatite nucleation under supersaturated conditions, more soluble Ca-P species may affect apatite formation according to the Ostwald step rule (Morse and Casey, 1988; Nancollas et al., 1989). Thus other Ca-P minerals, such as OCP and amorphous calcium phosphate (ACP), may represent significant sinks during early diagenesis and may influence soluble P levels. Incubation experiments with seawater indicated apatite formation by prior precipitation of metastable precursor phases, and a corresponding mechanism has been suggested for certain phosphorite deposits (Gunnars et al., 2004; Jahnke et al., 1983; Krajewski et al., 1994; Van Cappellen and Berner, 1988; Schenau et al., 2000). Apatite precursors are however difficult to identify, even in relatively simple mixtures without interfering sample matrices. Therefore, direct evidence for the sedimentary occurrence of OCP or ACP is scarce owing to methodological issues with respect to complex matrices.

Phosphorus analysis in sediment is generally performed by sequential extraction methods, which mainly separate the following P fractions: (i) exchangeable or loosely sorbed P, (ii) Al / Fe-bound P (Al / Fe-P), (iii) Ca-P, and (iv) organic P. Sequential extraction methods were developed for agricultural purposes (e.g. Hedley et al., 1982) and subsequently modified for sedimentary studies (e.g. SEDEX method, Ruttenberg, 1992). Phosphorus fractions were subdivided in some schemes in order to yield additional pools, such as CFAP (Ruttenberg, 1992) and Al-P (Kuo, 1996). To support the interpretation of these widely used extraction methods a complementary speciation method was recently developed to quantify OCP and different apatites (CONVEX; Fig. 2; Oxmann et al., 2008). This method combines a conversion of P compounds by batch titration with subsequent sequential extraction and was proposed to be used together with conventional methods.

To validate the specificity of the CONVEX method, Oxmann et al. (2008) conducted standard addition experiments using OCP, hydroxylapatite (HAP), strengite and variscite. For Ca-P, addition experiments revealed that OCP can be well distinguished from apatites due to its significantly higher solubility (Oxmann et al., 2008). Using CONVEX, we found that OCP was the major Ca-P species in alkaline sediments (Oxmann et al., 2008). Added HAP dissolved at lower pH and a third distinguishable compound dissolved upon further acidification. The latter was not identified due to lack of available standards but suggested to be fluorapatite (FAP). Carbonate fluorapatite remained also unverified by addition experiments despite its relevance. Theoretically, CFAP may be more or less soluble than HAP based on the apparent solubility product of Perrone et al. (2002; see Fig. 1). To demonstrate the applicability of the CONVEX method, especially for purposes of marine Ca-P analyses, we provide supplementary standardization by addition experiments in this study and show appropriate selective dissolution.

Our specific objectives were (i) to validate the specificity of the CONVEX method for FAP and CFAP, and (ii) to verify consistent differential dissolution of natural sedimentary species vs. added reference species. At the same time, this method refinement allowed for (i) a comparison of P species in similar sediments of temperate and tropical tidal flats in order to test cross-regional transferability, and (ii) an analysis of Ca-P solubilities in sediment including effects of mineral carbonate content on dissolution. Because CFAP is one of the most important species for marine P cycling and its solubility is largely unknown due to varying carbonate inclusion and incongruent dissolution (e.g. Perrone et al., 2002), we focused on an extended CFAP standard set from different localities.

2 Materials and methods

2.1 Sediment sampling, preparation and characterization

A sediment core was taken in a temperate tidal flat (Firth of Thames, New Zealand) using a polycarbonate corer (length 45 cm; diameter 9 cm). The sampling location (lat. −37.21321°, long. 175.38854°) was 180 m from the mangrove fringe and had an inundation duration of 130 days per year corresponding to a height of 67 cm above m.s.l. (mean sea level).

Sediment redox potential (Eh), pH and temperature were determined in situ at 5 cm intervals (sulfide resistant, sludge suitable SensoLyt SEA/PtA electrodes; Pt-100 temperature sensor; pH/Cond340i; pH 3310; WTW, Weilheim, Germany). Sensors were inserted through sealable corer inlets and connected to a computer using optoisolators (USB-isolator; Serial: 289554B; Acromag Inc., MI, USA). Data were recorded using the software MultiLab pilot (WTW, Weilheim, Germany) to determine final measurement values. Sediment from 30–35 cm was selected for subsequent analyses as this layer was less affected by tidal changes, runoff or bioturbation. Field moist subsamples were taken for salinity determination (wet sediment to deionized water ratio – 1 : 5; TetraCon 325 electrode; WTW, Weilheim, Germany) and for particle size measurement using laser diffractometry (sediment dispersed in 10 % sodium hexametaphosphate solution; Mastersizer 2000; Malvern Instruments Ltd., Malvern, UK). The remaining part of the sample was dried at 37 °C, ground with a planetary mill (< 300 μm mesh; PM 100; Retsch, Haan, Germany) and used for subsequent conversion–extraction analyses (see CONVEX method: Sect. 2.4). Basic P fractionation into Al / Fe-P and Ca-P was carried out according to Kurmies (1972) with minor modifications as described in Oxmann et al. (2008).

With an Eh of −98 mV and a pH of 7.467, the sediment was reduced and slightly alkaline and had clay, silt and sand fractions of 8, 34 and 58 %, respectively. Calcium phosphate bound P was the predominant fraction with a dry weight content of 5.51 μmol P g^{-1}, while the Al / Fe-P pool had an amount of 3.02 μmol P g^{-1}. Sediment moisture content was 58 % and salt concentration was 16.5 ‰. Results of CONVEX analysis of the sediment from the Firth of Thames were compared with those of tidal flat sediment from the Saigon River delta (Ca-P: 5.36 μmol P g^{-1}; Al / Fe-P: 5.90 μmol P g^{-1}; pH: 7.47; clay: 12 %, silt: 83 %; sand: 5 %; Oxmann et al., 2008) to test cross-regional transferability (see Sect. 3.1).

2.2 Preparation and identification of calcium phosphate reference minerals

In total, eleven Ca-P minerals were prepared for standard addition experiments. Their origin is given in Table 1. Specimens were acquired from mineral distributors, synthetic HAP was purchased from Sigma Aldrich (St. Louis, Missouri, USA) and OCP was synthesized as described in Christoffersen et al. (1990).

A large single FAP crystal was ground to powder and used without further preparation. All CFAP specimens with visually distinguishable mineral attachments had characteristics of either (i) numerous small needle-like or small hexagonal crystals attached to a matrix, or (ii) spheres of a botryoidal cluster embedded into a matrix (Table 1). To generate pure mineral samples, crystals were picked and spheres were isolated from the matrix under a binocular microscope. Received fragments were again inspected to avoid any significant contamination and subsequently ground. Cryptocrystalline collophane could not be purified and in this case the entire specimen was ground. Correspondingly, all bones of a fresh Australasian snapper (*Pagrus auratus*) were ground after they were cleaned from attached tissue with a knife and dried at 40 °C. Heating at elevated temperatures, boiling in water or chemical clean-up was avoided to prevent changes to the crystalline structure or carbonate content. Because collophane and bone specimens were not purified, their carbonate content was not determined. In contrast, care was taken during purification of other samples for precise quantification of their mineral carbonate content.

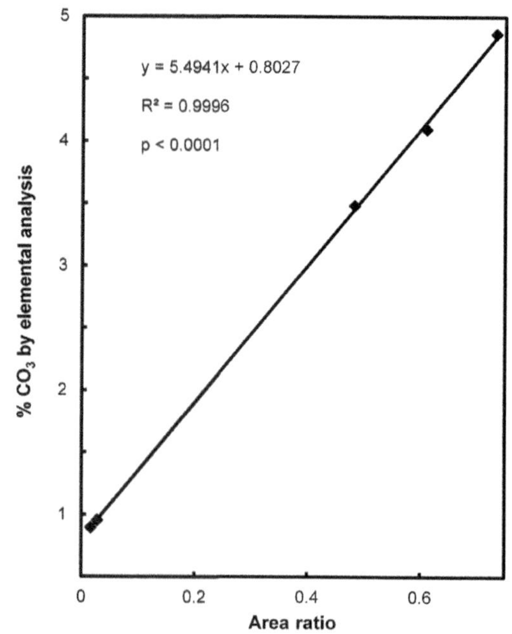

Figure 2. FTIR calibration for carbonate content determination of calcium phosphate minerals. Ratio of the extinction of the FTIR carbonate band to phosphate band (areas, $E_{1380-1440}/E_{550-590}$) vs. carbonate content (% by weight) determined by elemental analysis. For calibration, data of five minerals with known CO_3 content were used (cf. Table 1, minerals with performed elemental analysis). Contents of other minerals listed in Table 1 were calculated using the indicated correlation (method after Featherstone et al., 1984 with modification).

Mineral analyses were performed using X-ray diffraction (XRD; PW 1130 HV generator; automation-modified 1050/25 goniometer fitted with PW 1752 curved graphite crystal monochromator and PW 1965 proportional detector; Philips Analytical, Amsterdam, the Netherlands; Cu-tube, 40 kV, 20 mA), Raman spectroscopy (RM1000 confocal micro-Raman system; Renishaw, Gloucestershire, UK; laser excitation at 488 nm, 26 mW) and Fourier transform infrared spectroscopy (FTIR; Spectrum One FT-IR Spectrometer; Perkin Elmer, MA, USA; 4 cm^{-1} spectral resolution). Synthesized OCP was additionally confirmed using X-ray absorption near-edge structure spectroscopy (XANES; P K-edge; see Oxmann, 2014; spectrum collected by Kruse et al., 2010). Briefly, crystals and spheres were pure, highly crystalline and correctly identified. In contrast, snapper bones and collophane were found to be impure and less crystalline. Despite impurities, both samples had relatively high P contents. Snapper bone contained more than 50 % of the P content of pure minerals and collophane approximately five-fold less than pure minerals as determined by photometric determination (see Sect. 2.4).

Table 1. Basic properties, performed analyses and carbonate content (% by weight) of Ca-P reference minerals used for standard addition experiments.

ID-Mineral[a]	Origin	Type/appearance	Characterization	Purity[b]	CO_3
MS-CFAP	Moculta Phosphate Quarry, Angaston, Australia	Var. Staffelite, botryoidal, yellow-brown	XRD, FTIR, Raman, EA	PS	4.4
MS-CFAP	Staffel, Limburg, Germany	Var. Staffelite, botryoidal, brown	XRD, FTIR, Raman, EA	PS	4.7
KC-CFAP	Zheleznyi Mine, Kola Peninsula, Russia	Var. Staffelite, botryoidal, blue-green	XRD, FTIR, Raman, EA	PS	3.4
KC-CFAP	Holmbush Mine, Callington, Cornwall, UK	Var. Holmbushite, needle-like crystals	XRD, FTIR, Raman	PC	3.4
CW-CFAP	Geevor Mine, Pendeen, Cornwall, UK	Small hexagonal crystals, brown	XRD, FTIR, Raman	PC	2.3
CW-CFAP	Levant Mine, Pendeen, Cornwall, UK	Small hexagonal crystals, brown	XRD, FTIR, Raman	PC	2.1
BL-CFAP	Bonny Lake, Florida, USA (with HAP and FAP)	Var. Collophane, cryptocrystalline	XRD, FTIR, Raman	BS	ND
DU-FAP	Cerro de Mercado Mine, Durango, Mexico	Single large hexagonal crystal, green	XRD, FTIR, Raman, EA	PC	1.0
SB-AP	*Pagrus auratus* bones, New Zealand	Nanocrystalline apatite inclusion	XRD, FTIR, Raman	BS	ND
ST-HAP	Synthetic	Crystalline powder	XRD, FTIR, Raman, EA	S	0.9
ST-OCP	Synthetic	Crystalline powder	XRD, FTIR, Raman, XANES	S	0.8

[a] The upper six CFAP samples were pooled into three composite samples according to their similar CO_3 content (see minerals with equal ID; MS-CFAP: 4.5 %; KC-CFAP: 3.4 %; CW-CFAP: 2.2 % pooled CO_3 content). [b] Samples were prepared from pure spheres (PS), pure crystals (PC) or synthetic origin (S) and confirmed by XRD, FTIR and Raman or from bulk samples (BS) including other components. Var.: variety. EA: elemental analysis. ND: not determined.

2.3 Carbonate content of Ca-P reference minerals

In order to determine mineral carbonate content, FTIR and elemental analysis were used. Calibration of FTIR by apatite samples with known carbonate contents is an accurate method for carbonate quantification, which uses the extinction ratio of the FTIR carbonate to phosphate band. For FTIR, 2 mg of sample was ground with 200 mg of potassium bromide and pressed in a mechanical press to form pellets for spectroscopy. Analysis was carried out as described by Featherstone et al. (1984) with two modifications: (i) carbonate contents of apatite reference samples were determined using an elemental analyzer (TruSpec CNS; LECO Corp., MI, USA) instead of preparing blended apatite standards with barium carbonate, and (ii) area ratios ($E_{1380-1440}/E_{550-590}$; Omnic Software Suite; Thermo Scientific, MA, USA) were used instead of extinction ratios at specific wave lengths. For extinction ratio calibration, data of five minerals with known carbonate content were used (cf. Table 1). Contents of other minerals listed in Table 1 were calculated using the correlation given in Fig. 2.

Results of elemental analysis correlated with those of FTIR at a significance level of 0.0001. Hexagonal crystals from the UK had lowest carbonate contents and spheres of botryoidal clusters from Germany and Australia had highest contents (Table 1). Carbonate contents of natural FAP, purchased HAP and synthesized OCP were comparable and much lower than those of natural CFAP (Table 1). To verify the effect of mineral carbonate content on sedimentary CFAP solubility by a lower number of addition experiments, CFAP samples were partly pooled, excluding collophane. The upper six CFAP samples in Table 1 were pooled into three composite samples according to similar CO_3 inclusions of each pooled pair (4.4 and 4.7 %; 3.4 and 3.4 %; 2.1 and 2.3 %).

2.4 CONVEX method

The CONVEX method is shown in Fig. 3. For the conversion procedure (1A/1B), sediment subsamples of 1 g were incubated for seven days in 10 mL pH-adjusted 0.01 M $CaCl_2$ using two wash steps with 0.01 M $CaCl_2$ after two days. Subsequently, subsamples were centrifuged and pH was measured in the clear supernatant solution with the instrumental set-up described above but using an InLab Micro electrode (Mettler Toledo GmbH, Greifensee, Switzerland). Soluble P analyses of supernatant solutions were not performed because corresponding DIP concentrations were already shown (i) to alter with pH in a typical manner for different sediment and soil types, (ii) to vary inversely with total inorganic P, and (iii) to be insignificant with respect to solid-phase levels within a certain pH range despite large P transfers between Al / Fe-P and Ca-P (see Murrmann and Peech, 1969; Oxmann et al., 2008; and Sect. 3.5). For sequential extraction, dried sediment pellets were extracted by a series of steps (2A-3C) that enable an accurate separation of inorganic P into Al / Fe-P (mineral and adsorbed P) and Ca-P (mineral and residual P). Total inorganic P was calculated by summing Al / Fe-P and Ca-P. Sediment subsamples without prior incubation served as controls while differences between spiked and unspiked incubations yielded pH-dependent losses for each added standard mineral.

For addition experiments, minerals were weighed into small polyethylene caps using an ultra-micro balance (XP6U; Mettler Toledo GmbH, Greifensee, Switzerland) and then added to the sediment subsamples before incubation. The separate addition ensured homogeneity of added P among subsamples as opposed to a possible mineral addition to a larger sediment amount before subsample weighing. Care was taken to maintain experimental conditions and to minimize sample loss during centrifugation and decanting steps. For example, tubes were briefly centrifuged (3 min, 3000 *g*) and shaken once before each centrifugation step (15 min, 3000 *g*) in order to remove remaining particles from

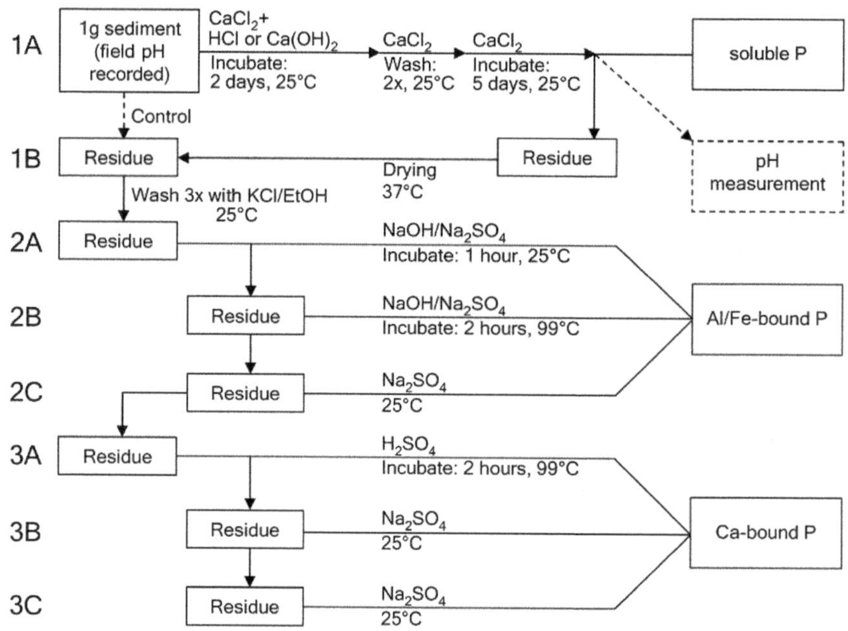

Figure 3. CONVEX scheme for the quantification of phosphorus species in sediment. 1A/1B: conversion procedure, 2A–3C: extraction procedure. For addition experiments, reference minerals were added to the sediment subsamples before incubation. Adapted from Oxmann et al. (2008).

tube walls with clean supernatant solution. This procedure ensured minimal loss by subsequent decanting. Phosphate concentrations in solutions obtained by sequential extraction were determined according to Murphy and Riley (1962) using a UV-Vis spectrophotometer (Cintra 2020; GBC Scientific Equipment, Dandenong, Australia). A detailed description of the CONVEX method is given in Oxmann et al. (2008).

2.5 Data analysis

To quantify mineral loss for all standard addition experiments at equal pH values, data sets for Ca-P vs. pH were interpolated (0.001 pH intervals) using piece-wise cubic hermite interpolating polynomial (Octave, version 3.2.4; Eaton et al., 2008). Based on maximum changes in curve slopes ($\Delta c/\Delta \text{pH}$) for unspiked sediment (Sect. 3.2), differential dissolution of sediment FAP, CFAP and OCP occurred at pH ranges of ca. 3.0–3.5, 3.5–7.0 and 7.0–8.1, respectively. Selective losses of sediment species (sediment FAP, CFAP, OCP) and total species (added species plus corresponding sediment species) at different pH values were calculated according to

$$\Delta c_{\text{loss}} = \Delta c_{\text{initial}} - \Delta c_{\text{actual}}, \tag{1}$$

where $\Delta c_{\text{initial}}$ (initial concentration of each species) equals the difference between Ca-P concentrations at the beginning (upper cutoff pH values of 3.5, 7.0 and 8.1 for FAP, CFAP and OCP) and end (lower cutoff pH values of 3.0, 3.5 and 7.0 for FAP, CFAP and OCP) of each selective loss, and Δc_{actual} (actual concentration of each species after acidification-induced losses; actual pH < upper cutoff pH) equals the difference between Ca-P concentrations at actual and lower cutoff pH values (see example of calculation for OCP loss at pH 7.5 in Fig. 5a). For calculation of selective losses of added reference species (FAP, CFAP, OCP), $\Delta c_{\text{initial}}$ and Δc_{actual} were defined by the difference between Ca-P concentrations of spiked and unspiked incubations at upper cutoff pH values and actual pH values. The difference in Ca-P content ($\Delta c_{\text{initial}}$) between spiked and unspiked incubations at upper cutoff pH values of 3.5, 7.0 and 8.1 for FAP, CFAP and OCP (average of 2.04 μmol P g^{-1}, relative standard deviation of 3.1 %) closely matched the quantity of spiked P (2 μmol P g^{-1}) for each addition experiment. Hence, added FAP, CFAP and OCP were not significantly dissolved during incubations at pH values of 3.5, 7.0 and 8.1, respectively, and were quantitatively recovered by the Ca-P fraction at these pH values. Results from Eq. (1) were subsequently used to calculate loss percentages of sediment species, of total species and of added reference species for different pH values (Table 2). Based on results of standard addition experiments a slightly lower final pH (6.9) was preferable to calculate OCP contents (see Sect. 3.3).

Table 2. Loss (%) of added reference species (FAP, CFAP, OCP), total species (added species plus corresponding sedimentary species) and sedimentary species (sediment FAP, CFAP, OCP) at varying pH as determined by the CONVEX method. See Table 1 for abbreviations of added standard minerals.

Final pH	DU-FAP	Total FAP	Sediment FAP	MS-CFAP	KC-CFAP	CW-CFAP	Total CFAP[a]	Sediment CFAP	ST-OCP	Total OCP	Sediment OCP
3.0	**98**	**99**	**100**								
3.2	40	48	76								
3.5	**0**	**0**	**0**	**94**	**91**	**84**	**96**	**100**			
4.0				*79*	*55*	*40*	70	82			
4.5				*47*	*32*	*21*	53	70			
5.0				31	23	18	43	58			
5.5				18	19	16	32	43			
6.0				8	10	15	20	27			
6.5				2	6	13	10	14			
6.8				1	4	8	6	7			
6.9				0	3	5	4	4	90	98	105[b]
7.0				**0**	**0**	**0**	**0**	**0**	**81**	**91**	**100**
7.1									68	78	88
7.5									35	28	20
8.0									6	3	0
8.1									**0**	**0**	**0**

Values in bold indicate cutoff pH values for distinguishing Ca-P species (FAP: 3.0–3.5; CFAP: 3.5–7.0; OCP: 7.0–8.1). Italics: largest differences in CFAP dissolution due to different CO_3 substitution in the pH range 4–4.5 (MS-CFAP: 4.6 %; KC-CFAP: 3.4 %; CW-CFAP: 2.2 % CO_3). [a] Among the different CFAP phases, KC-CFAP was used to calculate the total CFAP loss. [b] Indicates a possible overestimate if pH 6.9 is used for distinguishing between OCP and CFAP.

3 Results and discussion

3.1 Sedimentary Ca-P speciation

For both unspiked sediments, the slopes of curves change at pH values of approximately 7, 3.5 and 3 (Fig. 4a and b) due to differential dissolution of Ca-P species, as proven by standard addition experiments (see Sect. 3.2). Octacalcium phosphate dissolved within a narrow pH range above pH 7, CFAP within a large pH interval of ca. 3.5–7 and FAP within a small pH range below pH 3.5. Due to these dissolution ranges and given species concentrations, incremental acidification-induced losses of OCP, CFAP and FAP per pH (ΔOCP, CFAP, FAP / ΔpH) resulted in a steep, a less steep and a steep slope, respectively (cf. unspiked sediment in Fig. 4a and b).

Species contents of unspiked sediments are indicated by columns at the right of Fig. 4a and b. Carbonate fluorapatite in the sample from the Firth of Thames was largely elevated (43 %) based on a lower amount of FAP (8 %) and residual P (12 %) if compared to the sample from the Saigon River delta (CFAP: 25 %; FAP: 18 %; residual P: 20 %). Thus, species analysis showed different P compositions despite similar sediment properties of the two sediments compared (similar Ca-P content, reduced, slightly alkaline). However, OCP contents of both sediments were equal and with a value of 37 % surprisingly high. These findings generally agree with a kinetic control of species compositions because OCP precipitates in natural systems more rapidly (days to weeks; Bell and Black, 1970) than apatite (ten to some thousand years; Schenau, 2000; Jahnke et al., 1983; Gulbrandsen et al., 1984). Hence, more soluble species might better mirror ongoing processes, whereas sparingly soluble apatite minerals obviously reflect varying conditions over time due to slow precipitation. In accordance with such a kinetic control, OCP has been consequently proposed as a precursor for apatite formation (Gunnars et al., 2004; Krajewski et al., 1994).

The occurrence of OCP supports suggestions that authigenic apatite is formed through an OCP precursor mechanism. The precursor hypothesis agrees with earlier reported kinetic barriers but tends to contrast with the often suggested strong undersaturation of seawater with respect to CFAP. If direct nucleation of apatite is kinetically inhibited in marine environments by magnesium (Golubev et al., 1999; Gunnars et al., 2004; Martens and Harriss, 1970), the ambient water must be supersaturated with respect to a precursor to facilitate apatite formation by a less magnesium susceptible pathway. For example, in long-term incubation experiments apatite precipitated from natural brackish seawater only in case of supersaturation with respect to OCP (Gunnars et al., 2004). Under these conditions, the morphology of the precipitates (spherulites with OCP-like crystals) provided supporting evidence that OCP was a precursor to the observed formation of apatite. The liquid phase must be several times supersaturated with respect to apatite, though, in order to attain supersaturation with respect to OCP or ACP. When the calculation of the solubility of Ca-P phases under experimental conditions (Fig. 1) is modified to obtain solubilities for seawater of normal composition (5–25 °C; see e.g. Appendix in Atlas, 1975), pore-water appears to be generally undersaturated with respect to OCP. Due to widely differing solubility products (see e.g. HAP in Fig. 1) and the challenging application to multi-ion systems of high ionic strength, the uncertainty associated with such thermodynamic models is, however, relatively large. Incubations of P-spiked natural seawater on the other hand indicated that homogeneous precipitation of non-apatitic Ca-P is generally possible within some days under naturally occurring pore-water P concentrations (see kinetic considerations in Atlas, 1975).

Conversely, it is difficult to explain the occurrence of a highly soluble Ca-P mineral with rapid dissolution kinetics, such as OCP, when ambient water is undersaturated even with respect to a much more insoluble Ca-P species like CFAP. However, an undersaturation of bottom water with respect to marine apatite is difficult to reconcile even with the occurrence of comparatively stable ancient CFAP because the latter did not dissolve despite exposure to undersaturated conditions for millions of years (Sheldon, 1981). Accordingly, Atlas (1975) inferred that prolonged apatite persistence requires at least saturation of sediment with respect to apatite. It is even more remarkable that recent studies suggest CFAP formation in sinking particulate matter throughout the water column (Faul et al., 2005; Lyons et al., 2011) using extractions of Ruttenberg's SEDEX method (1992). Although protective coatings and microenvironments may be considered for the prevented dissolution of ancient CFAP and potentially occurring precipitation in the water column, apparent undersaturation is frequently not confirmed by field observations (Sheldon, 1981). Furthermore, a systematic comparison of P K-edge XANES fingerprints from reference materials and marine sediment particles provided direct evidence for the occurrence of OCP in marine surface sediment (Oxmann, 2014).

However, a suggested precursor pathway depends more on continuously high pH values and elevated DIP and calcium concentrations compared to direct CFAP nucleation. For example, Ca-P formation through an OCP precursor mechanism was detected after approximately two years of incubation in calcite-seeded brackish seawater using a DIP concentration of 500 μM, which is slightly higher than the range reported for interstitial water of Baltic Sea sediments (Gunnars et al., 2004). In addition, low dissolved magnesium levels may be favourable for OCP precipitation because, possibly due to less adsorption of foreign ions caused by lattice water in OCP, the growth of OCP seems to be much less but not completely uninfluenced by magnesium (e.g. Bell and Black, 1970; Johnsson and Nancollas, 1992). Yet, equilibrium calculations by means of pore water data imply that most interstitial waters of the Baltic Proper are supersaturated for long periods, at least with respect to apatite (Carman and Rahm, 1997), and thus agree with observed P transformations into apatite (van Beusekom and De Jonge, 1997).

In view of the proposed but not unambiguous OCP precursor mechanism, Gunnars et al. (2004) highlighted the difficult in situ identification of authigenic apatite. However, OCP detection may pose an even larger problem because OCP is not quantifiable in sediment using any conventional method. Furthermore, OCP is not included in operationally defined fractions of common sequential extraction methods. Calcium phosphate fractions of these methods are consequently incompletely defined and those definitions may result in overestimates of other fractions and the exclusion of OCP. Possible misinterpretations might be not insubstantial in view of OCP contents detected in the present study. Overall, the proposed method offers a good opportunity to further evaluate the issues mentioned above by field or experimental studies. Results for the sediments investigated support suggestions of step-wise formation of apatite via OCP. However, further analyses of samples from other regions are required to determine spatial OCP distributions; to clarify whether successive or direct nucleation is favoured in environmental systems; and to identify factors that may facilitate or inhibit one or the other pathway at different locations.

3.2 Solubility of calcium phosphates in sediment: verification of differential dissolution

Results of Ca-P addition experiments (Figs. 4, 5 and 6) show mineral species successively dissolved by acidification according to their theoretical solubility, agree with thermodynamic considerations and verify the applicability of the CONVEX method. Moreover, solubility of added CFAP increased with its carbonate content, which is in line with previous studies (Chien and Black, 1976; Jahnke, 1984; LeGeros and LeGeros, 1984).

The OCP and HAP addition experiments showed that the CONVEX method can distinguish OCP from HAP in sediment. Almost complete dissolution of sedimentary and added OCP at pH 7 (cf. ST-OCP in Fig. 4a and b) verified OCP quantification by the Ca-P loss at this pH because added HAP was not significantly dissolved at or above pH 7 (cf. ST-HAP in Fig. 4a and b). Addition experiments with HAP showed Ca-P loss above pH 7 but this loss was caused by the sedimentary OCP content. The detected OCP presence only at pH values above ca. 7 corroborates results of soil column incubations by Bell and Black (1970), which confirmed OCP formation only in the alkaline range. Precipitation of OCP in the alkaline range also agrees with DIP concentrations in supernatant solutions of pH batch experiments of soils and sediments because soluble P followed the OCP isotherm (Murrmann and Peech, 1969; Oxmann et al., 2008).

The CFAP addition experiments verified that the CONVEX method can also distinguish CFAP from OCP (differential dissolution of CFAP and OCP in Fig. 4a) despite the apparently high solubility of CFAP (cf. Fig. 1; based on apparent solubility products, CFAP could be more soluble than HAP). In fact, all incubations with various CFAP minerals

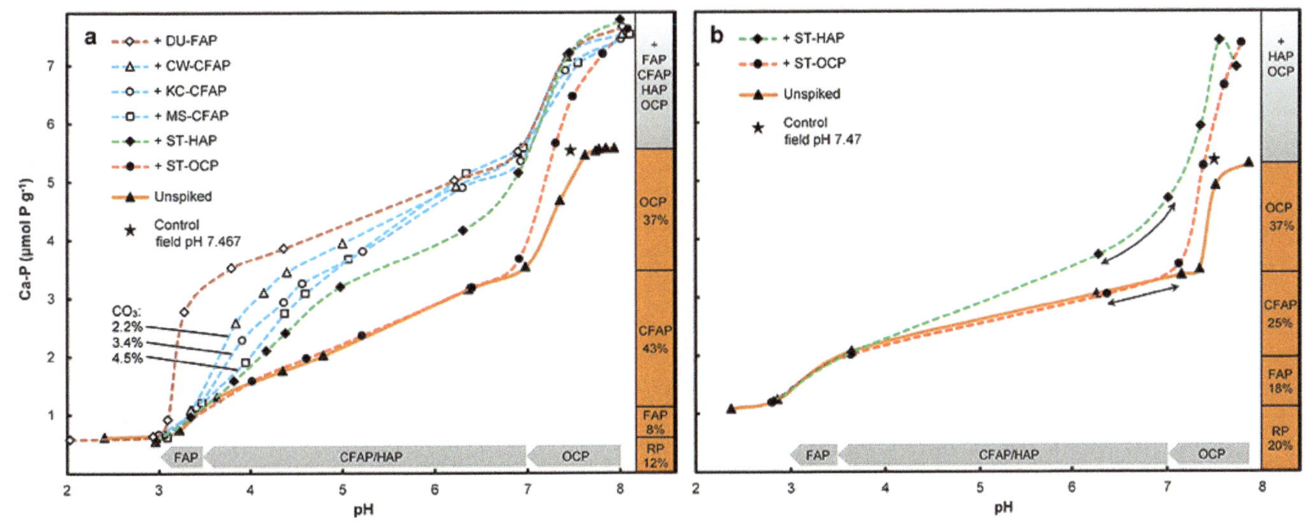

Figure 4. Differential Ca-P dissolution in sediment at varying pH as determined by subsample incubation of sediment (unspiked and with reference mineral additions) and subsequent fractionation (CONVEX method). **(a)** Standard addition experiments with natural and synthetic Ca-P using sediment from the Firth of Thames, New Zealand. **(b)** Standard addition experiments with synthetic Ca-P using sediment from the Saigon River delta, Vietnam (adapted from Oxmann et al., 2008). Columns indicate sedimentary Ca-P species composition in % (CFAP fraction may contain some HAP; see text) and standard additions. Note the greater extent of CFAP dissolution with increasing carbonate substitution. Note the increased slope of HAP addition for both sediments in the pH range 6.3–6.9, which was absent for unspiked sediment (arrows for visualization in **b**) and other additions (cf. **a**) except collophane (cf. Fig. 6). Control: fractionation without prior incubation. RP: residual P. See Table 1 for abbreviations and origins of added standard materials.

revealed a reduced loss compared to HAP spiked incubations within the pH range 3.5–7. Therefore, natural CFAP minerals with differing carbonate content were even more stable in the sedimentary environment than synthetic HAP and consequently did not bias determined OCP contents. Lower solubility of CFAP in comparison to HAP is in agreement with reported HAP conversion into more stable CFAP or increased HAP dissolution in the water column and near-surface sediments (e.g. Burdige, 2006; Schenau and De Lange, 2000). In contrast to the separation of CFAP or HAP from OCP, quantitative discrimination between HAP and CFAP is not possible without further methodological optimization because both apatites dissolved within the same pH range. This finding agrees with similarities between the calculated solubilities of CFAP and HAP (Fig. 1). Yet, the presence of significant quantities of HAP can be verified because approximately half of added HAP dissolved in the pH range 6.3–7 (cf. ST-HAP in Fig. 4a and b) while CFAP did not. Thus, incubations with HAP additions show a steeper curve within that interval compared to unspiked sediment incubations (illustrated by curved and straight arrows in Fig. 4b, respectively) and all FAP and CFAP additions shown in Fig. 4a (cf. DU-FAP, CW-CFAP, KC-CFAP, MS-CFAP). Loss of sediment Ca-P within the range of ca. 3.5–7 was previously ascribed to HAP based on concurrent HAP dissolution during corresponding addition experiments (Oxmann et al., 2008). However, as unspiked incubations and CFAP additions did not show an increased curve slope in the pH range 6.3–7, contrary to HAP additions, sedimentary Ca-P loss in the pH range 3.5–7 was likely caused by a predominance of sediment CFAP. Yet presence of some HAP in the unspiked sediments could not be ruled out.

Standard addition with igneous FAP from Mexico (cf. DU-FAP in Fig. 4a) demonstrated that Ca-P dissolution below pH 3.5 is caused by FAP and therefore showed that the CONVEX method is able to distinguish FAP from other Ca-P phases. All other apatites almost completely dissolved at ca. pH 3.5, whereas added FAP dissolved below this value. The observed significantly lower FAP solubility is consistent with notable differences between solubilities of laboratory synthesized FAP and CFAP detected by Jahnke (1984). Jahnke (1984) and other investigations (Chien and Black, 1976; LeGeros and LeGeros, 1984) did not only point to solubility differences between pure FAP and CFAP but also showed that increasing carbonate content strongly increases the solubility of CFAP minerals with different degrees of carbonate substitutions. Regarding the effect of crystal carbonate content on CFAP solubility, increased carbonate substitution also caused a greater extent of CFAP dissolution in the present study (cf. CW-CFAP, KC-CFAP, MS-CFAP in Fig. 4a; Table 2).

Due to this dependency on carbonate substitutions and naturally occurring carbonate traces in FAP (Table 1), separation of sedimentary FAP from CFAP could have been insufficient. Presence of both, FAP with low carbonate content and CFAP with varying extent of carbonate substitution, could

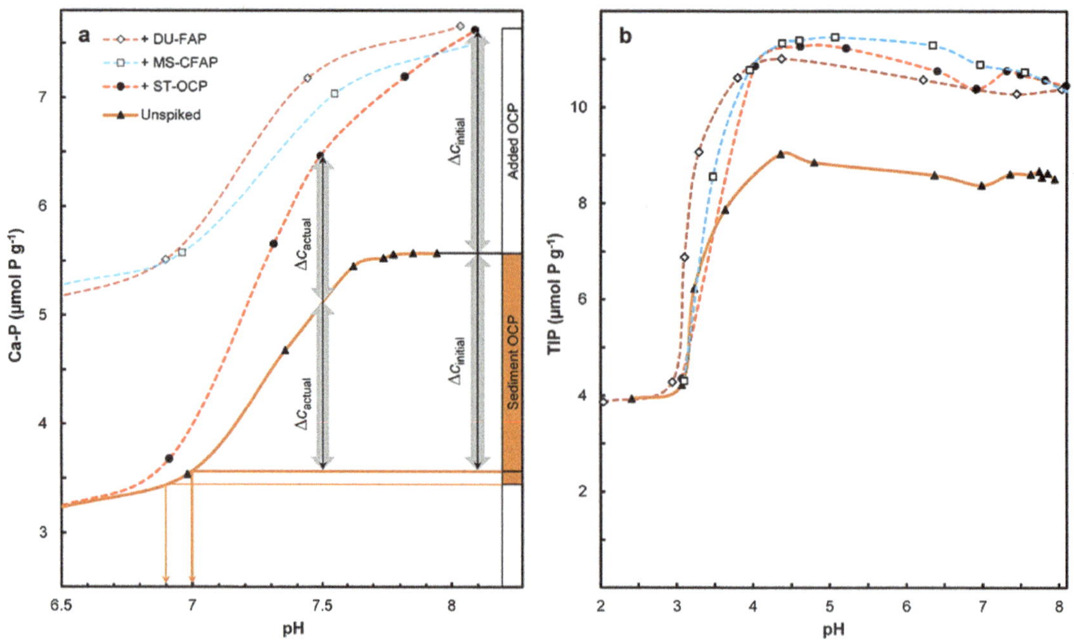

Figure 5. **(a)** Calculation of Ca-P species dissolution by acidification-induced losses of the Ca-P fraction as indicated for OCP. Differential OCP dissolution occurred at ca. pH 7 (thick orange arrow) and a slightly lower pH (6.9; thin orange arrow) was preferable to calculate the OCP content (see text). Curves were fitted by piece-wise cubic hermite interpolating polynomial to calculate selective losses of species ($\Delta c_{\text{loss}} = \Delta c_{\text{initial}} - \Delta c_{\text{actual}}$) at different pH values. Grey arrows denote initial and actual concentrations (exemplified for pH 7.5) of added OCP (+ ST-OCP) and sediment OCP (unspiked). Superimposed black arrows denote corresponding concentrations of total OCP (added OCP plus sediment OCP). Calculated loss percentages for different P species at different pH values are given in Table 2. Standard addition of FAP and CFAP is shown for comparison (+ DU-FAP; + MS-CFAP; insignificant dissolution in the range of OCP dissolution). **(b)** Acidification-induced changes of total inorganic P (TIP = Ca-P + Al / Fe-P) for unspiked sediment and standard additions of OCP, FAP and CFAP (same symbols as in **a**).

have resulted in a significant concurrent dissolution. However, there is a pronounced slope change for unspiked sediment at ca. pH 3.5, and the slope height in the pH range 3–3.5 obviously discriminates a certain amount of less soluble apatites from more soluble apatites. The ratio of slope heights is not constant for both sediments indicating that this is not an acidification-caused constant feature similarly related to all sedimentary apatites. Sediment from the Saigon River delta contained less CFAP but more FAP than sediment from the Firth of Thames. Finally, the slope change for unspiked sediment agrees with FAP and CFAP addition experiments. In conclusion, there must be certain mineral features that permit discrimination of these sedimentary apatites in the sedimentary environment in addition to the clear separation of added reference species.

However, the different mineral features that permit discrimination of FAP and CFAP are attributable to the different origins of sedimentary apatites. In contrast to authigenic CFAP, FAP is in general derived from detrital sources and delivered to marine environments by continental weathering. The pH at which differential dissolution of FAP and CFAP occurred during our addition experiments agrees approximately with a pH of 4, at which best separation of FAP and CFAP is attained in acetate buffer solutions or dilute mineral acids (Ruttenberg, 1992). Therefore, the CONVEX method distinguishes detrital from authigenic apatite within the sediment matrix. Whether this discrimination and hence the large solubility differences between sediment FAP and CFAP were primarily attributable to different carbonate contents remained unclear. Addition experiments showed that a rather small variation in carbonate content between DU-FAP and CW-CFAP (1.0 and 2.2 %, respectively) coincided with large solubility differences, whereas a larger variation between CW-CFAP and MS-CFAP (2.2 and 4.5 %, respectively) corresponded to less pronounced solubility differences (cf. Table 2). As detrital FAP and authigenic CFAP can contain various impurities, which are dependent on depositional and post-depositional conditions (Guidry and MacKenzie, 2003), other factors may also contribute to their strongly differing dissolution characteristics.

3.3 Congruency of species dissolution and possible sources of error

Based on maximum changes in curve slopes ($\Delta c/\Delta\text{pH}$) for unspiked sediment, complete dissolution of sediment OCP, CFAP and FAP was assumed to take place at pH values of 7.0, 3.5 and 3.0, respectively (Table 2). More than 80 % of

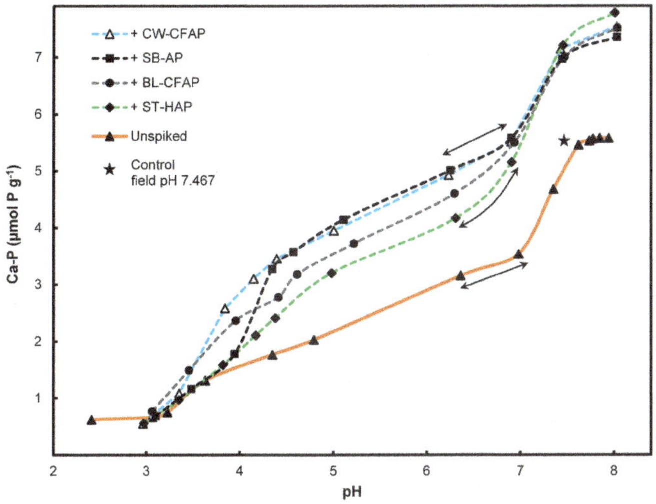

Figure 6. Differential Ca-P dissolution in sediment at varying pH as determined by subsample incubation of sediment (unspiked and with reference mineral additions) and subsequent fractionation (CONVEX method). Cryptocrystalline collophane (BL-CFAP) dissolved similar to HAP (ST-HAP) in the range 6.3–6.9 (see increased slope, arrows). Analysis was performed using sediment from the Firth of Thames, New Zealand. Control: fractionation without prior incubation. See Table 1 for origins of added standard materials.

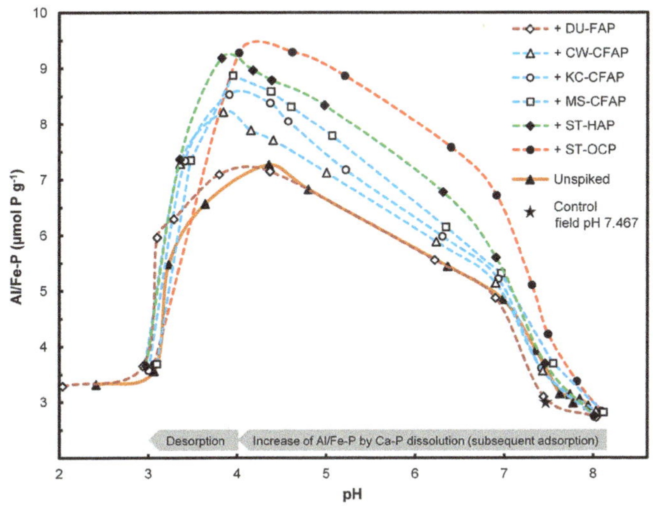

Figure 7. Increase of Al / Fe-P due to calcium phosphate dissolution (subsequent adsorption; pH 4–8) and decrease of Al / Fe-P due to desorption (pH < 4) as determined by subsample incubation of sediment (unspiked and with reference mineral additions) and subsequent fractionation (CONVEX method). Note the mirror-like reflection in the pH range 4–8 when compared to Fig. 2a (except for unspiked sediment; see text). Analysis was performed using sediment from the Firth of Thames, New Zealand. Control: fractionation without prior incubation. See Table 1 for origins of added standard materials.

added OCP dissolved at pH 7 (value in bold in Table 2; Fig. 5a for an example of calculation). However, CFAP only sparingly dissolved at slightly lower pH values (see loss percentages of CFAP at pH > 6.5) thereby minimizing a concurrent dissolution of CFAP and OCP. Thus, a slightly lower pH of 6.9 was preferable to calculate OCP contents (Fig. 5a). At this pH, the total OCP loss was approximately 100 %, loss of added OCP was 90 %, CFAP loss was lower than 5 % and sediment OCP might be overestimated by 5 % (Table 2).

In general more than 90 % of added CFAP and FAP dissolved at pH 3.5 and 3.0, respectively. Largest differences in CFAP dissolution due to varying carbonate inclusion were observed in the pH range 4 to 4.5 (Table 2, numbers in italic). Further, the dissolution of the CFAP sample with the highest degree of carbonate substitution (MS-CFAP: 4.6 %) was best correlated with the dissolution of sediment CFAP ($R^2 = 0.92$ for the values shown in Table 2). This correlation was progressively less pronounced with decreasing carbonate substitution of the other CFAP samples (KC-CFAP: 3.4 %, $R^2 = 0.85$; CW-CFAP: 2.2 %, $R^2 = 0.73$) and may therefore indicate a relatively high degree of carbonate substitution of authigenic apatite of the unspiked sediment sample. However, despite the effect of carbonate substitution on CFAP dissolution, different degrees of carbonate substitution did not significantly change the dissolution range of CFAP (Fig. 4a). All in all, dissolution ranges of added FAP, CFAP and OCP were almost congruent with the three dissolution ranges of sediment Ca-P (Table 2) verifying that standards were well suited to explain these dissolution ranges.

Regarding potential transformations, incubations at in situ pH did not significantly change P compositions. For instance, OCP was not generated by desorption of adsorbed P because Ca-P and Al / Fe-P contents of controls without incubation (cf. Sect. 2.4.; Fig. 3) closely matched those of sediment incubated at in situ pH (Figs. 4 and 7). Hence, OCP did not form by P desorption (or any other non Ca-P source) because total Ca-P did not increase by incubation. Furthermore, one Ca-P species did not transform to another. If anything, one would expect extremely slow HAP formation by OCP hydrolysis (e.g. Gunnars et al., 2004) and certainly not OCP formation by apatite dissolution. None of this took place as shown by standard additions. Finally, controls without incubation were not biased by OCP formation during sequential extraction because the employed method was developed and validated to eliminate OCP precipitation (Kurmies, 1972; removal of Ca using three wash steps with KCl/EtOH prior to the alkaline extraction; Fig. 3).

However, control values fit even better to Ca-P concentrations of incubations at higher pH (see Ca-P contents closer to pH 8 in Fig. 4a and b) and were used for calculation of the most soluble species (OCP). Incubations with a final pH similar to field pH showed slight Ca-P dissolution because the adjusted pH was initially somewhat lower than the final pH and drifted subsequently upwards. Therefore, the final pH did not precisely reflect the pH of observed loss. However, this shift is not significant because added Ca-P minerals (i) were subjected to the same conditions as sediment Ca-P, (ii) explained characteristic Ca-P dissolution ranges of unspiked incubations, and (iii) were measurable in the sediment matrix by differential dissolution.

3.4 Solubility of collophane and fish bone apatite in sediment

Collophane and fish bone specimens were also tested by standard addition experiments because various substitutions and crystal imperfections may result in increased solubility of collophane and bone apatite. Although biogenic apatite is likely an insignificant P sink on a global scale, it can constitute a considerable P fraction in some sediments. If so, biogenic apatite might bias the quantification of precursor phases due to its relatively high solubility compared to other apatites (see Schenau et al., 2000; Schenau and De Lange, 2000).

Dissolution of nanocrystalline snapper bone apatite (SB-AP) was comparable to that of crystalline CFAP with lowest carbonate inclusion (CW-CFAP) at higher pH but matched synthetic HAP (ST-HAP) dissolution at ca. pH 4 (Fig. 6). Bone apatite is structurally similar and frequently termed as HAP but also shows considerable chemical and structural differences (Wopenka and Pasteris, 2005). This may explain why the dissolution behaviour of the bone apatite was comparable to both, crystalline CFAP and synthetic HAP. Bone apatite contains typically ca. 7 % carbonate, little or a complete lack of hydroxyl substitutions and other ions may be incorporated to adjust for required properties. For example, incorporation of fluorine into apatite reduces its solubility in organisms. However, bone apatite did neither bias the determination of OCP nor the detection of detrital apatite due to absent additional Ca-P loss above pH 6.9 (loss only caused by sediment OCP) and pronounced Ca-P loss at ca. pH 4. Likewise, OCP presence in the bone sample can be ruled out by the missing additional loss in the alkaline pH range. An OCP precursor during bone apatite formation has been proposed in the past but a recent publication concluded that OCP has not been detected hitherto in bone (Rey et al., 2009).

Cryptocrystalline collophane (BL-CFAP) seemed to include an apatite mixture because it dissolved in agreement with HAP, CFAP and FAP at high, medium and low pH, respectively. Thus, incubations with collophane only showed increased dissolution comparable to HAP (increased dissolution marked by curved arrow in Fig. 6) within the pH range 6.3–7 among all incubation experiments. This is in agreement with expected varying collophane compositions including presence of HAP. Thus, collophane incubations indicate that significant contents of HAP are likely detectable. This conclusion additionally suggests minor contents of HAP in the unspiked sediments because they did not show increased dissolution within the pH range 6.3–7 (straight arrow for unspiked sediment in Fig. 6).

3.5 Increase of Al / Fe-P due to calcium phosphate dissolution

The pH-dependent Ca-P variations were generally reflected by Al / Fe-P contents because P liberated by Ca-P dissolution was subsequently adsorbed and determined by the Al / Fe-P fraction (Fig. 7). Thus, Al / Fe-P mirrored Ca-P in the pH range 4–8 for each individual addition experiment with an accompanied sequence inversion among all addition experiments (cf. Figs. 7 and 4a). In consequence of the almost quantitative adsorption (ΔAl / Fe-P $\approx$ ΔCa-P), total inorganic P contents remained largely unchanged in the pH range 4–8 as shown in Fig. 5b for unspiked sediment and sediment with FAP, CFAP and OCP additions. These findings are in agreement with previous observations (Oxmann et al., 2008) and demonstrate that most pH-induced sedimentary P dissolution processes cannot be observed by total inorganic P variations or reverse soluble P levels due to almost quantitative solid-phase P transfers. Data of Al / Fe-P and Ca-P for unspiked sediment were also mirrored in the pH range 4–8. Because they always involved lowest concentrations relative to other incubations (see Fig. 5b) they naturally did not follow sequence inversion.

Below ca. pH 4, less P was adsorbed despite large Ca-P losses (Fig. 7). The accompanied overall loss of sediment inorganic P below pH 4 (Fig. 5b) is in turn reflected by increased soluble P (see Oxmann et al., 2008). Very little free orthophosphate generated by Ca-P dissolution was adsorbed below pH 3 resulting in an almost complete liberation of added P from the sedimentary environment. Thus, even the relatively low original sedimentary Al / Fe-P content ($3.02\,\mu$mol P g^{-1}) increased just marginally ($3.31\,\mu$mol P g^{-1}) due to adsorption in the strongly acidic range (Fig. 7).

4 Conclusions

We found that OCP dissolved above ca. pH 7 and its quantification was not biased by FAP, CFAP, HAP or biogenic apatites. Thus, the CONVEX method is able to differentiate OCP from the apatite group. Carbonate fluorapatite dissolved together with HAP in the pH range 3.5–7. However, HAP dissolved to a greater extent at equal pH and was therefore more soluble than CFAP. Below pH 3.5, FAP began to dissolve and was therefore distinguishable from other apatites, verifying that the CONVEX method also distinguishes detrital from authigenic apatites. Solubility of CFAP increased with carbonate content, particularly in the pH range 4–4.5. Bone apatite and collophane dissolved in a similar manner to different identified reference apatites.

Overall, this study demonstrated that the CONVEX method can be used for experimental evaluation of calcium phosphate solubilities in the sedimentary environment. Modifying the conversion procedure by adjusting the incubation time or varying the solution composition (e.g. different ionic strengths), could provide an understanding of dissolution kinetics and solubilities under different environmental conditions. Moreover, the method performs accurately for quantification of the most important sedimentary Ca-P species. Further analysis of sediments of different textural properties suggests that the method is robust towards differing characteristics of sample matrices (Oxmann and Schwendenmann, unpublished data). Generally, obtained results are not biased by common matrix interferences because the method of standard addition eliminates matrix effects.

The two coastal sediment samples differed in their FAP and CFAP content despite similar physical–chemical properties. However, contribution of OCP to total Ca-P was equal for both sediments and, with a value of 37 %, surprisingly high. The detected OCP occurrence supports suggestions of apatite formation by an OCP precursor mechanism thereby pointing to the importance of addressing corresponding compositions of species and their geographic distributions in future studies.

Acknowledgements. We thank Michel Nieuwoudt (School of Chemical Sciences) and John Wilmshurst (School of Environment) for assistance with FTIR, Raman and XRD analyses at The University of Auckland. Special thanks to John Robertson for providing access to grinding and high precision weighing facilities at the Auckland University of Technology. Thanks also to John Rakovan (Miami University) for making the FAP specimen available and to Melissa Bowen and Charles Hyland for providing valuable comments on an earlier version of the manuscript and assisting with language editing. The project was funded by the German Research Foundation through a research fellowship granted to J. F. Oxmann under the code OX 54/2-1.

Edited by: M. Hoppema

References

Avnimelech, Y.: Phosphorus and calcium carbonate solubilities in Lake Kinneret, Limnol. Oceanogr., 28, 640–645, 1983.

Atlas, E. L.: Phosphate equilibria in seawater and interstitial waters, Ph.D. thesis, Oregon State University, Corvallis, 1975.

Atlas, E. L. and Pytkowicz, R. M.: Solubility behavior of apatites in seawater, Limnol. Oceanogr., 22, 290–300, 1977.

Baker, M. J., Blowes, D. W., and Ptacek, C. J.: Laboratory development of permeable reactive mixtures for the removal of phosphorus from onsite wastewater disposal systems, Environ. Sci. Technol., 32, 2308–2316, 1998.

Barrow, N. J., Bowden, J. W., Posner, A. M., and Quirk, J. P.: Describing the effects of electrolyte on adsorption of phosphate by a variable charge surface, Aust. J. Soil Res., 18, 395–404, 1980.

Bar-Yosef, B., Kafkafi, U., Rosenberg, R., and Sposito, G.: Phosphorus adsorption by Kaolinite and Montmorillonite: I. Effect of time, ionic strength, and pH, Soil Sci. Soc. Am. J., 52, 1580–1585, 1988.

Baturin, G. N. (Ed.): Principal features of the marine geochemistry of disseminated phosphorus, in: Developments in Sedimentology, Elsevier B. V., Amsterdam, 1981.

Bell, L. C. and Black, C. A.: Transformation of dibasic calcium phosphate dihydrate and octacalcium phosphate in slightly acid and alkaline soils, Soil Sci. Soc. Am. Proc., 34, 583–587, 1970.

Bentor, Y. K. (Ed.): Phosphorites: The unsolved problems, in: Marine Phosphorites: Geochemistry, occurrence, genesis, SEPM Special Publication, 29, 3–18, 1980.

Bray, J. T., Bricker, O. P., and Troup, B. N.: Phosphate in interstitial waters of anoxic sediments: Oxidation effects during sampling procedure, Science, 180, 1362–1364, 1973.

Burdige, D. J. (Ed.): Geochemistry of marine sediments, Princeton University Press, New Jersey/Oxfordshire, 2006.

Carman, R. and Rahm, L.: Early diagenesis and chemical characteristics of interstitial water and sediments in the deep deposition bottoms of the Baltic Proper, J. Sea Res., 37, 25–47, 1997.

Chen, Y. S. R., Butler, J. N., and Stumm, W.: Kinetic study of phosphate reaction with aluminum oxide and kaolinite, Environ. Sci. Technol., 7, 327–332, 1973.

Chien, S. H. and Black, C. A.: Free energy of formation of carbonate apatites in some phosphate rocks, Soil Sci. Soc. Am. J., 40, 234–239, 1976.

Christoffersen, M. R., Christoffersen, J., and Kibalczyc, W.: Apparent solubilities of 2 amorphous calcium phosphates and of octacalcium phosphate in the temperature-range 30–42 °C, J. Cryst. Growth, 106, 349–354, 1990.

Crosby, C. H. and Bailey, J.: The role of microbes in the formation of modern and ancient phosphatic mineral deposits, Front. Microbiol., 3, 241, doi:10.3389/fmicb.2012.00241, 2012.

Daesslé, L. W., Camacho-Ibar, V. F., Carriquiry, J. D., and Ortiz-Hernández, M. C.: The geochemistry and sources of metals and phosphorus in the recent sediments from the Northern Gulf of California, Cont. Shelf Res., 24, 2093–2106, 2004.

Eaton, J. W., Bateman, D., and Hauberg, S.: GNU Octave Manual Version 3, Network Theory Ltd, UK, 2008.

Einsele, W.: Über die Beziehungen des Eisenkreislaufs zum Phosphatkreislauf im eutrophen See, Arch. Hydrobiol., 29, 664–686, 1936.

Emerson, S. and Widmer, G.: Early diagenesis in anaerobic lake sediments II. Thermodynamic and kinetic factors controlling the formation of iron phosphate, Geochim. Cosmochim. Ac., 42, 1316–1316, 1978.

Faul, K. L., Paytan, A., and Delaney, M. L.: Phosphorus distribution in sinking oceanic particulate matter, Mar. Chem., 97, 307–333, 2005.

Featherstone, J. D. B., Pearson, S., and LeGeros, R. Z.: An infrared method for quantification of carbonate in carbonated apatites, Caries Res., 18, 63–66, 1984.

Gaudette, H. E. and Lyons, W. B.: Phosphate geochemistry in nearshore carbonate sediments: a suggestion of apatite formation, SEPM Special Publication, 29, 215–225, 1980.

Golterman, H. L.: Phosphate release from anoxic sediments or "What did Mortimer really write?", Hydrobiologia, 450, 99–106, 2001.

Golubev, S. V., Pokrovsky, O. S., and Savenko, V. S.: Unseeded precipitation of calcium and magnesium phosphates from modified seawater solutions, J. Cryst. Growth, 205, 354–360, 1999.

Guidry, M. W. and MacKenzie, F. T.: Experimental study of igneous and sedimentary apatite dissolution: Control of pH, distance from equilibrium, and temperature on dissolution rates, Geochim. Cosmochim. Ac., 67, 2949–2963, 2003.

Gulbrandsen, R. A., Roberson, C. E., and Neil, S. T.: Time and the crystallization of apatite in seawater, Geochim. Cosmochim. Ac., 48, 213–218, 1984.

Gunnars, A., Blomqvist, S., and Martinsson, C.: Inorganic formation of apatite in brackish seawater from the Baltic Sea: an experimental approach, Mar. Chem., 91, 15–26, 2004.

Hedley, M. J., Stewart, J. W. B., and Chauhan, B. S.: Changes in inorganic and organic soil phosphorus fractions by cultivation practices and by laboratory incubations, Soil Sci. Soc. Am. J., 46, 970–976, 1982.

Hutchison, K. J. and Hesterberg, D.: Dissolution of phosphate in a phosphorus-enriched ultisol as affected by microbial reduction, J. Environ. Qual., 33, 1793–1802, 2004.

Jahnke, R. A.: The synthesis and solubility of carbonate fluorapatite, Am. J. Sci., 284, 58–78, 1984.

Jahnke, R. A., Emerson, S. R., Roe, K. K., and Burnett, W. C.: The present day formation of apatite in Mexican continental margin sediments, Geochim. Cosmochim. Ac., 47, 259–266, 1983.

Johnsson, M. S. and Nancollas, G. H.: The role of brushite and octacalcium phosphate in apatite formation, Crit. Rev. Oral Biol. Med., 3, 61–82, 1992.

Jourabchi, P., Van Cappellen, P., and Regnier, P.: Quantitative interpretation of pH distributions in aquatic sediments: A reaction-transport modeling approach, Am. J. Sci., 305, 919–956, 2005.

Kazakov, A. V.: The phosphorite facies and the genesis of phosphorites, in: Geological investigations of agricultural ores, Transactions of the Science Institute of Fertilizers and Insecto-Fungicides, Leningrad, 95–113, 1937.

Krajewski, K. P., Van Cappellen, P., Trichet, J., Kuhn, O., Lucas, J., Martinalgarra, A., Prevot, L., Tewari, V. C., Gaspar, L., Knight, R. I., and Lamboy, M.: Biological processes and apatite formation in sedimentary environments, Eclog. Geol. Helvet., 87, 701–745, 1994.

Kruse, J., Negassa, W., Appathurai, N., Zuin, L., and Leinweber, P.: Phosphorus speciation in sequentially extracted agro-industrial by-products: Evidence from X-ray absorption near edge structure spectroscopy, J. Environ. Qual., 39, 2179–2184, 2010.

Kuo, S.: Phosphorus, Methods of Soil Analysis: Chemical Methods. Part 3, WASA and SSSA, Madison, 869–919, 1996.

Kurmies, B.: Zur Fraktionierung der Bodenphosphate, Die Phosphorsäure, 29, 118–149, 1972.

LeGeros, R. Z. and LeGeros, J. P.: Phosphate minerals in human tissues, Phosphate minerals, Springer Verlag, Berlin, 1984.

Lindsay, W. L., Vlek, P. L. G., and Chien, S. H.: Phosphate minerals, in: Minerals in soil environments, edited by: Dixon, J. B. and Weed, S. B., Soil Sci. Soc. Am., Madison, 1089–1130, 1989.

Lyons, G., Benitez-Nelson, C. R., and Thunell, R. C.: Phosphorus composition of sinking particles in the Guaymas Basin, Guf of California, Limnol. Oceanogr., 56, 1093–1105, 2011.

Martens, C. S. and Harriss, R. C.: Inhibition of apatite precipitation in the marine environment by magnesium ions, Geochim. Cosmochim. Ac., 84, 621–625, 1970.

Morse, J. W. and Casey, W. H.: Ostwald processes and mineral paragenesis in sediments, Am. J. Sci., 288, 537–560, 1988.

Morse, J. W. and Cook, N.: The distribution and form of phosphorus in North Atlantic Ocean deep-sea and continental slope sediments, Limnol. Oceanogr., 23, 825–830, 1978.

Mortimer, C. H.: The exchange of dissolved substances between mud and water in lakes, J. Ecol., 29, 280–329, 1941.

Mortimer, C. H.: Chemical exchanges between sediments and water in Great Lakes – Speculations on probable regulatory mechanisms, Limnol. Oceanogr., 16, 387–404, 1971.

Murphy, J. and Riley, J. P.: A modified single solution method for the determination of phosphate in natural waters, Anal. Chim. Ac., 27, 31–36, 1962.

Murrmann, R. P. and Peech, M.: Effect of pH on labile and soluble phosphate in soils, Soil. Sci. Soc. Am. Proc., 33, 205–210, 1969.

Naidu, R., Syersy, J. K., Tillman, R. W., and Kirkman, J. H.: Effect of liming on phosphate sorption by acid soils, Eur. J. Soil Sci., 41, 165–175, 1990.

Nancollas, G. H., LoRe, M., Perez, L., Richardson, C., and Zawacki, S. J.: Mineral phases of calcium phosphate, Anat. Rec., 224, 234–241, 1989.

Oh, Y.-M., Hesterberg, D. L., and Nelson, P. V.: Comparison of phosphate adsorption on clay minerals for soilless root media, Commun. Soil Sci. Plan., 30, 747–756, 1999.

Oxmann, J. F.: Technical Note: An X-ray absorption method for the identification of calcium phosphate species using peak-height ratios, Biogeosciences, 11, 2169–2183, doi:10.5194/bg-11-2169-2014, 2014.

Oxmann, J. F., Pham, Q. H., and Lara, R. J.: Quantification of individual phosphorus species in sediment: A sequential conversion and extraction method, Eur. J. Soil Sci., 59, 1177–1190, 2008.

Oxmann, J. F., Pham, Q. H., Schwendenmann, L., Stellman, J. M., and Lara, R. J.: Mangrove reforestation in Vietnam: The effect of sediment physicochemical properties on nutrient cycling, Plant Soil, 326, 225–241, 2010.

Parfitt, R. L., Atkinson, R. J., and Smart, R. S. C.: The mechanism of phosphate fixation by iron oxides, Soil Sci. Soc. Am. J., 39, 837–841, 1975.

Parrish, J. T., Zeigler, A. M., Scotese, C. R., Humphreville, R. G., and Kirschvink, J. K.: Proterozoic and Cambrian phosphorites – specialist studies, Early Cambrian palaeogeography, palaeoceanography and phosphorites, in: Phosphate deposits of the world. Volume I, Proterozoic and Cambrian phosphorites, edited by: Cook, P. J. and Shergold, J. H., Cambridge University Press, Cambridge, 280–294, 1986.

Perrone, J., Fourest, B., and Giffaut, E.: Surface characterization of synthetic and mineral carbonate fluoroapatites, J. Colloid Interf. Sci., 249, 441–452, 2002.

Pevear, D. R.: The estuarine formation of United States Atlantic Coastal Plain phosphorite, Econ. Geol., 61, 251–256, 1966.

Puigdomenech, I.: Windows software for the graphical presentation of chemical speciation, in: 219th ACS National Meeting, Abstracts of Papers, Vol. 1, Am. Chem. Soc., Washington, DC, 2000.

Rey, C., Combes, C., Drouet, C., and Glimcher, M. J.: Bone mineral: update on chemical composition and structure, Osteoporosis Int., 20, 1013–1021, 2009.

Ruttenberg, K. C.: Development of a sequential extraction method for different forms of phosphorus in marine sediments, Limnol. Oceanogr., 37, 1460–1482, 1992.

Ruttenberg, K. C. and Berner, R. A.: Authigenic apatite formation and burial in sediments from non-upwelling continental margin environments, Geochim. Cosmochim. Ac., 57, 991–1007, 1993.

Schenau, S. J. and De Lange, G. J.: A novel chemical method to quantify fish debris in marine sediments, Limnol. Oceanogr., 45, 963–971, 2000.

Schenau, S. J., Slomp, C. P., and De Lange, G. J.: Phosphogenesis and active phosphorite formation in sediments from the Arabian Sea oxygen minimum zone, Mar. Geol., 169, 1–20, 2000.

Schulz, H. N. and Schulz, H. D.: Large sulfur bacteria and the formation of phosphorite, Science, 307, 416–418, 2005.

Sheldon, R. P.: Ancient marine phosphorites, Annu. Rev. Earth Pl. Sc., 9, 251–284, 1981.

Spiteri, C., Cappellen, P. V., and Regnier, P.: Surface complexation effects on phosphate adsorption to ferric iron oxyhydroxides along pH and salinity gradients in estuaries and coastal aquifers, Geochim. Cosmochim. Ac., 72, 3431–3445, 2008.

Tribble, J. S., Arvidson, R. S., Lane III, M., and Mackenzie, F. T.: Crystal chemistry, and thermodynamic and kinetic properties of calcite, dolomite, apatite, and biogenic silica: applications to petrologic problems, Sediment. Geol., 95, 11–37, 1995.

Tyler, G.: Phosphorus fractions in grassland soils, Chemosphere, 48, 343–349, 2002.

Van Beusekom, J. E. E. and De Jonge, V. N.: Transformation of phosphorus in the Wadden Sea: Apatite formation, Deutsche Hydrographische Zeitschrift, 49, 297–305, 1997.

Van Cappellen, P. and Berner, R. A.: A mathematical model for the early diagenesis of phosphorus and fluorine in marine sediments; apatite precipitation, Am. J. Sci., 288, 289–333, 1988.

Wopenka, B. and Pasteris, J. D.: A mineralogical perspective on the apatite in bone, Mater. Sci. Eng., 25, 131–143, 2005.

16

Impact of a 30 % reduction in Atlantic meridional overturning during 2009–2010

H. L. Bryden[1], **B. A. King**[2], **G. D. McCarthy**[2], **and E. L. McDonagh**[2]

[1]National Oceanography Centre Southampton, University of Southampton, Empress Dock, Southampton, UK
[2]National Oceanography Centre, Southampton, Empress Dock, Southampton, UK

Correspondence to: H. L. Bryden (h.bryden@noc.soton.ac.uk)

Abstract. The Atlantic meridional overturning circulation comprises warm upper waters flowing northward, becoming colder and denser until they form deep water in the Labrador and Nordic Seas that then returns southward through the North and South Atlantic. The ocean heat transport associated with this circulation is 1.3 PW, accounting for 25 % of the maximum combined atmosphere–ocean heat transport necessary to balance the Earth's radiation budget. We have been monitoring the circulation at 25° N since 2004. A 30 % slowdown in the circulation for 14 months during 2009–2010 reduced northward ocean heat transport across 25° N by 0.4 PW and resulted in colder upper ocean waters north of 25° N and warmer waters south of 25° N. The spatial pattern of upper ocean temperature anomalies helped push the wintertime circulation 2010–2011 into record-low negative NAO (North Atlantic Oscillation) conditions with accompanying severe winter conditions over northwestern Europe. The warmer temperatures south of 25° N contributed to the high intensity hurricane season in summer 2010.

1 Introduction

In monitoring the Atlantic meridional overturning circulation (AMOC) at 25° N within the Rapid programme (Cunningham et al., 2007), we observe the Gulf Stream flow through Florida Straits, wind-driven Ekman transport in the surface layer and mid-ocean geostrophic flow between the Bahamas and Africa (Fig. 1). After 5 years of small variability in annual average overturning of 18.6 Sv, the overturning circulation slowed by 30 % during 2009–2010 to 12.8 Sv: the net northward warm water flow above 1000 m decreased by 5.6 Sv and the net southward flow of cold water below 1000 m decreased by 5.6 Sv (McCarthy et al., 2012). Based on the estimated uncertainty in annual-average overturning of 1.5 Sv (Cunningham et al., 2007), this change of 5.6 Sv is significant. The observed change is also extraordinary with respect to variability in the Atlantic overturning circulation found in coupled climate model control runs (McCarthy et al., 2012). Here we quantify the strength and duration of the AMOC slowdown and show that the reduction in ocean heat transport is the main cause of changes in heat content north and south of 25° N. We combine the publicly available time series of the AMOC (http://www.rapid.ac.uk/rapidmoc), an analysis of ocean heat transport following established methods (Johns et al., 2011), a new climatology of North Atlantic temperature and salinity variations based on Argo profiles (ftp://ftp.noc.soton.ac.uk/pub/bak/mapping) and ship-based estimates of air–sea heat exchange (Berry and Kent, 2009; available at http://noc.ac.uk/science-technology/earth-ocean-system/atmosphere-ocean/noc-surface-flux-dataset) as well as model-based estimates of air–sea heat flux (Dee et al., 2011).

2 Size of the event

To define the strength and duration of the slowdown, we take the first 4.75 years (April 2004–December 2008) of Rapid observations to represent the "normal" AMOC for all components. During the first 5 years , the annual average AMOC values (1 April to 31 March) were 17.8, 19.9, 19.5, 18.0, 17.5, showing little interannual variability about a mean of 18.6 Sv (Smeed et al., 2014). Because the "normal" period is not an integral number of years, we first remove the seasonal

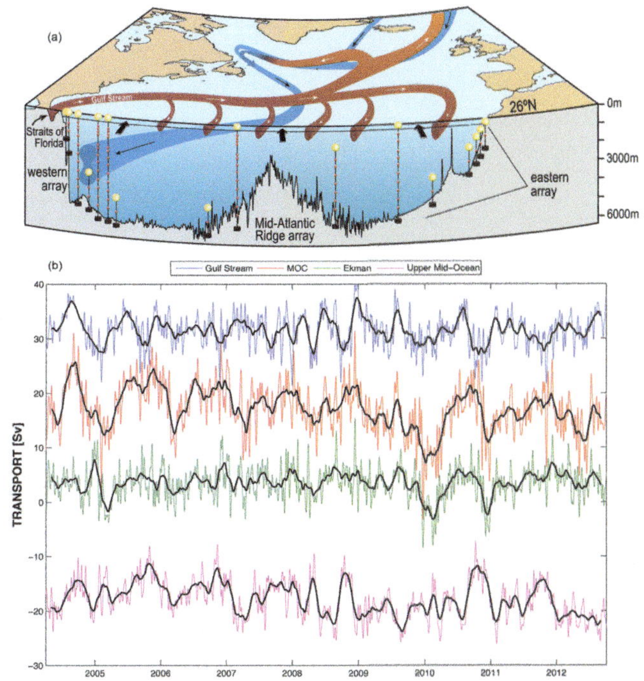

Figure 1. **(a)** Schematic of Rapid monitoring system for the Atlantic meridional overturning circulation (AMOC) at 26° N. **(b)** Rapid time series transports across 26° N since April 2004. Gulf Stream transport is derived from electromagnetic cable measurements in Florida Straits. Ekman transport is derived from ERA Interim wind estimates. Upper mid-ocean transport is derived from geostrophic velocity profiles from moored instruments across the Atlantic Ocean. The MOC transport is the sum of Gulf Stream, Ekman and upper mid-ocean transports. Black lines are 3-month low pass values.

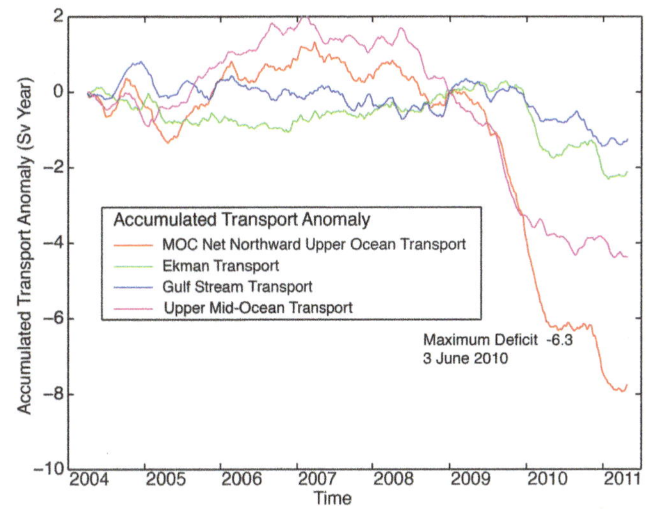

Figure 2. Accumulated Transport Anomaly during the Rapid event starting in early 2009. The normal AMOC is defined to be the average for each component for the first 4.75 years of the array, April 2004 to December 2008, where a seasonal cycle has been subtracted from each component prior to calculating the normal AMOC. The anomaly transport at each time for each component is defined as the difference between the instantaneous transport and the "normal" transport. Anomalous transport for each component is then accumulated over time. A local maximum accumulated transport deficit of 6.3 Sv years is achieved on 3 June 2010.

cycle from each component and then accumulate with time the deviation in transport from the 2004–2008 average for each component of the AMOC (Gulf Stream, Ekman, upper mid-ocean and total northward upper ocean transport 0–1030 m). Cumulative values are given in units of Sv years (Fig. 2) so that the size of the slowdown can be quantified as a change in AMOC transport (in Sv) and a duration (in years). The accumulated transport deficit reaches 6.3 Sv years in June 2010, it levels out through the remainder of 2010, and later in 2011 it increases as part of a long-term decline in the AMOC as documented by Smeed et al. (2014). Here we concentrate on the 2009–2010 event and conclude that the AMOC slowdown of 5.6 Sv persisted for about 1.1 years.

The cumulative values naturally oscillate with small amplitude about zero for the first 5 years which define the "normal" AMOC. In early 2009, the mid-ocean recirculation representing the southward flow of thermocline waters across the basin appears to strengthen; later in 2009 and early in 2010 the northward Gulf Stream and Ekman transports weaken. The increased southward recirculation starts earlier, lasts longer and accounts for more than half of the slowdown in the AMOC. For the overall slowdown of 6.3 Sv years, the increased southward thermocline circulation accounts for 3.8 Sv years or 61 % of the slowdown; the reduced northward Ekman transport accounts for 1.7 Sv years or 27 % of the slowdown; and the reduced Gulf Stream flow accounts for 0.8 Sv years of 12 % of the slowdown. The reduced northward flow of upper waters across 25° N is balanced by a reduced southward flow below 1000 m. The reduced southward transport of cold waters occurs mainly in the Lower North Atlantic Deep Water below 3000 m depth (McCarthy et al., 2012). To the extent that the northward Gulf Stream and Ekman transports are not very different from normal, we conclude that during the event the southward flow compensating for the northward Gulf Stream and Ekman transports is due to more thermocline recirculation and less deep overturning circulation: more recirculation, less overturning (McCarthy et al., 2012).

The overturning circulation in which warm upper waters flow northward and cold deep waters flow southward across 25° N is the principal mechanism transporting heat northward: 1.33 PW of northward ocean heat transport for an overturning circulation of 18.6 Sv (Johns et al., 2011). Because an increase or decrease in overturning of 1 Sv is accompanied by an increase or decrease in heat transport of 0.07 PW, a reduction in AMOC of 5.6 Sv is expected to result in decreased northward heat transport of 0.4 PW. We calculate the heat transport using methods described in Johns et al. (2011), remove the annual cycle, and accumulate it relative to the

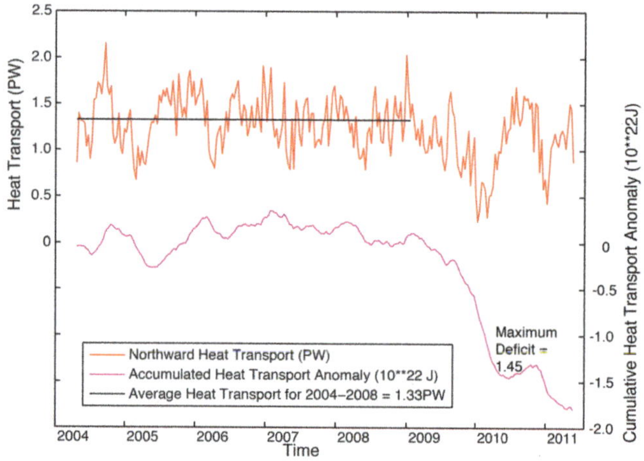

Figure 3. Accumulated northward heat transport deficit during the Rapid event starting in April 2009. Normal AMOC heat transport is defined as the average heat transport for the first 5 years of the array, April 2004 to December 2008, after removing a seasonal cycle in heat transport. The anomaly transport at each time is defined as the difference between the instantaneous transport and the "normal" transport. Anomalous heat transport is then accumulated over time and a local maximum accumulated heat transport deficit of 1.45×10^{22} J is achieved in June 2010.

"normal" heat transport estimated for the first 4.75 years of Rapid observations. The accumulated heat transport deficit reaches a maximum of 14.5×10^{21} J in June 2010 (Fig. 3). The uncertainty in heat transport is principally due to the uncertainty in overturning transport. For an event lasting 1.1 years, the error in accumulated heat transport deficit is 3.6×10^{21} J (1.1 years × 1.5 Sv × 0.07 PW per Sv).

3 Impact of the slowdown on ocean heat content

To examine the effects of reduced northward heat transport during 2009–2010, we analyse Argo float profiles in the region 10–45° N, 60–20° W relative to the Hydrobase seasonal climatology (http://www.whoi.edu/science/PO/hydrobase) so that the seasonal cycle should be removed. Anomaly potential temperature and salinity profiles at 20 dbar intervals from 10 to 1990 dbar are objectively analysed every 10 days on a 2.5° × 2.5° latitude, longitude grid (ftp://ftp.noc.soton.ac.uk/pub/bak/mapping). We then calculate heat content anomaly by vertically integrating from 1000 m depth to the surface the potential temperature anomaly (times density and specific heat). The heat content anomaly is noisy in space and time due to the fact that the Argo array of profiling floats has a typical horizontal scale of 300 km by 300 km so the array is not eddy-resolving. As a consequence, an Argo profile in a warm or cold eddy can have a large effect on the heat content value. Here we spatially average heat content anomaly north of 25° N (25–45° N, 60–20° W) and south of 25° N (10–25° N, 60–20° W)

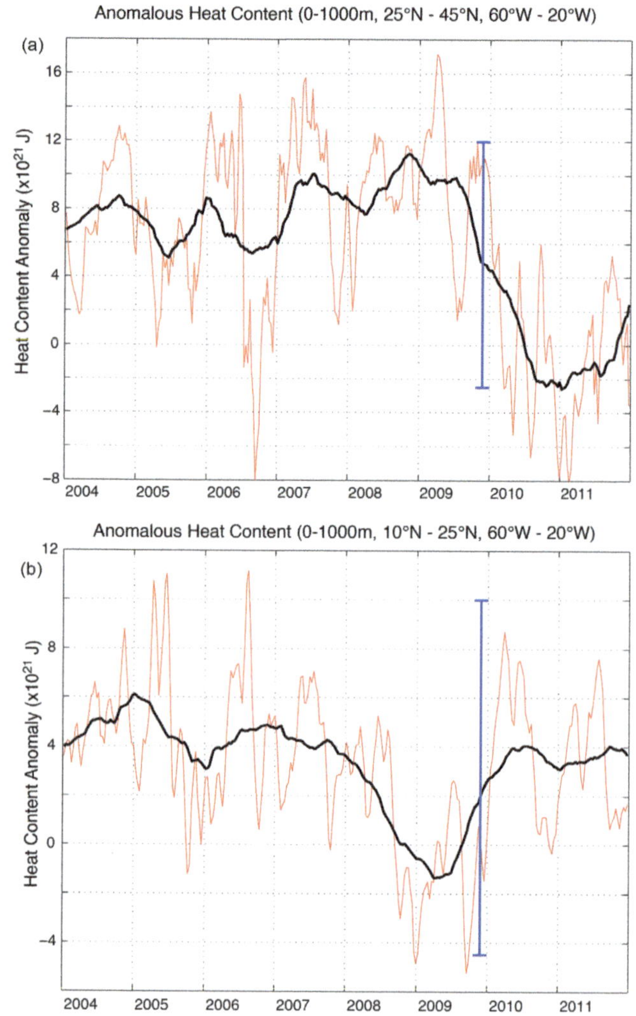

Figure 4. Anomalous heat content **(a)** north and **(b)** south of 25° N. Heat content anomaly is defined as the deviation in temperature from Hydrobase seasonal climatology multiplied by density and specific heat and integrated vertically from 0 to 1000 m depth. Heat content anomaly is then integrated over 60 to 20° W and over latitude bands 25 to 45° N (north) and 10 to 25° N (south). The red curves represent 30-day average heat content and the black curves are 12-month running mean values. The blue bar represents a change in heat content of 14.5×10^{21} J.

and we also smooth the time series with a 12-month running mean (Fig. 4). The objective analysis procedure yields error estimates for heat content anomaly at 30-day intervals that amount to 2×10^{21} J for the northern or southern region (see the Appendix). Considering the 50-day heat content anomaly values to be independent, the error in 12-month running mean heat content is 0.7×10^{21} J.

Heat content anomaly is generally positive: the Argo period since 2004 is warmer than the historical period represented in the Hydrobase climatology. Heat content anomaly north of 25° N shows a sharp decrease in 2009–2010.

Remarkably, the decrease in heat content anomaly occurs at the same time as the reduced heat transport and the magnitude of the decrease in heat content (13.5×10^{21} J) is almost the same as the reduced ocean heat transport (14.5×10^{21} J) across 25° N. South of 25° N, heat content anomaly increases during 2009–2010 contemporaneous with the decrease in ocean heat transport, but only by about 5.5×10^{21} J. Unfortunately, there are nearly no Argo floats in the Gulf of Mexico during 2004–2010 so we cannot account for any heat content changes for this substantial region immediately south of 25° N. Overall, the reduced ocean heat transport across 25° N during 2009–2010 is accompanied by decreased ocean heat content anomaly north of 25° N and increased ocean heat content anomaly south of 25° N.

To examine the vertical structure of the change in heat content anomaly, we estimate the temperature change for the regions north and south of 25° N for the periods winter (DJF) 2009–2010, summer (JJA) 2010 and winter (DJF) 2010–2011. Based on the objective mapping, the error in the upper ocean temperature profile is about 0.04 °C for these 3-month averages north or south of 25° N (see the Appendix). The cooling north of 25° N reaches a maximum of 0.8 °C at 50 m depth in summer 2010 (Fig. 5a). During the winters north of 25° N, maximum cooling of 0.6 to 0.7 °C occurs at the sea surface (Fig. 5a). South of 25° N, the warming reaches a maximum of 0.8 °C at 50 dbar in summer 2010 while the warming during the winters is a maximum of 0.5 °C at 50 dbar (Fig. 5b). Thus, for the 2009–2010 event where the accumulated heat transport deficit across 25° N peaks in June 2010, the cooling north of 25° N and warming south of 25° N peak in summer 2010. As might be expected, wintertime cooling and warming are smaller in magnitude: the accumulated heat transport deficit in winter 2009–2010 has not yet reached its maximum and the deficit in winter 2010–2011 has been reduced from its maximum by a small uplift in heat transport across 25° N during July–December 2010.

As suggested by the profiles of temperature anomaly in Fig. 5, the vertical structure for changes in heat content anomaly is different for the regions north and south of 25° N. North of 25° N, the changes in heat content anomaly during the event penetrate down to 1000 dbar and they decrease only slowly down from the surface, amounting to 3.6×10^{21} J for the interval 0–200 dbar, decreasing to 2.6×10^{21} J for 400–600 dbar and to 1.2×10^{21} J for 800–1000 dbar. South of 25° N, the changes in heat content anomaly during the event are concentrated in the upper ocean, amounting to 3.4×10^{21} J for the interval 0–200 dbar, decreasing to 0.6×10^{21} J for 400–600 dbar and to 0.1×10^{21} J for 800–1000 dbar. The heat content anomaly changes (and hence the temperature anomaly changes) in the interval 0–200 dbar are similar in magnitude but opposite in sign north and south of 25° N during the event.

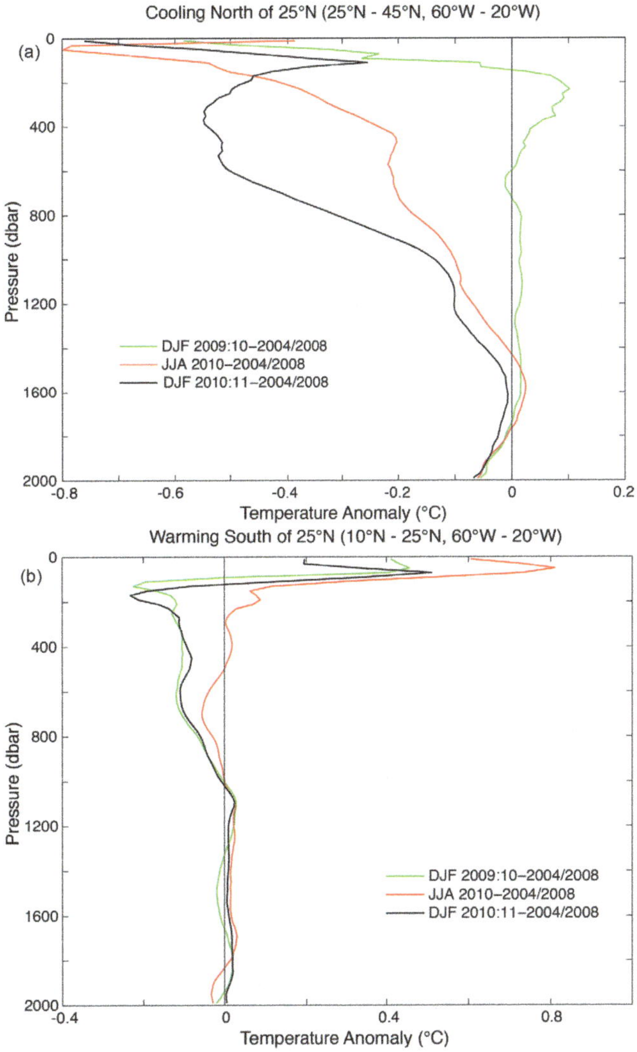

Figure 5. Vertical structure of temperature change **(a)** north and **(b)** south of 25° N during winter (DJF) 2009–2010, summer (JJA) 2010 and winter (DJF) 2010–2011. The temperature change is estimated as the difference from the average vertical temperature profiles for the period April 2004 to December 2008 in the north and in the south.

4 Role of air–sea heat flux anomalies in heat content changes

Traditionally, changes in ocean heat content are attributed to changes in air–sea heat exchange. For example, the cold sea surface temperatures north of 25° N in winter 2009–2010 have been attributed (Taws et al., 2011) to anomalously cold wintertime conditions associated with a negative North Atlantic Oscillation (NAO) atmospheric circulation. In contrast, here we find that over a broad area the anomalies in air–sea heat exchange during the event are small compared to the changes in heat content anomaly or in ocean heat transport deficit during the 2009–2010 event.

To examine the contribution of air–sea heat exchange during the Rapid event, we analyse the NOC monthly air–sea flux climatology (Berry and Kent, 2009; http://noc.ac.uk/science-technology/earth-ocean-system/atmosphere-ocean/noc-surface-flux-dataset) and the ERA Interim air–sea flux climatology (Dee et al., 2011). We generate average monthly air–sea heat flux for the decade from 1999–2008 at each 1° × 1° grid point for the region 10–45° N, 60–20° W and then subtract these average monthly values to create anomaly air–sea heat fluxes. We use the 10-year (1999–2008) baseline fluxes rather than the shorter 2004–2008 time period in an effort to remove the sizeable seasonal cycle in heat flux. To match the heat content analysis, we average the anomaly air–sea heat fluxes north of 25° N and south of 25° N and accumulate the anomalies since 1999 (Fig. 6). The pattern of anomaly air–sea heat fluxes is similar for the NOC and ERA Interim analyses with a small decline in the exchange during 2009–2011 north of 25° N in both analyses and a larger decline in both analyses south of 25° N during 2009–2011. A decline in anomaly air–sea flux represents anomalous cooling of the ocean.

North of 25° N where ocean heat content anomaly decreased by 13.5 × 10^{21} J, anomaly air sea heat fluxes from NOC contributed a cooling of about 0.9 × 10^{21} J during the Rapid event from January 2009 to June 2010 compared with a cooling of 1.4 × 10^{21} J for ERA Interim. Such cooling may well be associated with the extreme negative NAO event in winter 2009–2010, but the air–sea heat flux contributes only about 10 % to the observed reduced heat content anomaly north of 25° N. The anomalous air–sea flux cooling north of 25° N during the event of about 1.2 × 10^{21} J is the same size as the reduction in heat content anomaly in the upper 0–60 dbar and is a factor of 3 smaller than the reduced heat content anomaly of 3.6 × 10^{21} J in the upper 200 dbar, which we consider a reasonable layer thickness over which air sea fluxes have direct effect. Thus, even for the upper ocean, the reduction in heat content anomaly is larger than the anomalous air–sea flux cooling during the event. Hence we conclude that the cold upper ocean temperatures over the northern subtropics north of 25° N in winter 2009–2010 are primarily due to the slowdown in the AMOC and not to air–sea fluxes associated with the negative NAO conditions. In winter 2010–2011, cold surface temperatures north of 25° N are due to continuing effects of the slowdown in the AMOC that persisted until June 2010 and their subsequent re-emergence in winter (Taws et al., 2011; the overall pattern in seasonal upper ocean temperature anomalies during the period December 2009 to February 2011 is beautifully illustrated in their Fig. 1).

South of 25° N where ocean heat content anomaly increased by 5.5 × 10^{21} J during the Rapid event, anomaly NOC air–sea heat fluxes contribute a cooling of about 3.2 × 10^{21} J from January 2009 to June 2010 and ERA Interim anomaly heat fluxes contribute a cooling of 1.5 × 10^{21} J. It is likely that warmer upper ocean temperatures during 2009–2010 led to the reduced warming by air–sea exchange (anomalous

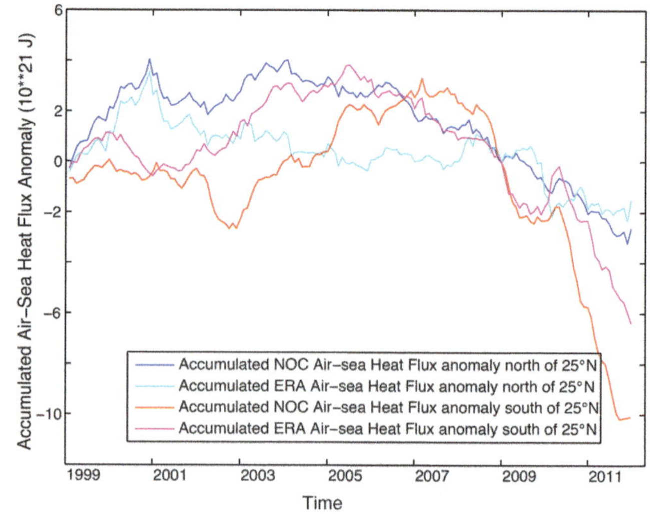

Figure 6. Accumulated anomalous air–sea heat flux versus 1999–2008 average for the NOC air–sea and ERA-Interim heat flux climatologies. North of 25° N represents the area 25–45° N, 60–20° W; south of 25° N represents the area 10–25° N, 60–20° W. Normal air–sea heat flux is defined as the averaged flux for each month over the 10-year period January 1999 to December 2008. The anomaly air–sea flux is then the difference between the monthly flux and the normal monthly flux. Anomaly heat fluxes are then accumulated over time from January 1999.

cooling) for the tropical region between 10 and 25° N. The increase in ocean heat content anomaly in the upper ocean (0–200 dbar) of 3.4 × 10^{21} J during the Rapid event is the same size but of opposite sign to the anomalous air–sea heat flux cooling of order 2.5 × 10^{21} J, suggesting that the effects of the reduction in northward heat transport during the event were partially compensated by an adjustment in air–sea heat exchange. Overall for the region south of 25° N, the reduced northward ocean heat transport across 25° N of 14.5 × 10^{21} J is larger than the sum of heat content anomaly increase and anomaly air–sea heat loss of 7 to 9 × 10^{21} J. As stated earlier, the region considered here south of 25° N does not include the Gulf of Mexico and Caribbean Sea where there were not enough Argo floats to estimate heat content change.

We do not attempt a full heat budget analysis for either region north or south of 26° N because there is no boundary section of observational quality as good as the Rapid 26° N section with which to measure ocean heat transport divergence. In a careful and impressive analysis of the northern subtropical region between 26 and 41° N (Cunningham et al., 2013), the ocean heat transport convergence was found to be not statistically different from zero within error bars, even for the large Rapid event of 2009–2010, due to uncertainties in boundary transports. Thus there is no point in performing a full heat budget analysis when heat transport divergence cannot be estimated to be different from zero for observed transports. In point of fact, we have a firm physical basis for maintaining that heat is conserved for any oceanic region: heat content changes must be balanced by the combination of divergence in heat transport across the boundaries and the air–sea heat flux into or out of the region. Ocean state estimates (e.g. Wunsch and Heimbach, 2013) effectively use heat content changes and air–sea fluxes to derive a time varying ocean circulation that maintains heat balance within errors for every ocean region. This is the inverse problem: use the principle of heat conservation along with the time history of heat content from Argo profiles and air–sea fluxes to derive the time-varying circulation that is necessary to conserve heat. Here we concentrate on the forward problem: how does the time-varying circulation modify the distribution of temperature in the North Atlantic Ocean. During the slowdown of the AMOC in 2009–2010, northward ocean heat transport across 25° N decreased, causing temperatures in the northern subtropics to decrease substantially (as has also been demonstrated by Cunningham et al., 2013) and temperatures in the tropics to increase. Air–sea fluxes contributed little to the observed temperature changes. The slowdown in the overturning circulation produced the spatial pattern of cooler waters north of 25° N and warmer waters south of 25° N that peaked in summer 2010.

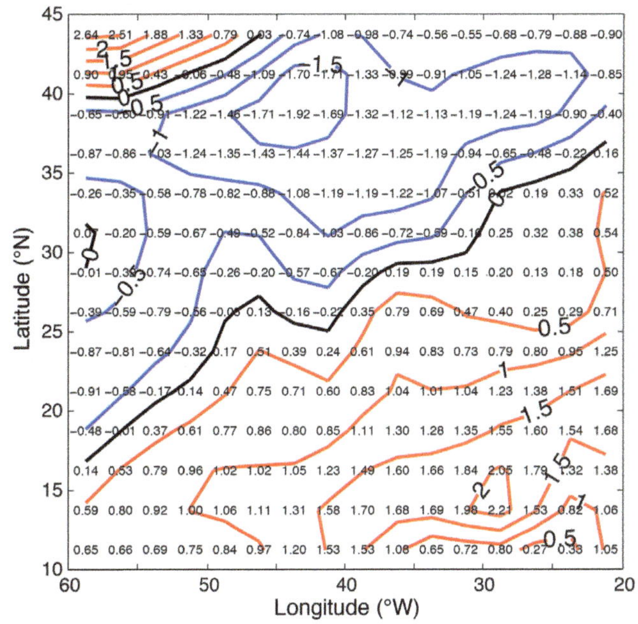

Figure 7. Temperature anomaly at 50 m depth averaged for May–July 2010 at the end of the slowdown in the AMOC.

5 Response of the atmosphere to the rapid event

At the end of the Rapid event in summer 2010, the temperature anomaly pattern at 50 m depth derived from the Argo analysis (Fig. 7) shows the impact of the 5.6 Sv slowdown in the overturning circulation over 14 months on the upper ocean temperature patterns that interact with the atmosphere. Temperatures south of 25° N are as much as 2 °C warmer in the tropical regions extending southwestward from Africa; maximum warming is at 15° N 30° W. Temperatures north of 25° N are more than 1.5 °C colder in the northern central subtropics; maximum cooling occurs at 40° N 42.5° W. The contour of no change in 50 m temperature anomaly does not exactly follow 25° N but stretches from 14° N 60° W (near Barbados) to 40° N 20° W (towards northwestern France). In this temperature pattern one can see the dipole anomaly pattern that has been suggested to re-emerge in winter 2010–2011 to affect the winter atmospheric circulation and produce a second winter of extreme NAO negative conditions (Taws et al., 2011) Also one can see the warm upper ocean tropical temperatures along the traditional hurricane trajectory westward from Africa across the tropical Atlantic (Emmanuel, 1987, 2005; Goni et al., 2009) that arguably contributed to the intensity of the 2010 Atlantic hurricane season that was the strongest since 2005.

Model studies (Cassou et al., 2007; Buchan et al., 2014) have shown that the upper ocean temperature distribution associated with a severe NAO-negative winter pushes the atmospheric circulation in the succeeding winter towards stronger negative NAO conditions. In models, the NAO-negative temperature distribution in winter is a tripole pattern consisting of warm anomalies in the tropics south of 25° N, cold anomalies in the northern subtropics north of 25° N and warm anomalies in the Labrador Sea (Taws et al., 2011). Here we attribute the dipole pattern of warm tropical waters and cold subtropical waters (two-thirds of the tripole pattern) to the 30 % slowdown in the AMOC. In a seasonal forecast for the winter of 2010–2011 over northwestern Europe, it was found that the distribution of upper ocean temperature anomalies in October–November 2010 was the key factor allowing severe winter conditions to be successfully forecast 3 months in advance (Maidens et al., 2013). On this basis we argue that the AMOC slowdown created an upper ocean temperature distribution that helped push the atmospheric circulation into record-low NAO negative states in both the winters of 2009–2010 and 2010–2011 with consequent effects on UK winter weather (Osborn, 2011). At the culmination of the event in summer 2010, the slowdown also led to very warm upper ocean temperatures in the tropical region south of 25° N potentially providing energy for developing hurricanes (Goni et al., 2009). The AMOC slowdown began in spring 2009 and persisted through June 2010, providing a 6 to 12-month advance warning for the anomalous thermal structure of the tropical and subtropical Atlantic Ocean in winter 2009–2010, summer 2010 and winter 2010–2011.

6 Discussion

The slowdown in the AMOC in 2009–2010 was unanticipated and its magnitude was exceptional compared with interannual variability in coupled climate models. Its origin remains uncertain 5 years on: there have been suggestions that wind stress curl variability 2 to 3 years earlier may have generated increased southward gyre flow associated with the slowdown (Duchez et al., 2014) or that ocean–atmosphere boundary conditions, partially set by the ocean thermal state, could generate the event (Roberts et al., 2013). Such efforts representing attempts to show that the atmosphere must drive all ocean variability do not yet explain why the slowdown also shows up in a reduction in the southward flow of Lower North Atlantic Deep Water below 3000 m depth, but not in a change to the southward flow of upper North Atlantic Deep Water between 1000 and 3000 m depth. The cooling south of 25° N (Fig. 4) preceding the warming during the 2009–2010 event suggests to us that the slowdown started earlier further south. At this point in time the evidence is that the ocean exhibits modes of variability that are not explained by recent atmospheric forcing: the ocean has a mind of its own.

The surface temperature distribution in the Atlantic during December 2009 has previously been linked with strongly negative NAO conditions associated with severe winter weather in the UK (Taws et al., 2011; Jung et al., 2011). Here we show that the pattern of cooler waters north of 25° N and warmer waters south of 25° N was generated not by anomalous air–sea fluxes but by an event in the AMOC during which it was 30 % below its "normal" value for about 14 months. The slowdown in the AMOC reduced northward ocean heat transport across 25° N by 0.4 PW, decreasing temperatures north of 25° N and increasing temperatures south of 25° N. To the extent that the upper ocean temperature spatial pattern nudged the atmospheric circulation into a strong NAO-negative state during winters 2010–2011 (Buchan et al., 2014) and 2009–2010, the anomalous upper ocean temperatures were the result of the slowdown of the AMOC, as quantified by Rapid observations at 25° N. With ongoing technological developments, the prompt delivery of these observations to the forecasting community could prove to have significant societal impact.

Appendix A: Error in temperature and heat content from Argo profiles

We make an estimate of the uncertainty in the heat content of a region from the uncertainty of individual points mapped by optimal interpolation (OI). First note that instrument error (order 0.01 °C) is small compared with sampling error and can be ignored. The best available estimate of the uncertainty of temperature at an OI grid point is given by the OI mapping error. This takes account of the variability of data that were included in the mapped estimate, and the number and spatial distribution of those data.

The heat content of a region is the sum of the individual gridded estimates. To estimate the uncertainty in heat content, we need to know the quality of individual estimates, and the number of statistically independent estimates being accumulated. We examine the median over space and time, as a representative value, of the temperature mapping error. This is of order 0.2 °C at the surface, 0.17 °C over the upper 300 m, 0.12 °C from 300 to 1400 m and 0.08 °C below 1400 m. Conservatively, we use 0.2 °C as a representative error over all depths, regions and time.

The OI mapping is done with a length scale of 500 km in latitude and longitude and we assume that the mapped values are independent between 700 km × 700 km areas, equal to 49×10^{10} m^2. In order to estimate the uncertainty in heat content of such a region for Fig. 4, we multiply the temperature uncertainty (0.2 °C) by the depth 1000 m, by density 1000 kg m^{-3} and by heat capacity 4000 J kg^{-1} °C^{-1} to estimate an uncertainty in heat content for an independent 700 km × 700 km area of 3.9×10^{20} J. The region 25–45° N, 60–20° W is 16.5 times larger than this independent region, so the uncertainty in heat content for the northern region is 1.6×10^{21} J (equal to $\sqrt{16.5} \times 3.9 \times 10^{20}$ J). The region 10–25° N, 60–20° W is 14.4 times larger than the independent region so the uncertainty in heat content for the southern region is 1.5×10^{21} J. Conservatively, we take the error for either northern or southern region to be 2×10^{21} J for the 30-day average values in Fig. 4. We estimate integral timescales for the heat content time series to be 25 days so that 50-day heat content values are independent of each other and the uncertainty in 12-month running mean heat content is 0.7×10^{21} J (equal to 2×10^{21} J divided by $\sqrt{365/50}$).

For the temperature profiles in Fig. 5, the uncertainty in seasonally averaged temperature anomaly is about 0.04 °C (equal to 0.2 °C divided by $\sqrt{16.5}$ or $\sqrt{14.4}$ and divided by $\sqrt{90/50}$). The temperature anomalies in Fig. 6 have an uncertainty of 0.15 °C (equal to 0.2 °C divided by $\sqrt{90/50}$ for the 90-day averaging).

Acknowledgements. This work formed the basis for H. L. Bryden's Nansen Medal Lecture at the European Geosciences Union (EGU) Meeting, April 2013. H. L. Bryden thanks the EGU for awarding him the Nansen Medal and for supporting his presentation in Vienna. We thank Natural Environment Research Council (NERC) for long-term support for the analysis of the Rapid observations. B. A. King and E. L. McDonagh are supported by NERC under the MONACO project of Rapid-WATCH. G. D. McCarthy is supported by NERC under the continuing Rapid-WATCH monitoring programme. Observations in the Rapid array along 26° N are funded by the Natural Environment Research Council (UK) and National Science Foundation (USA) and are freely available from www.rapid.ac.uk/rapidmoc. As part of the Rapid monitoring, Florida Current transport estimates are funded by the National Oceanic and Atmospheric Administration and are available from www.aoml.noaa.gov/phod/floridacurrent. Argo profiles are freely available from www.argodatamgt.org. NOC air–sea heat fluxes are freely available from http://noc.ac.uk/science-technology/earth-ocean-system/atmosphere-ocean/noc-surface-flux-dataset and ERA-Interim air–sea fluxes were obtained from the ECMWF Data Server.

Edited by: J. M. Huthnance

References

Berry, D. I. and Kent, E. C.: A new air-sea interaction gridded dataset from ICOADS with uncertainty estimates, B. Am. Meteorol. Soc., 90, 645–656, 2009.

Buchan, J., Hirschi, J. J.-M., Blaker, A. T., and Sinha, B.: North Atlantic SST anomalies and the cold North European weather events of Winter 2009/10 and December 2010, Mon. Weather Rev., 142, 922–932, 2014.

Cassou, C., Deser, C., and Alexander, M. A.: Investigating the impact of re-emerging sea surface temperature anomalies on the winter atmospheric circulation over the North Atlantic, J. Climate, 20, 3510–3526, 2007.

Cunningham, S. A., Kanzow, T., Rayner, D., Baringer, M. O., Johns, W. E., Marotzke, J., Longworth, H. R., Grant, E. M., Hirschi, J. J.-M., Beal, L. M., Meinen, C. S., and Bryden, H. L.: Temporal Variability of the Atlantic Meridional Overturning Circulation at 26.5° N, Science, 317, 935–938, 2007.

Cunningham, S. A., Roberts, C. D., Frajka-Williams, E., Johns W. E., Hobbs, W., Palmer, M. D., Rayner, D., Smeed, D. A., and McCarthy, G.: Atlantic meridional overturning circulation slowdown cooled the subtropical ocean, Geophys. Res. Lett., 40, 6202–6207, doi:10.1002/2013GL058464, 2013.

Dee, D. P., Uppala, S. M., Simmons, A. J., Berrisford, P., Poli, P., Kobayashi, S., Andrae, U., Balmaseda, M. A., Balsamo, G., Bauer, P., Bechtold, P., Beljaars, A. C. M., van de Berg, L., Bidlot, J., Bormann, N., Delsol, C., Dragani, R., Fuentes, M., Geer, A. J., Haimberger, L., Healy, S. B., Hersbach, H., Hólm, E. V., Isaksen, L., Kållberg, P., Köhler, M., Matricardi, M., McNally, A. P., Monge-Sanz, B. M., Morcrette, J.-J., Park, B.-K., Peubey, C., de Rosnay, P., Tavolato, C., Thépaut, J.-N., and Vitart, F.: The ERA-Interim reanalysis: Configuration and performance of the data assimilation system, Q. J. Roy. Meteorol. Soc., 137, 553–597, 2011.

Duchez, A., Hirschi, J. J.-M., Cunningham, S. A., Blaker, A. T., Bryden, H. L., de Cuevas, B., Atkinson, C. P., McCarthy, G. D., Frajka-Williams, E., Rayner, D., Smeed, D., and Mizielinski, M. S.: A new index for the Atlantic Meridional Overturning Circulation at 26° N, J. Climate, in press, 2014.

Emmanuel, K. A.: The dependence of hurricane intensity on climate, Nature, 326, 483–485, 1987.

Emmanuel, K. A.: Increasing destructiveness of tropical cyclones over the past 30 years, Nature, 436, 686–688, 2005.

Goni, G., Demaria, M., Knaff, J., Sampson, C., Ginis, I., Bringas, F., Mavume, A., Lauer, C., Lin, I. I., Ali, M. M., Sandery, P., Ramos-Buarque, S., Kang, K., Mehra, A., Chassignet, E., and Halliwell, G.: Applications of satellite-derived ocean measurements to tropical cyclone intensity forecasting, Oceanography, 22, 190–197, 2009.

Johns, W. E., Baringer, M. O., Beal, L. M., Cunningham, S. A., Kanzow, T., Bryden, H. L., Hirschi, J. J.-M., Marotzke, J., Meinen, C. S., Shaw, B., and Curry, R.: Continuous, array-based estimates of Atlantic Ocean heat transport at 26.5° N, J. Climate, 24, 2429–2449, 2011.

Jung, T., Vitart, F., Ferranti, L., and Morcrette, J.-J.: Origin and predictabiity of the extreme negative NAO winter of 2009/10, Geophys. Res. Lett., 38, L07701, doi:10.1029/2011GL046786, 2011.

Maidens, A., Arribas, A., Scaife, A. A., MacLachlan, C., Peterson, D., and Knight, J.: The influence of surface forcings on prediction of the North Atlantic Oscillation regime of Winter 2010–11, Mon. Weather Rev., 141, 3801–3813, doi:10.1175/MWR-D-13-00033.1, 2013.

McCarthy, G., Frajka-Williams, E., Johns, W. E., Baringer, M. O., Meinen, C. S., Bryden, H. L., Rayner, D., Duchez, A., Roberts, C., and Cunningham, S. A.: Observed interannual variability of the Atlantic meridional overturning circulation at 26.5° N, Geophys. Res. Lett., 39, L19609, doi:10.1029/2012/GL052933, 2012.

Osborn, T. J.: Winter 2009/2010 temperatures and a record–breaking North Atlantic Oscillation index, Weather, 66, 19–21, 2011.

Roberts, C. D., Waters, J., Peterson, K. A., Palmer, M. D., McCarthy, G. D., Frajka-Williams, E., Haines, K., Lea, D. J., Martin, M. J., Storkey, D., Blockley, E. W., and Zuo, H.: Atmosphere drives recent interannual variability of the Atlantic meridional overturning circulation at 26.5° N, Geophys. Res. Lett., 40, 5164–5170, 2013.

Smeed, D. A., McCarthy, G. D., Cunningham, S. A., Frajka-Williams, E., Rayner, D., Johns, W. E., Meinen, C. S., Baringer, M. O., Moat, B. I., Duchez, A., and Bryden, H. L.: Observed decline of the Atlantic meridional overturning circulation 2004–2012, Ocean Sci., 10, 29–38, doi:10.5194/os-10-29-2014, 2014.

Taws, S. L., Marsh, R., Wells, N. C., and Hirschi, J.: Re-emerging ocean temperature anomalies in late-2010 associated with a repeat negative NAO, Geophys. Res. Lett., 38, L20601, doi:10.1029/2011GL048978, 2011.

Wunsch, C. and Heimbach, P.: Dynamically and kinematically consistent global ocean circulation and ice state estimates, in: Ocean circulation and climate: A 21st century perspective, edited by: Siedler, G., Griffies, S., Gould, J., and Church, C., Chapter 21, 553–579, Elsevier, doi:10.1016/B978-0-12-391851-2.00021-0, 2013.

17

Hydrographic situation during cruise M84/3 and P414 (spring 2011) in the Mediterranean Sea

D. Hainbucher[1], A. Rubino[2], V. Cardin[3], T. Tanhua[4], K. Schroeder[5], and M. Bensi[3]

[1]CEN, Institut für Meereskunde, University of Hamburg, Bundesstraße 53, 20146 Hamburg, Germany
[2]Università Ca' Foscari di Venezia, Dorsoduro, 2137 – C.A.P. 30121, Italy
[3]OGS – Istituto Nazionale di Oceanografia e di Geofisica Sperimentale, Borgo Grotta Gigante, 42/c, 34018 Sgonico, Trieste, Italy
[4]GEOMAR Helmholtz Centre for Ocean Research Kiel, Department of Chemical Oceanography, Düsternbrooker Weg 20, 24105 Kiel, Germany
[5]Istituto di Scienze Marine – ISMAR, Consiglio Nazionale delle Ricerche (CNR), Arsenale – Tesa 104, Castello 2737/F, 30122 Venice, Italy

Correspondence to: D. Hainbucher (dagmar.hainbucher@zmaw.de)

Abstract. Aspects of hydrography and large-scale circulation observed in the Mediterranean Sea during the M84/3 and P414 cruises (April and June 2011, respectively) are presented. In contrast to most of the recent expeditions, which were limited to special areas of the basin, these two cruises, especially the M84/3, offered the opportunity of delineating a quasi-synoptic picture of the distribution of the relevant physical parameters along a section extending through the whole Mediterranean, from the Lebanese coast up to the Strait of Gibraltar. The foci of our analysis are the observed water mass properties and velocity fields. The first are investigated through T–S diagrams and an optimum multiparameter (OMP) analysis and the results are discussed also in the context of recently identified modes of variability; the second are studied by comparing the velocity fields observed using a vessel-mounted Acoustic Doppler Current Profiler and those calculated from the observed density fields. Overall, a distribution of temperature, salinity and geostrophic velocities emerges, which is far from that observed before the beginning of the so-called "Eastern Mediterranean Transient", a major climatic shift in the hydrography and circulation of the Mediterranean Sea which began at the end of the 1980s. The picture which emerges helps to further address the complexity of long-term evolution of hydrography and large-scale circulation of the Mediterranean Sea.

1 Introduction

The Euro-African Mediterranean Sea is a marginal sea, as it is separated from the Atlantic Ocean through the narrow Strait of Gibraltar. It consists of two sub-basins, the western (WMed) and the eastern (EMed) Mediterranean, which communicate through the broad (145 km) and shallow (maximum depth at its western sill is 550 m) Sicily Channel. This represents a natural barrier for the inter-basin circulation of intermediate and deep waters. As a consequence of the different bathymetry and the different forcings which exist between the two sub-basins, the WMed and EMed show distinct differences both in their hydrographic conditions and circulation. Different attempts, based on observational evidence, to schematize surface as well as intermediate and deep paths of circulation of the Mediterranean Sea have been made in the past (see e.g. the POEM Group, 1992; Malanotte-Rizzoli et al., 1997, 1999; Robinson et al., 2001; Roether et al., 2007; Schroeder et al., 2012).

The surface layer in the WMed is dominated by the inflow of relatively low-salinity Atlantic Water (AW) through the Strait of Gibraltar. This water mass is subject to different, mostly wind-driven currents that coexist with a complex, mostly cyclonic path of circulation. The AW flows from the WMed along the Algerian coast and bifurcates at the Sicily Strait. One branch, accounting for about 1/3 of the total water transport, deviates eastwards, entering the Tyrrhenian Sea whereas the other branch, accounting for about 2/3 of the transport, enters the eastern basin (Pierini and Rubino, 2001; Molcard et al., 2002). The intermediate layer is occupied by the Levantine Intermediate Water (LIW), originated in the EMed, whose core is found typically between 200 and 400 m depth. The abyssal water mass in the WMed, produced by shelf and open-ocean convection, is the Western Mediterranean Deep Water (WMDW), which is formed in the Gulf of Lion (see e.g. Leaman and Schott, 1991).

In the EMed, the AW, the Levantine Surface Water (LSW) and the Ionian Surface Water (ISW) dominate in near-surface layers, approximately down to 100 m depth. In the intermediate layer the LIW is clearly associated with a salinity maximum which is found typically between 50 and 600 m depth; in the eastern Mediterranean in this depth horizon Cretan Intermediate Water (CIW) can also be found. Note that the depth variability of such salinity maximum is largely regional: generally, it increases westward. LIW exerts a large influence on the oceanography and climatology of the Mediterranean Sea: formed in the Levantine Sea and in the Rhodes Gyre (Özturgut, 1976), it is able to flow through the Sicily Channel into the WMed and thus represents the intermediate water mass connection between the two basins. Additionally, more regionally confined water masses, like the Cretan Intermediate Water (CIW) can be found in the intermediate layer of the EMed (Robinson et al., 2001). The Adriatic Deep Water (AdDW) and water of Aegean origin named Cretan Deep Water (CDW) (Schlitzer et al., 1991) are deep water masses and are produced in the two deep water formation regions of the EMed. Thus, the predominant water mass of the bottom layers filling the abyssal plains of the Ionian and Levantine basins is the Eastern Mediterranean Deep Water (EMDW), being a mixture of AdDW, CDW and shallower water masses.

The EMed has gone through dramatic changes in hydrography and circulation during the last decades. The most drastic of those changes is known as the Eastern Mediterranean Transient (EMT; see Roether et al., 1996). Together with the occurrence of anomalous meteorological, oceanic, and hydrological phenomena, the source of the EMDW shifted abruptly from the Adriatic Sea to the Aegean Sea (see e.g. Roether et al., 1996; Klein et al., 1999; Lascaratos et al., 1999). As a consequence, the characteristics of the EMDW changed dramatically and resulted in a new, rapidly and still evolving thermohaline circulation. The deep parts of the Ionian basin were filled by the dense waters of Aegean origin and the isopycnals were lifted up by several hundred metres (Klein et al., 1999). Additionally, the northern Ionian upper circulation changed from anticyclonic to cyclonic in 1997, whereupon Borzelli et al. (2009) suggested that this phenomenon was associated with the filling up of the Ionian bottom layer with Cretan Sea Overflow Water (CSOW) following the EMT. Gačic et al. (2010) generalized this assumption and suggested that the redistribution of Ionian water masses, resulting from changes in the thermohaline properties of waters entering the basin, can sustain inversions of the upper-layer circulation. They propose a feedback mechanism between variations in the properties of waters from the southern Adriatic Sea and the Ionian circulation which they name BiOS (Adriatic–Ionian Bimodal Oscillating System).

Concerning the basin's long-term stability in the WMed, one has to note that, since 1960, the basin has experienced a marked long-term warming and salinification trend in its deep water (Béthoux et al., 1990). Possibly, such a trend has started well earlier, but the accuracy of the available observations does not allow a statistically significant reconstruction. These variations have been mostly attributed to changes in regional climate. Furthermore, two conditions led to an event which is known as the Western Mediterranean Transition (WMT): a significant increase both in temperature and salinity in the intermediate and deep layers occurred in recent years, which was identified as a propagating signal from the eastern basin to the western one originating from the EMT (Schroeder et al., 2006). The winters of 2004/2005 and 2005/2006 were extreme with a high rate of heat loss, less precipitation and persistent winds. This led to strong deep convection increasing also temperature and salinity in the WMed's deep layers up to a depth of about 1600 m (Lopez-Jurado et al., 2005; Schroeder et al., 2008). These changes became noticeable by an uplift of the isopycnals and by sharp inversions in the temperature–salinity diagrams.

In this study we present aspects of hydrography and large-scale circulation observed in the Mediterranean Sea during cruises M84/3 and P414 (April and June 2011, respectively) on a section extending from the eastern parts of the Mediterranean to the Strait of Gibraltar in the west. Special attention is paid to identify differences between the eastern and western basins, whilst elsewhere in this special issue (Cardin et al., 2014), aspects of the long-term variability in hydrography and circulation are presented which complement this investigation, particularly as far as the evolution of the thermohaline properties of the EMed during the last three decades is concerned. Similarly, in a study using transient tracers, Schneider et al. (2014), report on changes in ventilation of the Mediterranean Sea. Our focus here is a discussion of the observed water mass properties analysed through T–S diagrams and through an optimum multiparameter (OMP) analysis and a comparison between observed acoustic Doppler current profiler (ADCP) velocities and computed geostrophic velocities. Quasi-synoptic data like the ones presented here are rare and hence very valuable, although they refer to just one single section and cannot be used to elucidate three-dimensional structures of circulation and hydrography. The distributions of salinity, temperature and density along the section occupied by the M84/3 cruise have been reported by Tanhua et al. (2013b).

2 Data and methods

Since the EMT started at the end of the 1980s several surveys have been carried out in the Mediterranean to monitor the evolution of the basin's hydrography, circulation and biogeochemistry (POEM, 1987; POEM-BC, 1991; LIWEX, 1995; BOUM, 2008). However, most of them were restricted to special areas. METEOR cruise M84/3 constitutes an exception, as it is one of very few recent cruises (another one was the BOUM cruise, see Moutin et al., 2012), carried out after the appearance of the EMT, which ran on a section across the whole Mediterranean Sea with the goal of producing a quasi-synoptic view of the distribution of relevant physical and chemical properties in both basins. Furthermore, it reoccupied relevant stations of previous surveys, most importantly several of the stations from the METEOR survey M5/6 in 1987, which provided the last view of the situation before the EMT. POSEIDON cruise P414 (2011) was designed to complete M84/3, with stations located especially in the southern Adriatic and northern Ionian Sea. One major goal of that cruise was to observe the overflow of dense water of Adriatic origin in the Ionian basin, with the purpose of monitoring the effects of the thermohaline changes observed in the Adriatic on the EMDW (Bensi et al., 2013a). Both cruises were designed to follow the lines of the GO-SHIP Program (Global Ocean Ship Based Hydrographic Investigation Program, www.go-ship.org), which aims at creating a globally coordinated network of sustained hydrographic sections as part of the global ocean/climate observing system. Besides acquiring data for a nearly synoptic picture of the water mass property distribution across the whole Mediterranean Sea, the data collected by these two cruises can be used to determine changes in the circulation and in the distribution of properties and thereby fill in the gaps in the knowledge of the Mediterranean system.

More information about the cruises can be found in Tanhua et al. (2013a) and Hainbucher (2012).

2.1 Conductivity–temperature–depth (CTD)/rosette and hydrographic samples

Altogether, 61 full depth standard hydrographic stations were sampled during cruise M84/3 (Fig. 1a) and 33 full depth standard hydrographic stations during cruise P414 (Fig. 1b). A SeaBird SBE911plus CTD-O2 probe equipped with dual sensors of temperature, salinity and oxygen was employed attached to a SeaBird carousel 24 (12) bottle water sampler during the M84/3 (P414) together with an altimeter. Additionally, a fluorometer sensor was installed on the CTD during the M84/3 cruise and a digital reversing thermometer was attached to the probe for calibration purposes on both cruises.

At almost all stations water samples for dissolved oxygen (DO) were taken at different depths throughout the whole water column. Dissolved oxygen from bottle samples were analysed on board using a Winkler potentiometric method.

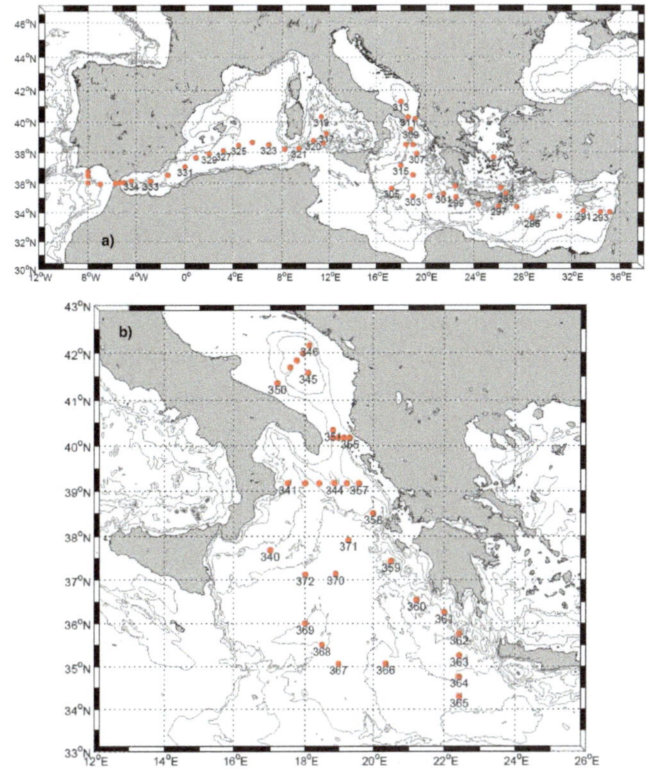

Figure 1. **(a)** CTD stations (red dots) of METEOR cruise M84-3 during April 2011. Depth contour lines are 200, 500, 1000, 2000, 3000, 3500 and 4000 m. **(b)** CTD stations (red dots) of POSEIDON cruise P414 during June 2011. Depth contour lines are 200, 500, 1000, 2000, 3000, 3500 and 4000 m.

A detailed description of oxygen sampling and distribution during cruise M84/3 can be found in Tanhua et al. (2013a, b). Dissolved oxygen is relevant in the discussion here only in the context of the applied OMP analysis.

From three depth levels, depending on the vertical profile of the stations, water samples were taken also for salinity analysis for calibration purposes and they were analysed onboard using a Guildline Autosal Salinometer.

Temperature, salinity and pressure data were post-processed by applying Seabird software and MATLAB routines. Spikes were removed, 1 dbar averages calculated, and temperature, salinity, and dissolved oxygen corrected with a regression analysis which fits the downcast profiles with the temperature, salinity and DO. Since the corrections applied to these parameters were small, the data quality can be considered excellent. Overall accuracies are within expected ranges: 0.002 °C for temperature, 0.003 for salinity and 2 % of saturation for dissolved oxygen. Additionally, we calculated potential temperature with the appropriate MATLAB routines. (We always refer to potential temperature when we use the term temperature in our discussions and figures.)

2.2 Acoustic Doppler current profiler (ADCP) velocity measurements

A vessel mounted ADCP (Teledyne RD Instruments) 38 kHz (with bin size of 32 m) was used during the M84-3 cruise. On P414 cruise we used an ADCP (Teledyne RD Instruments) 75 kHz ADCP (with bin size of 8 m). Both worked in narrow-band mode. The data were post-processed with the software package ossi14 (ocean surveyor sputum interpreter), developed by the Leibniz Institute of Marine Sciences, Kiel, which also corrects for the misalignment angle. An objective quality control of the data was not performed as no data for comparison or correction (like additional lADCP data) were available.

2.3 Quantification of water mass fractions

The mixing of water masses and their spatial distributions in the western and eastern Mediterranean Sea were determined using the OMP analysis developed by Tomczak (1981) and described in detail in Tomczak and Large (1989). To apply this method we assume that all hydrographic parameters are conserved and therefore not influenced by biogeochemical processes. We apply this assumption also for dissolved oxygen as our analysis is conducted over a limited region. However, we have to keep in mind that consumption along the pathways can play a role due to the rather long turnover times in the Mediterranean Sea which in the deep waters may amount to decades. Another problem is that the OMP method assumes steady-state situations while the period under consideration is characterized by a transient process (EMT, WMT). With these restrictions, although we believe they can add valuable information to our analyses, our OMP results have to be considered as tentative.

The OMP method is essentially based on establishing a linear system of n equations from observations of $n-1$ parameters and from the equation of mass conservation, and by deriving the contribution of the n water types to a given water sample as the exact solution to the n equations. For our analyses the method bears two challenges: the determination of parameter weights and the determination of water types. As described in Cardin et al. (2011) and in Kovačević et al. (2012), who applied the method to the Adriatic Sea and to the Ionian Sea, respectively, the weight for each parameter was derived as follows: the variance of each parameter of the considered water masses was calculated within the relevant vertical water layer and for the relevant stations. The largest variance of the same parameter among the considered water masses is represented by v2. The variance (v1) of the overall mean (the average of the considered different mean values associated with the different water masses) is determined. The ratio between v1 and v2 represents the weight, i.e. the relevance of each parameter used in the calculation. Some additional adjustments are made to equate salinity and mass conservation weights to that of temperature after normalization

Table 1. Water masses and their characteristics for (a) the WMed and (b) the EMed. Acronyms used to define water masses: AW (Modified Atlantic Water), LSW (Levantine Surface Water), LIW (Levantine Intermediate Water), WMDWn (Western Mediterranean Deep Water new), WMDWo (Western Mediterranean Deep Water old), AdDW (Adriatic Deep Water), CDW (Cretan Deep Water).

(a) Western Mediterranean Sea			
Water Mass	Temperature (°C)	Salinity	Oxygen (μmol kg^{-1})
AW	16.39	36.381	230
LIW	14.01	38.726	164
WMDWn	12.905	38.485	185
WMDWo	12.870	38.462	179
(b) Eastern Mediterranean Sea			
Water Mass	Temperature (°C)	Salinity	Oxygen (μmol kg^{-1})
AW	15.20	38.562	214
LSW	16.85	39.217	216
LIW	16.33	39.180	213
AdDW	13.25	38.730	197
CDW	14.43	39.040	197

(to sum of 100) of the values. The weights reflect the degree of conservativeness of each parameter, measurement accuracy and other processes which may render some parameters less reliable than others (Tomczak and Large, 1989). Therefore, temperature and salinity (and mass conservation) always obtain the largest weights. For our OMP analysis we additionally added dissolved oxygen, as this parameter possesses the same vertical resolution (1 m after averaging) as temperature and salinity due to the fact that it was measured through CTD sensors. Note that, because high weights are obtained for temperature and salinity, other parameters unavoidably exert a smaller influence on the OMP results. The results for weights came up as follows: temperature, salinity and conservation of mass equals 33 and oxygen equals 1. In fact, the effect of dissolved oxygen is negligibly small reflecting thereby the delicate assumption that in our case we regard dissolved oxygen to be a conservative property.

In the Mediterranean Sea the water mass definitions are broadly diversified and the definition of their characteristics differs often to some extent, presumably due to location and time when the water mass was found. Ideally, properties of water types are defined in the source region and reasonably back in time, which certainly is not possible for most hydrographical investigations due to lack of data. Therefore, in our study water mass properties were determined with the help of T–S diagrams, profiles and the knowledge about typical water mass distributions in the Mediterranean Sea. The characteristics are listed in Table 1. Depending on season and location, as already stated, the properties of the water masses can vary distinctly. Table 2 specifies temperature and salinity values found in the literature for the relevant water masses (Roether et al., 1998; Millot, 1999; Lascaratos et al., 1999; Theocharis et al., 1999; Gertman et al., 2006; Schroeder et

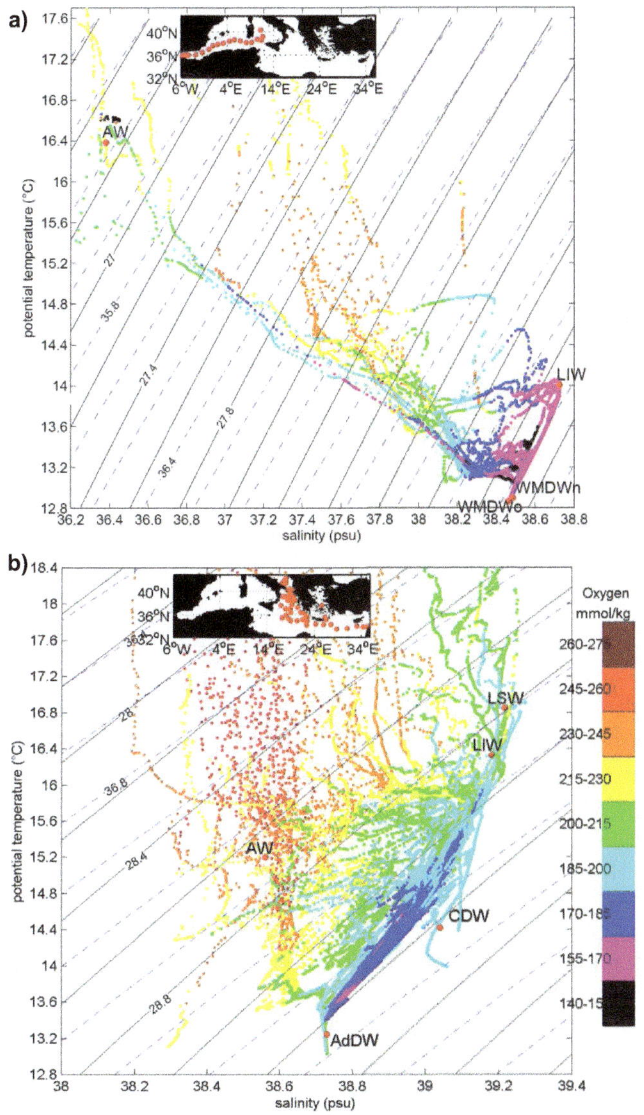

Figure 2a. T–S diagrams of the western **(a)** and eastern **(b)** Mediterranean Sea. Colours represent the concentration of oxygen in μmol kg^{-1}. All depth levels from 0 m to bottom are taken into account. The locations of water mass sources are indicated by red dots (AW = Modified Atlantic Water, LSW = Levantine Surface Water, LIW = Levantine Intermediate Water, WMDW = Western Mediterranean Deep Water, AdDW = Adriatic Deep Water, ADW = Cretan Deep Water). The inset panels show the location of considered stations. Note that the scales of T and S of the two panels are not identical. Data are from cruise M84-3 and P414 in spring 2011.

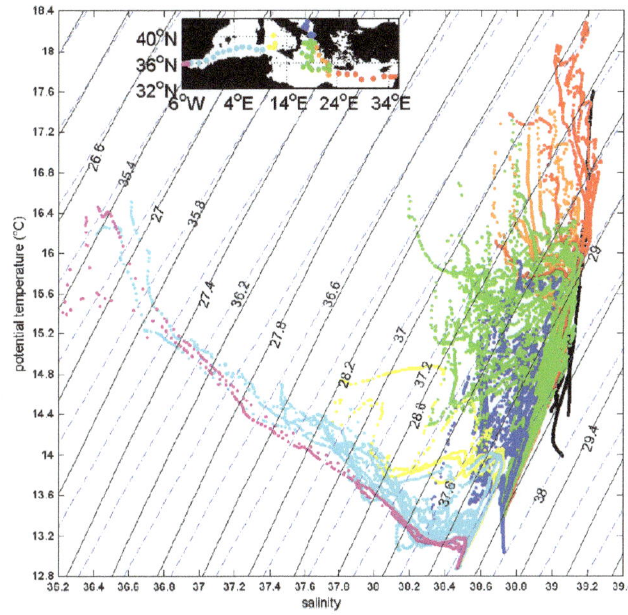

Figure 2b. T–S diagram of all CTD stations taken during cruise M84-3 and P414 in spring 2011. Values from 50 m depth down to the bottom are considered. Colours show different regions of the Mediterranean Sea. Black dots: Aegean Sea, red dots: Levantine Sea, green dots: Ionian Sea, orange dots: Hellenic Trench, blue dots: Adriatic Sea, yellow dots: Tyrrhenian Sea, cyan dots: remaining western Mediterranean Sea, magenta dots: Strait of Gibraltar. Isolines for potential density $\sigma_0 = \rho_0$ (referenced to sea surface) $- 1000\,\mathrm{kg\,m^{-3}}$ (black continuous line) and for potential density $\sigma_2 = \rho_{2000}$ (referenced to 2000 m) $-1000\,\mathrm{kg\,m^{-3}}$ (blue dashed line) are given. Inset panel shows the location of the different stations.

al., 2006; Kovačević et al., 2012; Bensi et al., 2013a, b). Note that the values presented in Table 1 refer to the year 2011, and might not be valid for other years. For instance, for a water mass analysis of years prior to the EMT the values of temperature and salinity in Table 1 are too high. Again here, the possible weakness of our OMP analysis, i.e. the fact that we are not in the presence of a steady-state situation, is evident.

In the WMed we generally find four major water masses (Fig. 2a, panel a): Modified Atlantic Water (AW) in the surface layer, LIW in the intermediate layer, and old and new WMDW from mid-depths down to the bottom. Our definition for AW in the WMed results from the T–S profiles of station 334, which is close to the entrance of the Strait of Gibraltar, where we find a pronounced salinity minimum in the surface layer. The definition of LIW for the WMed is given by the T–S profiles of station 320, which is close to the Channel of Sicily, where we find the salinity maximum at about 350 m depth. As is known, WMDW is a mixture of water masses resulting from deep convection in the Gulf of Lion. Old and new WMDW were determined by considering the values encountered in the hook-like T–S structures of Fig. 6b and c: new WMDW is slightly warmer and saltier than older WMDW, and more oxygenated. Thus, for the OMP analysis we focus on these four main water masses.

For the EMed we also revert to the basic source water masses: AW, LSW for the surface layers, LIW for the intermediate layers and AdDW and CDW for the bottom layers (Fig. 2a, panel b). Note again that here we choose the characteristics of the year 2011 for our analysis but the pre-EMT

Table 2. Water masses and their characteristics as found in the literature for (a) the WMed and (b) the EMed (Roether et al., 1998; Millot, 1999; Lascaratos et al., 1999; Teocharis et al., 1999; Gertman et al., 2006; Schroeder et al., 2006; Kovačević et al., 2012; Bensi et al., 2013a, b). Ranges of properties are given if possible. Acronyms used to define water masses: AW (Modified Atlantic Water), LSW (Levantine Surface Water), LIW (Levantine Intermediate Water), WMDWn (Western Mediterranean Deep Water new), WMDWo (Western Mediterranean Deep Water old), AdDW (Adriatic Deep Water), CDW (Cretan Deep Water).

(a) Western Mediterranean Sea		
Water Mass	Temperature (°C)	Salinity
AW	> 14	> 36.5
LIW	> 13.1	> 38.47
WMDWn	12.9	38.48
WMDWo	12.75–12.80	38.44–38.46
(b) Eastern Mediterranean Sea		
Water Mass	Temperature (°C)	Salinity
AW	17.7–23.29	37.38–38.10
LSW	17.0–28.0	39.0–39.4
LIW	12.85–14.2	38.37–38.80
AdDW	13.00–13.60	38.663–38.72
CDW	13.66–14.2	38.80–38.95

salinity for AdDW was distinctly lower. Much of the old AdDW is still present in the EMed and therefore our choice might underestimate the CDW.

AW was defined according to the profiles of stations 304/305 where we found the salinity minimum at around 100 m depth. The locations of these stations were closest to the Sicily Channel where AW enters the EMed. For LSW we considered the T–S characteristics at the surface of station 287, which is located in the Aegean Sea. Note that LSW is usually predominant during summer with much higher temperature values than those found during these cruises at the beginning of the summer season. LIW was defined according to station 296 located in the centre of the Levantine basin, where we found the salinity maximum at $\sim$ 170 m depth. We assume that the EMDW, which we also found during our cruises, results from the mixing of AdDW and CDW and therefore it is not taken into account for the OMP analyses. AdDW is defined according to the profiles collected in the Strait of Otranto and CDW according to the profiles collected in the Kasos Strait.

2.4 Geostrophic velocities

Using pressure, temperature and salinity data, geostrophic velocities have been calculated for cross sections through the western and eastern Mediterranean Sea. The reference layer of no motion was set to 1000 m. The use of deeper layers of no motion did not change the results substantially. Note that, in general, the distance between stations was too large to resolve any mesoscale features in the velocity distributions. These are mostly limited by the Rossby radius of deformation, which is of the order of 10–15 km in the Mediterranean Sea (Robinson et al., 2001).

3 Analysis and discussion

The general state of the circulation in the Mediterranean Sea for spring 2011 is shown exemplarily by the map of the Absolute Dynamic Topography (ADT) averaged for April 2011 (Fig. 2c). A variety of sub-basin-scale gyres can be identified for the eastern as well for the western Mediterranean Sea. A complete discussion of these features for the EMed can be found elsewhere in this issue (Cardin et al., 2014). The ADT shows a host of very complex, highly spatially variable circulation features in the basin. Such a complexity can be only partially captured by our in situ measurements.

3.1 Water mass properties and vertical distribution

To define the water mass properties observed during the M84/3 and P414 cruises, potential temperature/salinity diagrams (T–S) constructed for different layer depths and regions are considered. Figure 2b gives an overview of the T–S characteristics in the whole Mediterranean Sea measured in April and June 2011, which can be considered as representative for the spring season. It highlights major differences existing between the EMed and WMed stratification. If we exclude a surface layer extending down to around 50 m, overall, in the EMed, values vary in a relatively small salinity range (between 38.4 and 39.1, i.e. $\Delta S = 0.7$) and in a relatively large temperature range (between 13.0 °C and 18.2 °C, i.e. $\Delta T = 5.2$ °C). In contrast, in the WMed the values vary in a much broader salinity range (from about 36.4 to 38.7, i.e. $\Delta S = 2.3$), but in a relatively narrow temperature range (from about 12.9 to 15.9 °C, i.e. $\Delta T = 3.0$ °C). Differences between the two sub-basins are encountered also in the water properties close to the bottom: in particular, the abyssal layer of the EMed is warmer and saltier than that of the WMed. In the former, salinity and temperature values range between 13.40–13.64 °C and 38.72–38.79, while in the latter they range between 12.87–13.29 °C and 38.46–38.58. As a consequence, bottom density reaches values $\sigma_0 \sim 29.30\,\mathrm{kg\,m^{-3}}$ ($\sigma_2 \sim 37.81\,\mathrm{kg\,m^{-3}}$) in the EMed and $\sigma_0 \sim 29.10\,\mathrm{kg\,m^{-3}}$ ($\sigma_2 \sim 37.75\,\mathrm{kg\,m^{-3}}$) in the WMed.

As expected, the LIW signal is stronger in the EMed than in the WMed (Fig. 3). Indeed, the LIW spreads westwards from its formation sites (in the northeastern Levantine Sea and Rhodes cyclonic gyre; see e.g. Ovchinnnikov, 1984; Malanotte-Rizzoli and Hecht, 1988; Lascaratos, 1993) and it is identified by a relative salinity maximum encountered at $\sim$ 50–100 m in the Levantine Sea or $\sim$ 200–600 m in the Ionian Sea and WMed, whereas in the lower depth also

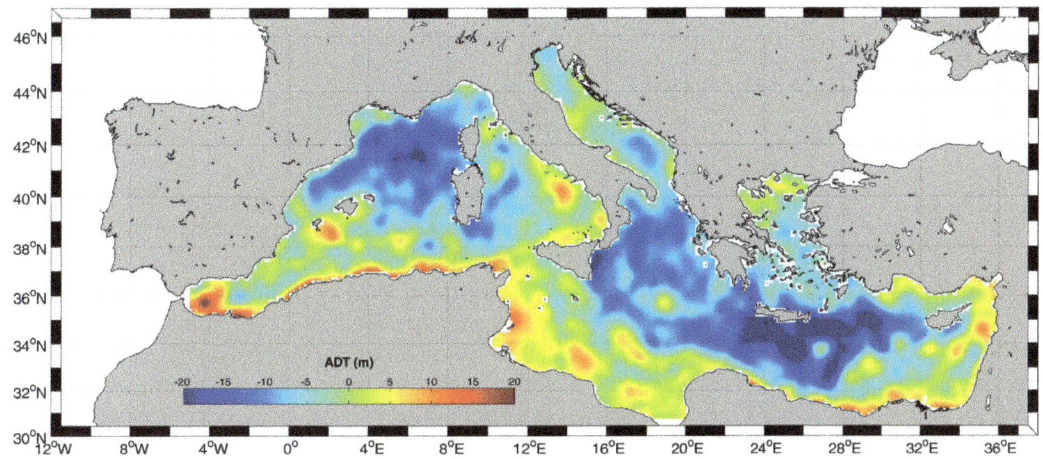

Figure 2c. Monthly absolute dynamic topography (ADT) map of the Mediterranean Sea for April 2011.

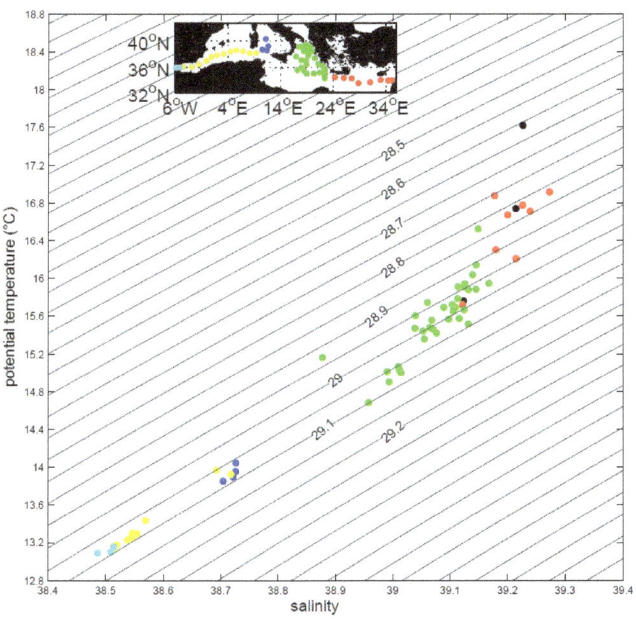

Figure 3. T–S diagram of maximum salinity and corresponding temperature in the layer 50–650 m, which shows the LIW core in the Mediterranean Sea. Colours show different regions. Black dots: Aegean Sea, red dots: Levantine Sea, green dots: Ionian Sea, blue dots: Tyrrhenian Sea, yellow dots: remaining western Mediterranean Sea, cyan dots: Strait of Gibraltar. Isolines for potential density $\sigma_0 = \rho_0$ (referenced to sea surface) $- 1000\,\mathrm{kg\,m^{-3}}$ (black continuous line) are given. Inset panel shows the location of the different stations.

CIW can be involved in the salinity maximum. The maximum reaches values of ~ 39.30 ($\sim 16.80°$ for temperature) in the Levantine Sea and reaches values of ~ 38.50 ($\sim 13.20\,°\mathrm{C}$ for temperature) at the Strait of Gibraltar, after a strong dilution along its way has occurred. Nevertheless, the density differences are small along the whole Mediterranean (from 28.85 to 29.05 increasing westward) due to the compensation between the thermohaline parameters.

During the two cruises the densest water ($\sigma_0 > 29.30\,\mathrm{kg\,m^{-3}}$, $T_{\mathrm{Pot}} \approx 13.95\,°\mathrm{C}$ and $S \approx 39.06$) was found (Fig. 4) in the Aegean Sea (St. 288), and it corresponded to the CDW. However, we did not observe an overflow of such water outside its formation region (see Cardin et al., 2014). The densest water of Adriatic origin ($\sigma_0 < 29.27\,\mathrm{kg\,m^{-3}}$, $T_{\mathrm{Pot}} \approx 13.06\,°\mathrm{C}$ and $S \approx 38.73$, St. 313), instead, was less dense than the CDW, but able to flow out from its formation region through the Strait of Otranto, where σ_0 was still larger than $29.2\,\mathrm{kg\,m^{-3}}$, $T_{\mathrm{Pot}} \approx 13.28\,°\mathrm{C}$ and $S \approx 38.72$ (St. 352–353).

According to Roether et al. (2007) the anomalously abundant and dense water outflow of Aegean origin produced during the EMT caused a noticeable increase of temperature and salinity throughout the EMed deep waters. Especially in the Levantine basin and in the Hellenic Trench, this increase could be detected as a long-lasting, pronounced T–S inversion (Roether et al., 2013). Indeed, during M84/3 we still can find this inversion in a broad area of the Levantine basin (Fig. 5a). There, T–S values are distributed closely together, with their minima being located at about 1000 m depth. In the deeper stations (excluding the easternmost three stations) a T–S maximum is reached at about 2200 m depth, followed by a steady T and S decrease. Those stations located south of the eastern Mediterranean ridge (Fig. 5a, blue, cyan orange, magenta) do not show such strong decreases in temperature and salinity toward the bottom. The T–S distribution is similar, but the minima are found at about 1300–1700 m depth and the maxima at about 1800–2100 m depth. The salinity difference between the extreme values is much larger in the Levantine than in the eastern Hellenic Trench. Compared to the distributions found by Roether et al. (2007) during the EMT, the shape of the T–S diagrams changed slightly. During the latter the intermediate maxima were not found in the

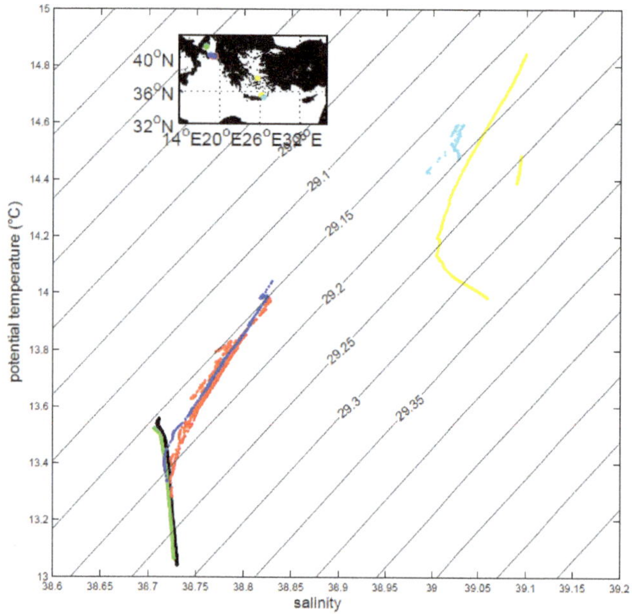

Figure 4. T–S diagram in the areas where dense waters are produced (Aegean and Adriatic Seas). Only values from 600 m to bottom are considered. Yellow dots: Aegean Sea stations of M84-3. Cyan dots: Kasos Strait station of M84-3. Black dots: Adriatic Pit stations of P414. Green dots: Adriatic Pit stations of M84-3 and red dots: Strait of Otranto, stations of P414. Inset panel shows the location of the different stations.

Levantine basin, but T and S increased steadily in the deep layer, after reaching their minimum. Only in the Hellenic Trench were shapes characterized by intermediate minima and maxima found. Also the depth of the inversions changed. While during the EMT the inversions took place in a layer between 1000–1700 m, this feature can now be found in a wider layer.

During the EMT, this kind of shape of the T–S diagram ("peaking") was prominent in the Ionian Sea only in the northern part of the eastern Mediterranean ridge, the principal entrance for water of Aegean origin (Roether et al., 2007). Remarkably, now it can be found in the whole Ionian Sea (Fig. 5b), but it is characterized by a much less pronounced range between minima and maxima due to presumably ongoing mixing. In general, the T–S range between stations is small. The inversions are also less pronounced in the Ionian Sea than in the Levantine and Hellenic Trench. The depths of the inversions lay between 1200 and 1900 m, the corresponding density level being $\sigma 0 = 29.18\ \mathrm{kg\,m^{-3}}$ ($\sigma 2 = 37.88\ \mathrm{kg\,m^{-3}}$), which corresponds to the same density level encountered in the Levantine basin. For the deep stations an additional near-bottom minimum can be found at about 3000 m depth. Salinity is 38.73 in P414, and 38.72 in M84/3 (station 306). Note that salinity increases close to the bottom, while temperature remains constant at 13.42 °C. This minimum corresponds to a branch of old EMDW of Adriatic origin (Bensi et al., 2013a). The increase of salinity further down to the bottom is certainly due to AdDW of very recent origin and leads to the highest density ever observed in the Ionian Sea (Roether et al., 2013).

The stations in the western Hellenic Trench show a more scattered distribution (Fig. 5c). It seems that, especially the P414 stations 362 (black) and 361 (red) and the M84-3 station 300 (cyan) are under the influence of the Antikythera outflow in the upper layers. We note an increase in salinity from April 2011 (M84/3) to June 2011 (P414), which confirms the more direct response of this region to the outflow. In the deepest station (blue) we find a minimum in T and S close to the bottom which is certainly due to advection from the Ionian Sea (Roether et al., 2007).

Since 2005 the deep waters of the WMed experienced significant physical changes (Schroeder et al., 2006; Font et al., 2009). Probably, the major change corresponds to an abrupt increase in the deep heat and salt content. This increase appears as a "hook" in the deep T–S diagrams which arises from the interaction between old deep water, new deep water formed by open-sea convection and dense shelf cascading water (Salat et al., 2009). The data collected during the M84/3 cruise clearly show the presence of this hook-like pattern (Fig. 6b and c). In these data, the minimum is located at a depth of about 1200–1600 m, with temperature values of about 12.87–12.89 °C and salinity values of about 38.46–38.48. The differences between the minimum values and the values at the bottom are small: in T and $S <= 0.01$ °C, psu respectively. The new WMDW is still missing in the Tyrrhenian Sea (Fig. 6a), because of the sill at 1900 m that separates this sub-basin from the rest of the WMed. Here, the T–S diagrams show an almost linear decrease from mid-depths to the bottom. Bottom values are higher ($T \sim 13$ °C, $S \sim 38.5$) than in the "hook" shape case even when depths of more than 1600 m are reached.

3.2 Determination of water mass fractions by OMP

Due to the availability of just three parameters (T, S, O2) and to the fact that dissolved oxygen has a very small weight we had to make several separate OMP runs for the western Mediterranean to calculate the fractions of the four relevant water masses. In particular we determined the relation between AW–LIW–WMDWnew, AW–LIW–WMDWold and LIW–WMDWnew–WMDWold.

In all considered OMP cases (Fig. 7a) AW is encountered as a shallow near-surface lens throughout the whole section. Also the appearance of LIW in the intermediate water column from about 200 to 1000 m with a distinct core at about 500 m in the eastern side of the section (Fig. 7b) can be observed in all OMP cases. All different OMP combinations also reveal WMDW more or less exclusively in the layers beneath 1000 m, with WMDWnew dominating the deeper layers. Only the combination LIW–WMDWnew–WMDWold could resolve the new WMDW from its old version.

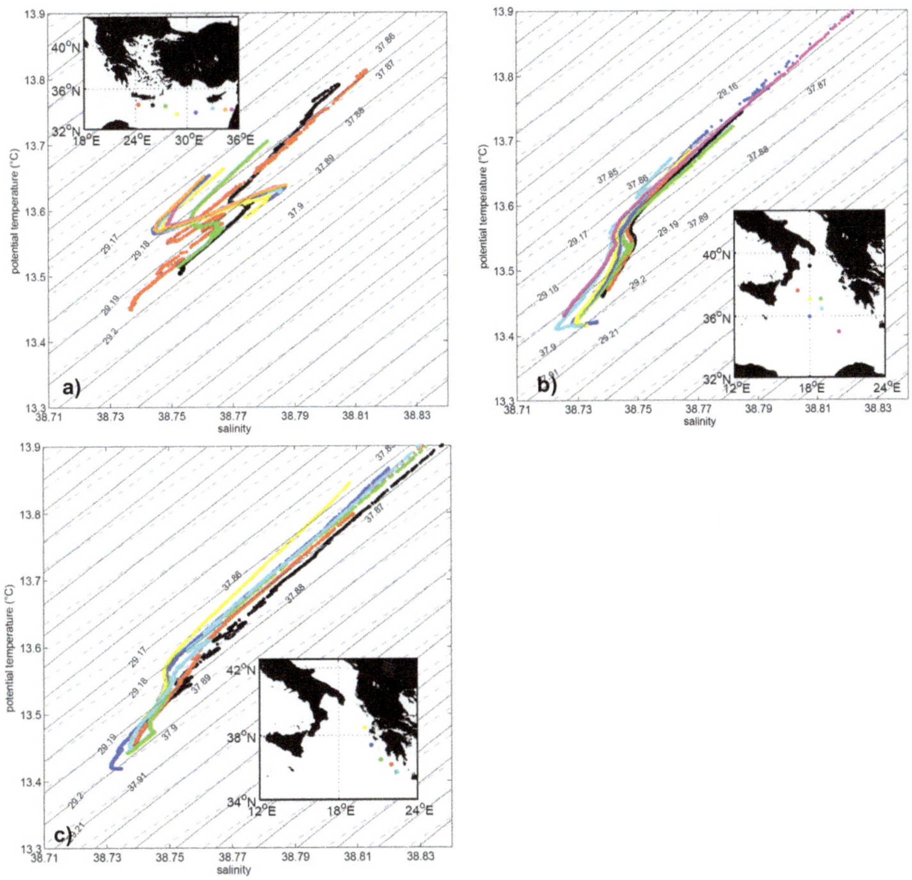

Figure 5. *T–S* diagrams of M84-3 and P414 CTD stations in **(a)** the Levantine basin along the eastern Hellenic Trench and south of the eastern Mediterranean ridge, **(b)** in the Ionian basin (for clarity only selected CTD stations are shown) and **(c)** in the western part of the Hellenic Trench. Values from 800 m depth down to the bottom are taken into account. The inset panels show the location of the different stations.

In the case of the EMed we also considered three parameters for our OMP analyses. Again, to compensate for this limitation we carried out different runs to determine the relative importance of the water masses. One run was focused on the upper 500 m of the water column. Here we combined AW with LSW and LIW. All other runs considered the water column from 500 m to the bottom. We combined LIW with AdDW and CDW, AW with LIW and AdDW. LSW is dominant in the eastern and central part of the section in the upper 200 m (Fig. 8a) while LIW is predominant in the intermediate layers throughout the section forming different cores at depths between 100 and 300 m (Fig. 8b). The different OMP runs for the deep water masses AdDW and CDW produce qualitatively more or less the same results. AdDW fills the basins from 1000 m to the bottom (Fig. 8d), but with decreasing percentages towards the east where CDW contributes to about 30 % of the deep water (Fig. 8c).

Summarizing our results, we can conclude that, for the EMed, the hydrographic situation which we found in the upper depths during the two cruises reveals well known situation with LSW at the surface and LIW in the intermediate layers. The situation beneath is dominated by water masses originating from the Adriatic Sea, which confirms the assumption that after the EMT, and in particular after 2003, AdDW returned to be the main water mass for deep water formation in the EMed (Hainbucher et al., 2006; Rubino and Hainbucher, 2007; Bensi et al., 2013a). The high amount of AdDW which we found even in the Levantine basin in 2011 together with the fact that the turnover times in the EMed are longer than decades queries the overall importance of AdDW for the deep water formation in the EMed over longer timescales. This is supported by a recent study by Roether et al. (2013) that found no evidence of EMT-like events during the observational record.

3.3 Geostrophic and mean ADCP velocities

There are remarkable differences between the geostrophic velocities of the WMed and EMed. Peak velocities (maximum and minimum) are distinctly higher in the WMed than in the EMed, but the structures do not reach as deep and are not as wide as in the EMed. In the former (Fig. 9a)

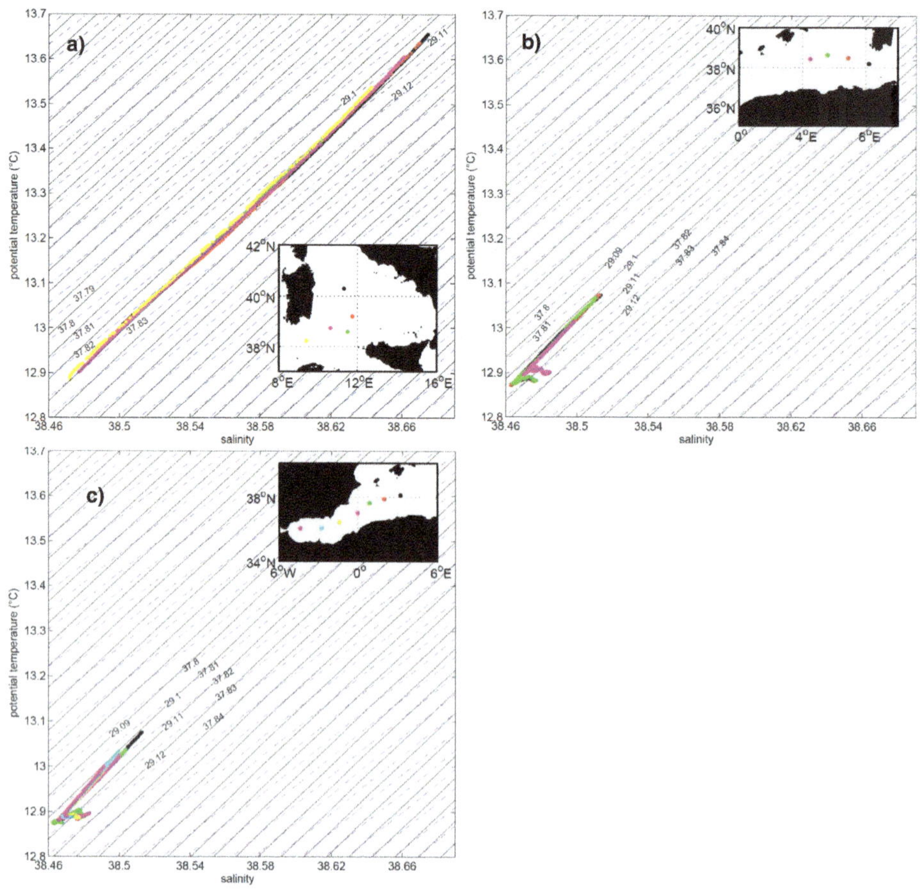

Figure 6. T–S diagrams of M84/3 stations in **(a)** the Tyrrhenian Sea, **(b)** the western Mediterranean Sea between the Balearic Islands and Sardinia and **(c)** the western Mediterranean Sea between Gibraltar and the Balearic Islands. Values from 800 m depth down to the bottom are taken into account. The inset panels show the location of the different stations.

the features reach down to about 200 m depth, are of the order of 100 km width, and reach maximum amplitudes of 12–15 cm s^{-1}. Differently, maximum velocities in the EMed are 5 cm s^{-1} and the features widths are about 300–400 km (Fig. 10a). These structures might be related to sub-basin-scale gyres, this assumption being supported by the range of the features and the changing appearance of positive and negative amplitudes. In the EMed the velocity feature is correlated with the Mersa-Matruh anticyclone (Tanhua et al., 2013b, this issue Fig. 2).

In order to compare the ADCP velocities with the results of the geostrophic calculations, the ADCP data collected between two consecutive CTD stations were averaged. The along-track velocities were determined and the velocities across the sections are presented in Figs. 9b and 10b. In the EMed, due to data gaps and also to the fact that often the track between CTD stations was not direct, only seven significant CTD stations could be taken into account for applying the method.

Altogether, large velocity amplitudes are achieved mainly in the upper 500 m of the water column in both basins. Beneath this water depth current speeds are around 1–2 cm s^{-1}. For the WMed, geostrophic velocities and mean ADCP velocities agree well. Structures and amplitudes are similar and the ageostrophic fraction (Fig. 9c) determined by subtracting the geostrophic velocity from the ADCP velocity is comparably low. Based on this result, we conjecture that during the measurements large-scale velocities in the upper part of the ocean were mainly determined by geostrophy in the WMed, while in the EMed the situation was opposite. Again, the largest velocities were reached in the area of the Mersa-Matruh anticyclonic gyre. These far exceed the calculated geostrophic velocity with maximum amplitudes of more than 12 cm s^{-1} (Fig. 10c), indicating that the ageostrophic fraction in the EMed circulation areas exerted a larger role in shaping the observed velocity distribution.

4 Summary and conclusion

In the EMed the effects induced by the EMT on the deep layers (increase in temperature and salinity due to the massive production of CDW) were still visible during our cruises

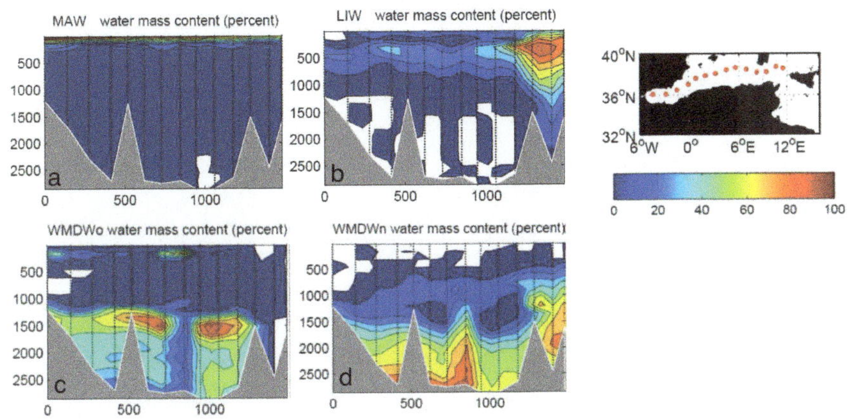

Figure 7. Water mass fractions in percent for the western Mediterranean Sea. Panels **(a, b)** show Modified Atlantic Water and Levantine Intermediate Water for the OMP case AW–LIW–WMDW$_n$. Panel **(c)** shows old Western Mediterranean Deep Water for the case AW–WMDW$_n$–WMDW$_o$ and panel **(d)** shows new Western Mediterranean Deep Water for the case LIW–WMDW$_n$–WMDW$_o$. The y axes indicate depth in dbar and x axes specify distance in km. The right panel shows the location of considered stations. White patches are areas for which the OMP method is not valid.

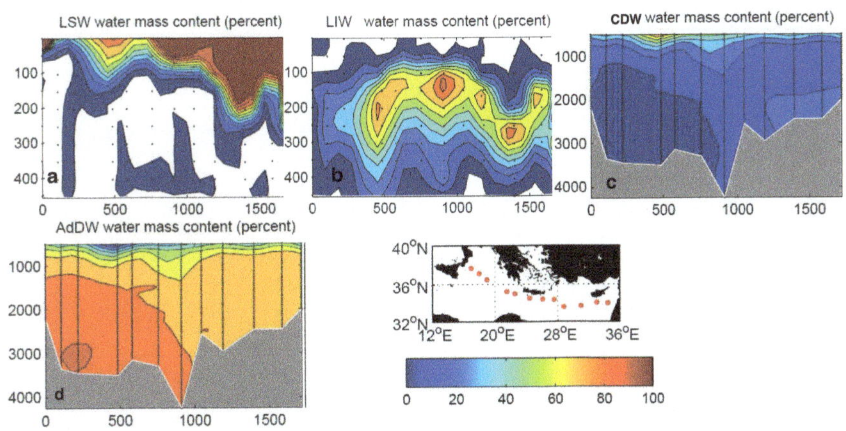

Figure 8. Water mass fractions in percent for the eastern Mediterranean Sea. **(a)** Levantine Surface Water from surface to 500 m. **(b)** Levantine Intermediate Water from surface to 500 m. Panels **(a, b)** result from the OMP run which combines AW, LSW and LIW in the upper 500 m. **(c)** Cretan Deep Water from 500 m to bottom. **(d)** Adriatic Deep Water from 500 m to bottom. Panels **(c, d)** result from an OMP combination of AW, AdDW and CDW from 500 m down to the bottom. Lower right panel: location of section and colour scale for percentage. Left axis of panels: depth in dbar; lower axis: distance in km. White patches are areas for which the OMP method is not valid.

(2011). T–S diagrams show that the AdDW returned to represent the main source for the EMDW. However, it became much saltier and warmer than before the EMT. Our analyses show that this water mass represents also the dominating water filling the deeper basins of the EMed, in the Levantine also CDW added during the peak years of the EMT is still present (Roether et al., 2013). CDW of more recent outflows is limited to a shallower layer (approximately around 2000 m) in the proximity of Kasos Strait. In the upper layer of the EMed, LIW is prevailing in the intermediate layers, while AW dominates in the western part of the EMed and LSW in the eastern regions. The velocity distribution derived by the ADCP measurements and geostrophic calculations indicates that in the EMed the ageostrophic component of the velocity is stronger than in the WMed. Large velocity amplitudes ($> 10\,\mathrm{cm\,s^{-1}}$) can only be found in the upper 500 m of the water column.

In the WMed the stratification in the deep layers is dominated by the "hook" shape of the T–S distribution, indicating a transition in bottom waters between old and new WMDW. One exception is represented by the stratification found in the Tyrrhenian Sea, where temperature and salinity did not reach the range where the "hook" typically occurs. Here, temperature and salinity are distinctly higher than further west and are influenced by strong downward mixing of Sicily Strait overflow (Gasparini et al., 2005; Roether and Lupton, 2011). In the deep layers of the rest of the WMed, WMDW is filling the basin, but from our results we cannot discern whether new or old WMDW dominates. In the upper layers, AW is predominant in the surface layers, whereas LIW is filling

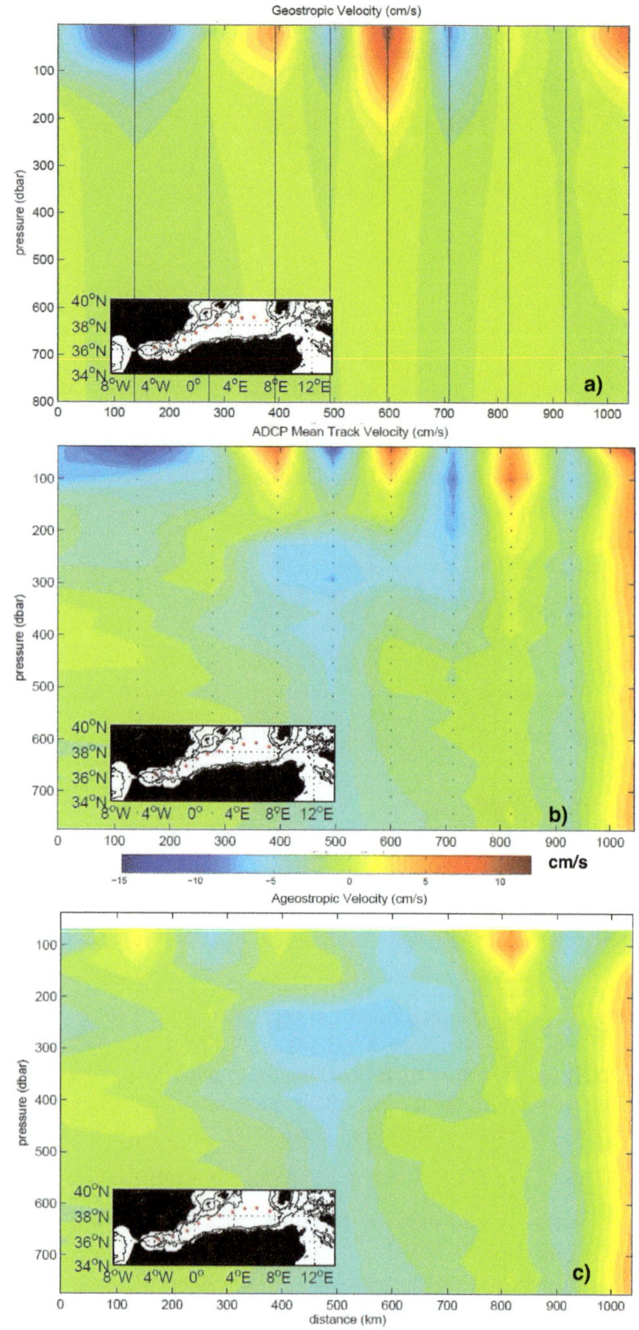

Figure 9. Velocity distribution on a section through the western Mediterranean Sea. The inset panels show the location and nodes of the section. Left axis of panels: depth in dbar; lower axis: distance in km. **(a)** Geostrophic velocity determined between the CTD stations. **(b)** Mean ADCP velocity across the section between adjoining CTD stations. Means are calculated between two CTD stations. **(c)** Ageostrophic velocity determined by subtracting the geostrophic velocity from the mean ADCP velocity. Reference layer of no motion for geostrophic velocities is 1000 m.

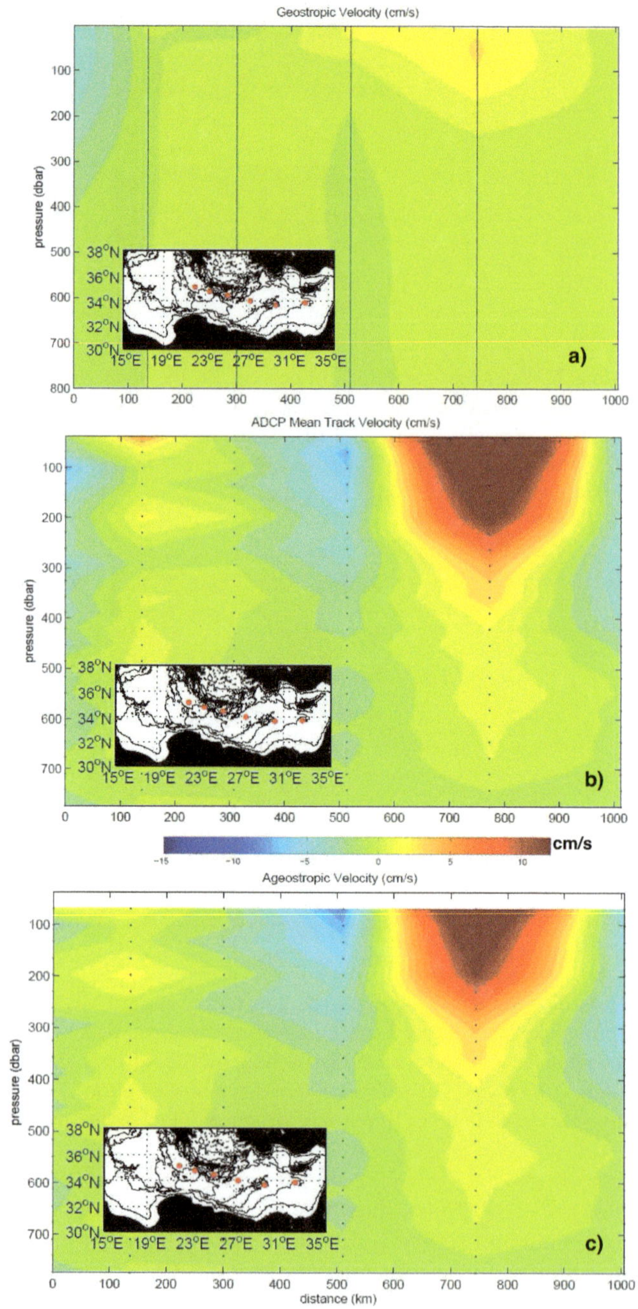

Figure 10. Velocity distribution on a section through the eastern Mediterranean Sea. The inset panels show the location and nodes of the section. Left axis of panels: depth in dbar; lower axis: distance in km. **(a)** Geostrophic velocity determined between the CTD stations. **(b)** Mean ADCP velocity across the section between adjoining CTD stations. Mean is the average between two CTD stations. **(c)** Ageostrophic velocity determined by subtracting the geostrophic velocity from the mean ADCP velocity. Reference layer of no motion for geostrophic velocities is 1000 m.

the intermediate depth levels with a core in the eastern part of the WMed. In contrast to the EMed, the velocity here is more geostrophically balanced. In both basins the observed structures are related to sub-basin-scale gyres – in the EMed we even can correlate the velocity feature we found with the Mersa-Matruh anticyclone.

The salinity range in the WMed is larger than in the EMed, certainly due to input of fresher water of Atlantic origin. On the other hand, the temperature range in both regions is of the same order. Bottom densities of the WMed are lower than in the EMed and are determined by lower salinity. In contrast, LIW is lighter in the EMed and the mean depth at which LIW flows deepens from east to west.

Overall, we can conclude that the hydrographic property distribution which we found during April–June 2011 in the western and eastern Mediterranean Sea characterizes a situation which is far from that observed before the EMT. The picture which emerges adds further information toward a deeper understanding of long-term evolution of hydrography and large-scale circulation of the Mediterranean Sea.

Acknowledgements. The authors want to thank the captains and crews on the research vessels RV *Meteor* and RV *Poseidon* for the excellent cooperation during the campaigns. Thanks go to all scientists involved in the field campaigns. Special thanks go to Norbert Verch for processing and calibrating the hydrographic data of both cruises.

The *Meteor* cruise M84/3 was supported by a grant from the Deutsche Forschungsgemenschaft – Senatskommission für Ozeanographie (DFG), and from a grant from the DFG; TA 317/3-1. The *Poseidon* cruise 414 was supported by funding from the University of Hamburg.

Edited by: N. Kress

References

Bensi, M., Rubino, A., Cardin, V., Hainbucher, D., and Mancero-Mosquera, I.: Structure and variability of the abyssal water masses in the Ionian Sea in the period 2003–2010, J. Geophys. Res., 118, 1–13, doi:10.1029/2012JC008178, 2013a.

Bensi, M., Cardin, V., Rubino, A., Notarstefano, G., and Poulain, P. M.: Effects of winter convection on the deep layer of the Southern Adriatic Sea in 2012, J. Geophys. Res.-Ocean, 118, 6064–6075, doi:10.1002/2013JC009432, 2013b.

Bethoux, J. P., Gentili, B., Raunet, J., and Taillez, D.: Warming trend in the Western Mediterranean deep water, Nature, 347, 660–662, 1990.

Borzelli, G. L. E., Gačic, M., Cardin, V., and Civitarese, G.: Eastern Mediterranean Transient and reversal of the Ionian Sea circulation, Geophys. Res. Lett., 36, L15108, doi:10.1029/2009GL039261, 2009.

Cardin, V., Bensi, M., and Pacciaroni, M.: Variability of water mass properties in the last two decades in the South Adriatic Sea with emphasis on the period 2006–2009, Cont. Shelf Res., 31, 951–965, 2011.

Cardin, V., Civitarese, G., Hainbucher, D., Bensi, M., and Rubino, A.: Thermohaline properties in the Eastern Mediterranean in the last three decades: is the basin returning to the pre-EMT situation?, Ocean Sci. Discuss., 11, 391–423, doi:10.5194/osd-11-391-2014, 2014.

Font, J., Béranger, K., Bryden, H., Budillon, G., Fuda, J. L., Gaèiæ, M., Gascard, J. C., Packard, T., Puig, P., Roether, W., Salat, J., Salusti, E., Schroeer, K., Theocharis, A., and van Haren, H.: Executive summary of CIESM workshop 38 "Dynamics of Mediterranean deep waters", in: CIESM, 2009, Dynamics of Mediterranean Deep Waters, N_38 in CIESM Workshop Monographs, edited by: Briand, F., Monaco, 132 pp., 2009.

Gačic, M., Borzelli, G. L. E., Civitarese, G., Cardin, V., and Yari, S.: Can internal processes sustain reversals of the ocean upper circulation? The Ionian Sea example, Geophys. Res. Lett., 37, L09608, doi:10.1029/2010GL043216, 2010.

Gasparini, G. P., Ortona, A., Budillon, G., Astraldi, M., and Sansone, E.: The effect of the Eastern Mediterranean Transient on the hydrographic characteristics in the Strait of Sicily and in the Tyrrhenian Sea, Deep Sea Res. Pt. I, 52, 915–936, 2005.

Gertman, I., Pinardi, N., Popov, Y., and Hecht, A.: Aegean Sea water masses during the early stages of the Eastern Mediterranean Climatic Transient (1988–1990), J. Phys. Oceanogr., 36, 1841–1859, 2006.

Hainbucher, D.: Cruise Report RV POSEIDON Cruise P414, available at: http://www.geomar.de/fileadmin/content/zentrum/ze/fs/Poseidon_Berichte_2011_PDF/POS414_Hainbucher.pdf (last access: 12 December 2013), 2012.

Hainbucher, D., Rubino, A., and Klein, B.: Water mass characteristics in the deep layers of the western Ionian Basin observed during May 2003, Geophys. Res. Lett., 33, L05608, doi:10.1029/2005GL025318, 2006.

Klein, B., Roether, W., Manca, B. B., Bregant, D., Beitzel, V., Kovacevic, V., and Luchetta, A.: The large deep water transient in the Eastern Mediterranean, Deep-Sea Res. Pt. I, 46, 371–414, 1999.

Kovačević, V., Manca, B. B., Ursella, L., Schroeder, K., Cozzi, S., Burca, M., Mauri, E., Gerin, R., Notarstefano, G., and Deponte, D.: Water mass properties and dynamic conditions of the Eastern Mediterranean in June 2007, Prog. Oceanogr., 104, 59–79, 2012.

Lascaratos, A.: Estimation of deep and intermediate water mass formation rates in the Mediterranean Sea, Deep-Sea Res. Pt. II, 40, 1327–1332, 1993.

Lascaratos, A., Roether, W., Nittis, K., and Klein, B.: Recent changes in deep water formation and spreading in the Eastern Mediterranean Sea: a review, Progr. Oceanogr., 44, 5–36, 1999.

Leaman, K. D. and Schott, F.: Hydrographic structure of the convection regime in the Golfe du Lion, J. Phys. Oceanogr., 23, 575–598, 1991.

López-Jurado, J.-L., González-Pola, C., and Vélez-Belchi, P.: Observation of an abrupt disruption of the long-term warming trend at the Balearic Sea, western Mediterranean Sea, in summer 2005, Geophys. Res. Lett., 32, L24606, doi:10.1029/2005GL024430, 2005.

Malanotte-Rizzoli, P. and Hecht, A.: Large-scale properties of the eastern Mediterranean: a review, Oceanol. Acta, 11, 323–335, 1988.

Malanotte-Rizzoli, P., Manca, B. B., Ribera d'Alcala, M., Teocharis, A., Bergamasco, A., Bregant, D., Budillon, G., Civitarese, G., Georgopoulos, D., Michelato, A., Sansone, E., Scarazzato, P., and Souvermezoglou, E.: A synthesis of the Ionian Sea hydrography, circulation and water mass pathways during POEM-Phase I, Prog. Oceanogr., 39, 153–204, 1997.

Malanotte-Rizzoli, P., Manca, B. B., Ribera d'Alcala, M., Theocharis, A., Brenner, A., Budillon, G., and Ozsoy, E.: The Eastern Mediterranean in the 80s and in the 90s: the big transition in the intermediate and deep circulations, Dynam. Atmos. Oceans, 29, 365–395, 1999.

Millot, C.: Circulation in the Western Mediterranean Sea, J. Marine Syst., 20, 423–444, 1999.

Moutin, T., Van Wambeke, F., and Prieur, L.: Introduction to the Biogeochemistry from the Oligotrophic to the Ultraoligotrophic Mediterranean (BOUM) experiment, Biogeosciences, 9, 3817–3825, doi:10.5194/bg-9-3817-2012, 2012.

Molcard, A., Gervasio, L., Griffa, A., Gasparini, G. P., Mortier, L., and Özgökmen, T. M.: Numerical investigation of the Sicily Channel dynamics: density currents and water mass advection, J. Marine Syst., 36, 219–238, 2002.

Ovchinnikov, I. M.: Intermediate Water formation in the Mediterranean Sea, Okeanologiya, 24, 217–225, 1984.

Özturgut, E.: The Source and Spreading of the Levantine Intermediate Water in the Eastern Mediteranean, Saclant ASW Research Center Memorandum Sm-92, La Spezia, Italy, 45 pp. 1976.

POEM Group: General circulation of the eastern Mediterranean, Earth-Sci. Rev., 32, 285–309, 1992.

Pierini, S. and Rubino, A.: Modeling the oceanic circulation in the area of the Strait of Sicily: the remotely forced dynamics, J. Phys. Oceanogr., 31, 1397–1412, 2001.

Robinson, A. R., Leslie, W. G., Theocharis, A., and Lascaratos, A.: Mediterranean Sea Circulation Encyclopedia of Ocean Sciences, Academic Press, 1689–1706, 2001.

Roether, W. and Lupton, J. E.: Tracers confirm downward mixing of Tyrrhenian Sea upper waters associated with the Eastern Mediterranean Transient, Ocean Sci., 7, 91–99, doi:10.5194/os-7-91-2011, 2011.

Roether, W., Manca, B. B., Klein, B., Bregant, D., Georgopoulos, D., Beitzel, V., Kovacevic, V., and Lucchetta, A.: Recent changes in Eastern Mediterranean deep waters, Science, 271, 333–335, 1996.

Roether, W., Klein, B., Beitzel, V., and Manca, B. B.: Property distributions and transient-tracer ages in Levantine Intermediate water in the Eastern Mediterranean, J. Marine Syst., 18, 71–87, 1998.

Roether, W., Klein, B., Manca, B. B., Theocharis, A., and Kioroglou, S.: Transient Eastern Mediterranean deep waters in response to the massive dense-water output of the Aegean Sea in the 1990s, Prog. Oceanogr., 74, 540–571, 2007.

Roether, W., Klein, B., and Hainbucher, D.: The Eastern Mediterranean transient: evidence for similar events previously?, in: The Mediterranean Sea: Temporal Variability and Spatial Patterns, edited by: Borzelli, G. L. E., AGU monographs, 2013.

Rubino, A. and Hainbucher, D.: A large abrupt change in the abyssal water masses of the eastern Mediterranean, Geophys. Res. Lett., 34, L23607, doi:10.1029/2007GL031737, 2007.

Salat, J., Emelianov, M., and Puig, P.: From bottom water (Lacombe, 1985) to New-WMDW since 2005, Possible shifts on open sea deep convection, in: CIESM, 2009, Dynamics of Mediterranean Deep Waters, N_38 in CIESM Workshop Monographs, edited by: Briand, F., Monaco, 132 pp., 2009.

Schlitzer, R., Roether, W., Oster, H., Junghans, H.-G., Hausmann, M., and Michelato, A.: Chlorofluoromethane and oxygen in the eastern Mediterrancah, Deep-Sea Res., 38, 1531–1551, 1991.

Schneider, A., Tanhua, T., Roether, W., and Steinfeldt, R.: Changes in ventilation of the Mediterranean Sea during the past 25 year, Ocean Sci., 10, 1–16, doi:10.5194/os-10-1-2014, 2014.

Schroeder, K., Gasparini, G. P., Tangherlini, M., and Astraldi, M.: Deep and intermediate water in the Western Mediterranean under the influence of the Eastern Mediterranean Transient, Geophys. Res. Lett., 33, L21607, doi:10.1029/2006GL027121, 2006.

Schroeder, K., Ribotti, A., Borghini, M., Sorgente, R., Perilli, A., and Gasparini, G. P.: An extensive western Mediterranean deep water renewal between 2004 and 2006, Geophys. Res. Lett., 35, LI8605, doi:10.1029/2008GL035146, 2008.

Schroeder, K., Schroeder, K., Garcìa-Lafuente, J., Josey, S. A., Artale, V., Buongiorno Nardelli, B., Carrillo, A., Gacic, M., Gasparini, G. P., Herrmann, M., Lionello, P., Ludwig, W., Millot, C., Özsoy, E., Pisacane, G., Sánchez-Garrido, J. C., Sannino, G., Santoleri, R., Somot, S., Struglia, M., Stanev, E., Taupier-Letage, I., Tsimplis, M. N., Vargas-Yáñez, M., Zervakis, V., and Zodiatis, G.: Circulation of the Mediterranean Sea and its variability, in: The Climate of the Mediterranean Region: From the Past to the Future, edited by: Lionello, P., Elsevier Insights, Amsterdam, 2012.

Tanhua, T., Hainbucher, D., Cardin, V., Álvarez, M., Civitarese, G., McNichol, A. P., and Key, R. M.: Repeat hydrography in the Mediterranean Sea, data from the *Meteor* cruise 84/3 in 2011, Earth Syst. Sci. Data, 5, 289–294, doi:10.5194/essd-5-289-2013, 2013a.

Tanhua, T., Hainbucher, D., Schroeder, K., Cardin, V., Álvarez, M., and Civitarese, G.: The Mediterranean Sea system: a review and an introduction to the special issue, Ocean Sci., 9, 789–803, doi:10.5194/os-9-789-2013, 2013b.

Theocharis, A., Nittis, K., Kontoyiannis, H., Papageorgiou, E., and Balopoulos, E.: Climatic changes in the Aegean Sea influence the Eastern Mediterranean thermohaline circulation (1986–1997), Geophys. Res. Lett., 26, 1617–1620, 1999.

Tomczak, M.: A multiparameter extension of temperature/salinity diagram techniques for the analysis of non-isopycnal mixing, Prog. Oceanogr., 10, 147–171, 1981.

Tomczak, M. and Large, D. G. B.: Optimum multiparameter analysis of mixing in the thermocline of the Eastern Indian Ocean, J. Geophys. Res., 94, 16141–16149, 1989.

18

Heat loss from the Atlantic water layer in the northern Kara Sea: causes and consequences

I. A. Dmitrenko[1], S. A. Kirillov[1], N. Serra[2], N. V. Koldunov[2], V. V. Ivanov[3,4], U. Schauer[5], I. V. Polyakov[4], D. Barber[1], M. Janout[5], V. S. Lien[6], M. Makhotin[3], and Y. Aksenov[7]

[1]Centre for Earth Observation Science, University of Manitoba, Winnipeg, Canada
[2]Institute of Oceanography, University of Hamburg, Hamburg, Germany
[3]Arctic and Antarctic Research Institute, St. Petersburg, Russia
[4]International Arctic Research Center, University of Alaska Fairbanks, Fairbanks, AK, USA
[5]Alfred Wegener Institute for Polar and Marine Research, Bremerhaven, Germany
[6]Institute of Marine Research, Bergen, Norway
[7]National Oceanography Centre Southampton, Southampton, UK

Dedicated to the memory of our colleague Klaus Hochheim, who tragically lost his life in the Arctic expedition in September 2013

Correspondence to: I. A. Dmitrenko (igor.dmitrenko@umanitoba.ca)

Abstract. A distinct, subsurface density front along the eastern St. Anna Trough in the northern Kara Sea is inferred from hydrographic observations in 1996 and 2008–2010. Direct velocity measurements show a persistent northward subsurface current ($\sim 18\,\mathrm{cm\,s^{-1}}$) along the St. Anna Trough eastern flank. This sheared flow, carrying the outflow from the Barents and Kara seas to the Arctic Ocean, is also evident from shipboard observations as well as from geostrophic velocities and numerical model simulations. Although we cannot substantiate our conclusions by direct observation-based estimates of mixing rates in the area, we hypothesize that the enhanced vertical mixing along the St. Anna Trough eastern flank favors the upward heat loss from the intermediate warm Atlantic water layer. Modeling results support this hypothesis. The upward heat flux inferred from hydrographic data and model simulations is of $O(30–100)\,\mathrm{W\,m^{-2}}$. The region of lowered sea ice thickness and concentration seen both in sea ice remote sensing observations and model simulations marks the Atlantic water pathway in the St. Anna Trough and adjacent Nansen Basin continental margin. In fact, the sea ice shows a delayed freeze-up onset during fall and a reduction in the sea ice thickness during winter. This is consistent with our results on the enhanced Atlantic water heat loss along the Atlantic water pathway in the St. Anna Trough.

1 Introduction

The warm Atlantic water (AW) plays a role in reducing the sea-ice cover in the Northern Hemisphere through upward heat loss (Polyakov et al., 2010). The AW heat loss impacts the ice cover along the Arctic Ocean Eurasian continental margins (Polyakov et al., 2010), but is significantly stronger over the regions of the AW inflow into the Arctic Ocean through the Fram Strait (the Fram Strait branch of AW inflow – FSBW) and in the Barents Sea (the Barents Sea branch of AW inflow – BSBW). These are areas where the inflowing AW with temperatures $T \sim 5\,°\mathrm{C}$ occupies the surface layer, directly affecting the ocean–sea-ice–atmosphere interface (Sirevaag and Fer, 2009; Årthun et al., 2012; Ivanov et al., 2012). Over the last decades, observations indicate an increase of AW transport into the Arctic through the Barents Sea (e.g., Skagseth et al., 2008), which impacts the observed sea ice reduction (e.g., Schlichtholz, 2011; Årthun et al., 2012). The AW transport into the Arctic through the Fram Strait shows similar patterns and impacts (e.g., Ivanov et al., 2012; Alexeev et al., 2013). Model simulations of future projections show analogous regularities (e.g., Koenigk and Brodeau, 2013).

While transiting along the continental slope of Svalbard (red arrows in Fig. 1) and through the Barents Sea (blue arrows in Fig. 1), the AW gradually cools and deepens, becoming isolated from the surface. In the area of the St. Anna Trough (hereinafter ST) in the northern Kara Sea, the submerged BSBW with temperatures slightly above 0 °C merges with a fraction of the warm intermediate FSBW ($T \sim 2.0$–2.5 °C), which enters the western ST steered by the topography (Hanzlick and Aagaard, 1980; Schauer et al., 2002a) – Figs. 1 and 2. Hereafter we refer to this fraction of the FSBW as ST-circulating FSBW (SFSBW).

The SFSBW return flow to the Arctic Ocean along the ST eastern slope is colder and less saline than the FSBW (Schauer et al., 2002b), which indicates mixing and heat loss along the SFSBW pathway in the ST, as suggested by Hanzlick and Aagaard (1980). This allows speculations that a fraction of the SFSBW heat is released upwards with potential implication for the ocean–sea-ice–atmosphere interface. This paper addresses the issue of SFSBW modification in the ST with a special focus on water dynamics and mixing occurring along the ST eastern flank. We suggest that vertical mixing promoted by the interaction of the SFSBW with the Barents and Kara seas outflow can favor an enhanced AW heat loss, resulting in a consistent delay of freeze-up onset in fall, sea ice thinning in winter, and earlier spring retreat.

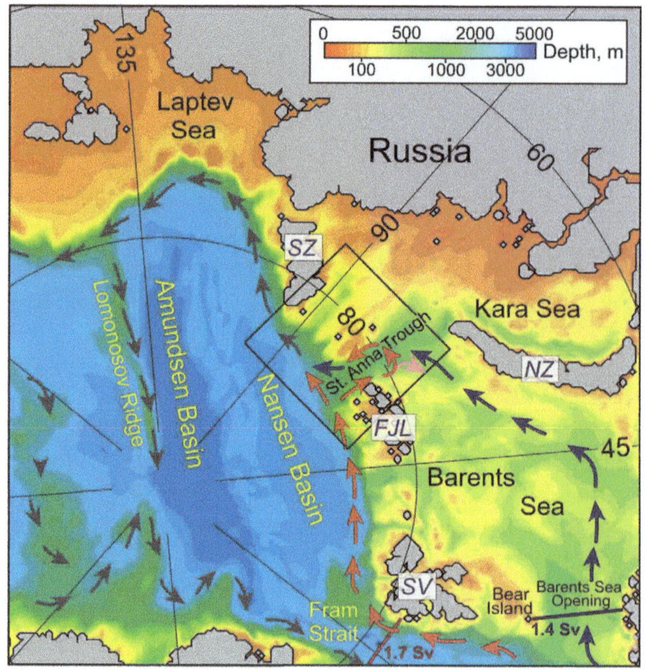

Figure 1. Map of the inflow region of Atlantic water (AW) to the Arctic Ocean. Arrows trace the Atlantic water (AW) pathways; red and blue arrows show the Fram Strait and Barents Sea branches, respectively, of the AW inflow into the Arctic Ocean in accordance with Rudels et al. (1994). The pink arrow shows a fraction of the Fram Strait branch water flowing west along the shallower trough between Novaya Zemlya and Franz Josef Land. Dark red and blue numbered lines in Fram Strait and Barents Sea Opening, respectively, show sections used for verifying the simulated volume transport of AW. Dark blue number indicates the simulated volume transport of AW (defined as $T > 3$ °C) through the Barents Sea Opening. Dark red number shows the simulated volume transport of AW (defined as $T > 2$ °C) in the core of the West Spitsbergen Current. The black rectangle encloses the region enlarged in Fig. 2. SV – Svalbard, FJL – Franz Josef Land, NZ – Novaya Zemlya, SZ – Severnaya Zemlya.

2 Data and methods

Oceanographic CTD transects across the ST (Fig. 2) were taken from the icebreaker *Kapitan Dranitsyn* along $\sim 81°$ N (24 October 2008) and along ~ 81 and 82° N (22–23 August 2009 and 2–3 September 2009, respectively), and from the RV *Nikolay Evgenov* along $\sim 81°$ N (22 September 2010). These sections were sampled using a shipboard SBE19 + CTD (in 2009–2010) and Lockheed Martin Sippican Expendable Bathythermographs – XBTs and Expendable Conductivity, Temperature, and Depth Profilers – XCTDs (in 2008). These data were complemented by oceanographic stations occupied across the ST along $\sim 82°$ N in July 1996 during the ARKXII cruise of the RV *Polarstern* (Schauer et al., 2002b). In addition, we used the velocity data collected from a conventional mooring equipped with Teledyne RD Instruments 300 kHz Workhorse Sentinel Acoustic Doppler Current Profilers (ADCPs) measuring velocity through the depth range of 134–218 m and 376–468 m from 24 August 2009 to 22 September 2010. The mooring was deployed at the eastern slope of the ST at 81°01′ N, 73°02′ E (Fig. 2) in ~ 520 m water depth.

We use gridded satellite data of ice concentration (from the Integrated Climate Data Center at the University of Hamburg, Germany), ice thickness (from the Jet Propulsion Laboratory, USA), and ice drift (from the Center IFREMER of Brest, France). The sea ice concentration is derived from the Advanced Microwave Scanning Radiometer for EOS – AMSR-E (2005–2010) and Special Sensor Microwave Imager/Sounder – SSMIS (2011) (Spreen et al., 2008). The sea ice thickness is from the Ice, Cloud, and land Elevation Satellite (ICESat) (Kwok et al., 2009). The sea-ice drift is derived from the 89 GHz brightness temperature of the AMSR (Ezraty et al., 2007). The spatial grid resolution for ice concentration, thickness and drift is 6.25, 25 and 31.25 km, respectively. According to Kwok and Cunningham (2008), the uncertainty of the ICESat sea ice thickness is ~ 0.5 m based on the assessment with data from submarines and upward-looking sonars. Comparisons of the satellite derived sea ice drift data with buoys for a 3-day period show almost no bias; however, the standard deviation for the ice drift velocity and direction is 6.7 km and 35°, respectively (Ezraty et al., 2007). Although the uncertainty of individual scans of satellite measurements is large, the spatial and temporal

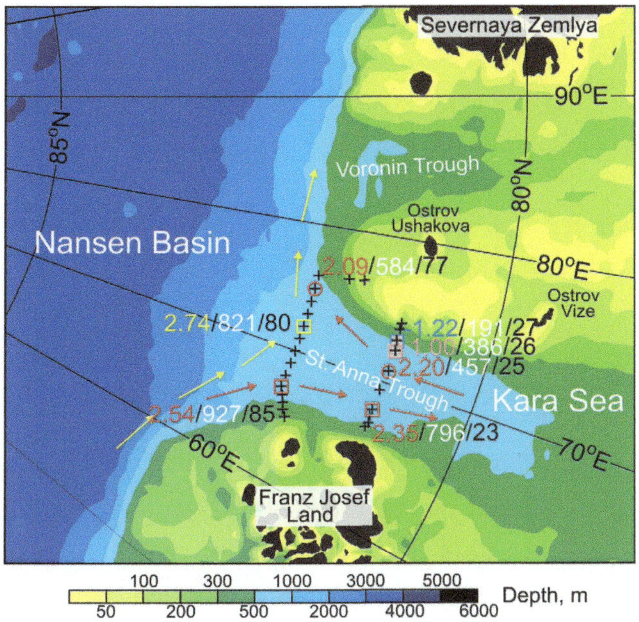

Figure 2. Map of the northern Kara Sea showing the St. Anna Trough (ST). Arrows show the Fram Strait branch of the AW inflow into the Arctic Ocean, which recirculates in the ST (red arrows, SFSBW) and follows the continental margin (yellow arrows, FSBW). Crosses depict the positions of CTD stations taken in September 2009 at two sections crossing the ST at ~ 81 and 82° N. Red squares and circles identify stations taken through the core of the SFSBW inflow and outflow to/from the ST, respectively. The yellow square identifies a station taken through the core of the FSBW boundary current. The pink square depicts the mooring position. The pink and blue squares with gray shading identify stations used for estimates of uncertainty in the vertical heat flux due to spatial undersampling of the AW jet. The first (yellow/red) number shows the FSBW/SFSBW core temperature (in °C) in September 2009. The first pink and blue numbers show temperature at 110 m. The second (white) number is heat content (in MJ m^{-2}) computed relative to the freezing temperature between 30 to 90 m depth. The third (black) number denotes the station number.

averaging reduces the associated errors down to about 5 cm and 0.07 km day^{-1} for the sea thickness and drift, respectively. The surface air temperature for the ST area used here is from the National Centers for Environmental Prediction (NCEP) reanalysis (Kalnay et al., 1996).

In this study we further use ocean velocity, sea ice thickness and concentration, and vertical heat fluxes simulated with the coupled ocean–sea-ice MIT general circulation model (Marshall et al., 1997; Adcroft et al., 2004). The model covers the Atlantic region north of 30° S including the Nordic seas and the Arctic Ocean. It has 50 levels in the vertical, with resolution varying from 10 m in the upper ocean to 550 m in the deep ocean. It was initialized from rest and with the annual mean temperature and salinity from the World Ocean Atlas (WOA) 2005 monthly climatology (Boyer et al., 2005). Bottom topography was interpolated from the ETOPO2 database (Smith and Sandwell, 1997).

The model was forced at the surface by fluxes of momentum, heat and freshwater computed using bulk formulae and the 1948–2010 6-hourly atmospheric state from the NCEP RA1 reanalysis (Kalnay et al., 1996). At the southern open boundary the model was forced by a 1° resolution global model solution, and at Bering Strait a barotropic net inflow of 0.9 Sv was prescribed. The model sea surface salinity was relaxed to the WOA2005 monthly climatology to prevent long-term drifts. In the area of interest, the Barents and Kara seas including the ST, the horizontal resolution is 7 km.

The vertical mixing parameterization follows the K-profile parameterization (KPP) formulation by Large et al. (1994). Background coefficients of vertical diffusion and viscosity are both $10^{-5}\,m^2\,s^{-1}$. Biharmonic diffusion and viscosity represent unresolved eddy mixing, both with coefficients of $5 \times 10^9\,m^4\,s^{-1}$. Further details on the model setup and an application can be found in Serra et al. (2010).

The model is capable of realistically reproducing the AW inflow into the Arctic Ocean through the Barents Sea Opening (BSO) and Fram Strait. The vertical structure of temperature and velocity at BSO (Fig. 3a) is in overall agreement with that shown in Skagseth (2008). The AW and the Bear Island Trough outflow temperatures (Fig. 3c) are also consistent with observations. Furthermore, the simulated volume transport of AW (defined as $T > 3$ °C) of about 1.7 Sv (Figs. 1 and 3e) agrees well with the 1.8 Sv value given in Skagseth et al. (2008).

In the Fram Strait, the model performs reasonably well in terms of the vertical structure of mean temperature and velocity (Fig. 3b). The West Spitsbergen Current (WSC) has a realistic structure, and the temperature of the AW (defined as $T > 2$ °C – Fig. 3d) is in agreement with the time series presented in Beszczynska-Moeller et al. (2012). The long-term mean (1997–2009) simulated volume transport of AW in the core of the WSC at Fram Strait of about 1.4 Sv is directly comparable with the 1.3 Sv presented in Beszczynska-Moeller et al. (2012) based on measurements by an array of moorings in Fram Strait over the period 1998–2010 (Figs. 1 and 3f). The maximum mean (2000–2009) temperature in 150–800 m shows that there is a well-defined inflow of AW, which turns to the right after entering the Fram Strait and gradually loses heat on the way (Fig. 4).

The model's ability to adequately simulate the AW dynamics, and thus to elaborate the relationships between the AW dynamics and ice conditions over the ST region, is briefly discussed elsewhere (Dmitrenko et al., 2012). They also reported on a fair model capability in reproducing the basic hydrography for the downstream area over the Laptev Sea continental margin.

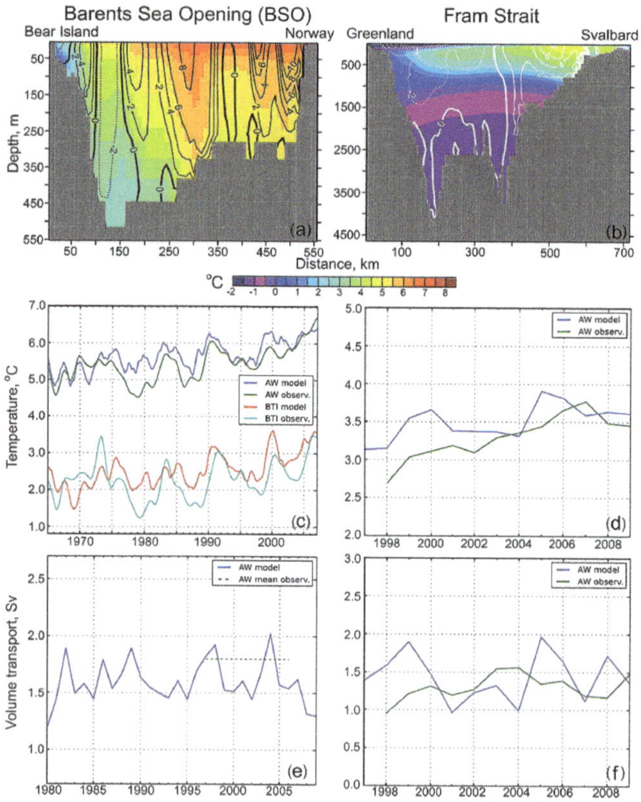

Figure 3. Simulated and observed characteristics of the AW inflow into the Arctic Ocean through the (left) Barents Sea Opening from Bear Island to Norway (BSO) and (right) Fram Strait. The first row shows the simulated mean temperature (in color) and velocity (contour lines) in **(a)** the Fram Strait from 2002 to 2008, as in Beszczynska-Moeller et al. (2012) and **(b)** the BSO from 2003 to 2005, as in Skagseth (2008). Positive velocities in **(a)** and **(b)** are to the east and north, respectively. **(c)** Temperature in the AW layer (blue – model, green – observations) defined as the mean between 72–73° N and 50–200 m and in the outflow water in the Bear Island Trough (red – model, cyan – observations) at 73.5° N, defined as the mean from 300 to 450 m. Definitions and observational data are from Skagseth (2008). **(d)** The AW mean temperature in the core of the West Spitsbergen Current (WSC) from the model (blue) and observations (green), defined as waters with $T > 2$ °C between and 8°40′ E. Definition and observational data are from Beszczynska-Moeller et al. (2012). **(e)** The AW (defined as $T > 3$ °C) volume transport through the BSO as simulated by the model (blue) with the 1997–2006 mean from observations (green) superimposed (Skagseth et al., 2008). **(f)** The AW volume transport in the WSC (AW defined as in **d**), from the model (blue) and observations (green) by Beszczynska-Moeller et al. (2012).

Figure 4. The maximum mean (2000–2009) simulated temperature (°C) in 150–800 m.

3 Results

First, we focus on the patterns of the CTD profiles taken in 2009, which captured the cross-trough temperature maxima in the intermediate water layer conditioned by the AW inflow into the Arctic Ocean. We suggest that these temperature maxima trace the SFSBW pathway in the ST (red arrows in Fig. 2). Second, we discuss the density structure across the ST with a main focus on the eastern flank, where the ST outflow is conditioned by the interaction between the Barents Sea outflow and the SFSBW. Third, we focus on the sheared flow along the ST eastern slope, revealed from directly measured, inferred (using geostrophy) and simulated velocities.

The 2009 survey shows the SFSBW core temperatures to decrease continuously along the SFSBW pathway from 2.54 °C (station 85) to 2.35 °C (station 23) over the ST western flank, and to 2.20 °C (station 25) and 2.09 °C (station 77) over the ST eastern flank (Figs. 2, 5a and b). The same pattern was found by Hanzlick and Aagaard (1980) based on five CTD transects crossing the ST from ~79 to 82° N in 1966, except that in the 1960s the temperatures were about 1 °C lower. For the SFSBW inflow, the overlaying water layer from 50 to ~100 m exhibits well-defined staircases indicating double diffusive upward heat flux from the warm AW layer (Fig. 5a and b). In contrast, the SFSBW outflow shows heavily eroded double diffusive staircases in the upper AW layer. In addition, over the eastern flank of ST, the halocline and the upper SFSBW (15–80 m) are fresher and less dense relative to the western flank by ~0.3–1.0 psu (Fig. 5a and b), and 0.3–0.7 kg m^{-3}, respectively. At the same time, we observe a deepening of the thermocline and, to a lesser extent, the halocline. All these patterns are in agreement with those revealed by the CTD profiles taken in 1996 (Fig. 5c) and 2008 (not shown; note that no hydrographic information for the western ST is available for 2010). The 1996 CTD profiles taken in July show a relatively cool and saline surface layer with lower/higher salinity/temperature vertical

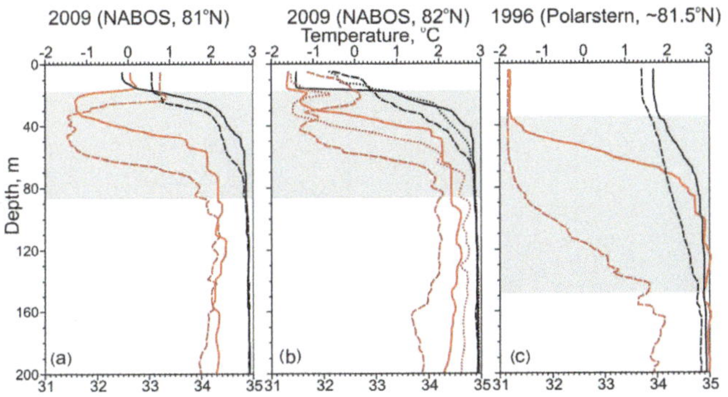

Figure 5. Vertical profiles of temperature (red, °C) and salinity (black, psu) in the upper 200 m layer taken in July 1996 and September 2009 through the core of SFSBW inflow (solid line, **a** – station 23, **b** – station 85, **c** – 81°27′ N, 65°51′ E) and outflow (dashed line, **a** – station 25, **b** – station 77, **c** – 81°17′ N, 72°01′ E) to/from the ST. **(b)** The dotted line depicts the CTD profile taken in September 2009 through the core of the FSBW boundary current (station 80). For station positions in 2009 see Fig. 2. Gray shading highlights the outflow from the ST modified by vertical mixing.

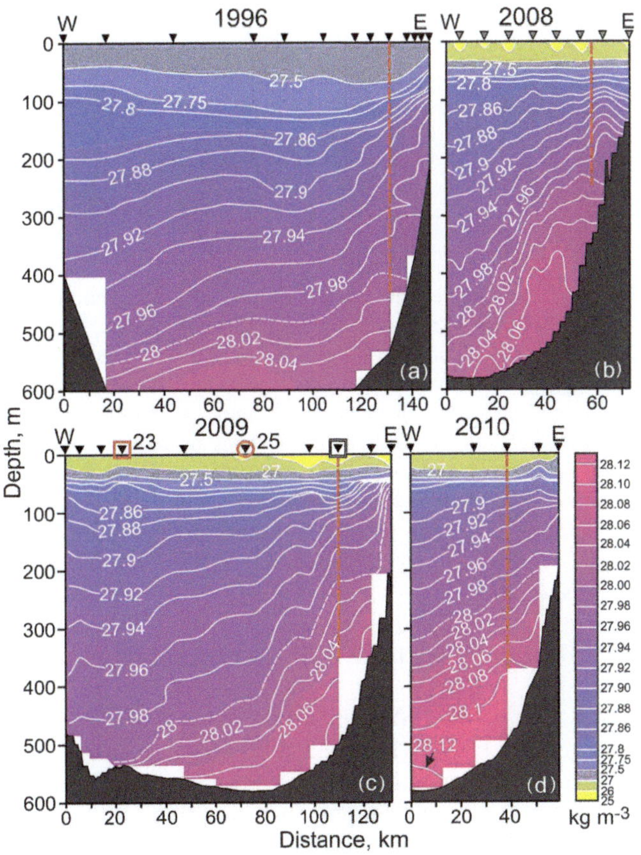

Figure 6. The 10 m binned cross-trough sections of potential density (σ–zero, kg m^{-3}) along 81° N for **(a)** July 1996, **(b)** September 2008, **(c)** September 2009 and **(d)** October 2010. Black and gray triangles on the top identify positions of CTD and XCTD stations, respectively. Following Fig. 2, the red square and circle and the white square identify stations taken through the core of the SFSBW inflow and outflow to/from the ST and mooring position, respectively, with their reference numbers on the top. Red dashed lines identify locations of the geostrophic velocity profiles shown in Fig. 7a. The white dashed line highlights the σ–zero = 28 kg m^{-3}, which roughly corresponds to the core of the cooler fraction of BSBW (Dmitrenko et al., 2008a, 2009).

gradients comparing to the 2009 profiles taken in late August to early September (Fig. 5). This difference implies the role of seasonal variability attributed to the sea ice melting and solar heating. The higher temperature of the inflowing SFSBW in 1996 compared to 2008 is due to the warm AW anomaly entering the Arctic Ocean through the Fram Strait in the 1990s (e.g., Schauer et al., 2002b). We also note that in 2009 the intermediate temperature maximum ($T_{\max} = 2.74$ °C) in the central part of the ST mouth at 82° N (station 80, Figs. 2 and 5b) exceeds that of the inflowing SFSBW by 0.2–0.4 °C.

The observations in 1996 and 2008–2010 consistently show denser water at the eastern flank of the ST below the halocline layer deeper than 50 m (Fig. 6). In 1996, over the ST eastern flank shallower than ~300 m, the density front extended throughout the water column (Fig. 6a). At 150 m, the potential density gradient across the eastern flank along 81° N ranged from 0.09×10^{-2} kg m^{-3} km^{-1} in 2010 (Fig. 6d) to 0.18×10^{-2} kg m^{-3} km^{-1} in 2008–2009 (Fig. 6b and c). At 300 m, the cross-slope density gradient in 2008–2010 was more uniform with ~0.16–0.18×10^{-2} kg m^{-3} km^{-1} (Fig. 6b–d). For comparison, in 1996, the density gradient across the eastern flank was ~0.22×10^{-2} and 0.09×10^{-2} kg m^{-3} km^{-1}, at 150 and 300 m, respectively (Fig. 6a). Moreover, in 1996 the surface layer is denser (saltier), and consequently the surface layer stratification is weaker (Fig. 5a).

The mean velocity profile derived from the 2009–2010 year-long ADCP velocity measurements at the eastern flank shows a relatively stable, nearly barotropic northward flow aligned to ~2° in 140–220 m and to ~11° in 380–460 m, which is consistent with the orientation of the ST eastern flank. The flow strength slightly increases with depth from 17 ± 7 cm s^{-1} at 140 m to 18 ± 7 cm s^{-1} at 200 m (Fig. 6b), 21 ± 5 cm s^{-1} at 380 m, and 23 ± 5 cm s^{-1} at 460 m (not shown). This is in agreement with the ship-board ADCP observations in 1996 showing the northward sub-surface flow of ~25–30 cm s^{-1} (Fig. 7b, see Schauer et al., 2002a, for more details). In addition, the velocity profile in Fig. 7b demonstrates a velocity difference of ~20 cm s^{-1} in the upper 90 m. The sheared flow is also evident from the geostrophic velocity profiles at the mooring position in the ST eastern flank derived from dynamic heights relative to the 0 dbar pressure surface using the potential density profiles from the nearest CTD stations (Figs. 6 and 7a). The simulated velocities also demonstrate a northward sheared flow along the ST eastern flank, but twice as large as the geostrophic velocities (Figs. 7c, d, and 8).

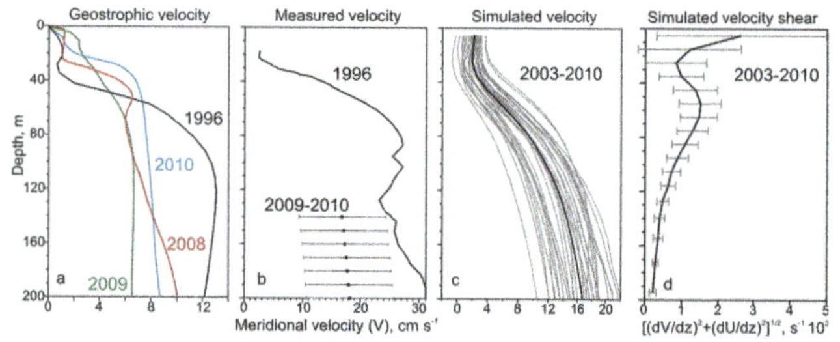

Figure 7. Vertical velocity profiles for the northward current over the ST eastern flank: **(a)** geostrophic, **(b)** measured and **(c)** simulated for each grid node over the ST eastern flank between 78 and 81° N (gray lines, averaged for 2003–2010 using annual mean velocity profiles) with their mean (black line). **(d)** Vertical profile of the simulated velocity shear at the mooring position averaged for 2003–2010 using daily mean velocity profiles. **(b, d)** Error bars show 1 standard deviation from the mean.

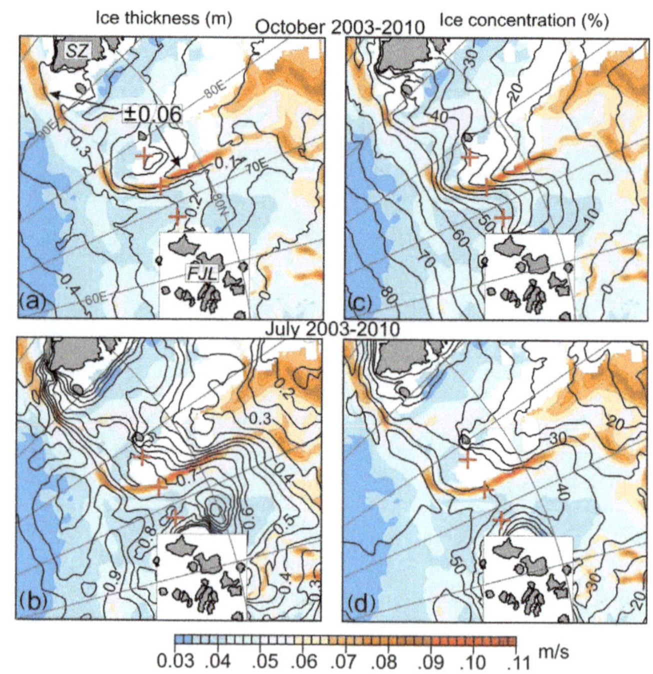

Figure 8. The 7-year mean (2003–2010) simulated velocity vertical difference (ΔV) in the 0–75 m layer ($m\,s^{-1}$) in the northern Kara Sea with overlaid simulated mean **(a, b)** sea-ice thickness (m) and **(c, d)** concentration (%) for **(a, c)** October and **(b, d)** July. Crosses depict locations along 81° N where the measured June–November time series of sea-ice concentration are compiled in Fig. 12. **(a)** The arrowed number shows ±1 standard deviation for the ΔV maximum over the ST eastern flank and Kara Sea continental margin obtained using monthly mean velocity data.

4 Discussion

4.1 Heat loss from the AW in the ST: causes

Expanding on the AW circulation scheme suggested by Rudels et al. (1994) (Fig. 1), we have demonstrated that a significant portion of the heat associated with the SFSBW in the ST disappears en route through the ST. In this context, our observations confirm earlier findings by Hanzlick and Aagaard (1980). This concept, however, requires explanation for the AW intermediate temperature maximum of 2.74 °C observed in the ST mouth at 82° N (station 80) that exceeds the temperature of the SFSBW inflow and outflow by 0.2 and 0.65 °C, respectively (Fig. 5a and b).

The current view on the AW circulation in the Eurasian Basin, as proposed by Rudels et al. (2004), implies that the FSBW could potentially feed all three temperature maxima along 82° N recorded at stations 77, 80, and 85 (Fig. 2). However, the formation of each maximum is due to different FSBW branches. That is, if the ST on-slope temperature maxima are associated with SFSBW, then the AW temperature maximum in the ST mouth could be attributed to the boundary current of the FSBW flowing along the Siberian continental slope (yellow arrows in Fig. 2). The double diffusive staircases noticeable in station 80 (Fig. 5b) provide indication of attributing the intermediate temperature maximum of 2.74 °C to the core of the FSBW that follows the Siberian continental slope rather than to the SFSBW. In contrast to the SFSBW inflow (stations 23 and 85), the CTD profiles taken in the SFSBW outflow (stations 25 and 77) show heavily eroded double diffusive staircases (Fig. 5a and b). The ST outflow also exhibits strong modification at the upper AW interface, which is likely caused by vertical mixing and interaction with cooler and fresher surface water of Barents Sea origin. In contrast, the temperature and salinity profiles in the ST mouth (station 80) are only slightly modified at the upper AW interface and maintain double diffusive staircases that are almost disrupted in stations 25 and 77 (Fig. 5a and b).

The enhanced turbulent diffusivity usually results in substantial degradation of the thermohaline staircases, and the turbulent mixing associated with shear instability nearly disrupts the double-diffusive staircases (e.g., Melling et al., 1984; Dmitrenko et al., 2008a). Over the ST eastern flank, turbulent mixing between the upper SFSBW and overlaying cooler and fresher Barents and Kara Sea waters may be inferred from the velocity shear across the upper SFSBW interface (Figs. 7 and 8). The cross-trough potential density distribution in Fig. 6 is also consistent with a sheared geostrophic flow above the eastern ST (Fig. 7a). Under this scenario, the SFSBW inflow (double-diffusive staircases present) has not been affected by enhanced turbulent mixing, while the SFSBW outflow (no double-diffusive staircases present) has experienced turbulent mixing during propagation along the eastern ST. The FSBW boundary current at the central ST (station 80) has been partly affected by turbulent mixing, but double-diffusive staircases remain recognizable.

The sheared flow along the eastern ST in the upper 100 m layer is evident from direct current observations (Fig. 7b), geostrophic velocity calculations (Fig. 7a), and model simulations (Figs. 7c, d and 8). However, the difference between measured and geostrophic velocities indicates that the SFSBW flow along the eastern ST is not entirely density driven. In this context, Kirillov et al. (2012) recently showed that wind forcing over the northern Kara Sea is important in facilitating the SFSBW outflow from the ST. The gradient Richardson number (*Ri*) provides a strong constraint for identifying turbulence produced by shear instability. It is defined as $Ri = N^2\ S^{-2}$, where N is the Brunt–Väisälä frequency, and S is the flow vertical shear $S = [(\mathrm{d}U/\mathrm{d}z)^2 + (\mathrm{d}V/\mathrm{d}z)^2]^{1/2}$. Turbulent mixing can be assumed to occur for $Ri < 1$ (e.g., Polzin, 1996). Considering a velocity shear $S \sim 0.004\,\mathrm{s}^{-1}$ and the stratification in the 30–90 m layer observed in 1996 (Figs. 6d and 7b) results in $Ri = 3.7$, which significantly exceeds the criterion. Simulations show, however, that velocity shear across the surface layer might reach $0.007\,\mathrm{s}^{-1}$ (Fig. 7d). Assuming the higher $S \sim 0.007\,\mathrm{s}^{-1}$ induced by the SFSBW northward velocity $V \sim 40\,\mathrm{cm\,s}^{-1}$, as measured in January 2010 (not shown), the $O(Ri) = 1.4$ might possibly allow for turbulent mixing, taking into account reduced salinity (density) stratification in winter due to sea-ice formation. While it cannot be shown explicitly that shear instability is entirely responsible for the enhanced vertical mixing, it may play a role in facilitating upward fluxes from the AW over the ST eastern slope.

An important amount of the AW heat loss in the ST may be attributed to strong vertical mixing over the ST flanks and associated rough topography. Sundfjord et al. (2007) and Sirevaag and Fer (2009) concluded that the Arctic turbulent mixing is important along the boundaries and steep bottom topography. The enhanced vertical diffusion can be also associated with elevated tidal forcing over the sloping topography (e.g., Dewey at al., 1999). In the following, however, we show that the enhanced vertical heat loss occurs specifically over the ST eastern flank, the area where the SFSBW flows to the Arctic Ocean.

4.2 Heat loss from the AW in the ST: estimates

In 2009, the heat content of the SFSBW inflow over the western ST flank computed relative to the freezing temperature between 30 to 90 m depth decreased by $\sim 340\,\mathrm{MJ\,m}^{-2}$ upon reaching the eastern flank. For example, at 81° N the heat content between 30 to 90 m depth decreased from $796\,\mathrm{MJ\,m}^{-2}$ at station 23 (SFSBW inflow) to $457\,\mathrm{MJ\,m}^{-2}$ at station 25 (SFSBW outflow). We note, however, the opposite tendency of the heat content over the eastern ST between stations 25 and 77, showing increase from $457\,\mathrm{MJ\,m}^{-2}$ at station 25 to $584\,\mathrm{MJ\,m}^{-2}$ at station 77 (Fig. 2). This discrepancy seems to be attributed to a crude spatial resolution of our time snapshot CTD data, which are insufficient to resolve the relatively narrow jet of the AW outflow from the ST. At 81° N, the horizontal cross-slope temperature gradient between stations 26 and 27 at 110 m (depth of the temperature maximum at station 25 – Fig. 3a) is $0.18\,°\mathrm{C\,km}^{-1}$, and the horizontal cross-slope gradient of heat content is $16.25\,\mathrm{MJ\,m}^{-2}\,\mathrm{km}^{-1}$ (Fig. 2). We hypothetically extend this gradient further offslope to station 26 in order to obtain an estimate of uncertainty related to spatial undersampling of the relatively narrow AW jet between stations 25 and 26. This approach gives heat content estimates from $457\,\mathrm{MJ\,m}^{-2}$ to $629.75\,\mathrm{MJ\,m}^{-2}$ in ~ 4 km to 15 km off station 26, respectively (note that the distance between station 25 and 26 is ~ 16 km). This suggests the upper bound of potential heat content underestimate at station 26 to be $173\,\mathrm{MJ\,m}^{-2}$.

In the following we assume that all heat loss in the SFSBW occurs along the ST eastern flank downstream of the SFSBW confluence with the Barents Sea water outflow to the ST at $\sim 78°$ N. This assumption is based on the hypothesis that the SFSBW heat loss is primarily driven by the velocity shear over the ST eastern flank, in line with model simulations showing no velocity shear over the ST western flank (Fig. 8). This suggestion also implies neglecting the double diffusive heat loss from the AW over the ST western flank, which is justified by the typical double diffusive heat flux estimated in the Arctic Ocean to $0.6\,\mathrm{W\,m}^{-2}$ (Sirevaag and Fer, 2012). Moreover, this assumption is also based on the structure of the temperature profiles, showing that the major part of the deepening of the upper SFSBW interface occurs from 81° N on the western flank to 81° N on the eastern flank (Fig. 5). Furthermore, we neglect the lateral heat loss, particularly by a fraction of the SFSBW flowing west along the shallower trough between Novaya Zemlya and Franz Josef Land – Fig. 1 (e.g., Schauer et al., 2002a; Gammelsrød et al., 2009).

The travel time along the eastern ST flank between 78° N and the mooring location at 81° N is estimated using northward along-trough velocities simulated with the MIT model. At each grid node over the eastern flank, the simulated velocity profiles were averaged from 2003 to 2010 (Fig. 7c). The mean velocity profile computed from averaging all individual profiles over the eastern ST shows $\sim 10\,\mathrm{cm\,s}^{-1}$ northward velocity for the SFSBW core at ~ 90 m (Fig. 7c). This velocity implies a ~ 40-day travel time from 78 to 81° N along the ST eastern flank, the area where the heat content of the SFSBW was suggested to reduce by $\sim 340\,\mathrm{MJ\,m}^{-2}$. This is equivalent to an upward vertical heat flux from the SFSBW to the low halocline water of $\sim 100\,\mathrm{W\,m}^{-2}$. This rough estimate is similar to those reported by Sirevaag and Fer (2009) for the area north of Svalbard where the upward heat loss from the AW is among the largest throughout the Arctic Ocean (e.g., Steele and Morison, 1993). For comparison, over the Laptev Sea continental margins, Polyakov et al. (2012) estimated the heat flux across the upper AW interface of $\sim 8\,\mathrm{W\,m}^{-2}$. Microstructure measurements over the Laptev Sea continental shelf break reveal upward heat fluxes of $12\,\mathrm{W\,m}^{-2}$ (Lenn et al., 2011). An estimate of the uncertainty in the vertical heat flux owing to spatial undersampling of the AW jet at 81° N is obtained based on the heat content underestimate

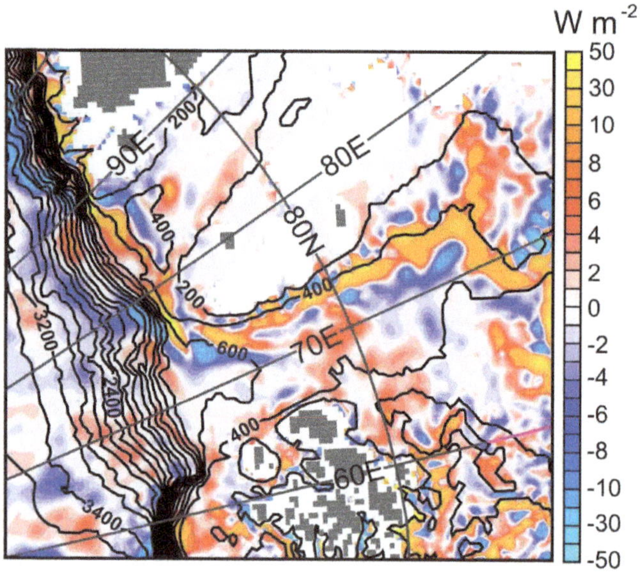

Figure 9. The 7-year mean (2003–2010) simulated vertical heat fluxes (W m^{-2}) across the upper SFSBW interface at 75 m showing enhanced heat loss over the ST eastern flank as well as over the Nansen Basin continental margin. Positive numbers indicate upward heat flux. The bathymetry is in meters. Yellow dashed rectangular depicts the area over the ST eastern flank where the simulated heat flux is averaged for compiling the time series shown in Fig. 10.

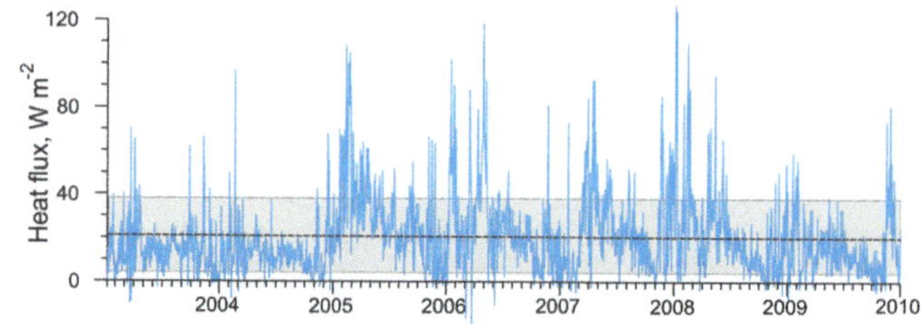

Figure 10. Time series (2003–2010) of daily mean simulated vertical heat fluxes (W m^{-2}) across the upper SFSBW interface at 75 m averaged over the ST eastern slope (the area depicted by the yellow dashed rectangle in Fig. 9). The black dashed line indicates the 2003–2010 mean (20.5 W m^{-2}) and the gray shading shows ± 1 standard deviation from the mean.

by 173 MJ m^{-2}, which reveals the heat flux lower bound of ~ 50 W m^{-2}.

Over the ST eastern flank, our model simulations show a mean upward heat flux across the upper SFSBW interface with values up to 30–50 W m^{-2} (Fig. 9) that is about half or less the heat loss compared with estimates inferred from CTD observations in the ST. The areal mean over the ST eastern flank is 21 ± 18 W m^{-2}. The difference between simulated and CTD-derived values indicates that the double diffusive heat loss over the ST western flank and the lateral heat loss might be also important. On the other hand, the model presents a large spatial and temporal variability in the vertical heat fluxes at the upper SFSBW interface, with values at times exceeding 80 W m^{-2} (Fig. 10). The daily averaged values of the upward heat flux in the model ST eastern flank (Fig. 10) present a seasonal cycle of about 20 W m^{-2}, showing a main maximum in early spring and a secondary maximum in early winter. Furthermore, strong interannual variability on the occurrence of the spring maximum can be noticed.

In view of such strong variability, the inconsistency to the observations might be, in part, attributed to unresolved temporal variability in the measurements. An incomplete mixing scheme in the model might also provide an explanation for model–observational discrepancies. Finally, the discrepancy between estimates derived from simulations and CTD data can also be explained by spatial undersampling of the AW jet over the ST eastern slope. The lower bound of heat flux at ~ 50 W m^{-2} retrieved from the CTD data is consistent with the 7-year mean simulated vertical heat fluxes (Fig. 9).

In summary, both the CTD data and the model simulation suggest the upward heat loss from the AW layer in the ST eastern flank, significantly exceeding that over the Siberian continental margin. The spatial undersampling results in relatively high uncertainty of our observationally based heat loss estimates. However, even at the lower bound of ~ 50 W m^{-2} the amount of heat transferred to the upper layer is capable of modifying the sea ice cover, as we discuss in the following section.

4.3 Heat loss from the AW in the ST: impact on sea ice

In what follows we test our hypothesis of enhanced SFSBW heat loss in the ST by using sea-ice remote sensing data. The fraction of the AW heat expended for heating the overlaying water depends on the characteristics of the Barents Sea outflow. However, any upward heat flux reaching the sea surface will eventually affect the ocean–sea-ice–atmosphere interface (e.g., through reduced sea-ice thickness, enthalpy flux to the atmosphere and delayed freeze-up). The winter mean (February–March) sea-ice thickness derived from five ICESat winter campaigns in 2004–2008 (Kwok et al., 2009) shows an area with negative anomalies of ~ 20 cm stretching from the ST mouth towards Severnaya Zemlya following the Nansen Basin continental margin (Fig. 11a; note that there is no available ICESat data south of 81° N). This is consistent with the simulated velocity difference in Fig. 8 and enhanced upward heat loss along the SFSBW pathway (Fig. 9). In February–March 2004–2008, the ice drift over the ST mouth is mainly westward with a mean velocity of ~ 4 km day^{-1} (Fig. 10b and c). The mean width of the ST eastern slope at the ST mouth is ~ 50 km. This yields a ~ 12.5-day pack ice residence time over the ST eastern slope at 82.5° N, where pack ice is exposed to the enhanced upward vertical heat flux from the SFSBW. The 20 cm sea-ice decrease in 12.5 days requires an upward heat flux of ~ 60 W m^{-2} to the ice–water interface, which is comparable to that obtained from the model simulations (Fig. 9).

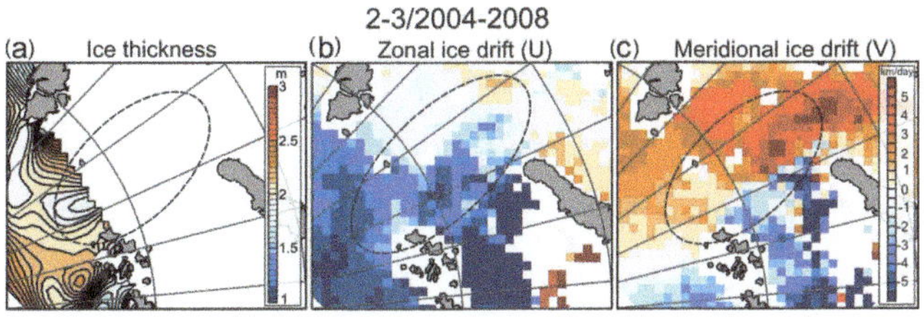

Figure 11. The 2004–2008 winter mean (February–March) **(a)** sea ice thickness (m) and **(b)** zonal and **(c)** meridional sea ice drift (km day^{-1}) derived from AMSR data. The dashed oval highlights the ST region.

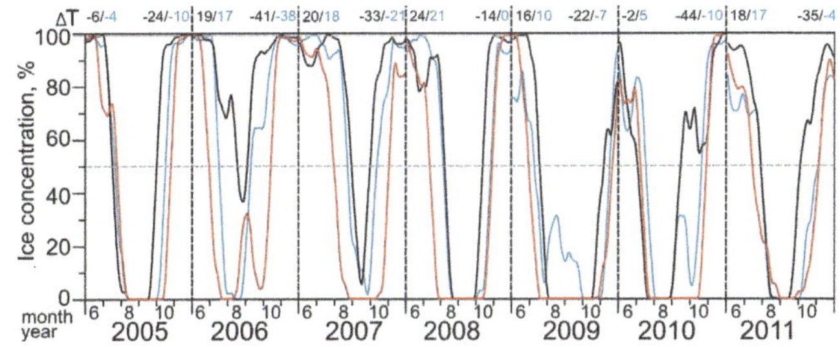

Figure 12. Observed June–November time series of sea ice concentration (%) at three locations along 81° N depicted with crosses in Figs. 8 and 12a: 68° E (black), 73° E – mooring position (red), and 78° E (blue). Black/blue numbers at the top show the time lag (ΔT, days) between 73° E and 68/78° E on reaching the 50 % sea ice concentration.

The time series of ice concentration across ST at 81° N reveals later freeze-up and earlier sea-ice decline at the eastern flank (73° E; mooring position), lagging the more western and eastern locations at 68 and 78° E, respectively, by ~0.5–2.5 months (Fig. 12). This is in agreement with simulated sea-ice concentrations showing consistent reduction over the ST eastern flank during freeze-up (October, contours in Fig. 8c) and melting (June, contours in Fig. 8d). For estimating the sea-ice growth during freeze-up, we use the semi-empirical method based on an analytical model known as the classical Stefan law (Stefan, 1891) that links the thermodynamic sea-ice growth to the cumulative surface air temperature below freezing. The accumulated sea-ice growth time series for 2005–2011 computed based on the surface air temperature from the NCEP reanalysis are shown in Fig. 13. For simplicity, we focus on satellite overpass times where the estimated thermodynamic sea-ice growth reaches ~20 cm, which roughly corresponds to the fall and winter sea-ice thickness anomaly over ST in Figs. 8a and 10a, respectively. In contrast to the expected thermodynamic sea-ice growth of ~20 cm, the ST eastern flank remains ice free even during the late fall at air temperatures far below freezing (~ −15 °C). For instance, on 25 October 2005 (Fig. 14a) the NCEP-derived daily mean air temperature was as low as −14 °C, which corresponds to a thermodynamic sea-ice growth of ~2 cm day^{-1} (Fig. 13). Melting this amount of ice requires ~70 W m^{-2}, a value approaching the upward heat flux inferred from the model simulation (Figs. 9 and 10). We note that this method overestimates the heat loss because Stefan's law does not account for solar insolation.

The observed ice-free area, stretching along the ST eastern flank (Fig. 14), is fully consistent with our assumption of a substantial AW heat loss along the eastern flank corroborated by the model simulation. Figure 14 is also in agreement with simulated sea ice thickness and concentration showing significant reduction over the ST eastern flank in October (Fig. 8a and c). Another remarkable feature of sea ice in the ST is that in 2005, 2007, 2008, 2009 and 2011 the area of reduced sea-ice concentration (45–85 %) was stretched along the ST eastern flank further north to the ST mouth at ~82° N where it turned east following the Nansen Basin continental slope (e.g., Fig. 14c). This is also in line with (i) the spatial distribution of sea ice thickness anomaly in Fig. 11a and (ii) the enhanced AW heat loss simulated along the Kara Sea continental slope (Fig. 9).

The BSBW, entering the Arctic Ocean along the ST, can also potentially impact the sea ice in this region (compare the BSBW pathway and configuration of the open water area in the ST – Figs. 1 and 13). At the Barents Sea exit between Novaya Zemlya and Franz Josef Land, Årthun and Schrum (2010) simulated a long-term mean surface heat flux from the ocean of ~10 W m^{-2}, which modified the ice cover distribution as shown in Årthun et al. (2012). However, en route to the ST, the warm core of the BSBW with temperatures above 0° and salinities between 34.8 and 35 psu is decoupled from the sea surface occupying the depths deeper than 70–100 m (Gammelsrød et al., 2009; Lien and Trofimov, 2013). Within the ST, the warm BSBW core submerges even below the SFSBW (e.g., Dmitrenko et al., 2008a, 2009). This suggests that the BSBW does not play a role in modifying the sea-ice conditions over the ST through direct heat loss.

5 Summary and concluding remarks

We have shown that the St. Anna Trough is one area of the Arctic Ocean where AW heat loss can modify the ocean–sea-ice–atmosphere interface affecting both the formation and decay of sea ice. In this sense, the St. Anna Trough is similar to an "ice bay" known as Whalers' Bay formed by inflowing Fram Strait branch of AW north of Svalbard (e.g., Rudels, 2010; Ivanov et al., 2012). The sheared St. Anna Trough outflow to the Arctic Ocean is evident from shipboard ADCP as well as from geostrophic velocities and a numerical model simulation. Although no clear evidence for the occurrence of shear instabilities could be obtained, we speculate that the enhanced upward heat flux in the St. Anna Trough is promoted by a vertical velocity shear along the St. Anna Trough eastern flank. We show that the zone of lowered sea ice thickness and concentration essentially marks the Fram Strait branch

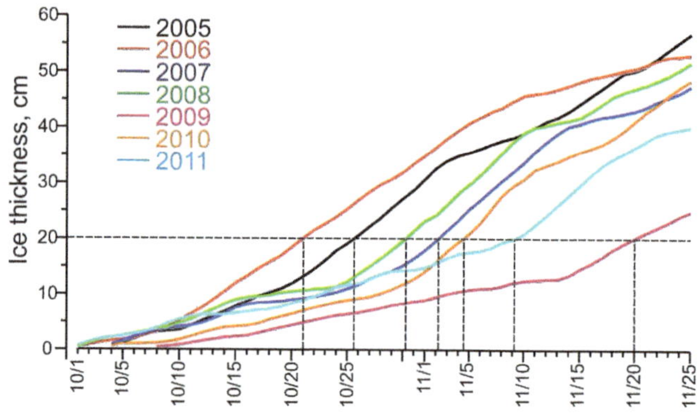

Figure 13. Time series of accumulated thermodynamic sea ice growth (cm) computed for October–November 2005–2011. Dashed lines indicate days in October–November when sea ice thickness reached 20 cm.

pathway in the St. Anna Trough and adjacent Nansen Basin continental margin from both sea-ice remote sensing observations and a model simulation.

Our analysis was significantly limited by the availability of velocity data, and our heat flux estimates based on CTD observations were necessarily illustrative. Our cross-trough CTD transects are time snapshots, and the crude spatial resolution is insufficient to resolve the jet of the AW outflow from the St. Anna Trough. We also note that our estimates of the heat flux at the ice–water interface using sea-ice remote sensing show a lower bound because water and sea-ice dynamics are not taken into account. Moreover, we note that heat loss from the St. Anna Trough recirculating Fram Strait branch is not entirely available for sea-ice melt, and its important fraction is likely to be consumed for modifying the overlaying water and atmospheric boundary layer. The lateral and downward heat loss were not considered and estimated. We also note that although Stefan's law is a simplification of the requisite physics it does provide first-order agreement in our observed surface temperature fluxes and the thermodynamic response of the sea ice.

Finally, our CTD observations cover a relatively short period of time, while over the last several decades the temperature of the FSBW has shown sustainable tendency to rise with several significant warm pulses in the 1990s (e.g., Schauer et al., 2002b) and 2000s (e.g., Dmitrenko et al., 2008b). In fact, our conclusions are based on measurements taken in 1996 and 2008–2010 during the warm pulses of the AW inflow into the Arctic Ocean through the Fram Strait. These conditions represent relatively well the modern tendency of FSBW temperature to rise. At the same time, the physical mechanisms of the upward heat loss in the ST are not obvious, in the light of the ongoing warming with a possible increased ocean heat transport from the Atlantic into the Arctic Ocean (e.g., Koenigk and Brodeau, 2013). It seems that they depend not only on the AW temperature in the SFSBW but also on the vertical velocity shear controlled by

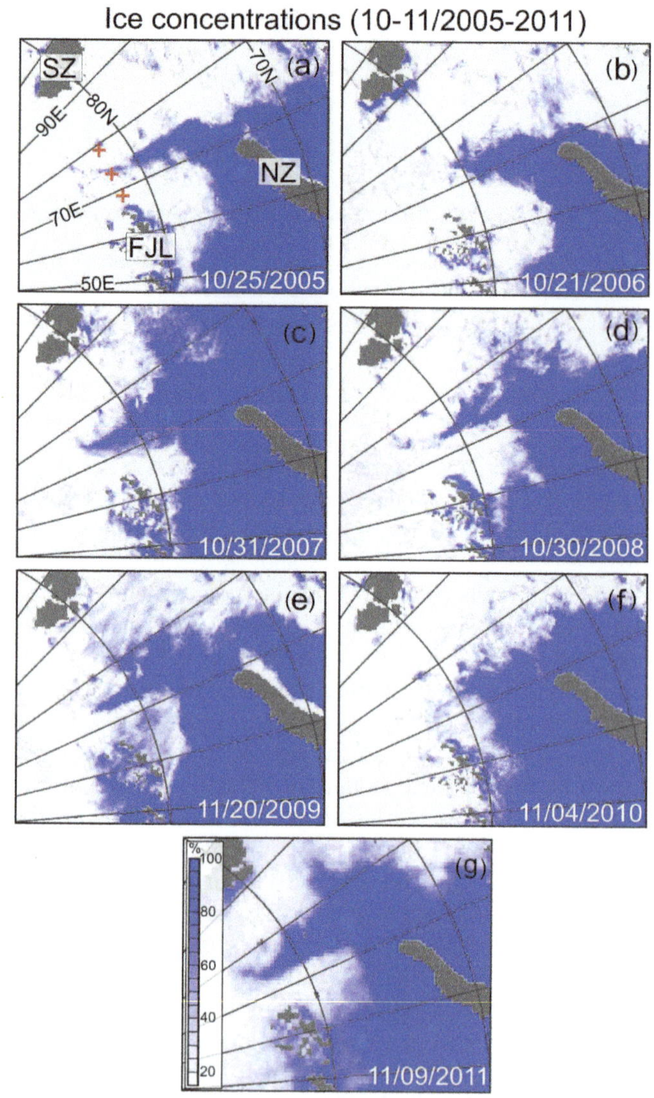

Figure 14. (a–f) The AMSR-E and **(g)** SSMI sea ice concentrations (%) in October–November 2005–2011. Crosses in **(a)** depict locations along 81° N where the June–November time series of sea ice concentration shown in Fig. 12 are compiled.

the horizontal density gradient across the ST eastern flank (Figs. 6 and 8). In turn, this gradient is conditioned by the density difference between the SFSBW and the Barents Sea outflow to the Arctic Ocean through the St. Anna Trough.

The deficiencies in our analyses clearly define a necessity for further research in this area to quantify the efficiency of shear instability in facilitating the heat loss from the St. Anna Trough recirculating Fram Strait branch, involving advanced field experiments and further analyses of model simulations.

Acknowledgements. This study was supported by the Canada Excellence Research Chair (CERC) program, the Canada Research Chairs (CRC) program, and the National Sciences and Engineering Research Council of Canada (NSERC) grant RGPIN-2014-03606.

The 2008–2010 oceanographic observations in the St. Anna Trough were conducted under the working frame of the NABOS project with support from NOAA and NSF, the "System Laptev Sea" project funded through the German BMBF, and project "Arctika" (Russia). The numerical simulations were performed at the DKRZ, Hamburg, Germany, in the frame of the German BMBF project "Nordatlantik" (WP 4.1) and the European Commission FP7 collaborative project "MONARCH-A" (FP7-Space-2009-1 contract 242446). AARI co-authors of this study were supported by the EU FP7 "Arctic Climate Change Economy and Society" (ACCESS) project and RFBR grant 13-05-41443. We acknowledge Lars Kaleschke (University of Hamburg, Germany) for providing the sea ice concentration data through the Integrated Climate Data Center (ICDC) at the University of Hamburg, Germany. Yan Waddington and Torben Klagge have rendered valuable assistance in performing mooring deployment and recovery in the St. Anna Trough.

Edited by: M. Meier

References

Adcroft, A., Campin, J.-M., Hill, C., and Marshall, J.: Implementation of an atmosphere-ocean general circulation model on the expanded spherical cube, Mon. Weather Rev., 132, 2845–2863, doi:10.1175/MWR2823.1, 2004.

Alexeev, V. A., Ivanov, V. V., Kwok, R., and Smedsrud, L. H.: North Atlantic warming and declining volume of arctic sea ice, The Cryosphere Discuss., 7, 245–265, doi:10.5194/tcd-7-245-2013, 2013.

Årthun, M. and Schrum, C.: Ocean surface heat flux variability in the Barents Sea, J. Marine Syst., 83, 88–98, 2010.

Årthun, M., Eldevik, T., Smedsrud, L. H., Skagseth, Ø., and Ingvaldsen, R. B.: Quantifying the influence of Atlantic heat on Barents Sea ice variability and retreat, J. Climate, 25, 4736–4743, 2012.

Beszczynska-Möller, A., Fahrbach, E., Schauer, U., and Hansen, E.: Variability in Atlantic water temperature and transport at the entrance to the Arctic Ocean, 1997–2010, ICES J. Mar. Science, 69, 852–863, doi:10.1093/icesjms/fss056, 2012.

Boyer, T., Levitus, S., Garcia, H., Locarnini, R. A., Stephens, C., and Antonov, J.: Objective analyses of annual, seasonal, and monthly temperature and salinity for the World Ocean on a 0.25° grid, Int. J. Climatol., 25, 931–945, doi:10.1002/joc.1173, 2005.

Dewey, R., Muench, R., and Gunn, J.: Mixing and vertical heat flux estimates in the Arctic Eurasian Basin, J. Marine Syst., 21, 199–205, doi:10.1016/S0924-7963(99)00014-7, 1999.

Dmitrenko, I. A., Kirillov, S. A., Ivanov, V. V., and Woodgate, R. A.: Mesoscale Atlantic water eddy off the Laptev Sea continental slope carries the signature of upstream interaction, J. Geophys. Res., 113, C07005, doi:10.1029/2007JC004491, 2008a.

Dmitrenko, I. A., Polyakov, I. V., Kirillov, S. A., Timokhov, L. A., Frolov, I. E., Sokolov, V. T., Simmons, H. L., Ivanov, V. V., and Walsh, D.: Toward a warmer Arctic Ocean: Spreading of the early 21st century Atlantic Water warm anomaly along the Eurasian Basin margins, J. Geophys. Res., 113, C05023, doi:10.1029/2007JC004158, 2008b.

Dmitrenko, I. A., Bauch, D., Kirillov, S. A., Koldunov, N., Minnett, P. J., Ivanov, V. V., Hölemann, J. A., and Timokhov, L. A.: Barents Sea upstream events impact the properties of Atlantic water inflow into the Arctic Ocean: Evidence from 2005–2006 downstream observations, Deep-Sea Res. Pt. I, 56, 513–527, 2009.

Dmitrenko, I. A., Kirillov, S. A., Ivanov, V. V., Rudels, B., Serra, N., and Koldunov, N. V.: Modified halocline water over the Laptev Sea continental margin: Historical data analysis, J. Climate, 25, 5556–5565, 2012.

Ezraty, R., Girard-Ardhuin, F., and Croizé-Fillon, D.: Sea ice drift in the central Arctic using the 89 GHz brightness temperatures of the advanced microwave scanning radiometr, User's Manual, IFREMER, Brest, France, 2007.

Gammelsrød, T., Leikvin, Ø., Lien, V., Budgell, W. P., Loeng, H., and Maslowski, W.: Mass and heat transports in the NE Barents Sea: Observations and models, J. Marine Syst., 75, 56–69, 2009.

Hanzlick, D. and Aagaard, K.: Freshwater and Atlantic Water in the Kara Sea, J. Geophys. Res., 85, 4937–4942, 1980.

Ivanov, V. V., Alexeev, V. A., Repina, I., Koldunov, N. V., and Smirnov, A.: Tracing Atlantic water signature in the Arctic sea ice cover east of Svalbard, Adv. Meteorol., 2012, 201818, doi:10.1155/2012/201818, 2012.

Kalnay, E., Kanamitsu, M., Kistler, R., Collins, W., Deaven, D., Gandin, L., Iredell, M., Saha, S., White, G., Woollen, J., Zhu, Y., Leetmaa, A., Reynolds, R., Chelliah, M., Ebisuzaki, W., Higgins, W., Janowiak, J., Mo, K. C., Ropelewski, C., Wang, J., Jenne, R., and Joseph, D.: The NCEP/NCAR 40 Year Reanalysis Project, B. Am. Meteorol. Soc., 77, 437–471, 1996.

Kirillov, S. A., Dmitrenko, I. A., Ivanov, V. V., Aksenov, E. O., Makhotin, M. S., and de Quevas B. A.: The influence of atmospheric circulation on the dynamics of the intermediate water layer in the eastern part of the St. Anna Trough, Dokl, Earth Sci., 444, 630–633, 2012.

Koenigk, T. and Brodeau, L.: Ocean heat transport into the Arctic in the twentieth and twenty-first century in EC-Earth, Clim. Dynam., 42, 3101–3120, doi:10.1007/s00382-013-1821-x, 2013.

Kwok, R. and Cunningham, G.: ICESat over Arctic sea ice: estimation of snow depth and ice thickness, J. Geophys. Res., 113, C08010, doi:10.1029/2008JC004753, 2008.

Kwok, R., Cunningham, G. F., Wensnahan, M., Rigor, I., Zwally, H. J., and Yi, D.: Thinning and volume loss of the Arctic Ocean sea ice cover: 2003–2008, J. Geophys. Res., 114, C07005, doi:10.1029/2009JC005312, 2009.

Large, W. G., McWilliams, J., and Doney, S. C.: Ocean vertical mixing: a review and a model with a nonlocal boundary layer parameterization, Rev. Geophys., 32, 363–403, 1994.

Lenn, Y. D., Rippeth, T. P., Old, C. P., Bacon, S., Polyakov, I., Ivanov, V., and Hölemann, J.: Intermittent intense turbulent mixing under ice in the Laptev Sea continental shelf, J. Phys. Oceanogr., 41, 531–547, 2011.

Lien, V. S. and Trofimov, A. G.: Formation of Barents Sea Branch Water in the Northeastern Barents Sea, Polar Res., 32, 18905, doi:10.3402/polar.v32i0.18905, 2013.

Marshall, J., Adcroft, A., Hill, C., Perelman, L., and Heisey, C.: A finite-volume, incompressible Navier Stokes model for studies of the ocean on parallel computers, J. Geophys. Res., 102, 5753–5766, 1997.

Melling, H., Lake, R. A., Topham, D. R., and Fissel, D. B.: Oceanic thermal structure in the western Canadian Arctic, Cont. Shelf. Res., 3, 233–258, 1984.

Polyakov, I. V., Timokhov, L. A., Alexeev, V. A., Bacon, S., Dmitrenko, I. A., Fortier, L., Frolov, I. E., Gascard, J.-C., Hansen,

E., Ivanov, V. V., Laxon, S., Mauritzen, C., Perovich, D., Shimada, K., Simmons, H. L., Sokolov, V. T., Steele, M., and Toole, J.: Arctic Ocean warming contributes to reduced polar ice cap, J. Phys. Oceanogr., 40, 2743–2756, 2010.

Polyakov, I. V., Pnyushkov, A. V., Rember, R., Ivanov, V. V., Lenn, Y. D., Padman, L., and Carmack, E. C.: Mooring-based observations of double-diffusive staircases over the Laptev Sea slope, J. Phys. Oceanogr., 42, 95–109, 2012.

Polzin, K. L.: Statistics of the Richardson number: Mixing models and fine structure, J. Phys. Oceanogr., 26, 1409–1425, 1996.

Rudels, B.: Constraints on exchanges in the Arctic Mediterranean – do they exist and can they be of use?, Tellus, 62A, 109–122, doi:10.1111/j.1600-0870.2009.00425.x, 2010.

Rudels, B., Jones, E. P., Anderson, L. G., and Kattner, G.: On the intermediate depth waters of the Arctic Ocean, in: The Polar Oceans and Their Role in Shaping the Global Environment: The Nansen Centennial Volume, edited by: Johannessen, O. M., Muench, R. D., and Overland, J. E., Geophys. Monogr. Ser., 85, AGU, Washington DC, 33–46, 1994.

Schauer, U., Loeng, H., Rudels, B., Ozhigin, V. K., and Dieck, W.: Atlantic Water flow through the Barents and Kara Seas, Deep-Sea Res. Pt. I, 49, 2281–2298, 2002a.

Schauer, U., Rudels, B., Jones, E. P., Anderson, L. G., Muench, R. D., Björk, G., Swift, J. H., Ivanov, V., and Larsson, A.-M.: Confluence and redistribution of Atlantic water in the Nansen, Amundsen and Makarov basins, Ann. Geophys., 20, 257–273, doi:10.5194/angeo-20-257-2002, 2002b.

Schlichtholz, P.: Influence of oceanic heat variability on sea ice anomalies in the Nordic Seas, Geophys. Res. Lett., 38, L05705, doi:10.1029/2010GL045894, 2011.

Serra, N., Käse, R., Köhl, A., Stammer, D., and Quadfasel, D.: On the low-frequency phase relation between the Denmark Strait and the Faroe-Shetland Channel dense overflows, Tellus A, 62, 530–550, 2010.

Sirevaag, A. and Fer, I.: Early spring oceanic heat fluxes and mixing observed from drift stations north of Svalbard, J. Phys. Oceanogr., 39, 3049–3069, 2009.

Sirevaag, A. and Fer, I.: Vertical heat transfer in the Arctic Ocean: The role of double-diffusive mixing, J. Geophys. Res., 117, C07010, doi:10.1029/2012JC007910, 2012.

Skagseth, Ø.: Recirculation of Atlantic Water in the western Barents Sea, Geophys. Res. Lett., 35, L11606, doi:10.1029/2008GL033785, 2008.

Skagseth, Ø., Furevik, T., Ingvaldsen, R. B., Loeng, H., Mork, K. A., Orvik, K. A., and Ozhigin, V.: Volume and heat transports to the Arctic Ocean via the Norwegian and Barents Seas, in: Arctic Subarctic Ocean Fluxes: Defining the Role of the Northern Seas in Climate, edited by: Dickson, R., Meincke, J., and Rhines, P., Springer, New York, 45–64, 2008.

Smith, W. H. F. and Sandwell, D. T.: Global seafloor topography from satellite altimetry and ship depth soundings, Science, 277, 1957–1962, 1997.

Spreen, G., Kaleschke, L., and Heygster, G.: Sea ice remote sensing using AMSR-E 89 GHz channels, J. Geophys. Res., 113, C02S03, doi:10.1029/2005JC003384, 2008.

Steele M. and Morison J.: Hydrography and vertical fluxes of heat and salt northeast of Svalbard in autumn, J. Geophys. Res., 98, 10013–10024, 1993.

Stefan, J.: Über die Theorie der Eisbildung, insbesondere über Eisbildung im Polarmeere, Annalen der Physik und Chemie, Wiedemann-Annalen, 42, 269–286, 1891 (in German).

Sundfjord, A., Fer, I., Kasajima, Y., and Svendsen, H.: Observations of turbulent mixing and hydrography in the marginal ice zone of the Barents Sea, J. Geophys. Res., 112, C05008, doi:10.1029/2006JC003524, 2007.

Permissions

All chapters in this book were first published in Ocean Science, by Copernicus Publications; hereby published with permission under the Creative Commons Attribution License or equivalent. Every chapter published in this book has been scrutinized by our experts. Their significance has been extensively debated. The topics covered herein carry significant findings which will fuel the growth of the discipline. They may even be implemented as practical applications or may be referred to as a beginning point for another development.

The contributors of this book come from diverse backgrounds, making this book a truly international effort. This book will bring forth new frontiers with its revolutionizing research information and detailed analysis of the nascent developments around the world.

We would like to thank all the contributing authors for lending their expertise to make the book truly unique. They have played a crucial role in the development of this book. Without their invaluable contributions this book wouldn't have been possible. They have made vital efforts to compile up to date information on the varied aspects of this subject to make this book a valuable addition to the collection of many professionals and students.

This book was conceptualized with the vision of imparting up-to-date information and advanced data in this field. To ensure the same, a matchless editorial board was set up. Every individual on the board went through rigorous rounds of assessment to prove their worth. After which they invested a large part of their time researching and compiling the most relevant data for our readers.

The editorial board has been involved in producing this book since its inception. They have spent rigorous hours researching and exploring the diverse topics which have resulted in the successful publishing of this book. They have passed on their knowledge of decades through this book. To expedite this challenging task, the publisher supported the team at every step. A small team of assistant editors was also appointed to further simplify the editing procedure and attain best results for the readers.

Apart from the editorial board, the designing team has also invested a significant amount of their time in understanding the subject and creating the most relevant covers. They scrutinized every image to scout for the most suitable representation of the subject and create an appropriate cover for the book.

The publishing team has been an ardent support to the editorial, designing and production team. Their endless efforts to recruit the best for this project, has resulted in the accomplishment of this book. They are a veteran in the field of academics and their pool of knowledge is as vast as their experience in printing. Their expertise and guidance has proved useful at every step. Their uncompromising quality standards have made this book an exceptional effort. Their encouragement from time to time has been an inspiration for everyone.

The publisher and the editorial board hope that this book will prove to be a valuable piece of knowledge for researchers, students, practitioners and scholars across the globe.

List of Contributors

J. Yi
State Key Laboratory of Resources and Environmental Information System, Institute of Geographic Sciences and Natural Resources Research, Chinese Academy of Sciences, Beijing 100101, China

Y. Du
State Key Laboratory of Resources and Environmental Information System, Institute of Geographic Sciences and Natural Resources Research, Chinese Academy of Sciences, Beijing 100101, China

Z. He
State Key Laboratory of Tropical Oceanography (LTO), South China Sea Institute of Oceanology, Chinese Academy of Sciences, Guangzhou 510301, Guangdong, China
College of Ocean and Earth Sciences, Xiamen University, Xiamen 361005, China

C. Zhou
State Key Laboratory of Resources and Environmental Information System, Institute of Geographic Sciences and Natural Resources Research, Chinese Academy of Sciences, Beijing 100101, China

Wei Song
Physical Oceanography Laboratory, Ocean University of China, Qingdao, China
State Key Laboratory of Tropical Oceanography, South China Sea Institute of Oceanology, Chinese Academy of Sciences, Guangzhou, China

Jian Lan
Physical Oceanography Laboratory, Ocean University of China, Qingdao, China

Qinyan Liu
State Key Laboratory of Tropical Oceanography, South China Sea Institute of Oceanology, Chinese Academy of Sciences, Guangzhou, China

Dandan Sui
State Key Laboratory of Tropical Oceanography, South China Sea Institute of Oceanology, Chinese Academy of Sciences, Guangzhou, China

Lili Zeng
State Key Laboratory of Tropical Oceanography, South China Sea Institute of Oceanology, Chinese Academy of Sciences, Guangzhou, China

DongxiaoWang
State Key Laboratory of Tropical Oceanography, South China Sea Institute of Oceanology, Chinese Academy of Sciences, Guangzhou, China

F. Zheng
International Center for Climate and Environment Science (ICCES), Institute of Atmospheric Physics, Chinese Academy of Sciences, Beijing 100029, China

J. Zhu
International Center for Climate and Environment Science (ICCES), Institute of Atmospheric Physics, Chinese Academy of Sciences, Beijing 100029, China

S.-Q. Zhou
State Key Laboratory of Tropical Oceanography, South China Sea Institute of Oceanology, 164 West Xingang Road, Haizhu District, Guangzhou 510301, China

L. Qu
State Key Laboratory of Tropical Oceanography, South China Sea Institute of Oceanology, 164 West Xingang Road, Haizhu District, Guangzhou 510301, China
University of Chinese Academy of Sciences, Beijing, 100049, China

Y.-Z. Lu
State Key Laboratory of Tropical Oceanography, South China Sea Institute of Oceanology, 164 West Xingang Road, Haizhu District, Guangzhou 510301, China
University of Chinese Academy of Sciences, Beijing, 100049, China

X.-L. Song
State Key Laboratory of Tropical Oceanography, South China Sea Institute of Oceanology, 164 West Xingang Road, Haizhu District, Guangzhou 510301, China
University of Chinese Academy of Sciences, Beijing, 100049, China

M. Menna
Istituto Nazionale di Oceanografia e Geofisica Sperimentale - OGS, Trieste, Italy

P.-M. Poulain
Istituto Nazionale di Oceanografia e Geofisica Sperimentale - OGS, Trieste, Italy

D. Mignac
Graduate Program in Geophysics, Physics Institute and Geosciences Institute, Federal University of Bahia, Salvador, Brazil

Oceanographic Modeling and Observation Network, Center for Research in Geophysics and Geology, Federal University of Bahia, Salvador, Brazil

C. A. S. Tanajura
Oceanographic Modeling and Observation Network, Center for Research in Geophysics and Geology, Federal University of Bahia, Salvador, Brazil
Physics Institute, Federal University of Bahia, Salvador, Brazil
Ocean Sciences Department, University of California, Santa Cruz, CA, USA

A. N. Santana
Oceanographic Modeling and Observation Network, Center for Research in Geophysics and Geology, Federal University of Bahia, Salvador, Brazil

L. N. Lima
Oceanographic Modeling and Observation Network, Center for Research in Geophysics and Geology, Federal University of Bahia, Salvador, Brazil

J. Xie
Institute of Atmospheric Physics, Chinese Academy of Sciences, Beijing, China

P. Malanotte-Rizzoli
Massachusetts Institute of Technology, Cambridge, MA, USA

V. Artale
ENEA, Agenzia nazionale per le nuove tecnologie, l'energia e lo sviluppo economico sostenibile, Centri ricerca Frascati e Roma, Rome, Italy

G. L. Borzelli-Eusebi
Istituto Nazionale di Oceanografia e di Geofisica Sperimentale - OGS, Trieste, Italy

S. Brenner
Department of Geography and Environment, Bar llan University, Ramat Gan 52900, Israel

A. Crise
Istituto Nazionale di Oceanografia e di Geofisica Sperimentale - OGS, Trieste, Italy

M. Gacic
Istituto Nazionale di Oceanografia e di Geofisica Sperimentale - OGS, Trieste, Italy

N. Kress
National Institute of Oceanography, Israel Oceanographic & Limnological Research, Haifa, Israel

S. Marullo
ENEA, Agenzia nazionale per le nuove tecnologie, l'energia e lo sviluppo economico sostenibile, Centri ricerca Frascati e Roma, Rome, Italy

M. Ribera d'Alcalà
Stazione Zoologica Anton Dohrn, Naples, Italy

S. Sofianos
University of Athens, Division of Environmental Physics, Ocean Physics and Modeling Group, Athens, Greece

T. Tanhua
GEOMAR Helmholz Center for Ocean Research, Kiel, Germany

A. Theocharis
Institute of Oceanography, Hellenic Centre for Marine Research, Sounio Ave., P.O. Box 712, Athens, Greece

M. Alvarez
Instituto Espanol de Oceanografia, Calle del Corazon de Maria 8, Madrid 28020, Spain

Y. Ashkenazy
Department of Solar Energy & Environmental Physics, The Jacob Blaustein Institutes for Desert Research

A. Bergamasco
CNR ISMAR, Istituto di Scienze Marine, Venice, Italy

V. Cardin
Istituto Nazionale di Oceanografia e di Geofisica Sperimentale - OGS, Trieste, Italy

S. Carniel
CNR ISMAR, Istituto di Scienze Marine, Venice, Italy

G. Civitarese
Istituto Nazionale di Oceanografia e di Geofisica Sperimentale - OGS, Trieste, Italy

F. D'Ortenzio
Laboratoire d'Oceanographie de Villefranche (LOV), CNRS and Université Pierre et Marie Curie, Paris 06, UMR7093,Villefraneche-sur-mer, France

J. Font
Institut de Ciencies del Mar - CSIC, Barcelona, Spain

E. Garcia-Ladona
Institut de Ciencies del Mar - CSIC, Barcelona, Spain

J. M. Garcia-Lafuente
Physical Oceanography Group University of Malaga, Malaga, Spain

A. Gogou
Institute of Oceanography, Hellenic Centre for Marine Research, Sounio Ave., P.O. Box 712, Athens, Greece

M. Gregoire
Universite de Liege - Laboratoire d'Oceanologie - Centre MARE, Liege, Belgium

D. Hainbucher
University of Hamburg - Institute of Oceanography, Hamburg, Germany

H. Kontoyannis
Institute of Oceanography, Hellenic Centre for Marine Research, Sounio Ave., P.O. Box 712, Athens, Greece

V. Kovacevic
Istituto Nazionale di Oceanografia e di Geofisica Sperimentale - OGS, Trieste, Italy

E. Kraskapoulou
Institute of Oceanography, Hellenic Centre for Marine Research, Sounio Ave., P.O. Box 712, Athens, Greece

G. Kroskos
Institute of Oceanography, Hellenic Centre for Marine Research, Sounio Ave., P.O. Box 712, Athens, Greece

A. Incarbona
Universita di Palermo, Dipartimenti di Geologia & Geodesia, Palermo, Italy

M. G. Mazzocchi
Stazione Zoologica Anton Dohrn, Naples, Italy

M. Orlic
University of Zagreb, Faculty of Science Andrija Mohoroviviv Geophysical Institute, Zagreb, Croatia

E. Ozsoy
Middle East Technical University, Institute of Marine Sciences, Erdemli, Turkey

A. Pascual
Instituto Mediterraneo de Estudio Avanzados, IMEDEA (CSIC-UIB), Balearic Islands, Spain

P.-M. Poulain
Istituto Nazionale di Oceanografia e di Geofisica Sperimentale - OGS, Trieste, Italy

W. Roether
University of Bremen, Bremen, Germany

A. Rubino
Universita Ca' Forscari di Venezia, Venice, Italy

K. Schroeder
CNR ISMAR, Istituto di Scienze Marine, Venice, Italy

J. Siokou-Frangou
Institute of Oceanography, Hellenic Centre for Marine Research, Sounio Ave., P.O. Box 712, Athens, Greece

E. Souvermezoglou
Institute of Oceanography, Hellenic Centre for Marine Research, Sounio Ave., P.O. Box 712, Athens, Greece

M. Sprovieri
CNR IAMC, Unita Operativa di Capo Granitola, Trapani, Italy

J. Tintoré
Instituto Mediterraneo de Estudio Avanzados, IMEDEA (CSIC-UIB), Balearic Islands, Spain

G. Triantafyllou
Institute of Oceanography, Hellenic Centre for Marine Research, Sounio Ave., P.O. Box 712, Athens, Greece

D. E. Gwyther
Institute for Marine and Antarctic Studies, University of Tasmania, Private Bag 129, Hobart, Tasmania 7001, Australia
Antarctic Climate & Ecosystems Cooperative Research Centre, University of Tasmania, Private Bag 80, Hobart, Tasmania 7001, Australia

B. K. Galton-Fenzi
Antarctic Climate & Ecosystems Cooperative Research Centre, University of Tasmania, Private Bag 80, Hobart, Tasmania 7001, Australia
Australian Antarctic Division, Channel Highway, Kingston, Tasmania 7050, Australia

J. R. Hunter
Antarctic Climate & Ecosystems Cooperative Research Centre, University of Tasmania, Private Bag 80, Hobart, Tasmania 7001, Australia

J. L. Roberts
Antarctic Climate & Ecosystems Cooperative Research Centre, University of Tasmania, Private Bag 80, Hobart, Tasmania 7001, Australia
Australian Antarctic Division, Channel Highway, Kingston, Tasmania 7050, Australia

K. O'Driscoll
Marine Research Group, School of Planning, Architecture & Civil Engineering, Queen's University Belfast, UK

B. Mayer
Institute of Oceanography, University of Hamburg, Bundesstrasse 53, Hamburg 20146, Germany

J. Su
Institute of Oceanography, University of Hamburg, Bundesstrasse 53, Hamburg 20146, Germany

M. Mathis
Institute of Oceanography, University of Hamburg, Bundesstrasse 53, Hamburg 20146, Germany

M. C. H. Tiessen
Royal Netherlands Institute for Sea Research, P.O. Box 59, 1790 AB Den Burg, Texel, the Netherlands

L. Fernard
Centre for Environment, Fisheries and Aquaculture Science (CEFAS), Pakefield Road, Lowestoft, Suffolk NR33 0HT, UK

T. Gerkema
Royal Netherlands Institute for Sea Research, P.O. Box 59, 1790 AB Den Burg, Texel, the Netherlands

J. van der Molen
Centre for Environment, Fisheries and Aquaculture Science (CEFAS), Pakefield Road, Lowestoft, Suffolk NR33 0HT, UK

P. Ruardij
Royal Netherlands Institute for Sea Research, P.O. Box 59, 1790 AB Den Burg, Texel, the Netherlands

H. W. van der Veer
Royal Netherlands Institute for Sea Research, P.O. Box 59, 1790 AB Den Burg, Texel, the Netherlands

T. S. Dotto
Laboratório de Estudos dos Oceanos e Clima, Instituto de Oceanografia, Universidade Federal do Rio Grande - FURG, Rio Grande, RS, 96203-900, Brazil

R. Kerr
Laboratório de Estudos dos Oceanos e Clima, Instituto de Oceanografia, Universidade Federal do Rio Grande - FURG, Rio Grande, RS, 96203-900, Brazil

M. M. Mata
Laboratório de Estudos dos Oceanos e Clima, Instituto de Oceanografia, Universidade Federal do Rio Grande - FURG, Rio Grande, RS, 96203-900, Brazil

M. Azaneu
Laboratório de Estudos dos Oceanos e Clima, Instituto de Oceanografia, Universidade Federal do Rio Grande - FURG, Rio Grande, RS, 96203-900, Brazil

I.Wainer
Laboratório de Oceanografia Física, Clima e Criosfera, Instituto Oceanográfico, Universidade de São Paulo - USP, São Paulo, SP, 05508-120, Brazil

E. Fahrbach
Stiftung Alfred-Wegener-Institut für Polar- und Meeresforschung in der Helmholtz-Gemeinschaft, Fachbereich Klimawissenschaften, Sektion Messende Ozeanographie, Postfach 120121 Bremerhaven, Germany

G. Rohardt
Stiftung Alfred-Wegener-Institut für Polar- und Meeresforschung in der Helmholtz-Gemeinschaft, Fachbereich Klimawissenschaften, Sektion Messende Ozeanographie, Postfach 120121 Bremerhaven, Germany

I. Grefe
School of Environmental Sciences, University of East Anglia, Norwich Research Park, Norwich, UK

J. Kaiser
School of Environmental Sciences, University of East Anglia, Norwich Research Park, Norwich, UK

M. Gunduz
Institute of Marine Sciences and Technology, Dokuz Eylul University, Izmir, Turkey

E. Özsoy
Institute of Marine Sciences, Middle East Technical University, Erdemli, Turkey

M. Duran-Matute
Dept. of Physical Oceanography, Royal NIOZ, P.O. Box 59, 1790 AB Den Burg (Texel), the Netherlands

T. Gerkema
Dept. of Physical Oceanography, Royal NIOZ, P.O. Box 59, 1790 AB Den Burg (Texel), the Netherlands

G. J. de Boer
Deltares, P.O. Box 177, 2600 MH Delft, the Netherlands

J. J. Nauw
Dept. of Physical Oceanography, Royal NIOZ, P.O. Box 59, 1790 AB Den Burg (Texel), the Netherlands

U. Gräwe
Leibniz Institute for Baltic Sea Research Warnemünde (IOW), Seestraße 15, 18119 Rostock, Germany

J. F. Oxmann
School of Environment, The University of Auckland, Auckland 1010, New Zealand

L. Schwendenmann
School of Environment, The University of Auckland, Auckland 1010, New Zealand

H. L. Bryden
National Oceanography Centre Southampton, University of Southampton, Empress Dock, Southampton, UK

B. A. King
National Oceanography Centre, Southampton, Empress Dock, Southampton, UK

G. D. McCarthy
National Oceanography Centre, Southampton, Empress Dock, Southampton, UK

E. L. McDonagh
National Oceanography Centre, Southampton, Empress Dock, Southampton, UK

D. Hainbucher
CEN, Institut für Meereskunde, University of Hamburg, Bundesstraße 53, 20146 Hamburg, Germany

A. Rubino
Università Ca' Foscari di Venezia, Dorsoduro, 2137 - C.A.P. 30121, Italy

V. Cardin
OGS - Istituto Nazionale di Oceanografia e di Geofisica Sperimentale, Borgo Grotta Gigante, 42/c, 34018 Sgonico, Trieste, Italy

T. Tanhua
GEOMAR Helmholtz Centre for Ocean Research Kiel, Department of Chemical Oceanography, Düsternbrooker Weg 20, 24105 Kiel, Germany

K. Schroeder
Istituto di Scienze Marine - ISMAR, Consiglio Nazionale delle Ricerche (CNR), Arsenale - Tesa 104, Castello 2737/F, 30122 Venice, Italy

M. Bensi
OGS - Istituto Nazionale di Oceanografia e di Geofisica Sperimentale, Borgo Grotta Gigante, 42/c, 34018 Sgonico, Trieste, Italy

I. A. Dmitrenko
Centre for Earth Observation Science, University of Manitoba, Winnipeg, Canada

S. A. Kirillov
Centre for Earth Observation Science, University of Manitoba, Winnipeg, Canada

N. Serra
Institute of Oceanography, University of Hamburg, Hamburg, Germany

N. V. Koldunov
Institute of Oceanography, University of Hamburg, Hamburg, Germany

V. V. Ivanov
Arctic and Antarctic Research Institute, St. Petersburg, Russia
International Arctic Research Center, University of Alaska Fairbanks, Fairbanks, AK, USA

U. Schauer
Alfred Wegener Institute for Polar and Marine Research, Bremerhaven, Germany

I. V. Polyakov
International Arctic Research Center, University of Alaska Fairbanks, Fairbanks, AK, USA

D. Barber
Centre for Earth Observation Science, University of Manitoba, Winnipeg, Canada

M. Janout
Alfred Wegener Institute for Polar and Marine Research, Bremerhaven, Germany

V. S. Lien
Institute of Marine Research, Bergen, Norway

M. Makhotin
Arctic and Antarctic Research Institute, St. Petersburg, Russia

Y. Aksenov
National Oceanography Centre Southampton, Southampton, UK